S0-CKR-013

Ultrastructure and the Biology of Plant Cells

Ultrastructure and the Biology of Plant Cells

Brian E. S. Gunning
Professor of Developmental Biology
Australian National University, Canberra

Martin W. Steer
Lecturer in Botany
The Queen's University of Belfast

Edward Arnold

First published 1975
by Edward Arnold (Publishers) Ltd.,
25 Hill Street,
London, W1X 8LL

ISBN: 0 7131 2494 6

This book is dedicated to Professor I. Manton

Filmset by Photoprint Plates Ltd, Rayleigh, Essex
Printed in Great Britain by Fletcher & Son Ltd, Norwich

Preface

Our subject in this book is the life of the plant cell. The approach is visual. We look at the cell and its components, and go on from there. We have tried to write for botanists who are concerned to see their subject treated in its biological context, for zoologists who would like to know something of plants, for physiologists and biochemists who would like to place their knowledge of processes and reactions in a structural background, and in general, for biologists who have no desire to be restricted by the traditional boundaries of any one branch of the study of life. Our aim is to present advanced information, exciting discoveries, and current problems and challenges at the frontiers of the subject, to set these against an introductory background, and to seek throughout to emphasize principles and generalizations. The material is selected with the hope that it will be of interest, useful, and comprehensible to student, teacher and research worker alike.

In order to promote general interest, usefulness, and comprehension, our presentation is restrictive in several senses. We use the minimum of technical jargon. Naturally, every subject has its own language, but the specialized vocabulary needed in order to find one's way around in the world of the cell is not large, and much is familiar from early education in biology. We introduce each new term the first time it is used in the text, and, for the more widely used words, include some background to the relevant history. We restrict the content of biochemistry to a minimum. There are no chemical formulae in the text. This is not because we undervalue the contribution of biochemistry: it is merely that we consider it possible to communicate the topics we wished to include through the use of relatively plain words, thereby allowing those who lack a thorough knowledge of chemistry to participate in the joys and excitements of discovery that exist in the subject, and that should as far as possible be available to all. We restrict the choice of topic and the depth of treatment. The book is not in any sense a compilation of facts for the research worker. Where factual matter is detailed, it is not so much for its own sake as to illuminate a principle or a generalization. We have tried to point to some of the things that are *not* known, as well as to some of those that are. At the risk of offending researchers whose work and discoveries may be acknowledged only indirectly, the list of references is presented not as an exhaustive bibliography, but as a means of enabling readers to gain an entry into the literature of the subject.

We thank all those, particularly our colleagues in the Botany Department of The Queen's University of Belfast, and Dr. B. Juniper at Oxford, who have helped us by making suggestions and correcting errors. Errors of fact and interpretation that remain are, of course, our own responsibility. Mrs. M. Pate, Mrs. L. Foster and Mrs. L. Steer have given invaluable assistance by typing innumerable drafts. As for the illustrations, we particularly thank our own students and those others who very kindly provided us with prints. Full acknowledgements are given in the appropriate Plate legends. Most of the printing was done, superbly, by Mr. D. Kernoghan. We also thank Mr. H. Turtle, who, never having heard of chloroplasts in his life before, nevertheless taught us much about them in the course of drawing Text Fig. 8 for us. Finally we are indebted to our undergraduate, schoolteacher, and research worker friends who gave us the idea that the book should be written.

B.E.S.G.
M.W.S.
1975

List of Plates

Contents

(Section and sub-section titles appear beneath the respective chapter headings in the text)

1 Aims and Methods

'To make a phenomenon visible is immensely to enlarge our powers of understanding it' (J. D. Bernal)*

1.1 Structure and Function

More than half of the human brain is concerned with the reception and processing of visual stimuli. With such an emphasis on the sense of sight it is no wonder we say that seeing is believing. More precisely, seeing aids understanding, and it is this direct benefit of the visual approach that we are about to exploit. The aim is to seek insight into the way plant cells function, by interpreting images of their structure.

The interpretation of visual images does not end with straightforward description of the observed structure. Although structural interpretation is essential, most biologists have come to regard it a mere preliminary to conceptual interpretation, and proceed to ask how structure might be related to function. The history of biology shows how generally this form of analysis has been employed, first in respect of easily observed features, and then repeatedly as newly invented instruments and techniques created opportunities to explore new realms of structure. The approach has contributed greatly to the current level of understanding of living organisms, not only by helping to explain known phenomena, but also by innovation. Many new fields of study have arisen out of hypotheses formulated during the conceptual interpretation of observed structures.

There are dangers in the search for meaningful relationships between structure and function. To suggest that structures are as they are for the express *purpose* of performing particular functions is now regarded as teleological, unjustified, and, in the sense that the finality of the conclusion inhibits further investigation, uninformative. Biological structures are better interpreted on an evolutionary basis than by assuming purposive design, in other words by considering that they have been shaped in the course of evolution by the natural selection and survival of those types which are in greatest harmony with the conditions under which they exist.

Some biologists take the view that conceptual interpretation of structure should be concerned only with mechanisms, such as how the structure is formed, how it operates, how it is controlled, and in general, with questions beginning with 'How . . . ?' Others go slightly further, to ask what might be the possible significance of the structure being as it is and where it is. Under the stimulus of such inquiries, testable hypotheses, based on observed structural features together with other background information, can be erected in attempts to explore the possible functions of a structure, and the manner in which it may be adapted in relation to its functions. It should be realized that structural features, like other categories of adaptation, can be extremely subtle, and are not always clearly recognizable as being of significance. It is not necessarily the biggest, or the most compact, or any of the more obvious manifestations of structure, that fit the organism best to the conditions of its existence.

There is a further stage in the mental processes by which visual images can be interpreted. It is to use the conceptual interpretations of what has been observed in order to make predictions concerning the as yet unobserved. George Haberlandt, a pioneer in the functional interpretation of botanical structure, was doing just this when he prophesied in his book *Physiological Plant Anatomy,* written in 1884, that

> 'Since the correlation of structure with function is always evident, as regards both the gross anatomy and the histology of the plant body, a similar connection between the morphological features of protoplasm and the functions of the individual protoplast would doubtless be demonstrable, *were it not that most of the structures concerned are of ultramicroscopic dimensions.*'†[303]

Haberlandt would be astonished to see the power and the diversity of the methods that are now available to test his prediction. From now on we shall rarely be outside the world of 'ultramicroscopic dimensions' that he wished to, but could not, penetrate. Many techniques have enabled the subject to grow, but few can have had such an impact as electron microscopy and the interpretation of electron microscope images.

* From Bernal J. D. (1969). Science in History, Vol. 3, *The Natural Sciences in our time*. Pelican Books, p. 782, lines 3 and 4. First published by C. H. Watts and Co. (1954).

† From Haberlandt, G. (1884). *Physiological Plant Anatomy*. Chapter 1. Taken from the reprint edition published by Today and Tomorrow's Book Agency, New Delhi (1965); based on the translation of the 4th German edition by M. Drummond, published by Macmillan (1914).

1.2 Electron Microscopy

The resolving power of the *electron microscope* is about one thousand times better than that of the optical microscope, though present techniques of specimen preparation, being imperfect, do not allow biologists to make full use of it. Nevertheless, in the past two decades this high performance has been exploited to such effect that revolutionary advances in our comprehension of living systems have been made. The three-dimensional architecture of cells and cell components has been magnified into the range of our ordinary senses. It has been possible in some cases to study the larger of the molecules found in cell structures and even to visualize some submolecular components.[304] One of the significant influences of electron microscopy has been to unify botany and zoology, for at high magnifications the distinctions between plant and animal cells seem unimportant in the face of the many features the two are seen to share. The consequence is that many research workers consider the properties of the *cell* to be as fundamental as those of the *organism* in which the cell is found. Although we concentrate on structure and function in *plant* cells, we shall be favouring the approach of the general biologist and drawing parallels with animal cells where appropriate. Much of what is written and illustrated in fact applies to both animals and plants.

Many of the recent spectacular advances in the subject of cell biology have stemmed from the production and careful inspection of electron micrographs, of which this book presents some examples (Plates 2–48). In order to make the most of the prodigious amount of information contained in such pictures, it is helpful to know something about electron microscopes and the way in which specimens are prepared for electron microscopy.

1.2.1 *Electron Microscopes*

The *transmission electron microscope* [903,695] produces an image of the specimen by passing a beam of electrons through it. Electromagnetic fields manipulate and focus the beam, and the magnified image can be viewed directly on a fluorescent screen or recorded by black and white photography. Because electrons are easily deflected, or scattered, they are given a path that is as nearly as possible collision-free by evacuating most of the molecules of air from within the body of the instrument. It follows that specimens to be placed in the microscope must (a) be strong enough to stand up to conditions of high vacuum, (b) be thin enough to transmit sufficient unscattered electrons to produce 'light' areas in the image, and (c) contain electron-scattering material which creates 'dark' areas in the image. *Thin* is a rather weak adjective in this context, and the prefix *ultra*, which pervades the literature of electron microscopy, is used. Sections and other types of specimen have to be *ultra-thin*, and the electron microscope reveals details of *ultrastructure* when they are examined. The units employed in descriptions of the world of ultrastructure are *nanometres* (symbol nm) and *micrometres* (symbol μm). One millimetre (mm) equals 1000 μm, and one μm equals 1000 nm. In another form,

$$1 \text{ nm} = 10^{-9} \text{ metre (m)},\ 1\ \mu\text{m} = 10^{-6} \text{ m},$$
and
$$1 \text{ mm} = 10^{-3} \text{ m}.$$

Instruments which impose slightly less stringent requirements on the specimen do now exist. For example, the *high voltage electron microscope*, in which the electron beam is accelerated by up to one million volts or more, allows objects up to two or three μm in thickness to be examined. In the *scanning electron microscope* the problems of transmission through the specimen do not arise, and usually the surface topography of the specimen is investigated.

1.2.2 *Magnification, Resolution, and True Dimensions*

Differences between the performance of the light and the electron microscope are immediately apparent in the Plates. It would be a simple matter to enlarge Plate 1 (light micrograph) by a factor of five until it had the same magnification as the first electron micrograph (Plate 2), but it is obvious that so doing would not yield comparable detail. Plate 1 does in fact include structures that are close to the limit of resolution imposed on the light microscope by the wave properties of light.

The true size of objects in a micrograph may be estimated using a simple and very useful rule of thumb. Since there are 1000 μm in 1 mm, a 1 μm object will appear to be 1 mm in size when the magnification is ×1000. Therefore, to place a scale marker representing 1 μm on any micrograph, simply draw a line as many mm long as there are thousands in the magnification. Having done this, it will be seen that Plate 1 contains some components (e.g. NE, the nuclear envelope) that are well below 1 μm in thickness while Plate 2 contains some (e.g. PM, the plasma membrane) that are so much thinner than 1 μm that nanometres are more convenient units. Scale marker lines enable the true dimensions of objects to be estimated at a glance, and precise measurements are equally easily obtained, thus:

$$\frac{\text{size in micrograph (mm)} \times 1000}{\text{magnification}} = \text{true size } (\mu\text{m})$$

or

$$\frac{\text{size in micrograph (mm)} \times 1\,000\,000}{\text{magnification}} = \text{true size (nm)}$$

1.2.3 *Specimen Preparation*

Cell biologists have so far used the conventional transmission electron microscope more than other types, and consequently have had to devise methods of preparing samples of cells and tissues in an ultra-thin form that can withstand evacuation.[160, 413, 437]

DIRECT PREPARATIONS Cells and tissues do not often assume an ultra-thin form. One exception of historical interest is the peripheral region of certain artificially cultured animal cells, which were amongst the first cells to have their internal structure examined electron microscopically. The simple technique of drying and staining whole cells was used. Nowadays the high voltage electron microscope is making a return to this simple method very profitable. Sub-cellular components extracted from cells may, however, be thin enough to be viewed in the conventional electron microscope after spreading them on a *support film*. This is an ultra-thin film of plastic or carbon, the electron microscopist's equivalent of the glass slide used for mounting specimens in light microscopy. The objects are dried and examined either directly or after staining with dissolved salts of heavy elements such as lead, uranium, osmium, tungsten, etc. The aim of the staining is to confer a range of electron-scattering powers on selected parts of the specimen so that the final image contains a variety of tone ranges, from dark, representing regions rendered *electron-dense* by the presence of the heavy elements, to light, representing *electron-transparent* regions that remained unstained. In *positive staining* the objects themselves (or parts of them) are stained and appear dark against a light background; in *negative staining* the objects are surrounded by stain so that they appear pale against a dark background.

ELECTRON DENSITY The terms 'dense' and 'transparent' are widely used in the literature of electron microscopy to describe regions of the image, along with variations such as 'less dense', 'more dense than . . .'. They have been borrowed from the literature of photography and light microscopy, and are to some extent misleading in that they imply that electrons behave in the manner of photons, able to penetrate an object at a given point, lose some energy, and pass on to contribute to the image of that point. Electrons which hit an electron microscope specimen may penetrate with little loss of energy, or at the opposite extreme they may be stopped completely. The arrangement of the apertures through which the electron beam passes is such that if an electron loses enough energy to become scattered, it will probably not reach the viewing screen. Thus, in practice, the specimen is likely to be either transparent or opaque to the individual electron, with few intermediate conditions. 'Dense', 'less dense', and the other descriptive terms that will be used in the text therefore refer not to the effect of the specimen upon the *individual* electron, but to variations across the specimen of the numbers of electrons that pass unchanged, relative to the numbers that do not. Solutions of heavy elements are used as stains in electron microscopy because the higher the atomic number of an atom, the greater the chance that it will deflect the incident electrons. Hence 'stained' areas transmit fewer electrons to the viewing screen and therefore appear dark.

SHADOW CASTING Another technique, *shadow casting,* developed in the early years of the electron microscope era, provides an image of the surface topography of the specimen. Objects such as small particles or ultra-thin fragments of plant cell wall are spread on a support film and, under conditions of high vacuum in a vacuum evaporator, sprayed with atoms of metal from a source placed to one side. All exposed surfaces accumulate an electron dense deposit, while those in the lee of projections remain uncoated. Just as objects seen in aerial photographs can be measured and identified from the size and shape of the shadow they cast, so it is with the shadow-cast electron microscope specimen. This procedure is somewhat limited by being applicable only to very small objects, but bulky specimens can be visualized if ultra-thin *replicas* of their surface are prepared, removed from the specimen, shadowed, and examined.

FREEZE ETCHING The *freeze-etching* procedure is a more recent and very important variation on the theme of shadow casting. It involves freezing cells or tissues—very rapidly so as to avoid distorting subcellular components—and then exposing internal surfaces by fracturing the ice. The fracture plane tends to follow lines of weakness, that is to say, regions where there was not much water to start with and which therefore have not produced strong ice. Cell membranes provide one such region, and so the exposed fracture surface usually jumps from one expanse of membrane to another. Details of the surface topography of these membrane surfaces and any other features of the fracture plane are then accentuated by momentarily raising the temperature (but still keeping it well below freezing point) and subliming off (etching) some ice. Finally, a replica of the surface is prepared, shadowed, and examined after the frozen specimen itself has been melted and removed. The method can give electron microscope images that are especially trustworthy in being representations of material that was alive at the moment of freezing, and not altered since.

ULTRA-THIN SECTIONING The need to obtain views of the interior of cells was, of course, recognized long before the freeze etching technique was devised. Following precedents set by light microscopists, attempts were made to use sectioned material. The first electron microscope view of at least one component of plant cells—the prolamellar body (Chapter 9)—was obtained by shadow casting a replica of a chemically etched, relatively thick section. This was a cumbersome procedure, however, and meanwhile, instruments *(ultra-microtomes)* and knife edges (made from glass or preferably diamond) that could cut *ultra-thin sections* were being developed, soon to become the tools used more than any others for preparing biological specimens for electron microscopy.

The technique is to cut, stain, and examine ultra-thin sections of fixed and embedded tissue. This statement summarizes the outcome of very many technical advances made over a long period of time and still being improved. It is only relatively recently that modern instrumentation has made the method a generally-available routine. Formerly it was surrounded by mystique and not many research workers could do it well. Each had his own ritual, believing that only thus could success be attained in the quest for perfect ultra-thin sections—smooth, not too thick, not too thin, no holes or scratches, and above all, clean, for at electron microscope magnifications particles or films of dirt can completely obscure the image. Justification for using the prefix 'ultra' in describing these thin sections is apparent when their actual dimensions are considered. They have to be about 50–75 nm thick, that is, a 1 mm slice of tissue could (in theory) be further sliced to yield about 15 000 ultra-thin sections.

FIXATION The specimen is first chemically *fixed* in a process that is equivalent in its aims to the rapid freezing step at the beginning of the freeze etching procedure.[560] The normal dynamic and changing state of the cell components is interrupted by the application of *fixatives* which rapidly kill the cell and, by forming chemical bridges, cross-link the constituent molecules into a three dimensional network rigid enough to stand up to the subsequent manipulations. Glutaraldehyde, formaldehyde, or mixtures of the two are good fixatives. Their mode of action can be demonstrated by adding some to a solution of a protein such as serum albumin: given suitable concentrations it is not many seconds before chemical cross-linking transforms the fluid solution into a solid mass. The first of the staining operations is carried out by transferring the tissue to a solution of osmium tetroxide. This highly reactive substance is itself a fixative, but it also imparts electron density to cell components by the insertion of osmium atoms. Some components take up more than others and hence appear darker in the final image—in other words the osmium tetroxide is a differential stain as well as a fixative.

EMBEDDING The fixed tissue is still not strong enough to be sectioned, so it is next dehydrated gradually and a plastic is introduced into the spaces formerly occupied by water. Epoxy resins are most frequently used nowadays. Once hardened in an oven they possess the necessary strength for ultra-thin sectioning, the sections are unaffected by being placed in a high vacuum, and since the resin itself lacks heavy elements, there is no background structure to detract from the image of the cell components, even though they are embedded in hard plastic. The sections are usually stained again prior to examination, and both general and chemically-specific staining methods exist, all based on the introduction of elements of high atomic number.

INTERPRETATION, AND VISUALIZATION OF DYNAMIC PROCESSES Most of our illustrations depict material prepared by fixing, embedding, and sectioning, and two limitations of the technique will be obvious. One is that the sections are so thin that it is hard to translate the two dimensional image into the three dimensional reality.[187] Some specific problems will be described in the following chapter, and a general analogy suffices here. Using a section 0.05 μm in thickness to investigate the architecture of a cell 20 μm in diameter is like trying to describe a house, its rooms, its cupboards, and all their contents down to 1 mm in size, by examining a two centimetre thick slice of the whole building. Obviously many such sections must be studied, and they should, if at all possible, be cut in known planes or in sequences from which three dimensional reconstructions can be made.

The other limitation is that the cells have to be killed. This need not mean that all cellular processes have to be inactivated, though very often the fixatives that preserve structural features best are those that totally prevent post-fixation enzymic activity. Some mild fixation procedures preserve structures adequately, and have the additional advantage of fixing enzyme molecules in place without totally destroying their ability to function. In favourable circumstances the enzymes can then be induced to reveal their location by supplying them with chemicals that they will act on to produce local, electron dense precipitates.

Although it is possible to preserve the activity of some enzymes during specimen processing, 'higher' functions of the cell are invariably lost: once fixed, the specimen is dead. It is therefore necessary to imagine dynamic life processes given only a series of static images, each one representing the moment in time when the cell was fixed. There are, however, several ways in which an impression of the time dimension can be gained. It is often possible to see how a structure alters with time by fixing samples at selected intervals. Alternatively, if the structure is represented many times and in many states in single samples of a cell or a tissue, a single fixation may preserve enough examples to allow micrographs to be taken and subsequently arranged in a series that describes the dynamic process. A source of error in this approach is that it is not always certain in which direction to view the series: unless there are obvious clues, either end could represent the beginning of the process.

It is even possible to detect sequential stages of some dynamic processes within single micrographs. This arises particularly where the processes are of the nature of assembly line systems. The main difficulty lies in recognizing that they are dynamic systems. As an example, a single still photograph taken in a factory might show all stages of the assembly of a product; but whereas the photographer can see the process in action and does not interfere with it, the electron microscopist has to render comparable processes in cells inactive before he can take his micrograph. While it is obvious to the photographer that his picture will represent a dynamic process, it is not so to the micrographer. This problem is encountered often, and the importance of studying living material with the light microscope before proceeding to fix samples for electron microscopy cannot be over-emphasized. Even if the structures involved are too small to be resolved, clues to the existence of dynamic processes, such as the accumulation of a product, may be visible.

Perhaps the most powerful method for studying dynamic processes that is available to the electron microscopist is the introduction of a label that can be followed through the processes by means of sequential fixations. A simple example is the feeding of cells with an electron-dense marker substance to permit description of the stages of its uptake and subsequent fate. The technique of *autoradiography* is a special, and most versatile, method of performing such studies. It has been used in experiments on the dynamic function of numerous cell components, and the principle of the method therefore needs to be described.

In autoradiography, radioactive substances essentially locate themselves by virtue of their emission of radioactive particles. Radioactive compounds that emit low energy beta particles are the most useful to the biologist, because these can be located with the greatest precision. Tritium, a radioactive isotope of hydrogen, has suitable emissions, and can be incorporated in place of hydrogen in a wide variety of molecules of the sort utilized by cells and found in them.

The technical problem of detecting a tritium-labelled compound in a cell or tissue can be overcome if the compound survives the processes of specimen preparation without moving, so that it is present in sections of the specimen. It is immaterial whether the sections are eventually to be viewed by light microscopy or by electron microscopy. Working in a dark room, the investigator coats the sections with a film of photographic emulsion, and still in darkness, leaves them for an exposure period of days, weeks, or even months, depending on how much radioactivity is present. Beta particles have much the same effect on photographic emulsions as has light, so when the section and the film superimposed on it are processed in the same manner as an ordinary photograph, radioactive areas in the section are rendered visible by darkening of the emulsion. Such areas are dark to the eye because they contain grains of silver: in the light microscope the grains appear as black particles, while the high atomic number of silver means that they are electron dense, and visible in the electron microscope as black particles or short lines. The most important part of the technique is that the section and the layer of film that contains the silver grains remain together throughout. The components of the cell or the tissue are visible under the grains, and so the structures that have become radioactive can be recognized.

The advantages of autoradiography can be exploited in several ways. The chemical nature of components can be studied by seeing which radioactive precursors they incorporate. The rate of incorporation can be studied by supplying labelled compounds for a range of times prior to fixation. The dynamic behaviour of components can be followed by first labelling them and subsequently fixing samples at intervals to see where and how rapidly the label moves. The latter procedure gives especially clear insight into dynamic processes that might otherwise be in the category described above as being almost undetectable to the electron microscopist, and several discussions in subsequent chapters will hinge upon results obtained in this way.

FACT OR ARTEFACT One final question must be raised concerning the rich detail to be found in electron micrographs of cells—is it meaningful?

Most workers are very conscious of the difficulty of deciding whether a structure has been altered, destroyed, or even created during specimen preparation procedures. Before accepting a value for, say, the molecular weight of a protein, biochemists like to prepare and purify the molecule by more than one method, and are reassured if consistent results are obtained. So it is with cell biologists, who place much reliance on comparisons made between living material viewed by light microscopy; chemically-fixed material viewed after ultra-thin sectioning (etc.); rapidly-frozen cells viewed after freeze etching; and components extracted from living cells and viewed by, for example, negative staining. It is also possible to bypass the procedures of fixation, dehydration and embedding, by very rapidly freezing portions of tissue and directly cutting ultra-thin sections of the frozen block. We shall be making a few such comparisons, and while each case has to be argued on its merits, it is a valid generalization that the structures we now proceed to examine do exist in life, in a form that can be visualized from their electron microscope images.

2 The Plant Cell and its Membranes: A General Survey

2.1 Introduction: The Cell

Concepts and definitions of the cell have changed greatly during the three centuries that have passed since Robert Hooke first used the word.[374] Looking at slices of cork, he saw small empty spaces delimited by walls, and applied to them the term that was already in use for similar constructions such as monks' cells and cells in a honeycomb. Although Hooke knew that in living material the spaces could contain 'juices', and despite recognition by himself and many others that a variety of components could be seen in the spaces, the wall came to be regarded as the structure by which a cell could be defined.[384] The 1830s saw this view reinforced as a consequence of the examination of a range of animal tissues by several investigators, all of whom likened the 'walls' (collagen layers) of hard epithelia and cartilage to the walls of plant tissues. One zoologist, Schwann, described cells in cartilage as resembling parenchymatous cells of plants, and his paper, written in 1839, is commonly regarded as the inception of the theory that all biological material is composed of cellular units.

The botanist Schleiden later pointed out to Schwann that the wall is not the only structure held in common by plant and animal cells, but that nuclei, established as consistent features of plant cells by Robert Brown earlier in the same decade, are present as well as walls. Schleiden's views on the role of the nucleus were quite wrong (Chapter 5), and the unifying concept of cells as nucleated, walled structures was also misleading. Already it was known from observations made on leaf cells, nerve cells, stamen hairs, and other materials, that a non-homogeneous substance occupies the compartment between the nucleus and the wall. Purkinje in 1840 and von Mohl in 1845 argued that this substance represents a stage in the development of a cell: apparently independently, they applied to it the word *protoplasm,* this being a theological term used at the time in referring to Adam, the 'first-formed', and so considered by them to be apt and suitable for use in the context of the development of cellular structure. Purkinje thought that plants differ from animals in that solid cell wall separates out of the fluid protoplasm more rapidly, and it was from this germ that more modern concepts have been elaborated. We now know that protoplasm (which includes the nucleus) is the basis of *all* cells, and that walls, formerly regarded as the unifying feature, are a product of the protoplasm of plant but not animal cells. In short, the emphasis has shifted from the walls which surround spaces, to the contents of the spaces. The word cell has been retained, but refers to the *protoplast* (consisting of protoplasm). The cell wall, where present, is relegated to the status of an extra-cellular product of cellular activity.

UNITY AND DIVERSITY All cells possess certain basic biochemical systems that synthesize carbohydrates, proteins, nucleic acids, and many other types of molecule. All have an outer surface that provides protection by excluding harmful material in the external environment, while at the same time permitting the controlled import and export of other substances (Chapter 3). All have a store of information where the hereditary material embodies in a chemical form instructions which guide the cells through the intricacies of their development and reproduction (Chapter 5). All have devices which provide chemical energy, to be utilized in the general maintenance of cellular integrity, and in syntheses leading to growth and development (Chapter 8). These attributes are fundamental to all living systems, and the structural and functional similarities of plant and animal cells stem from them.[596, 909] There are in addition cellular features in which the two kingdoms differ, mostly deriving from one major event in the evolution of living organisms—the development of a cell wall in the ancestors of plants. The consequences for plant cell structure and function were far-reaching (Chapter 3).

Plant cells vary in the extent to which different functions are developed, for, as with most other multicellular organisms, plants exhibit division of labour.[152, 153, 600] Specialized cells develop, related to the varied requirements of maintaining life and supporting growth and development—protection, mechanical support, synthesis or storage of food reserves, absorption, secretion, reproduction, cell division, and the humbler but vital role of connecting the more exotic tissues.

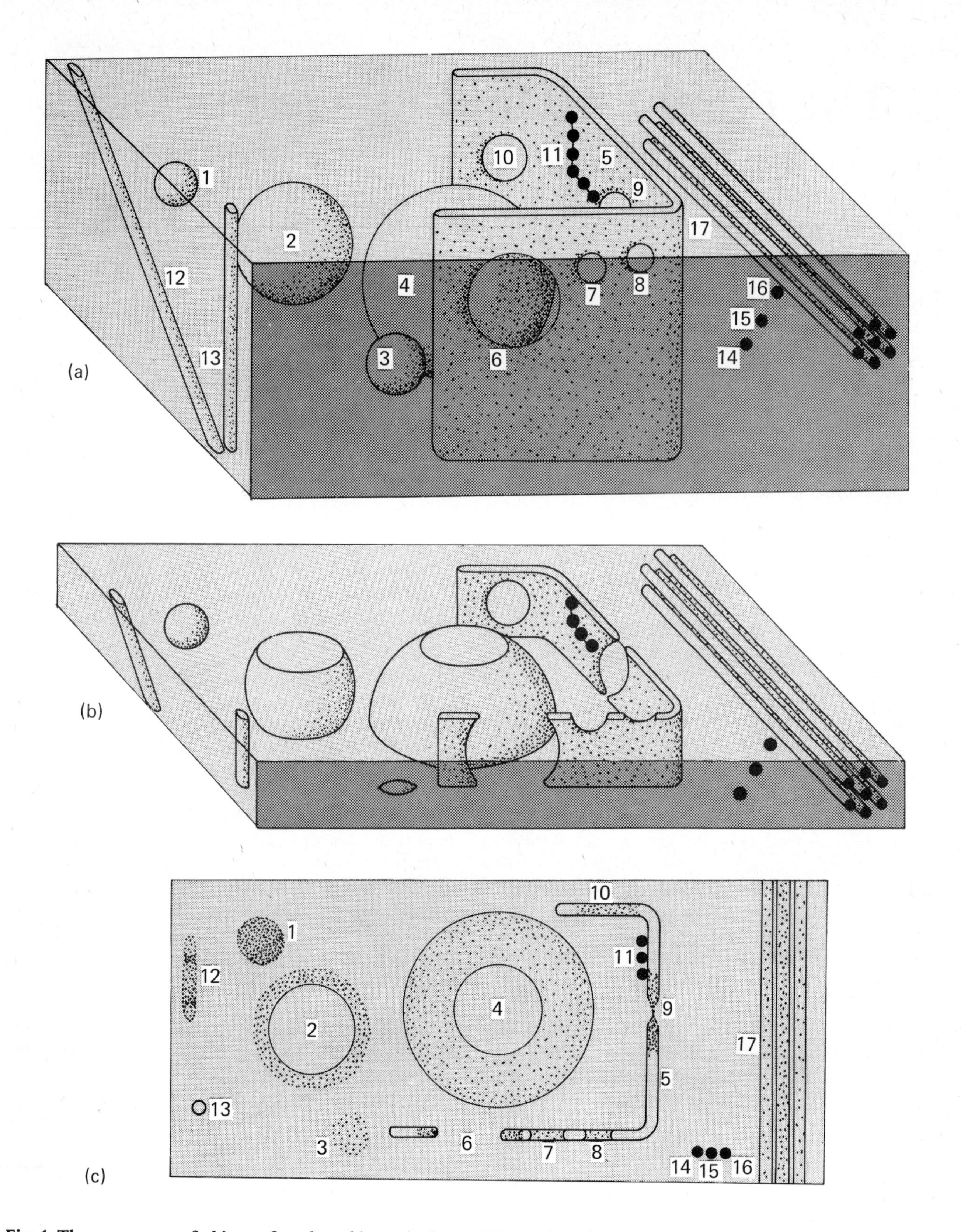

Fig. 1 The appearance of objects after ultra-thin sectioning and formation of a projected image. Various objects are viewed in the solid in (a). They are drawn so as to resemble vesicles of various sizes (1–4); a cisterna (5) with perforations (6–10) and an attached polyribosome (11); microtubules (12, 13); free ribosomes (14–16); and a bundle of rods or fibrils (17). Those portions that are included within the thickness of a section (b) generate the projected image (c).

Starting from the original material it is easy to see how the final projected image is produced. It is not so easy to visualize the 3-dimensional configuration of the original objects when, as is the situation in practice, only the final image is available. Some generalizations are as follows:

Dimensions: The projected image is strictly 2-dimensional, therefore 3-dimensional objects lose one dimension, and 2- and 1-dimensional objects lose one dimension if one of their dimensions happens to lie at right angles to the plane of the section (e.g. a 3-dimensional spherical vesicle (1) becomes a 2-dimensional circle; a 2-dimensional sheet of membrane like that bounding the cisterna (5) becomes a 1-dimensional line).

Orientations: The shape of the projected image depends on the orientation of the object with respect to the plane of the section. Thus the tubules (12) and (13) are identical except for orientation, but have different images. Membranes are seen as sharp black lines

Unlike animals, plants tend to restrict processes of cell multiplication to permanently embryonic regions termed *meristems,* and the zones between meristems and the nearby mature tissues contain cells in intermediate stages of maturation. A comparison of a juvenile and a mature stage illustrates the great precision and specificity with which cell differentiation takes place behind a meristem. Plate 49 shows the central portion of an unusually 'miniaturized' root at an early stage of development, and Plate 16a depicts the same cell types, distributed in exactly the same geometrical pattern, but in a mature part of the same root. Six different types of cell have matured in their own characteristic fashions and at their own characteristic rates, all starting from a population of comparatively uniform meristematic cells. In an organization of this sort there is clearly no such thing as a 'typical plant cell', but meristematic cells must at least contain a basic set of components. They alone have not, or have only just, started to diversify by maturation, so it is logical to use them in an introductory survey of 'the cell' (Plates 1–3, Fig. 2), before examining the constituent parts in detail.

Plate 1 is a light micrograph, showing cells in a broad bean root tip that was fixed in glutaraldehyde, dehydrated, embedded in plastic, sectioned at about 1 μm thickness, and the section stained by a combination of procedures chosen to reveal as many as possible of the cell components. Finer detail is visible in Plates 2 and 3, which are electron micrographs of ultra-thin sections of cells in other root tips. The final illustration in the introductory survey (Fig. 2) attempts to overcome the artificial two-dimensional impression created by the micrographs. It is a stylized three-dimensional interpretation of that mythical entity, the 'typical' plant cell. For the sake of clarity it is shown isolated from all the neighbours to which it should be joined. The components drawn within it are somewhat more symmetrical and simplified than they would be in life, also some have of necessity been enlarged in order to make them visible alongside their larger companions. They will be named and briefly described after some fundamental aids to the interpretation of ultrastructure have been considered.

2.2 The Appearance of Cell Membranes in Electron Micrographs

Every cell has a *membrane* which separates the cell contents from the external environment, and *eukaryotic* cells (cells with nuclei, as distinct from the *prokaryotic* bacteria and blue-green algae) possess in addition a series of internal *compartments* separated one from another by other membranes. Membranes serve to segregate and subdivide the biochemical machinery of the cell and to orient parts of it by providing surfaces where chemical reactions can take place. Before proceeding to look at specific structures it is worthwhile describing these vital cell membranes in very general terms, particularly in relation to the morphology of membrane-bound compartments and their appearance in electron micrographs.

MORPHOLOGY AND TERMINOLOGY Cell membranes are easily seen in the electron microscope, but because they are so thin they are visible in the light microscope only under favourable circumstances. The simplest type of organization is analogous to the skin of a balloon in that it is an unbroken expanse enclosing a compartment. The bounding membrane of the cell itself is of this nature, and other examples are found inside cells, delimiting *vacuoles* or *vesicles.* The latter term is usually reserved for very small compartments. Different conformations exist; thus vacuoles and vesicles are frequently spherical, but may be distorted in various ways. Two extreme cases are common, and receive special names. Elongation gives rise to *tubules,* which may or may not branch. *Cisternae* (singular *cisterna*) are markedly flattened, much as if partially inflated balloons were to be squashed between two surfaces to give compartments of great lateral extent, reduced thickness, and hence large surface to volume ratio. Cisternae may be flat, but more often have curved or irregular surfaces. They may come into contact and join with one another, or with vesicles or tubules, and may be lobed or branched. In the manner of ring-shaped balloons (or

---

only when one or both of the two knife cuts that produced the section slice through them. They are indistinct if they lie in or close to the plane of the section (3), and become more and more dense as they come closer to being perpendicular to the plane of the section (sloping sides of vesicles (4) and (2)).

Superimposed images: Even ultra-thin sections are of finite thickness, and everything that is within the section is projected. Objects that are in reality separate may be superimposed so as to appear close together, e.g. the three ribosomes (14), (15), and (16), at the bottom, middle, and top layers of the section, respectively, appear to be side by side. Some objects disappear through superimposition, e.g. the image shows only 3 of the 7 rods (17), and only 3 ribosomes in the polyribosome (11) instead of the 4 that are actually included in the section.

Section thickness: The thicker the section, the more material there is to contribute to the image. There is more membrane within the thickness of the section, and less chance of it all lying at right angles to the plane of the section: imagine, for instance, the effect of increasing or decreasing section thickness on the appearance of vesicles (3) and (4). The appearance of some structures depends upon whether they are entirely within the section (vesicle (1), pore (10)) or whether they have been cut through (vesicles (2) (3) (4), pores (6) (7) (8)).

Images of pores through cisternae are especially subject to these effects, the more so if their diameter is nearly the same as the section thickness. A pore that is slightly smaller than the thickness will, if it is perfectly centred, appear to have a narrow diaphragm traversing it (9). If the pore diameter is larger than the thickness (6), the cisterna is interrupted, but the gap is an underestimate of the pore diameter. Two pores that in reality are the same size may appear to be of different sizes if they lie at different levels ((7) and (8)). A pore that is entirely within the section has no clear cut edges and is only seen as a dense zone within the cisterna (10).

inner tubes from tyres) they may be pierced by anything from small pores to large windows.

No matter how extensively cisternae are deformed, interconnected, lobed, branched, perforated, or fenestrated, one rule is obeyed, namely that the membrane surface remains unbroken and continuous. Whereas the skin of a ruptured balloon is stable and remains torn, a broken cell membrane has a built-in tendency to reform a closed system or systems in which there are no free edges. This property of self-healing is very important. As far as is known, membrane-bound compartments are *always* enclosed and the contents are *never* in direct contact with the surroundings, except when certain membrane fusion processes liberate the contents of one compartment into another.

In all of the examples so far the compartmentation has been based on single expanses of membrane, but more complex situations exist in which one expanse is completely enclosed within another, giving three concentric spaces—the surroundings, the space *between* the two membranes, and the space *enclosed* by the inner of the two layers. In such cases the innermost space is described as being enclosed by a *double envelope*.

ORIENTATION AND APPEARANCE The lipid and protein molecules of which membranes are mainly composed take up considerable quantities of osmium and the other heavy element stains used in preparing ultra-thin sections for electron microscopy. Consequently membranes appear darker in the final image than their surroundings (unless they happen to enclose a very electron-dense substance). However, membranous expanses are only 6 to 10 nm in thickness, whereas ultra-thin sections are up to ten times thicker. It is therefore commonly found that membranes, occupying as they do only a small proportion of the bulk of the section, become obscured by other material. In fact their visibility and appearance depend greatly on their orientation with respect to the plane of the section. Just as a pane of glass is almost invisible when viewed at right angles to its surface, but becomes progressively more obvious if it is turned until it is seen edge-on, so membranes can be elusive or quite distinct. They are hard to see if they happen to lie in the plane of the section so that the electron beam 'looks' through them at right angles (unless they have particles or some other marks on their surface analogous to dirt on the pane of glass). If they lie at right angles to the plane of the section they are viewed edge-on and are at their most conspicuous, each one appearing as a dense black line. Intermediate orientations give less sharp images. This phenomenon recurs may times in the electron micrographs, and examples are specifically labelled in Plate 3a.

The whole range of possibilities can be visualized by imagining what a spherical vesicle, 1μm in diameter, would look like if viewed in ten successive, or 'serial' sections, each one 0.1 μm thick. The first section is tangential to the vesicle and merely grazes its surface. Any membrane that is included lies in, or very close to, the plane of the section and is therefore difficult to detect. By contrast, the fifth section slices through the diameter of the vesicle, and contains membrane lying at right angles to the plane of the section and appearing as a dark line circumscribing a circle of diameter 1 μm. Sections 2, 3 and 4 include progressively less oblique and therefore sharper and sharper views. A spherical vesicle, being symmetrical in three dimensions, is a relatively simple object, whereas the appearance in sections of tubules and cisternae can be so complex as to defy interpretation unless serial sections are available from which three-dimensional reconstructions (e.g. Plates 18b and 30c) can be built.

These and other aspects of the art of interpreting images of ultra-thin sections are explained in Fig. 1 and its caption.

2.3 The Components of the Cell

Plant tissues are composed of the non-living *extracellular* region and the living *protoplasm* of the cells proper. The former consists of *intercellular spaces* and *cell walls*. Each *protoplast* consists of *nucleus* (or sometimes several *nuclei*) and *cytoplasm*, and within these major subdivisions of the cell are the various membranous and non-membranous components with which subsequent chapters are concerned.[129, 473] The following list gives little more than a list of names and outline descriptions. Functions are not included. The numbers that follow each item indicate which of the first three Plates give the best views of the structure in question; the letters in brackets refer to labels on Fig. 2.

Cell wall: This is a thin structure in meristematic cells, but it can be very massive and elaborate (1, 2).

Plasma membrane: The bounding membrane of the protoplast, normally in close contact with the inner face of the cell wall (2, 3b).

Plasmodesmata: Narrow cytoplasmic channels, bounded by the plasma membrane, and interconneçting adjacent protoplasts through the intervening wall. The singular is *plasmodesma* (2, 3b) (PD).

Vacuole: Compared with the surrounding cytoplasm, these are usually empty looking spaces, spherical when small. They can, however, be very large, exceeding by many times the bulk of the cytoplasm itself (1) (V).

Tonoplast: The membrane that bounds a vacuole. Except for its position in the cell it looks very like the plasma membrane (3a, 3b).

Nuclear envelope: A cisterna wrapped around the contents of the nucleus (N). The space *between* the two membranous faces of the cisterna is the *perinuclear space* (1, 2, 3a).

Nuclear envelope pores: Perforations in the nuclear envelope, through which the cytoplasm may be in continuity with the contents of the nucleus (3a) (NP).

Chromatin: The genetic material of the cell, containing

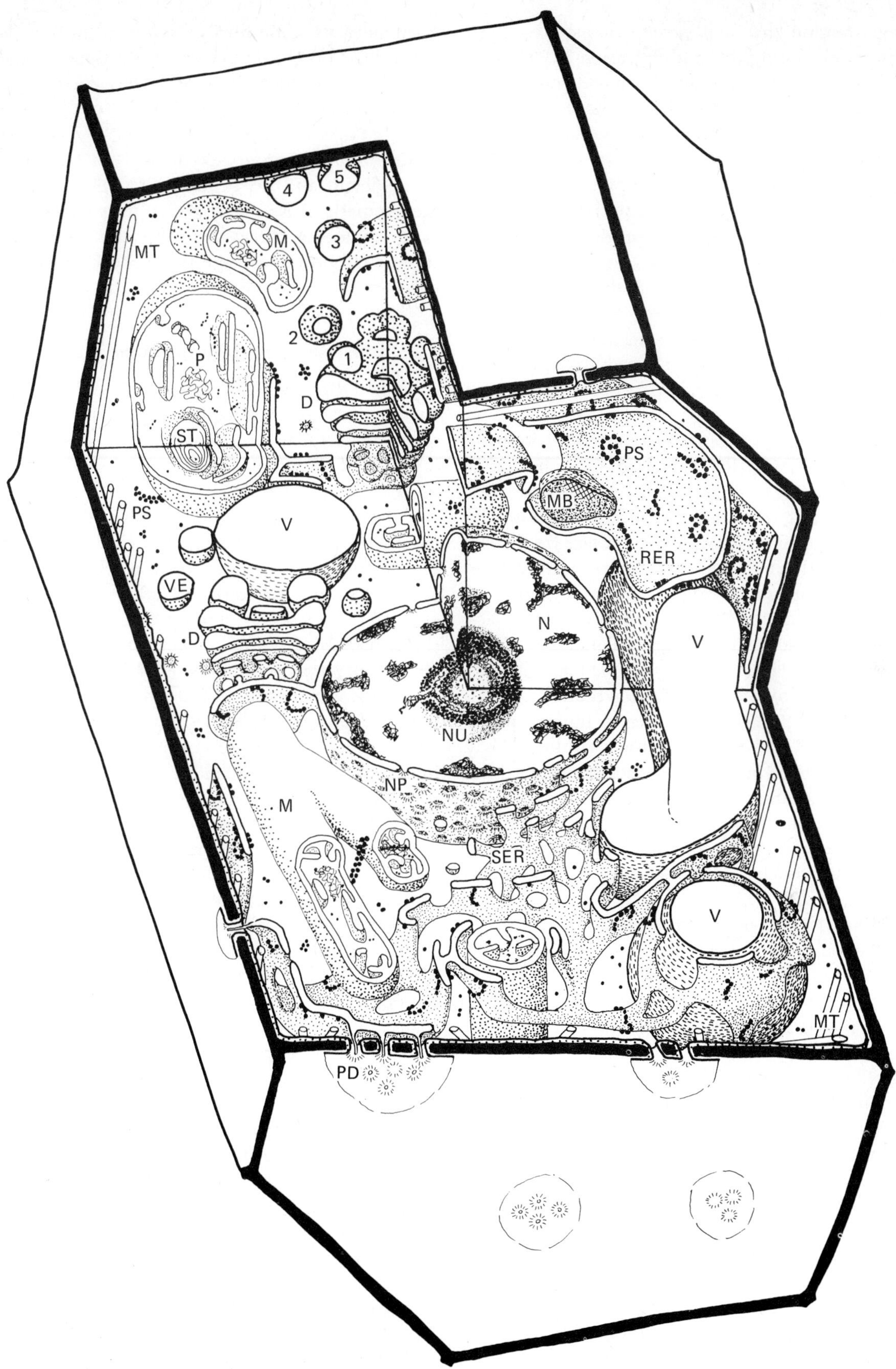

Fig. 2
Diagram of an undifferentiated cell cut open to show the three-dimensional structure of the principal components and their interrelationshps. For clarity they are not drawn to scale and some are illustrated by only a few examples (e.g. ribosomes, see Plate 2). They may be identified by the letters, which refer to those in parentheses in the list on pages 10 and 12.

the information that is passed from parent cell to daughter cell during the multiplication of cells and reproduction of the organism. It can exist in various forms, and during the division of nuclei it is condensed, so that discrete units, *chromosomes,* can then be recognized (1).

Nucleolus: A mass of fine threads and particles, largely a sequence of identical units of specialized genetic material together with materials produced from that genetic information (1, 2) (NU).

Nucleoplasm: Everything enclosed by the nuclear envelope falls in the category of nucleoplasm, just as objects outside it are constituents of the cytoplasm. The word is often, however, used to denote the ground substance in which the chromatin and nucleolus lie (1, 2).

Endoplasmic reticulum: Cisternae that ramify through the cytoplasm, occasionally connected to the outer membrane of the nuclear envelope. The inner face of the bounding membrane is in contact with the contents of the cisterna, and the outer face frequently bears attached ribosomes and polyribosomes (see below). Endoplasmic reticulum is described as 'rough', or granular (RER), and forms that lack ribosomes as 'smooth', or agranular (SER). The cisternae may or may not have visible contents, distending the cisternae when present in bulk (2, 3a, 3b).

Ribosomes: Small particles lying free in the cytoplasm or else attached to the endoplasmic reticulum. They may be aggregated in clusters, chains, spirals, or other *polyribosome* configurations (2, 3a).

Dictyosomes: The units of the *Golgi apparatus* of the cell. Each dictyosome consists of a stack of cisternae, and many small vesicles (VE, 1–5) are usually found nearby (2, 3a, 3b) (D).

Mitochondria: Components consisting of a compartment surrounded by a double envelope. The outer membrane of the double envelope is more or less smooth, but the inner is thrown into many folds—*mitochondrial cristae*—that project into the central compartment (2a, 3a) (M).

Plastids: This is a group name for a whole family of cell components, represented in the root-tip cells of the present survey by the simplest member, which is called the *proplastid* (P). Proplastids are usually larger than mitochondria, but like them have a double membrane envelope surrounding (in these examples) a fairly dense ground substance. *Starch grains* may be present in them (1, 3a) (ST).

Microbodies: These are bounded by a single membrane, and are distinguished from other vesicles by their size and dense contents (sometimes including a crystal) (2) (MB).

Microtubules: Except during cell division, these very narrow cylinders lie just inside the plasma membrane. The wall of the cylinder is made of protein and is *not* a cell membrane, though it superficially resembles one in its thickness and density (2, 3a) (MT).

Microfilaments: Fine fibrils, shown in some cases to be composed of a material resembling actin, one of the major constituents of muscle. They are not illustrated in Plates 1–3 (see Chapter 11).

Subsequent chapters examine these components in more detail, and since many of them are present in meristematic cells in only a very simple form, this will mean exploring specialized cell types as well, in order to seek out and assess significant aspects of the known diversity.

2.4 Substructure of Membranes

A general topic which if introduced now will provide a background for later chapters concerns the substructure of cell membranes—a complex, controversial and growing subject, its scope extending far beyond the outline given here.[78, 79, 82a, 769]

Cell membranes assume specific shapes, take up specific positions in the cell, develop specific features like perforations or surface coats of ribosomes, and to a certain extent have different affinities for the heavy elements used when staining ultra-thin sections. In short, no two types of cell membrane are completely alike. Chemical analysis of isolated membranes confirms this conclusion. All of them consist largely of lipid and protein molecules, but the proportions vary, and so does the precise composition. Some types of lipid may be present in all classes of membrane, indeed some lipid molecules can move from one membrane to another. Other lipids, and many of the proteins, are unique to specific membrane systems.

THE BIMOLECULAR LAYER The key to understanding the way lipid and protein molecules come together to create a membrane lies in their physical reactions with water molecules; reactions based on forces of attraction or repulsion. The majority of the lipids, and probably many of the proteins, are polarized molecules, the distribution of chemical groupings within them conferring high affinity for water on one pole (the *hydrophilic* end) and low affinity for water on another pole (the *hydrophobic* end). When in an aqueous environment such as the cell, polarized molecules like these tend to form aggregates with the hydrophobic portions internally located, mutually shielding one another so that the surrounding water is in contact only with the outwardly directed hydrophilic regions. Several types of aggregate are possible, and one of them is especially relevant because of its resemblance to cell membranes. Its construction can be envisaged in stages, first by picturing what happens when a small drop of polarized lipid is allowed to spread on the surface of water. The molecules line up side by side in an unbroken expanse with all the hydrophilic ends in the water and all the hydrophobic ends directed away from the water. Next consider a wire loop that is dipped into a suitable lipid in order to pick up a thin film, then immersed in water. In this situation, with water on *both* sides of the film, *two* molecular expanses of the sort formed at a water-air interface associate back to back to give a *bimolecular layer* with a hydrophobic *interior* and hydrophilic *surfaces*. Like this model system, genuine cell membranes are essentially thin films containing a high proportion of polarized molecules, immersed in an aqueous environment. After much controversy there seems now to be a consensus of opinion that they are indeed founded on a bimolecular layer of lipid.

A common variation on the bimolecular layer, easily made in the laboratory and also seen in cells which produce suitable lipids in bulk (e.g. Plate 3b), is the myelin figure, consisting of many bimolecular expanses superimposed on one another. The polarity of the constituent molecules is evident not only in their alignment, but also in their reaction with osmium tetroxide. The hydrophilic outer faces are electron-dense in the final image, so that each bimolecular layer appears as a dark-light-dark 'tramline' with the light centre representing the hydrophobic interior. The inserts in Plate 3b show that genuine cell membranes also have a tramline appearance, similar to that of bimolecular leaflets of lipid.

Some cell membranes are thicker than others, and all of them are thicker than a bimolecular layer of pure lipid (Plate 3b). Protein molecules dissolved in or attached to the lipid increase the overall thickness. Some float like micro-icebergs on one or other of the lipid surfaces, and some probably penetrate right through from one face to the other. Protein molecules give freeze etched membrane surfaces their characteristic knobbly appearance (e.g. Plates 4a, 17a, 25 and 33), the smooth regions between the knobs representing lipid. In many cases the knobs are in fact the *bottoms* of the 'icebergs', for, as stated on page 3, the fracture face that is observed in the freeze etching procedure represents regions in a frozen cell where the ice is weak. The *interior* of membranes is weak because, being hydrophobic, it contains comparatively few water molecules with which ice can be made. Consequently a frozen cell fractures preferentially along the *mid-line* of its membranes, between the two layers of lipid, so the knobs that are exposed are not protrusions from the membrane into the surrounding cytoplasm, but hydrophobic portions of proteins, anchored in the hydrophobic heart of the membrane.

DYNAMIC PROCESSES Membranes are subject to dynamic alterations. It is known that their gross morphology can change with time, as can the distribution and numbers of the particles seen by freeze etching. At a biochemical level, their constituent molecules are impermanent, continually being broken down and resynthesized. Models purporting to depict the molecular architecture of membranes must therefore be flexible enough to allow for such effects. One recent view is that the bimolecular layer of lipid can be regarded as a solvent, differing from conventional solvents mainly in that it is in an unusually thin form. The proteins of the membrane dissolve in the solvent, and lie on one or other face or in the interior. Just as molecules in ordinary solvents can move by diffusion, so can molecules in membranes. The diffusion coefficients of some of the lipid constituents are large enough to imply rates of molecular movement in the plane of the membrane of the order of micrometres per second. Protein molecules, being larger, and because they interact with the molecules of lipid 'solvent' around them, are less mobile. It is known, however, that at least some types can move in the plane of the membrane, and freeze etching studies that demonstrate dynamic changes in membranes, such as the appearance and disappearance of local congregations of particles, are consistent with the idea that proteins have restricted freedom to diffuse laterally, and hence that specialized areas can develop in localized regions of an expanse of membrane. It appears that both lipid and protein molecules are much less free to move from one face of the bimolecular layer to the other, a restriction which allows stable, asymmetric membranes to exist.

There are quite consistent differences in thickness between classes of membrane. In general the endoplasmic reticulum is the thinnest and the plasma membrane and tonoplast the thickest. The successive cisternae of dictyosomes are remarkable in that a transition from the thin to the thick category has been observed. The suspicion that dictyosomes may be devices which, amongst other functions, convert one type of membrane into another will be considered in more detail in Chapter 7.

The dimensional, chemical, and morphological heterogeneity of cell membranes should not be cause for surprise, considering the diversity of the functions they perform. But there is another reason for heterogeneity. If membranes were perfectly alike, it would presumably be as easy for them as it is for soap bubbles to fuse with one another. The compartments of the cell would coalesce into a disordered mixture of enzymes and substrates, many of them incompatible. A few minutes spent observing living cells by phase-contrast light microscopy should convince anyone that while certain types of fusion and coalescence can and do occur (vacuole with vacuole, mitochondrion with mitochondrion, etc.), others do not (vacuole with plasma membrane, mitochondrion with nucleus, etc.). Clearly, recognition systems have evolved and have been incorporated into cell membranes, limiting the options for membrane fusion reactions. Little is known about the nature of such recognition systems, except in the case of the highly specific antigens found on the plasma membrane of animal cells. However, they must exist, and the vital part they play in the life of the cell will be a recurring theme in later chapters.

Questions concerning the origin and growth of membranes are worth bearing in mind for the subsequent detailed descriptions of membranous cell components. Growth either involves the addition of molecules of the types that are already present, or, where the character of the membrane alters, of new types. Are the additional molecules made within the membranes themselves? If they are made elsewhere, how are they guided to their correct destinations? Do membranes have growing zones or can molecules be incorporated anywhere? Are the constituent molecules inserted in any special order? As will become apparent, not many such questions can be answered.

As to the origin of membranes, there are in the last analysis two contrasting possibilities. One is a *de novo* origin, where a suitable assemblage of molecules comes together to create a membrane anew. The other is that there is no *ultimate* origin of membranes, because new cells formed in meristems receive from their parents a full quota of all types, each one capable of growing, multiplying by fission or fragmentation, and being in turn passed on to the *next* generation of cells. The concept of structures arising from similar, pre-existing structures is familiar for its application, dating from Virchow in 1855, to cells themselves (*omnis cellula e cellula*), and such is the molecular complexity of cell membranes that intuitively it seems right to apply it to them too. Evidence to back this intuition will in fact be presented later in respect of certain membrane systems. However, membranes must have had a *de novo* origin sometime in the course of evolution, and they, or something not unlike them, can be made in the test tube. It may well be that the cell makes use of all possible modes of membrane formation, growth, and perpetuation.

2.5 Towards a Definition of the Cell

This chapter began by recounting how concepts and definitions of the cell had to be altered as new information accrued over the years. The advent of electron microscopy and the growth of other branches of biology have added immensely to the available body of knowledge, yet it is still difficult to define 'the cell'. Phrases such as 'the unit of life' or 'the unit of structure of living organisms' are commonly used, but are not very informative, though they express something of the properties of cells.

The case that there is no such thing as a 'typical' cell has already been stated (page 9), and it seems impossible to base a definition on purely morphological descriptions—such as of the cell components listed on pages 10–12, however painstakingly they are described. Individually the components are impermanent. They cannot function in isolation unless they are provided with all of the necessities normally available to them *in living cells.* When in the cell, they are able to interact and collaborate in a group existence, their collective activity tending to ensure both their own permanence and the permanence of the whole system, and endowing the system with a capacity to develop and reproduce as a unit. The collaboration and inter-dependence within the cell may be summarized by saying that a cell is greater than the sum of its components.

In multicellular organisms the environment of a cell contains other cells, all interdependent, and collaborating in a group existence just as do sub-cellular components within individual cells (Plate 49). It is possible, however, to devise culture media which will allow cells isolated from multicellular organisms to develop and reproduce. This cannot (yet!) be done with sub-cellular components: it is only at the level of integration seen in the whole cell that the breakthrough towards permanence and potential independence is achieved.

By placing emphasis on the mode of operation of the cell rather than on its morphology, it becomes possible to approach a definition. For instance, the cell can be stated to be the smallest independent system capable of survival, growth, and reproduction using raw materials and energy obtained from its environment. The statement is of the same nature as 'the cell is the unit of life', but with the substitution of operational descriptions for 'unit' and 'life'. It would be premature at this stage to delve deeper into problems of defining the cell: the topic will, however, be re-examined in the final Chapter.

3 The Outer Surface of the Cell—Cell Wall and Plasma Membrane

---

Botanists have for long used differences in the structure and staining reactions of cell walls as aids to the classification and definition of types of cell and tissue in plants. Indeed some texts, and most anatomical drawings, focus attention so exclusively upon walls that the student is in danger of being left with the impression that plants consist of nothing else. Since our primary concern is with the cell, the emphasis in the present book is in the opposite direction. In plants, just as in animals and bacteria, the outer limit of the cell is the plasma membrane, and the cell wall is an *extra-cellular* product, akin to comparable materials such as bone and collagen in animals and capsules in bacteria.

Walls have their own intrinsic importance and interest to the plant cell biologist, and a number of selected features is presented in this chapter. There are, however, additional reasons for including descriptions of walls early in the book. They serve to introduce many of the specialized cell types that will be met later in connection with other topics. Also, what we learn about walls tells us much about the life of the cells which they enclose. For instance, the wall is a long lasting testimony to the ability of the cell to organize and control the secretion of specific materials at specific times and in specific places. Again, the plant cell experiences its immediate environment through its wall, and the wall modifies environmental effects to the extent that it dominates the life of the cell and the plant. All of these themes will be developed. The wall will be considered first and the plasma membrane second. The two will then be viewed together as inseparable components of certain specialized wall-membrane complexes.

3.1 The Development of the Cell Wall: Introduction

All cells arise by division of pre-existing cells, and at the end of (almost) every division the progeny become separated by a newly synthesized partition, the *cell plate* (pp. 168–170 and Plate 48), which they share, and which will join on to the walls of the parent. Once the cell plate is fully developed, the cells that it separates can embark on their own developmental pathways. They deposit wall materials upon the shared partition, now called the *middle lamella,* which remains as, or is modified to become, a layer of intercellular cement. Walls that are deposited while the cells are enlarging are called *primary,* and there is either a gradual or an abrupt change to *secondary* walls, in general formed once the cell has enlarged. The successive stages of wall deposition create a time sequence rather like geological strata, the first-formed being adjacent to the middle lamella and the most recent being just outside the plasma membrane.

There are two fundamental modes of growth and development. *New* structures can be synthesized from precursors that are of molecular dimensions, and *pre-existing* structures can be modified. Cell wall development includes both processes, but the importance of the two varies from one wall to another. Some walls are direct derivatives of newly-synthesized cell plates, modified only by primary and secondary wall deposition. They are sometimes described as 'division walls'. Others have undergone such a long history of modification that their ultimate origin can hardly be discerned. An example of one such 'non-division wall' is the continuous wall that covers the whole outer surface of the plant. It is formed entirely by stretching and modification, the only stage at which a 'new' wall contributes to it being when the fertilized egg cell lays down its wall.

Maturation and specialization of cells bring both structural and chemical diversity to the walls. One wall can be utterly different from its neighbour just across a middle lamella. Within a single cell, the different walls, having different origins and histories, can differ in structure and composition. The same applies even within an individual wall, where different strata and different local areas can be specialized. Although the wall is non-living, it is far from being a static compartment where components, once inserted, remain for ever unchanged. They can be altered, or even digested and removed. Since the wall is non-living, its precision engineering is all the more remarkable. Many efforts are being made to discover how the living protoplast that is responsible for fashioning the wall is organized to control not only the synthesis and deposition of specific wall substances, but also their deposition in highly specific places and at specific times.

3.2 Fundamental Components: Microfibrils and Matrix

Many materials used in modern technology are composite systems, with mechanical and chemical properties of one component complementing those of the other. Carbon-fibre steel, fibreglass, and reinforced concrete resemble plant cell walls in being fibres embedded in an amorphous medium. They differ, however, in that they are static, whereas cell walls are dynamic structures, subject to alteration as the cells grow or respond to environmental stimuli. In cell walls the fibres are very small, and the word *microfibril* is used to convey their diminutive nature. The medium, or *matrix,* they lie in is a jelly-like assemblage of interlinked molecules, and of these we shall for the moment describe only those types known to occur in the primary and secondary walls of many, if not all, cells. Other types of matrix found in special locations will be encountered later, under the heading of secondary walls.[592, 703, 704]

CELLULOSE MICROFIBRILS Except in a few algae[670] and some fungi the microfibrils consist of cellulose.[670a] Like so many biological macromolecules, this is a *polymer,* made by joining together numerous relatively simple and identical units, *monomers,* in this case molecules of glucose. Glucose can be polymerized in various ways to produce starch, callose, or cellulose, depending on the nature and positioning of the chemical bonds that are used. Cellulose molecules are at least several hundreds of nanometres long, and consist of polymerized glucose units in a linear, unbranched chain. Cellulose microfibrils[137] are composed of such molecules linked together along their long axes in numbers sufficient to render the aggregate visible in the electron microscope. Microfibrils can be seen in ultra-thin sections[147] (Plates 14d, 42b), freeze etched specimens (Plate 4a), and best of all in portions of cell wall prepared by shadow casting after the matrix (which would obscure the microfibrils) has been extracted (Plate 4b, c).

Cellulose microfibrils are very long, and free ends are rarely seen. They range from about 3.5 nm to 10 nm in thickness, those of primary walls of higher plants being at the narrow end of this range, while even thicker examples are found in walls of various algal cells. Numerous long chain molecules of cellulose lie along individual microfibrils, and are in places held in relation to one another by hydrogen bonds in a precise crystalline arrangement. Elsewhere the molecules lie together, still hydrogen bonded, but more loosely, in non-crystalline arrangements. Individual cellulose molecules are long enough to lie along the microfibrils and link more than one crystalline zone via the intervening non-crystalline region or regions. Microfibrils are thus extremely resistant to stretching. If a wall is to stretch during growth of the cell, the mechanism must involve slipping of neighbouring microfibrils over one another, rather than stretching of the individual structures. This is not to say that microfibrils are unalterable structures—in fact it is known that the degree of polymerization of cellulose (equivalent to length of molecule) can alter during growth.[781]

THE WALL-MATRIX In view of the somewhat intractable mechanical properties of cellulose microfibrils, it has long been considered that the key to many of the properties of cell walls lies in the matrix, which is pictured as a network consisting of several types of macromolecule, bridging the spaces between microfibrils. Quantitatively, the most abundant matrix materials are, like cellulose, polysaccharides, that is, polymers of sugars or derivatives of sugars such as uronic acids (glucose and galactose, for instance, are sugars from which glucuronic acid and galacturonic acid are formed). Specialized proteinaceous material is also present in the wall matrix.

Analysis of the wall matrix in chemical terms, and hence also in structural terms, has proved to be extremely difficult.[592] The polysaccharide moiety has classically been regarded as being composed of two main fractions, the *hemicelluloses* and the *pectic substances.* The differences between the two are largely operational in that different procedures are used to extract them. Naturally, the fact that fractions *can* be distinguished implies that there are chemical differences. Further, within each major fraction there generally are chemically separable sub-fractions. Problems arise when attempts are made to put all of the available analytical data together. It is certain that the fractions and sub-fractions isolated by chemical extraction represent degradation products, and very uncertain whether the extracts are representative of the matrix as it is found in the wall of a living cell.

A great deal is known about the matrix of the primary wall of sycamore *(Acer)* cells that have been grown in artificial culture. It has proved possible to culture uniform batches of cells, to prepare samples of their walls, and to use enzymes to release fragments of matrix polysaccharides with far greater specificity than is obtained using chemical extraction. The results confirm much other work, and form the basis for a brief general account of the matrix and its structure.[39, 422, 808, 892]

HEMICELLULOSE In sycamore, the hemicellulose consists of glucose units together with units of the 5-carbon sugar, xylose. Like hemicelluloses from other sources, they are branched, non-crystalline molecules. Other sugars may occur, for instance there is mannose in wall matrix hemicelluloses of various gymnosperms. Of the greatest importance in relation to cell wall structure is the fact that the hemicellulose molecules can become linked by numerous hydrogen bonds to cellulose microfibrils, and by strong covalent bonds to certain of the pectic substances. It is because of the latter type of link that chemical extraction can yield fractions that have some of the properties of hemicellulose and some of pectic material.

PECTIC SUBSTANCES Pectic substances are not, as is often described, simple polymers of galacturonic acid units (so-called pectic acid), or of similar units with the acidic groups neutralized by esterification with methyl groups (so-called pectin). Such material can occur in some plants, but the general situation is considerably more complex.[2] The sub-fractions obtained from sycamore cells seem likely to be derived from a backbone polymer of galacturonic acid and rhamnose, with side branches containing arabinose and galactose. Work on other material shows that additional sugars can be present, and adds a further dimension of complexity: the composition of the pectic material is subject to dynamic alteration during the life of a cell.[592]

The acidic component of the wall matrix is a quantitatively significant trap for cations amongst the mineral salts taken in by plants, and it concentrates them to a remarkably high degree. Calcium salts such as *calcium pectate* have particular structural value because calcium ions are divalent and can therefore form strong bridges, or cross-links, between neighbouring acidic molecules. Experiments in which calcium is dissolved out of cell walls by the reagent ethylene diamine tetra-acetic acid testify to the exploitation of this phenomenon in plants: much of the quality of the middle lamella as an intercellular cement (referred to above) is due to it being impregnated with calcium pectate. If the calcium is extracted the tissue is liable to fall apart, indeed this is one way in which suspensions of cells can be obtained.

Linkage of hemicelluloses to cellulose microfibrils, and linkage of pectic material to hemicellulose have been mentioned above, and of equal significance is the fact that additional bonds extend the network of molecules in the wall matrix still further by linking on to the third major category of matrix material—that based on protein.

CELL WALL PROTEIN The primary wall contains an appreciable proportion of protein, and one which has unusual properties. It somewhat resembles collagen, the extra-cellular skeletal protein of animals, in its high content of the amino acid hydroxyproline, and in the absence of the amino acid tryptophane. It is not known what, if anything, lies behind this homology between extracellular proteins of the two kingdoms.[460]

MATRIX PLUS MICROFIBRILS The matrix emerges as a giant molecule, composed of proteinaceous, hemicellulosic, and pectic units all linked to one another, and in turn linked by hydrogen bonds to the microfibrillar part of the wall. It has been suggested that the number of the latter hydrogen bonds might regulate the extensibility of the wall, and hence the rate at which the cell enlarges (or the position in the wall at which extension occurs).[422] Biochemical systems which could, at any rate in theory, regulate the degree of hydrogen bonding do exist, and are being investigated to see whether the action of growth-stimulating hormone involves them.[320]

3.3 Primary Walls and Cell Enlargement

Microfibrils are deposited in the primary wall in specific patterns. They may be randomly arranged within the plane of the wall (= the plane of the paper in Plate 4b); they may be aligned in particular directions (Plate 4c); and they may be specifically sculptured around specialized areas of wall (Plates 4c and 14e).

Cell enlargement necessarily involves stretching of that part of the wall that has already been laid down. If the enlarging cell is in a tissue, it also involves overcoming the pressures exerted by the surrounding cells. These mechanical constraints influence the manner of microfibril deposition. As might be expected, the simplest pattern of microfibrils arises in the absence of mechanical constraints. In it, the microfibrils lie randomly in the plane of the wall. It is a pattern that develops in the wall of the more-or-less spherical single-celled alga *Chlorella,* and also in higher plant cells if they are in effect converted to unconstrained spherical unicells. This experiment can be done either by digesting away existing walls to allow naked protoplasts to escape into a liquid culture medium, round off, and start making new walls, or by withdrawing enough water from the protoplasts to make them shrink inwards, free of the surrounding wall, round off, and make a new wall within the old. The fact that naked protoplasts only seem to regenerate a wall if they contain a nucleus, emphasizes that processes of wall formation, even though they occur at the periphery of the cell, are subject to ultimate control by the nuclear genetic information, mediated by the cytoplasm (see p. 85).[131, 132]

Many types of *parenchyma* cells—the ground tissue of multicellular plants—are thin-walled and nearly spherical. A recent study of those in the flesh of the apple fruit has confirmed other reports of random microfibrils in their walls.[581]. Frequently, however, pressures exerted by the surrounding cells flatten the faces that are in contact, and so convert spheres to polyhedra. In such cases the flat faces retain the random pattern, but the edges of each face may become strengthened by ribs formed by microfibrils aligned in parallel. Plate 4c illustrates a rib. In general, random arrangements imply uniformity of strength in all directions (in the plane of the wall) and this in turn implies the ability to expand uniformly, as in spherical or iso-diametric polyhedral cells.

One of the most important of the other three-dimensional shapes of growing cells is the cylinder with flat ends, as exemplified in the zone of cell elongation behind the apical meristem of a root. Here the axis of each cylindrical cell lies along the axis of the root, and while the cells may not greatly increase in diameter, they do elongate, that is, the cylinders get longer. Their curved side walls (e.g. the *short* walls at the *top* and the *bottom* of the micrograph in Plate 2) therefore experience a much greater increase in surface area than do the end walls (e.g. the *long* walls at the *sides* of Plate 2). Any expansion undergone by the end walls is uniform in all directions in the plane of the wall, and correlating with this, they possess randomly oriented microfibrils. The side walls have a predominantly transverse mode of deposition, with the result that the microfibrils do not impede stretching of the cylinder, but, like the hoops round a barrel, restrict increases in girth. As in polyhedral parenchyma, tissue pressures may produce flattened faces up and down the walls of the cylinder, and again ribs composed of longitudinally oriented microfibrils may appear at the edges (Plate 4c).[704]

The phrase 'transverse mode of deposition' used above requires to be elaborated, as it conceals a fundamental aspect of the dynamics of cell enlargement. Cell elongation and primary wall deposition proceed hand in hand, and microfibril patterns that are laid down early on become progressively stretched out of their initial orientation. Predominantly transverse microfibrils become obliquely criss-crossed and eventually may end up closer to longitudinal than transverse. Meanwhile, the cell continues deposition on top of the microfibrils that are already in the wall. At any time during cell elongation the most recently formed microfibrils (just outside the plasma membrane) are the closest to transverse, and they become more and more oblique moving towards the middle lamella into older strata that have undergone more stretching. The sequence in depth in the wall represents a sequence in time.[704] These phenomena are illustrated in part in Plate 4b.

It is implicit in the above view of primary wall growth that expansion and deposition of wall material occurs uniformly all over the surface. This has in fact been confirmed in experiments utilizing radioactive raw materials to locate sites of incorporation. There are exceptions to the uniform type of wall growth, thus root hairs, pollen tubes, and fungal filaments grow only at the tip. Cells in a tissue do not extend as individuals, but in co-ordination as members of a commune; further, the members are interconnected by plasmodesmata throughout the period of extension. Neighbouring cells can evidently perform the delicate task of enlarging so precisely in step with one another that their plasmodesmata do not get sheared.

Clearly, the ability to enlarge, the three-dimensional shape that will develop, and the disposition of microfibrils, are all inter-related. We can catch glimpses of what may be a beautifully simple automatic regulatory system, in which an initial pattern of microfibrils creates a particular distribution of mechanical stresses in a wall that is subjected to forces of extension; the stresses then determine the orientation and distribution of new microfibrils, which in their turn determine the pattern of new stresses. . . .[290] However, there are still many unknown factors. The initial microfibrils and the stresses are like the chicken and the egg—which come first? How does the living protoplast perceive the stresses and translate them into patterns of deposition? How are sudden alterations initiated? We shall return to this subject in Chapter 11 in connection with cases of parallelism observed between microfibrils in the wall and microtubules in the cytoplasm just inside the plasma membrane, but cannot hope to provide full answers until much more is known about the synthesis of individual microfibrils and the nature and location of the enzymes involved (see also page 25).

3.4 Secondary Walls

Secondary walls are, by definition, formed mainly after growth has ceased. This does not mean that *all* cells make secondary walls once they stop growing. On the contrary, there are many examples of mature cells that remain thin-walled, apparently confining their wall synthesizing activities to maintenance, 'turnover', or perhaps modification of components. Others, however, produce distinctive secondary walls that reflect the function of the cells.

The microfibrils of secondary walls are conspicuous for the regularity with which they are aligned, largely because they are not subject to the distorting effects of cell enlargement. In elongated examples such as *fibres* they lie parallel to one another in helices wound around the long axis of the cell and at a set angle to it.[698] Some 'woody' cell types characteristically deposit two or three easily recognized layers of secondary wall. There may be many successive strata alternating in microfibril orientation.[704] Other specializations are achieved by varying the ratio of microfibrils to matrix materials. The hairs on cotton fruits illustrate one extreme. They are up to 95% cellulose. Examples of the other extreme include the thick walls of cells in some seeds (e.g. dates, or *Phaseolus* beans), where large quantities of matrix hemicellulose form a food reserve that is consumed during germination. Some tissues produce so much matrix carbohydrate that the middle lamellae are lost, the protoplasts becoming relatively isolated in masses of jelly-like mucilage. The alginic acid and agar-agar of brown and red seaweeds respectively, and the okra of *Hibiscus* pods are examples of commercial interest.

Finally, many secondary walls are specialized through their possession of chemical compounds that have not as yet been mentioned. In general these are non-microfibrillar materials that form either *encrusting* deposits on the cellulose microfibrils or *adcrusting* layers on the surface of the wall. Examples are illustrated in the following sections.

3.4.1 Lignin

Lignins are exceedingly complex polymers deposited on the surfaces of the microfibrils during secondary wall formation in a variety of cell types that at maturity are said to be *lignified*. The monomers are 3-carbon chains attached to 6-carbon phenols, i.e. phenylpropane units. Many such units polymerize in three dimensions to give large, amorphous molecules conferring much rigidity on the cell walls.[866a] The microfibrils resist stretching and the lignin resists compression.[705]

Exploitation of the properties of lignin for skeletal purposes may have been a vital factor in the evolutionary success and spread of the first land plants. It is produced by many types of cell, sometimes in bulk, sometimes in small and delicately placed quantities. One common misconception is that all lignified cells are dead at maturity, but this is far from being a valid generalization: it does, however, apply to *vessels* and *tracheids*, the two types of conducting element in the *xylem* tissue, functioning in the carriage of water and dissolved substances through the plant from regions of absorption to regions of water loss by evaporation. Between one-third and one-fifth of wood (depending on the species) is lignin, i.e. in general a higher quantity than the hemicellulose fraction, but a good deal less than the content of cellulose. Herbaceous plants have relatively small quantities of lignin, but as individuals their lignified cells are comparable to their more abundant counterparts in woody species.

DEVELOPMENT OF VESSELS Plates 5 and 6 are devoted to vessel 'elements' in the primary xylem of herbaceous plants, highlighting a lignified cell wall that forms by one of the most complex developmental sequences to be found in the plant kingdom. It exemplifies spatial, temporal and morphological precision in wall synthesis and breakdown.[829] Each xylem 'element' is derived from one cell, and a vessel consists of a file of elements joined end to end (Plate 5a). The life history of a vessel element culminates in sacrificial cell death. The living contents first of all manufacture a cell wall that is strong enough to withstand collapse, yet is permeable to water and solutes, and open-ended. Having played its part, the protoplast is then digested and disappears to leave an empty conduit.[598a]

The primary wall of a vessel element is not at all unusual, but in due course secondary deposits are laid down in discrete rings (Plate 5a), or in a helix (Plate 6d), or in a net (reticulum) with large open meshes (Plate 6a). Whatever the pattern, it is generated under the control of the protoplast, possibly exerted *via* groups of microtubules which take up a specific distribution along the primary wall over the sites of secondary band formation. (Plate 5b and e). How the microtubules themselves are directed into this strategic position is a mystery, as is the means by which they might influence the deposition of wall materials beneath them. Xylem elements forming at different times, and those in different tissues and different species, can have their own characteristic banding patterns, presumably specified by the genetic information in their nuclei. The corresponding walls in the secondary xylem are usually much smoother expanses of lignin-impregnated cellulose and hemicellulose (Plate 6d), perforated only by *pits* (see Plates 6d (insert at top right), 15c–e).[142, 292, 552]

The bands of thickening often arise back to back in neighbouring primary xylem elements (Plates 5b–d, 6c and 13b). Initially the bands are cellulosic with a hemicellulose matrix, but lignification soon begins deep within them at the middle lamella, and subsequently spreads throughout from this origin (Plates 5c and d).[336] Commencement of lignification in the older strata of walls seems to be a general phenomenon in lignified cells, and in other types the middle lamella or the primary wall at the corners of the cells are the initiation sites.[829] The enzyme peroxidase, which participates in lignification, is found in these sites.[342]

Meanwhile, the contents of the cell begin to degenerate (Plate 5b). Many digestive enzymes must be operating at this time, acting not only upon the cytoplasm but also upon the primary wall, which ultimately is digested everywhere *except* where it is impregnated with lignin.[597] Fragments of primary wall (including matrix) remain under the lignified bands (Plates 5d, 6c and d), but in unprotected zones all that survives are a few wisps of primary wall microfibrils (Plates 6c and d, and 13b). The process of wall dissolution is especially conspicuous and significant at the unlignified end walls of vessel elements (Plate 5f, g and h). Digestion of these leaves the functional vessel as a continuous pipe with no impediment to flow along its length, with reinforced walls, and with plenty of unlignified 'windows' suitable for transfer of water and dissolved nutrients between the lumen and the living cells (*xylem parenchyma* cells, plate 6d) in contact with it. Yet another cell wall specialization can develop at the xylem-xylem parenchyma interface. It is thought to facilitate loading and unloading processes, and since it concerns the plasma membrane as much as the wall it is described later, in Plate 13b and page 30.

DIGESTION OF WALL COMPONENTS Digestion of the contents and parts of the wall during the programmed development of a cell is a puzzling phenomenon in view of the hazards involved. How does the cell make and store the digestive enzymes until the correct moment for their release? Once released, what is to stop them from attacking the neighbouring tissues? The same questions arise in other situations where walls or wall components are broken down—during the formation of abscission layers in leaf or fruit stalks; during the ripening of fruits; during utilization of hemicellulose food reserves in seeds (see above); during the wholesale breakdown of food reserve tissues in, for example, a germinating grass seed; during the creation of reticulate patterns on the wall of spores in certain rust fungi, where a smooth wall is formed and is then etched enzymatically to produce a pattern of hollows.[331] Cellulose- and matrix-degrading enzymes are active not only in these situations and others like them, but also in the secretions of invading fungi. It seems that plants have evolved several ways of limiting the sphere of action of both their own digestive enzymes and the enzymatic armoury of their parasites.

As we have already seen, one strategy is to encrust the susceptible components of the wall with a non-susceptible coating of lignin. It is noteworthy that the native wall-degrading enzymes of higher plants include very few that are capable of breaking down lignin. Since their occurrence in a few situations in higher plants,[45] and in wood-rotting micro-organisms,[358] testifies to the biological feasibility of making such enzymes, their virtual absence would appear to represent an example of evolutionary restraint, thanks to which lignin can be both a skeletal and a protective substance in higher plant cell walls. Its efficacy in both functions has been illustrated by the survival, in an otherwise totally destroyed xylem element, of strong, lignin-covered, cellulosic and hemicellulosic, wall thickenings that include the underlying primary wall. By contrast, the matrix in the stretches of primary wall lying *between* the lignified bands on the side walls *is* digested, but only down to the level of the middle lamella (Plate 6c and d), thus raising speculations about the basis of the resistance of this layer.

Other strategies depend on placing inhibitors of wall-degrading enzymes in appropriate cell walls. Walls of the *aleurone cells* of germinating wheat seeds are thought to be protected from the enzymes that break down the neighbouring *endosperm* tissue by the presence of large amounts of a phenolic inhibitory substance.[250] Also, proteinaceous inhibitors of wall-degrading enzymes of fungi have been found in cell walls of a variety of plants.[3, 221]

3.4.2 Callose

Callose differs from cellulose in the nature of the bonds that join the constituent glucose monomers. It is probably a helical chain of glucose monomers, and its most noteworthy features as a cell wall component are the great speed with which plants can make it, and the delicacy with which it can be deposited in specific walls or parts of walls. It forms textureless clear areas in ultra-thin sections, being electron-transparent after the usual heavy metal staining procedures. Although it can be present in many cell types it is employed in two main and widely differing situations—*sieve elements,* and walls of reproductive cells.

CALLOSE IN SIEVE ELEMENTS Sieve elements may appear empty, but they retain a plasma membrane and their contents are interpreted as a highly specialized protoplast.[150b] They occur in the *phloem* tissue, generally but not invariably in association with *companion cells* (Plate 7) and perform the vital function of conducting sugars in high concentrations from sites of production to sites of utilization in the plant.[193] Two points are especially relevant in relation to the occurrence and function of callose. One concerns efficient communication between sieve elements, important in maintaining the flow of materials. They are in fact interconnected by pores in *sieve plates* (Plate 8), and each pore is a modified plasmodesma, having been enlarged by the local dissolution of wall materials and insertion of a lining of callose. Other plasmodesmata leading into sieve elements likewise include callose linings (Plates 7 and 14h). The other point concerns the unusual danger faced by interconnected sieve elements as a result of carrying a highly concentrated sugar solution. The high concentration means that there is a strong tendency for water to diffuse from the surrounding tissue into sieve elements, thus generating very high internal pressures. Water would be able to enter if the sieve element walls, which normally resist these pressures, were to be punctured. The sieve element sap, and its valuable sugar, would be flushed out, even from regions remote from the leak. However, it is found that injury triggers the deposition of plugs of callose at sieve plates, with such rapidity that it is in fact rather difficult for the electron microscopist to judge what sieve plates look like in life. The very manipulations involved in fixation may trigger callose-producing, injury-sealing systems—compare, for instance, Plates 8b and d.

CALLOSE AND REPRODUCTION Functions of callose in reproductive situations are not fully understood, but probably include other types of 'sealing'.[354, 702] Investments of callose effectively isolate developing *pollen grains* during and after the very critical early stage of being formed by meiosis (Plate 11a).[183, 284] Plasmodesmata do not penetrate these massive callose layers, so that each young grain is sealed off from the cells of the parent plant and from the other grains: having attained genetic individuality at meiosis, the grains perforce continue their development as individuals.

A more conventional sealing function is seen much later in reproductive development when a pollen grain germinates on the stigma of a flower and produces a *pollen tube.* There is only a limited amount of protoplasm in the grain, and it has to travel down the tube until the reproductive nuclei can fertilize the egg in the embryo sac. The tube exhibits tip growth (see section on primary walls) and the protoplasm stays close to the tip, progressively vacating the rearward portions. Periodically the advancing protoplasm cuts itself off from the empty tube behind it by laying down plugs of callose, thus ensuring that the apical growing region can remain turgid.[401]

A third type of callose-producing reaction, which it may just be permissible to include under the heading of 'sealing', is seen when a pollen grain lands on the stigma of a plant to which it is related, but incompatible. In such instances proteins exuding from the wall of the pollen grain (see page 22) stimulate the rapid formation of a pad of callose at those parts of the stigmatic cells which are in contact with the grain. It may not be true to say that the callose 'seals' and therefore prevents fertilization, but its production does seem to be at least part of an incompatibility reaction system.[166]

So little is known about callose and how the cell handles it that it is necessary to emphasize that there could be unsuspected functions, and indeed that the notion of 'sealing' may be misleading. Certainly it is not the only compound that plants use for sealing purposes. Certain algae manufacture and use special proteins in a manner analogous to the callose plugs of pollen tubes[103] (see page 61). Similarly, the pores of sieve plates can become clogged with fine strands of protein (Plate 8).[150b, 197]

3.4.3 Encrusting and Adcrusting Deposits Based on Hydrocarbons

Protection and waterproofing are the generally accepted functions for hydrocarbon type deposits *in* walls, and surface coats *on* walls.[533] Hydrocarbons, and alcohols, acids, esters, and oxygenated derivatives related to them are found particularly on the outer surface of the plant, whether a simple *epidermis,* a more complex *periderm* as in bark or cork, or the wall of pollen grains. The *endodermis* is one internal tissue layer where some of them occur (see p. 31–32), and while some of the internal intercellular spaces of plants have easily-wettable surfaces and can become waterlogged, others have hard-to-wet hydrocarbon coats and normally remain filled with gas.[776, 777]

WAX *Wax,* found on the surface of leaves (Plate 9a, b), fruits etc., as granules, rods, threads, scales, and flat or sculptured crusts,[31, 533] is the first non-polymeric wall component to be considered so far. It can only be visualized by procedures of specimen preparation that do not involve treatment with organic solvents. Thus scanning electron microscopy reveals it but conventional transmission electron microscopy of ultra-thin sections generally does not.

Complex mixtures of free molecules occur: apple fruit wax has more than fifty constituents; cabbage leaf more than thirty. Esters and paraffins predominate. In general the major components have many more carbon atoms than do the related compounds found in the lipid fraction of cell membranes, correlating with the tendency for wax to solidify and membranes to be pliable.[438] Some other aspects of wax formations are mentioned in the following section.

CUTIN AND SUBERIN Most exposed above-ground surfaces of plants possess a layer of *cuticle* lying on the outer face of a conventional microfibril-plus-matrix type of cell wall (Plates 9c-g, 10a and c). Wax, if present, lies predominantly outside the cuticle, but can intermingle with it—in general, mixtures of the various components are found at the wax-cuticle and cuticle-microfibril interfaces. Cuticles are composed largely of *cutin,* and the walls are described as being *cutinized.* The nomenclature of *suberized* corky deposits is similar, their main constituent being *suberin.* It was a suberized wall that led Robert Hooke to his great discovery of the cellular architecture of plant tissues (page 7).

Like 'lignin' and 'hemicellulose', the terms 'cutin' and 'suberin' refer to families of compounds. There is in fact considerable uncertainty regarding their chemical composition, arising in part from their resistance to being dissolved and decomposed—attributes annoying to the chemist but useful to the palaeobotanist and of considerable survival value to the plant. They are both polymers of long chain (16 and 18 carbon atoms) fatty acids, linked together and to other minor constituents by chemical bridges involving oxygen atoms. They differ in the relative proportions of their fatty acids, but few components are found exclusively in either one. One example of variation between species is that plants with thick cuticles tend to have a higher proportion of hydroxy-fatty acids in their cutin than plants with thin cuticles.[371] It may even be that the abundance of hydroxy groups is what makes the development of a thick cuticle possible, for each one of them represents an opportunity to build another molecule into the structure *via* an oxygen bridge.

DEPOSITION OF ADCRUSTING SUBSTANCES One of the difficulties in envisaging the process of wax, cutin, and suberin deposition on outer surfaces of walls is that to reach their final destination, these substances must migrate through the wall from their sites of synthesis in the protoplasts. The texture of the wall might be open enough to permit a slow flux of molecules towards the exterior. It has also been claimed that microchannels exist in both plant and insect cuticles, and that wax flows through them until solidification gives rise to the various species-specific surface forms.[220, 305, 857a] Certainly some young leaves can regenerate a wax 'bloom' after an original layer of wax has been wiped off.[307] Another proposal is that the final stages of polymerization of cutin and suberin take place outside the cell, stimulated by exposure to air. It is not really surprising that epidermal cells have been found to differ from deeper tissues in that they possess enzyme systems capable of making precursors of the surface coats.[439]

The species-specific morphology of the surface coat has been found in at least one case to derive from the physical properties of the major component.[396] The wax of Sitka spruce *(Picea sitchensis)* needles consists of tubes surmounting flat scales. The main constituent is N-nonacosanol-10-ol, and this compound will crystallize as tubes on the surface of artificial epidermal cell walls, simulated by porous discs through which the substance is slowly drawn by evaporation of solvent from the upper surface. In similar treatments the minor constituents of the wax give rise to flat scales.

Plates 9 and 10 illustrate a variety of cuticles. As with

all walls (except perhaps the most mucilaginous), the cells can produce specific patterns of deposition, exemplified by ridges in Plates 9c and d, by warts or knobs in Plates 9e and f, and by the cuticular coverings of the *guard cells* of *stomata,* seen in Plates 9a and d.

PERMEABILITY Plates 9g, 10a, and 10c serve to emphasize that cuticles are by no means impermeable structures and that they are not always external. They depict glands, and in one (Plate 9g) the secreted fluid is nectar, which despite being aqueous, can find its way through the massive cuticle (in this particular case into an internal, sub-stomatal cavity). The numerous strands of material penetrating from the microfibrillar wall into the cuticle may represent the escape routes. Most glands seem to be covered by cuticle, in some cases provided with definite pores.[498] Some even have a valve-like pressure release system that liberates secretory droplets at regular intervals.[213]

Quite apart from its permeability to secretions, the cuticle is not by any means a perfect skin. Water can be lost slowly through it ('cuticular transpiration')[294] and carbon dioxide can penetrate, at any rate through thin cuticles. Raindrops can wash both organic and inorganic molecules out of cuticularized leaves,[839] and in this connection a surface layer of wax formations plays an important part by reducing the wettability of the leaf.[370] Pictures of wax formations such as Plates 9a and b are much more than just academically interesting and visually elegant. They are of practical importance in trying to find out how best to wet leaves in order to aid penetration of the numerous chemicals used in modern crop husbandry.

SPOROPOLLENIN The final example of a protective outer wall layer takes to extremes the features of resistance to chemical attack, maturation outside the cell, and deposition in specific places. The substance *sporopollenin* is one of the most resistant materials known in the organic world, indeed it has been found in fossil deposits preserved from as early as the Pre-Cambrian.[87, 88] Nowadays we know it from the surfaces of spores and pollen grains (Plate 12a), and it is also present in the walls of some primitive algae (e.g. Plate 30b),[18] where it is so resistant to digestive enzymes that cells can pass through the digestive tract of a snail without loss of viability.

Sporopollenin is now known to be a polymer composed of derivatives of carotenoids.[87] These hydrocarbon pigments are found throughout the plant kingdom and are of interest to students of photosynthesis and other aspects of light absorption in plants, and to animal physiologists as the precursors of vitamin A and the visual pigments of the eye. Earlier reports that lignin is present in sporopollenin are now discounted.

While lignin, cutin, and suberin merely resist decay, sporopollenin is virtually non-biodegradable.[87] As such, it presents difficulties for the cells that make it, and the sporopollenin sculptures that adorn pollen grains seem all the more impressive. Special lipid surfaces are in fact prepared outside the protoplast at the site where sporopollenin is to be deposited. One case recently examined in detail (developing lily pollen) suggests extrusion of these 'parent lamellae' from or through the plasma membrane, with deposition commencing as soon as the surface become available externally.[165] The exact nature of the precursors and of the polymerization process is not yet known. Oxidation is involved, and enzymes are thought to participate in a form of ionic catalysis.[88]

SPOROPOLLENIN ON POLLEN AND ORBICULES Developing pollen is nourished by a tissue called the *tapetum* which either lines the anther sac (Plates 11c and 22) or else flows amongst the grains.[182] In lily,[355] the first sign of the elegantly sculptured pollen wall appears under the thick layer of callose that invests the young grain (see page 20 and Plate 11a). The callose later disappears and further deposition of sporopollenin completes the outer part of the wall. It is a rare example of a wall that is synthesized using precursors made not only in the cell it encloses (the grain itself) but also provided specifically by other cells (the tapetum). A more conventional inner wall stratum also develops.

Granular *orbicules* (sometimes called *Ubisch bodies*) are produced by the tapetal cells while the pollen grains are developing their mature wall (Plates 11b and 22c). The orbicule surface, like that of the 'parent lamellae' (above) is suitable for sporopollenin formation and it too becomes coated. As the anther and its pollen mature, orbicules are left on the inner face of the anther sac, where they may in due course have a role in pollen dispersal by forming an unwettable carpet to which the grains do not adhere (Plate 12b).

Mature pollen walls are so precisely and diagnostically sculptured (Plate 12a–c) that they can in most cases be used to identify the parent species, even after the grains have been lying in geological deposits for thousands of years. There is no finer example of the emergence of specific pattern in a cell wall, and it provides an opportunity for investigating how a cell can create an object of great complexity. Application of physical and chemical treatments at known stages of development to disrupt microtubules or displace cytoplasmic membrane systems reveals that the position of major features of the pollen wall is related to the plane of the previous cell division, and that a chain of events then ensues, like the working out of a computer programme, one step leading on to the next, and exerting progressively finer and finer control over the shaping of the wall (see legend to Plate 12).[356]

The biochemistry of pollen walls is as specific, or even more so, than their structure, for they contain proteins thought to function at the time of pollination by governing compatibility and incompatibility reactions (see p. 20). The grain will germinate if the 'recognition proteins' identify a correct mating between pollen and stigma. Millions of hayfever sufferers are sensitive to these pollen wall proteins and are painfully aware that they are very loosely held and diffuse out of the wall rapidly, both on the stigma of a flower that is being pollinated, and, unfortunately, in the human nose and eyes. A major site of pollen wall protein storage is the inner wall layer under the pores that many pollen grains possess (Plate 11b).[436]

3.5 The Plasma Membrane

ULTRASTRUCTURE The sub-structural complexity of the plasma membrane is evident both in ultra-thin sections, where the dark-light-dark 'tramline' appearance (Plates 3b, 16c) is usually interpreted in terms of a bimolecular layering within the membrane (pages 12–13), and in views given by freeze-etching (Plate 4a). Freeze-etching frequently (but not always)[343] reveals face views depicting the distribution of particles within the cleaved membrane. Particles 8–10 nm in diameter lie in a smooth matrix, the smooth expanses probably being the hydrophobic inner portion of the bimolecular layer of lipid, and the particles being proteinaceous. The particles may be scattered or they may be non-randomly distributed in local aggregates, short straight lines, or even in semi-crystalline arrays.[593, 690]

Like all other membranes in the cell, the plasma membrane is subject to concomitant synthesis and breakdown, i.e. to 'turnover' (page 68). Chapter 7 includes further information on this phenomenon, but there are observations which are relevant here, and which also demonstrate the dynamic nature of the plasma membrane. By freeze etching cells that are known to be in different stages of development, it has been noted that the distribution of particles and other topographical features can alter in a regular fashion throughout the cycle of growth and division.[789] Aggregates, lattice arrays, and folds appear and disappear. The fact that micro-topographical features may be transitory is no cause for surprise, in view of the fact that both lipid and protein molecules can diffuse within the plane of the membrane (page 13), but it is remarkable that temporal changes, such as those mentioned above, should be predictable and regular in their occurrence. Their origin may in part be due to rearrangements within the membrane, but could also be in part a consequence of the arrival of pre-formed expanses of membrane at specific times and places. Vesicles derived from the Golgi apparatus can, for instance, transfer a pattern of particles to the plasma membrane by membrane fusion.[426] The functional significance of the observed changes remains to be discovered, but as always with the plasma membrane, the two major areas deserving of special attention are the synthesis of cell wall material and the regulation of transport to and from the cytoplasm.

Selective staining procedures reveal that there is more to the plasma membrane than a bimolecular layer with included particles. Several independent observations have shown that there is, on the outer face of the membrane, a thin layer which contains polysaccharide, and which is probably equivalent to the rather more conspicuous *glycocalyx* of animal cell plasma membranes.[143, 706, 707] Work on animals and bacteria suggests that the glycocalyx consists of protein moieties anchored in the membrane, to which the externally located polysaccharide is attached. The glycocalyx is an important component of the recognition systems of animal cells, and its constituent polysaccharide-protein (glycoprotein) molecules may even extend right through the membrane, thereby providing a possible means by which signals could be transmitted from the exterior of the cell to the interior. This may or may not apply to plants, where at any rate vegetative cells may, due to the presence of the cell wall, have little or no need for an external recognition system, but where the need for a mechanism by which internal cell membranes can recognize the plasma membrane remains (see also pages 13 and 75–76).

It is extremely common for ultra-thin sections to include areas of plasma membrane that have been infolded in a variety of ways. Attempts have been made to classify such investigations, and a lengthy list of descriptive terms has emerged, for example, lomasomes, plasmalemmasomes, paramural bodies, and multivesicular bodies.[148, 531] In the latter, the membrane is seen to have given rise to numerous tubes and vesicles, and clusters of this nature occur not only at the plasma membrane, but also *apparently* free in the cytoplasm, and even in vacuoles.[507] Selective staining for polysaccharides suggests that the membrane in them is similar to that of the plasma membrane.[143] Some authors hold that they represent sites of uptake into the cell,[143] others that they are sites of export.[531] Others suspect that at least some of the structures within the general category of local infoldings of plasma membrane may be artefacts,[229] perhaps arising because the fixation process is not rapid enough to prevent the appearance of local excesses of membrane, due to unequal effects of fixation upon the normally balanced addition and removal processes that occur during turnover of the membrane. It is noticeable that in general, chemical fixation, followed by embedding and ultra-thin sectioning, yields plasma membranes that are considerably less smooth than those preserved by the very rapid freezing procedure employed in freeze-etching (compare Plate 4a with almost any ultra-thin section in the Plates).

DIFFERENTIAL PERMEABILITY It is obvious that living cells contain substances that do not freely leak out: equally there are substances that are not permitted free entry. Other substances are allowed restricted, or even relatively free passage. Although the cell wall can be impermeable (pages 30–32) it is in most cases the plasma membrane, and not the cell wall, that is the differentially permeable guardian of the protoplast.[840] Some of its selectivity is based on the physical capacity of combined lipid and protein layers to be differentially permeable; dissolved substances penetrating freely, poorly, or not at all, according to the particular types and combinations of chemical grouping they contain. In order to traverse a membrane, molecules must pass from the medium on one side and dissolve in the fabric of the membrane, and then diffuse out of it again into the aqueous environment on the other side. Equations that describe such processes can be applied to the diffusion of solutes across the plasma membrane. Some generalizations are that neutral molecules can pass more readily than cations or anions, and of course small molecules more readily than large. Trans-membrane transport via vesicles is a very different process, to be described later.

Water itself can penetrate with relative ease, as is easily shown by the simple experiment of placing cells in a solution of non-permeating solute, say sucrose.[804] In this situation the *water* on the two sides of the membrane is at different concentrations, and it tends to diffuse through the membrane until equality is attained. Matters can be expressed more correctly by making use of the concept of 'water potential'—a measure of the amount of energy in the molecules of water. Water tends to diffuse from regions of high water potential to regions of low water potential. The water potential is high when the water molecules are crowded together, as in dilute solutions or in pure water, where there are few or no solute molecules in amongst them, or when positive hydrostatic pressures concentrate them. It is low when the molecules are more spaced out, as when many solute molecules are interspersed amongst them, or when hydrostatic pressures are low or negative. Thus if the water potential of the sucrose solution outside the plasma membrane is lower than that inside the cell, water will diffuse outwards down the gradient of water potential, and the protoplast will accordingly shrink, i.e., undergo *plasmolysis*.

Volumes of water of the order of 10^{-5} cm^3 can pass per cm^2 of plasma membrane per second per unit driving force, expressed in atmospheres of pressure.[588] This may appear to be a very small flow rate, but simple arithmetic using realistic values for the surface area and volume of cells shows that it will in most cases cope with the passage of a volume equal to the volume of the cell itself in a matter of a few hundreds of seconds. It is more than ample for the rates of inflow needed to sustain cell growth. The permeability of the plasma membrane to water is so high that molecules undoubtedly pass both inwards and outwards even when the volume of the protoplast is not changing: there is, in other words, a dynamic equilibrium. Evidence for the existence of water-filled pores in plant cell membranes is not good,[156] and it seems more likely that molecules of water readily traverse even unperforated membranes. The major barrier to water movement across a membrane is the hydrophobic interior of the bimolecular layer, and it may be that thermal motions of the hydrophobic fatty acid chains of the membrane lipids open temporary passages large enough to admit small groups of water molecules, which then pass on as permitted by the random molecular oscillations that are known to occur within most membranes.[613]

A more biological type of selective permeability also exists, based on the presence of 'carrier' molecules in the membrane, each capable of recognizing a particular type of solute and transporting it across. The unique feature of these carrier-mediated solute transport systems is that, unlike diffusion, they can in favourable circumstances work against concentration gradients. They can absorb solutes from dilute solutions and concentrate them within the cell, or conversely they can be active in secreting substances from the cell. The term *solute pump* is often used to describe them, for like any other sort of pump, they need to be supplied with energy. In the context of the cell, this means that the solute pump molecules must be able to interact with the various energy-rich chemicals, including adenosine triphosphate, that are generated by cell metabolism. The plasma membrane of root cells is in fact known to contain an enzyme which is stimulated by the presence of ions (of the type taken in by roots from the soil) to break down adenosine triphosphate and release its energy.[479, 480] This ion-stimulated adenosine triphosphatase enzyme in the plasma membrane probably is a solute pump, or a vital part of one. The mode of operation of a solute pump can be pictured in several ways. Presumably a part of it has to be exposed on one face of the membrane, so that it can recognize and bind its solute molecule. There could then be a change in the conformation of the pump—and many proteins do change their shape when they bind other molecules—so that the bound solute becomes exposed and then released on the *opposite* face of the membrane. The requirement for energy could arise at almost any stage in the process, including that of recovery to the starting condition ready for another, similar, operation.

Diffusion of water and solutes presumably occurs all over the plasma membrane surface, though the flux may vary from one region to another. There are as yet no means of identifying 'fast' or 'slow' areas in either ultra-thin sections or face views. What is known about the sizes and numbers of solute pump molecules in plant and animal plasma membranes suggests that they occupy a fairly small fraction of the total surface area, and it is possible that they may be represented by or amongst the free and aggregated particles exposed to view by the freeze etching technique.

PINOCYTOSIS One remaining type of transport across the plasma membrane, is, however, relatively easily recognized. Solid material can come into contact with the outer face of the plasma membrane of animal cells, and in many cases the plasma membrane is capable of wrapping particles in small vesicles, which are then transported into the cytoplasm. This is the process of *pinocytosis*. It is a type of feeding mechanism not normally open to plant cells, where the microfibrils of the cell wall are closely enough meshed to form an effective filter, with open channels not much larger than 10 nm in average diameter. Plant cells of necessity subsist on dissolved mineral nutrients and gases small enough to percolate through these confined channels up to the plasma membrane, and presentation of larger materials to the plasma membrane is precluded. Although the opportunity for pinocytosis thus does not normally exist, the mechanism, remarkably, survives. If the cell wall is removed by attacking it with appropriate digestive enzymes, the naked protoplast, escaped from its normal confines, is found to be capable of pinocytotic uptake, just as in *Amoeba* and similar naked organisms, which habitually obtain much of their food by this process.[131]

In one very unusual botanical situation (the legume root nodule), nitrogen-fixing bacteria can, by a special infection mechanism, penetrate the cell wall to the plasma membrane, which envelops them prior to their entry into the cytoplasm.[277] This is essentially pinocytosis on a large size scale, and is called *endocytosis*. It is, of course, relatively common in animals.

Many authors have noted the existence of small cytoplasmic vesicles near the plasma membrane, and also of inpocketings of the membrane, and commented that their observations could be evidence that pinocytosis occurs more commonly in plant cells than is generally realized.[30, 141, 478] Unfortunately, the usual problems of how to interpret static electron microscope images are raised at this point. A vesicle seen close to the plasma membrane could be on its way *in* or on its way *out*. Vesicles can liberate their contents to the exterior, and because this is precisely the opposite of pinocytotic uptake it is described as *reverse pinocytosis*. It will be considered again in Chapter 7, and is illustrated in Plate 28.

THE PLASMA MEMBRANE AND THE CELL WALL A role for the plasma membrane in the synthesis of the cell wall has been suggested at various points in this chapter, but now that the time has come to develop the theme it must be confessed that the opportunities for research are many and the established facts few. There are only two mechanisms by which the cell wall can be made. One is extrusion of pre-formed materials from the cytoplasm, through the plasma membrane. The other is synthesis of wall components outside the plasma membrane under the control of the protoplast and using raw materials provided by it. Almost certainly both mechanisms occur. The former rests heavily on reverse pinocytosis and probably applies to the secretion of wall matrix polysaccharides (see Chapter 7). The latter implicates the plasma membrane, which is the only part of the protoplast that is in direct contact with the developing wall. Most discussions of the role of the plasma membrane relate to the problems posed by the shape, size, and orientation of cellulose microfibrils, and significant discoveries in this area can be anticipated now that the successful isolation of plant cell plasma membranes has been achieved. For instance, a preparation from onions has been found to polymerize glucose, and this activity is stimulated both in the plant and in the test tube by treatment with a plant regulating substance.[845]

One fruitful approach to the problems of the mechanism and control of microfibril deposition has been to compare the distribution of particles in the plasma membrane with the arrangement of the microfibrils that are being laid down in the wall. In some algal cells the two have been shown to be closely related.[701] Microfibrils have been seen in association with plasma membrane particles, and the particles do sometimes lie in linear arrays (Plate 4a).[591] It is therefore pertinent to ask whether the particles include enzyme systems for making cellulose, whether the local aggregates of particles (e.g. Plate 4a) provide any clue to the ability of cells to perform local synthesis of wall material, and whether cytoplasmic components can influence the distribution of the particles and hence the disposition of microfibrils.

None of the above questions can as yet be answered in full. Microfibrils of cellulose consist of uniformly thick, locally crystalline, bundles of very long cellulose molecules, each a linear chain of glucose units. The uniformity in thickness, and the complexity of the patterns in which they are sometimes seen to lie, suggest that microfibrils do not grow by addition or interpolation of pre-formed molecules of cellulose. Growth by the addition of glucose monomers to the exposed termini of the cellulose molecules at the end of a microfibril seems more feasible.[321] In contrast to newly secreted matrix polysaccharides, which might percolate between the microfibrils deep into a wall, the most recently synthesized microfibrils lie just outside the plasma membrane, though there are indications that the synthesizing system is not necessarily confined to the interface between wall and plasma membrane.

If these considerations and ideas are taken into account,[671] the picture that emerges is one in which cellulose-synthesizing enzymes extend a microfibril by a process of 'end-growth'. There would probably have to be numerous enzymes to cope with the numerous molecules of cellulose in the microfibril. A microfibril might be extended at a rate of the order of micrometers in minutes, and it could be that this growth is achieved by the activity of a cluster of enzymes that remains attached to the growing end; alternatively the extending microfibril might grow through a fairly continuous carpet of cellulose-synthesizing enzymes on or close to the outer face of the plasma membrane, and made available by the movements of random diffusion to all of the extending microfibrils in the wall.[694]

It follows from what is known of the disposition of microfibrils in the cell wall that there must also be a means of directing the syntheses into particular orientations, and further, that this orientation mechanism must be able to change with time. It is difficult to imagine how small particles or aggregates of enzymes could achieve the stability that would be necessary for directional syntheses. As already noted, they would probably be subject to random movements of diffusion, even if they are attached to or lie within the plasma membrane. Large aggregates or arrays in or on the membrane might, on the other hand, be sufficiently immobile. Alternatively, the cytoplasm just inside the plasma membrane might provide stabilizing and orienting influences in the form of membranes or microtubules—a possibility that will be reconsidered in Chapter 11.

It is apparent that the part played by the plasma membrane in synthesizing components of the cell wall is still conjectural. So too is its role in determining regional differentiation of the wall, but there is no doubt that it can influence cell wall morphology. For instance, in young lily pollen grains (Plate 12) the plasma membrane develops outward projections in a spatial arrangement upon which the polygonal disposition of wall sculpturing later rests.[164] The morphology of the membrane in effect predicts the morphology of the wall. As will be re-stated later in connection with the role of the endoplasmic reticulum in cell differentiation (page 66), it is presumed that the membrane is not of itself able to predict and then bring about the pattern in the cell wall. It merely mediates some stage or stages of an overall sequence that, since it eventually characterises the cell and the species, must be under the ultimate direction of genetic material, and must involve the co-ordinated activity of more cell components than the plasma membrane alone.

The necessity of a nucleus being present if an isolated protoplast is to lay down a wall upon its naked plasma membrane was mentioned on page 17. It now seems probable that studies of the mechanism of action of plant growth hormones will throw light on the nature of the interaction of nucleus, plasma membrane, and wall. Hormones such as indoleacetic acid very rapidly induce 'loosening' of the cell wall and so stimulate cell enlargement. The extremely short reaction times that are involved here suggest that the plasma membrane is stimulated directly by the hormone, with none of the delay that would accompany uptake and stimulation at an internal site. More sustained hormone action is known to require the activation of genes in the nucleus. It is now known that preparations of isolated plasma membrane will bind molecules of growth hormone, that glucose polymerization by enzymes in the membrane is enhanced by application of hormones, and that exposure to hormones releases a factor from the membrane that, in the test tube, is capable of promoting the activity of RNA polymerase, the enzyme needed during the first stages of the expression of genes.[312]

No doubt knowledge of the interactions of wall, plasma membrane, and nucleus will accrue rapidly. Already the plasma membrane is seen to be of central importance in controlling the traffic of materials in and out of the cell, whether molecule by molecule or by vesicles; in controlling the properties of the cell wall external to it; in mediating the reactions of the cell to hormones; and in guiding the morphogenesis of the cell in at least some instances.

3.6 Specialized Cell Wall—Plasma Membrane Complexes

Regulation of import and export is a necessity that faces all cells, and plants and animals have evolved remarkably similar specializations of their outer surfaces related to (a) the transport of solutes from cell to cell, (b) the intensive transport of solutes, in and out of the cell, and (c) restriction of solute movement in the non-living space between neighbouring cells. The botanical versions of these specializations, all highly flavoured by the existence of the cell wall, are now described.

3.6.1 Plasmodesmata

INTRODUCTION: SYMPLAST AND APOPLAST Of the numerous manifestations of 'division of labour' in the plant body, few distinctions are as fundamental as that between cells which can absorb nutrients directly from the environment, and those which, being internally located, cannot. It is a distinction that does not arise in some of the primitive algae which consist of simple or branched filaments of cells, or of single or double layered two-dimensional expanses of cells. There every cell is in contact with the surrounding pond, lake, or sea-water from which it derives its nutrients, and often the cells in the body of the plant are essentially identical copies, and behave as individuals. Any division of labour that exists is strictly limited. We may conjecture that the evolution of plants with more complex organizations of cells in three-dimensions had to await the appearance of structures which could cope with the nutritional requirements of internally-situated cells.

In the animal kingdom, evolutionary progress beyond the two layered organization (exhibited, for example, in the Coelenterates) must have been subject to a similar limitation. Animals, however, being soft, were free to exploit the potential of the spaces *between* cells as channels of transport, a trend which has culminated in elaborate vascular systems through which intercellular fluids are pumped to the vicinity of all cells, however remote they are from direct sources of nourishment. In plants, the presence of more-or-less rigid cell wall material in the spaces between cells precluded the evolutionary development of vascular systems based on *mechanical* pumping mechanisms. The vascular systems of higher plants doubtless evolved in stages. Judging by the survival to the present day of numerous simple algae and fungi which lack long distance transport systems such as the xylem and the phloem, but possess channels of short distance transport between neighbouring cells, the critical evolutionary step which allowed multicellular plants to begin to exploit the architectural possibilities of the third dimension, was the appearance of *plasmodesmata*.

Plasmodesmata are fine strands of cytoplasm, bounded by plasma membrane, connecting a living cell to its living neighbours through holes in the fabric of the intervening cell walls (Plate 14). It is thought that substances can move from cell to cell through them. To plant physiologists they represent a means whereby a plant is converted from a mere collection of individual cells to an interconnected commune of living protoplasts in which the activities of the different parts can be integrated and regulated. Cells and tissues remote from external sources can be nourished by diffusion or bulk flow through plasmodesmata, and materials can be passed to or from the long-distance transport tissues of the vascular system. One of the long distance transport systems, the sieve tubes of the phloem, is itself regarded as being an evolutionary extension of the possibilities opened up by the existence of intercellular connections, for the nutrients carried in bulk by the phloem pass from one sieve element to the next through the enlarged plasmodesmata of the intervening sieve plates (pages 20 and 66 and Plate 8).

It is probable that if any present day higher plant were to be examined in sufficient detail, every living cell would be found to be connected to its living neighbours by plasmodesmata. Their existence subdivides the plant body into two major compartments—the interconnected protoplasts, bounded by what is in effect the one continuous plasma membrane of the plant, and the non-living compartment external to the plasma membrane, and consisting of cell walls, intercellular spaces, and the lumen of dead structures such as xylem elements. The former compartment is known as the *symplast*, and the latter as the *apoplast*.

Whereas animals have exploited the apoplast for their vascular systems, plants have made greater use of the symplast. Even the apoplastic pathway followed by the transpiration stream is derived by sacrificial cell death of a part of the symplast (page 19). Some animal tissues do, however, have the equivalents of plasmodesmata. Thanks to the absence of cell walls, it is feasible for the plasma membranes of neighbouring animal cells to come into close contact with one another, and special 'low-resistance cell junctions' may occur in such places. The 'low resistance' refers to modifications of the plasma membranes which permit relatively free movement of small molecules from cell to cell.[493] More strikingly, structures which are much more akin to plasmodesmata develop in the cells (osteocytes) of the one animal tissue (bone) which resembles plant tissue in the presence of inflexible intercellular material.[663]

STRUCTURE OF PLASMODESMATA Plasmodesmata are illustrated specifically in Plates 14 and 15, and also appear in many other micrographs (especially Plates 2, 3b, 4a, 7, 10a, 27, 34b, c, 48f). They are tubular in side view (Plates 3b and 14a) and when the tubes are viewed end-on they appear as circular profiles (Plates 14b, d). At high magnification the part of the plasma membrane within a plasmodesma is seen to have the same tramline appearance as it does elsewhere in the cell (Plates 3b, 14g). Obviously, the length of the tube is related to the thickness of the wall. In some instances it elongates to keep pace with wall thickening processes, and in others the plasmodesmata lie in groups in the vicinity of which wall thickening is reduced or does not occur, so that a thin-walled, plasmodesmata-containing *pit* develops (Plate 15a and b).

In many cases the plasmodesmatal tubes, or canals, branch within the cell wall (Plates 7, 14f, g, h). The whole structure, with all the branches, can be called a compound plasmodesma. They occur, for instance, between sieve elements and their companion cells, a site which is of especial significance in relation to the loading and unloading of the sieve tubes.[302] Characteristically, several to many plasmodesmatal canals pass from the companion cell protoplast and converge in the centre of the wall, all funneling into a single canal which leads on into the sieve element (Plate 7, 14h). In other instances multiple canals may converge from both directions, meeting in a complex labyrinth in the centre of the wall (Plate 14f). Plate 14g illustrates a variation in which the individual canals do not converge to meet at a particular focal point, but nevertheless are connected laterally within the wall.

The lumen of an individual plasmodesmatal canal is narrow. The images in Plate 14 show that there would hardly be room for two ribosomes to sit side by side within it. Actual measurements of the diameter vary, but are generally in the range 30–50 nm. An attempt to assess the functional potential of such narrow channels will be made later, meanwhile, it is important to note that despite the narrowness, an internal structure is often to be seen along the axis of the channels.[98, 696] Where it is present it is visible both in side views (Plate 14a, c) and in transverse sections (end views, Plate 14d). It is not always present, however (Plates 3b, 27a). There is evidence, supported by pictures such as Plate 14a, that the axial structure is itself tubular, and a derivative of the endoplasmic reticulum. Sometimes it passes straight along the plasmodesmatal axis (Plate 14a) and in other instances it appears very convoluted (Plate 14c). It is not known whether the convolutions are genuine or an artefact resulting from imperfect fixation of what must be an extremely tenuous and delicate object. In some material the plasma membrane pinches in at the two opposite faces of the wall, leaving very little, if any, space between the plasma membrane and the axial structure (Plate 14c). This is not the case in other material (Plate 14a). The functional potential of these variations has to be evaluated later.

DISTRIBUTION OF PLASMODESMATA The probability that all living cells in higher plants are interconnected by plasmodesmata has already received mention. Two reservations should, however, be emphasized.

One is that to be interconnected, the neighbouring cells must both be alive. Thus, plasmodesmata occur between young, still living, xylem elements and their neighbours, but they are severed and disappear by the time the xylem matures and dies (e.g. Plate 6c).

The second reservation is that the neighbouring cells usually have to be members of the same individual plant. Where alternating generations of plants exist, one within tissues of the other, the two are not (or are very rarely) interconnected. Thus pollen mother cells lose the plasmodesmata that connect them to the parent generation (though they retain for a while very large cytoplasmic bridges,[352] too large for the word plasmodesma to be appropriate, that interconnect the population of mother cells, Plate 11a). Young pollen grains formed in groups of four by meiosis develop in isolation as individual members of the microgametophyte generation (Plate 11c). Embryo sacs are not usually connected to the sporophytic ovule tissue that is their host, and after fertilization has taken place, the zygote, and later the embryo, are not connected to the embryo sac or endosperm generations that surround them.[590]

Were it not for the fact that some exceptions exist, it might be concluded from the above that plasmodesmata cannot form between cells of genetically different individuals, perhaps because differences in the recognition systems of the plasma membranes preclude the fusion that is needed for plasmodesmatal continuity. The exceptions that render such a conclusion imperfect include observations of connections between certain specialized embryo sac cells and the surrounding nucellus tissues in *Capsella*;[757] the extraordinary case of the establishment of plasmodesmata between the invading absorptive cells of the parasitic flowering plant *Cuscuta* and the cells of the host plant;[177] and the existence of plasmodesmata between genetically different cell layers in graft-hybrids between *Cytisus* and *Laburnum*.[99] The first of these exceptions demonstrates that plasmodesmata can form between genetically differing cells within a species, and the second and third that the much wider genetical differences that exist between different genera and families are not insurmountable barriers.

DEVELOPMENT OF PLASMODESMATA It is thought that in the vast majority of cases, both the number of plasmodesmata per unit area of cell wall, and their distribution over the surface of the wall, are determined when the cell plate is being formed after cell division (Plate 48). The meshwork of microfibrils initiated at that time contains neatly circumscribed gaps at the plasmodesmatal sites, whether they are scattered as individuals all over the wall (Plate 4b), clustered in *primary pit fields* (Plate 4c–e), or in other non-random distributions (Plate 14b). We know nothing about the control mechanisms that must somehow specify not only the total number of plasmodesmata, but also whether they lie singly or in groups, and whether or not strands of endoplasmic reticulum are incorporated in them (Plate 48d, f). Different walls can differ markedly in these features. In a root tip cell the transverse walls have more plasmodesmata than the side walls (compare the upright (transverse) and horizontal (side) walls in Plate 2).[412] In complex vascular tissues where a single cell can have up to four different types of neighbouring cell, the intervening walls can all be quite different.[302]

Whilst it is widely accepted that the developing cell plate is the major site at which plasmodesmata originate, it is also recognized that they can develop in much more mature cell walls. The unusual case of the development of plasmodesmata between a host and a parasite,[177] mentioned above, is one case in point, for no cell division and no cell plate formation is involved: a cell of the parasite merely bulges into a cell of the host, and presumably the intervening cell wall (partly host, partly parasite) becomes dissolved in places to permit the two plasma membranes to establish continuity. The graft hybrids[99] provide another example. Here adjacent tissue layers are genetically different, being derived from the two partners that have become grafted together, and there is no question that any one parent cell could have divided to give rise to members of both adjacent layers. Like the host-parasite junction, the intervening wall can be classed as a 'non-division' wall. Plasmodesmata can develop in it, and again it is considered that the opposed cells can dissolve small holes in their walls; apparently this is not achieved with great precision, and the two partners do not always establish complete plasmodesmatal canals.

Since late development of plasmodesmata has been detected, it becomes a problem to know to what extent the process might occur in ordinary tissues. Yet other problems arise when considering how compound plasmodesmata develop.

FREQUENCY OF PLASMODESMATA Quantitative estimates of plasmodesmatal frequency are available for a considerable number of cells.[841] An example is shown in Plate 14b, where use of the method described in the caption yields a figure of about 3000 per 100 μm^2 of wall surface. This is a rather high value, probably reflecting the non-expanding state of this particular cell. Counts in the range 50 to 1500 per 100 μm^2 are more common (for example see the alternative (and less accurate) method of calculation presented in the caption to Plate 2). One exceptionally high value of about 5000 per 100 μm^2 has been recorded for the end walls of certain conducting cells in *Laminaria,* a large brown seaweed. Walls in cells with corresponding functions in conduction in higher plants also tend to have high frequencies, as in the stalk cells of secretory glands (Plate 10a).

If, as suggested above, most plasmodesmata are laid down in the cell plate stage of wall development, it follows that their numbers per unit area will diminish if the wall stretches during cell enlargement. This has in fact been documented for enlarging cells in the centre of root caps, where the plasmodesmata of the side walls become 'diluted' by a greater factor than those of the end walls, due to the unequal extent of wall-stretching in the two situations.[412]

FUNCTION OF PLASMODESMATA The concept of the symplast[841] was developed decades before any significant work had been carried out on the ultrastructure of plasmodesmata. Still earlier observations on the anatomical arrangements of cells in tissues had led pioneers in the art of relating structure to function to develop ideas such as the 'Theory of Expeditious Translocation'.[303] It was deduced from all of this background work that substances can pass from cell to cell. Now that much is known about plasmodesmata, the deduction still stands, but there are surprisingly few facts and figures to back it up.

Virus infections are thought to spread through the symplast, and several investigators have fixed material which proved to have virus particles within the lumen of plasmodesmata, apparently caught *in transit*.[431] The particles are, however, large in relation to the sort of solutes that physiologists believe to move through the symplast of normal plants, and it is just possible that plasmodesmata in the pathological state are abnormal. Certainly in normal plants ribosomes are never seen within the lumen, and there is no evidence of their movement from cell to cell, even though they are smaller than the viruses.

To date the only direct visual evidence that a low molecular weight solute can be present in a plasmodesmatal canal comes from studies in which the reagent silver acetate was added to the fixative used in preparing specimens for electron microscopy. Electron-dense silver chloride was precipitated where chloride ions were present, and plasmodesmata figured amongst the sites that contained electron dense deposits, both when the cells of salt-secreting glands were examined,[930] and when plants were pre-treated with solutions containing chloride.[848] It is presumed that the chloride found in the canals was moving through the symplast.

The remaining evidence is indirect. Measurement of electrical resistances, between cells suggests that open channels are present.[780] Diffusion of chloride along leaf tissues *other than* vascular tissues occurs at a rate of a few centimetres per hour, i.e. very much slower than the rate of transport in the phloem, but a rate that could be sustained by plasmodesmata, bearing in mind that the length of the diffusion pathway is not, thanks to the stirring effect of cytoplasmic streaming within the cells, the total length of the tissue sample, but only the sum of the lengths of all of the plasmodesmatal canals in all of the walls through which diffusion has to take place.

In the absence of more direct means of investigation, it has been necessary to resort to mathematical analysis of the functional potential of plasmodesmatal canals.[127, 841] It may or may not be valid to apply to plasmodesmata that are as narrow as 30–60 nm in diameter the formula that describes the flow of liquid through a pipe. The formula that describes diffusion of substances along concentration gradients has also been applied, bearing in mind that in diffusing through plasmodesmata from cell to cell, the canals add up to only about 1% of the total area of the intervening wall. The major conclusion that has been drawn is that tiny as plasmodesmata are, they are in theory likely to be able to cope with the sort of rates of movement of water and dissolved compounds that have been measured experimentally. In the absence of plasmodesmata, dissolved substances moving from cell to cell would have to traverse one plasma membrane, the intervening cell wall, and then the other plasma membrane. When this route is assessed in quantitative terms, it turns out, despite the relatively large surface areas that are available, to be considerably less efficient than the pathway offered by plasmodesmata of average size and frequency.

Unfortunately, the clarity of the outcome of mathematical assessments of plasmodesmatal function is clouded by a number of complications. For situations where the direction of transport is clearly one-way, as, for example, towards a secretory gland (Plate 10a) the conclusion may be valid, but what of other situations where the traffic appears to be two-way? For instance, at the endodermis of roots (see page 31), water and dissolved minerals flow inwards towards the vascular tissue, while sugar, etc., needed to sustain the cortical cells has to pass in the opposite direction. It seems that it may be necessary to search for means whereby the opposed fluxes could be segregated either in space to different plasmodesmata, or in time.[302]

The axial structures of plasmodesmata, described above, add to the complications. If the axial tubular structure is indeed an extension of the endoplasmic reticulum, what might be its role? Where it occurs, there could be two potential pathways for cell-to-cell transport, one inside the cisternae of endoplasmic reticulum, and one in the cytoplasmic annulus between the axial tube and the plasma membrane.[601] The former is so narrow that it is open to question whether it could be functional, and the latter, as already stated, can be severely constricted where the plasma membrane pinches in to come into close proximity to the axial strand.[599, 696]

These considerations are important, but at present are impossible to treat satisfactorily. They do, however, give rise to several intriguing speculations. The two potential pathways together admit the possibility that two-way traffic might occur in plasmodesmata. If the constriction of the plasma membrane effectively closes the cytoplasmic annulus to transport, then the symplast is reduced from a commune of interconnected protoplasts to a commune of interconnected cisternae of endoplasmic reticulum. Additionally the axial strand, connected as it seems to be to cytoplasmic cisternae of endoplasmic reticulum, could just conceivably act as a one-way valve to regulate the direction of transport.

It is suspected that as well as large fluxes of organic and inorganic solutes, trace quantities of growth regulators may move through the symplast. For instance, if the root cap is removed from a root tip, the root is no longer sensitive to gravitational stimulation, and if placed horizontally will not bend downwards during further growth. It may be that the key factor here is that the gravi-perceptive cells in the root cap (see page 118) cannot pass growth-regulating hormone back to the growing zone of the root unless the plasmodesmata at the root cap–root tip junction are intact. Replacing a severed root cap on the root tip does not restore the gravi-perception system.[411] The growth regulator indoleacetic acid displays a remarkable polarity in certain tissues, being much more mobile in one direction than the opposite (e.g., its movement in the coleoptile of a grass seedling is predominantly away from the apex). It is not clear how this polarity of transport is achieved, or indeed whether plasmodesmata are involved. Their existence, however, needs to be taken into account in any attempt to explain the basis of polar transport.

Experiments on the growth of tissues from free cells in artificial culture suggest that differentiation does not commence until divisions create groups of cells that are all interconnected by plasmodesmata: only then do individuals embark on their own developmental pathways.[891] It seems paradoxical that the onset of cellular individuality should depend upon the existence of connections between cells, the more so since in extreme cases, connections can lead to the *loss* of individual behaviour patterns. Thus in pollen mother cells and a few other situations in both plants and animals where there are large cytoplasmic bridges (Fig. 11a), the high degree of intercellular communication synchronizes the progress of the cells through cycles of nuclear division.[352] There must be a subtle dividing line between having as much cell-to-cell communication as is necessary, and having too much to permit individuality.

3.6.2 Transfer Cells

Most of the molecules that traverse the plasma membrane do so one by one, either by diffusion or by the activities of the 'carriers' in the membrane. The greater the surface area of the plasma membrane, the more space there is for trans-membrane diffusion and the more solute pump molecules that can be accommodated. Cells which seem to exploit this simple principle by enlarging the surface area of their plasma membrane are common in both plants and animals. They tend to occur in parts of organisms where there is intensive movement of solutes.

The commonest adaptation in animals is a fringe of *microvilli*—tubular extensions of the plasma membrane—lining those surfaces of the cells in contact with the solution from which solutes are absorbed. Animal cells develop a variety of other types of surface infoldings in relation to both absorptive and secretory processes.

The corresponding adaptation in plants (Plate 13a and b) is a system of inwardly projecting ingrowths of cell wall material.[301a–633] The plasma membrane follows the contours of this elaborate wall, and therefore has a larger surface area than would be the case in the same cell imagined to be devoid of wall ingrowths. Wall projections and plasma membrane together constitute a *wall-membrane apparatus* that can develop in a wide variety of plant cells, from secretory glands, to absorptive epidermal cells on the submerged leaves of water plants, to cells lining the xylem (as in Plate 13) and phloem and probably functioning in loading and unloading these channels of long-distance transport in the plant body. The name *transfer cells*, chosen with their likely function in mind, is used when referring to cells that have become specialized in this way, regardless of their anatomical position. Their cytoplasm is in some ways as specialized as their walls. Intensive trans-membrane solute transport requires correspondingly intensive production of energy, and this is likely to be related to the large population of mitochondria found in transfer cells (Plate 13b).

There is a functional relationship between plasmodesmata and transfer cells, in which these two specializations of the plasma membrane would seem to work in sequence. It is thought that in the case of transfer cells with an enlarged plasma membrane having *absorptive* functions, the solutes that are taken in across it can diffuse to underlying cells through plasmodesmata. In other words plasmodesmata might allow absorptive transfer cells to serve whole tracts of the symplast. In transfer cells where the enlarged plasma membrane has secretory functions, the plasmodesmata might provide symplastic supply lines for the solutes that are destined to be exported. Transfer cells emerge as cells which are specialized in relation to exchanges between the symplast and the apoplast.

3.6.3 Restriction of Transport through Cell Walls

The most abundant of all the constituents in cell walls has scarcely been mentioned as yet. It is water.[252] Hemicelluloses and pectic substances are especially hydrophilic and in a normal wall in its natural state the volume of the water is as great as or greater than that of the space occupied by microfibrils and matrix molecules. Consequently, dissolved molecules of the sort plants live on can diffuse through or along cell walls. Acid groups in the wall do impede the movement of ions, but if neutralized by salt formation, or if neutral molecules like sugars are being considered, non-encrusted walls emerge as very minor obstacles, compared to the plasma membrane, in the path of substances moving to or from the cytoplasm.[840] By contrast, lignified, cutinized, or suberized parts of walls have very much narrower and less permeable waterways.[771]

The open, watery texture of walls means that where there is no cuticle or cork, as in young roots, an external solution can permeate inwards. Conversely, the sap in the xylem can permeate outwards, since mature xylem elements have no plasma membranes to separate the sap from the surrounding walls (page 19). Considering these two facts in relation to the physiology of roots, the question arises, how is it that the xylem sap that moves upwards from root to shoot can contain a higher concentration of mineral salts and other solutes than does the soil solution?[188] Why do not solutes leak along the cell walls until the concentrations are equalized? Animal physiologists encounter the same type of problem. How is it that the extracellular fluid in blood serum is constantly different from extracellular fluids a few cells away in compartments such as the gut or kidney tubules?

The answer is that both kingdoms have evolved devices that restrict leakage through the gaps between plasma membranes of neighbouring cells. Animals simply fuse the two membranes together back to back, thereby forming the seals known as *tight junctions*.[493] The corresponding specialization in plants is not dissimilar. The neighbouring plasma membranes are fused to the intervening wall (for details see caption to Plate 16), and the permeability of the wall is reduced by impregnating it with what is usually described as suberin, despite the lack of modern chemical analyses to justify the use of the term. The structure is called the *Casparian strip*, after its discoverer, Caspari.[62a]

The Casparian strip is compared with a conventional wall and plasma membrane in Plate 16c, where the flattening of the plasma membrane gives an unrivalled view of its 'tramline' construction. Plate 16a and b shows where it occurs. It is a feature of one particular layer of cells, the *endodermis*, that surrounds the vascular tissues. It cements the endodermal cells into a continuous sheath, running as a band around each one of them in the walls along which solutes would otherwise leak from xylem to soil.

If the Casparian strip is truly as 'tight' a junction as is suspected, the only route across the endodermis is through the endodermal cell protoplasts, either by plasmodesmata or across the plasma membrane and cytoplasm.[699] In either case it is perfectly feasible for the apoplast (cell walls and xylem) *enclosed by* the endodermal cell layer to contain a fluid that differs from that in the apoplast *external* to the endodermis. Normally the fluid inside has the lower water potential of the two, because it is continually being enriched with solutes that were accumulated into the symplast from the soil solution by the activities of solute pumps, and passed through the symplast towards the vascular tissue of the root, where they pass from the symplast to the apoplast.[188] The endodermal protoplasts act as semi-permeable membranes between the internal and external solutions, and water diffuses across them, passing down its potential gradient from outside the endodermis to the region enclosed by it. This input of water creates the hydrostatic pressure (root pressure) that drives an upward flow of xylem sap from the root to the shoot.

An extreme version of this system, equally dependent on the Casparian strip, occurs in vascular bundles of some legume root nodules (Plate 13a).[302] Here the solutes that are being transported are products of nitrogen fixation. They are considered to diffuse *via* plasmodesmata from the cells containing nitrogen-fixing bacteria (referred to on page 25) up to and through the endodermal cells. The cells inside the endodermis are secretory transfer cells, and their wall ingrowths can be seen in Plate 13a. They secrete the incoming nitrogenous compounds into the apoplast of the vascular bundle, thereby lowering the water potential and so generating an inflow of water. As in ordinary roots, the solutes are flushed along the xylem.

Plate 13 illustrates what might be described as the primary stage of limitation of water movement through the apoplast. The Casparian strip, as already noted, allows differences in water potential to be maintained and to manifest themselves in the form of osmotic flows of water towards whichever compartment has the lower water potential. Secondary and tertiary stages of development are also recognized.[127, 318] In the former, a layer of suberin is deposited right around *all* of the walls of the endodermal cells. Given this deposit, water and solutes dissolved in it are not only prevented from moving *between* the cells of the endodermis (by the Casparian strip), but are prevented from reaching the plasma membranes of the endodermal cells. If the suberized layer is truly impermeable, these cells are now entirely dependent upon their plasmodesmata for their water and nutrient supplies. Furthermore, it would appear that the flow of water and solutes from the root cortex to the vascular heart of the root must also be restricted to the endodermal cell plasmodesmata. If sugars or other nutrients have to be passed from the vascular tissue to the cortex then the problem, mentioned on page 29, of bidirectional transport through plasmodesmata against a net flow of water arises.

A suberized layer occurs in the equivalent cell layer around the leaf veins of many grasses (Plate 34a–c), though in this situation it occurs on its own. It is not certain whether it serves the function of a Casparian strip or whether it is merely the equivalent of the 'secondary' suberized layer of the root endodermis. Its function in terms of the physiology of the leaf is obscure (see also page 108). By limiting the flow of water between the veins and the evaporating surfaces in the leaf to plasmodesmata (or to the smallest veins, where the suberized layer does not occur) the layer may help to ensure that water supplies are distributed to all parts of even very elongated grass leaves.[599] Again, however, problems of bidirectional transport arise, for products and intermediates of photosynthesis must pass into the veins, against the net flow of water.

The tertiary stage of endodermal cell wall development in roots is the deposition of thick layers on the radial walls and inner tangential wall.[127] Even these lignified deposits are pierced by plasmodesmata, and their function is probably merely skeletal.

Vascular tissue is not the only region of plants that becomes partially isolated by 'tight junctions'. Equivalent structures also occur very commonly around secretory organs, especially in the stalk cells of glands (e.g. Plate 10a and b).[498] It is presumed that the outside walls of the stalks are rendered impermeable, so that the outward flow of solutes and water that maintains the secretory activity of the gland must pass through the symplast. The abundance of plasmodesmata in the cross walls of stalk cells has already been noted. Further, because the impregnated portion of wall is continuous with the surrounding cuticle, the secreted product cannot percolate back into the tissue below the gland. Neither can the water that is present in the internal cell walls of the underlying tissue—so necessary to preserve cell turgidity—escape by uncontrolled evaporation into the open air, as distinct from internal evaporation and loss in a controlled manner *via* stomatal pores.

3.7 The Cell Wall: Some Generalizations

We have taken the view that the plasma membrane is the boundary of the living cell and that the wall is a non-living extra-cytoplasmic investment. In the sense that it cannot maintain or reproduce itself, the wall is certainly non-living, but it has nevertheless emerged as a lively and vital part of the plant. Some of the dynamic processes sited in walls have been mentioned but many others have had to be omitted. In particular the existence of cell wall enzymes has been virtually ignored, though functions in both the synthesis and breakdown of wall components and the transport of materials between wall and protoplast are likely.

It is appropriate to conclude by asking how many of the characteristics of plants depend upon the existence of cell walls. Naked protoplasts prepared from plant tissues by digesting away the cell walls are very sensitive and can only survive in culture media that match their own water potential. In dilute media (high water potential) they swell and burst because water diffuses into them along its potential gradient. It may be speculated that there was a very early stage of evolution characterized by cells as sensitive as naked protoplasts, superseded in due course by the evolution of osmoregulatory devices that coped with the problems of life in unbalanced media. This could well, in fact, have been the original selective advantage possessed by cells with cell walls. A protoplast enclosed within a wall cannot swell and burst. It can swell only until the stretching of its wall generates an inward 'wall pressure' that exactly counteracts the forces of swelling. A wall, in other words, would have allowed colonization of habitats where the concentration of solutes was low.

In general, plants still live in media of high water potential, at least, they accumulate solutes in their cells so that the water potential inside the plasma membrane is always lower than that outside (see also page 34). The consequent tendency for water to diffuse inwards keeps the protoplasts pressed against the walls, and this is one of the main sources of mechanical strength in plant tissues—compare turgid with wilted leaves. Given cell walls and adequate water, plants do not have to regulate the water potential of their apoplastic fluids (in cell walls, xylem sap, and fluid-filled intercellular spaces) in the way that is so vital for the survival of naked cells in higher animals.

One consequence of the evolution of a cell wall was that from then on solid food was in general unobtainable. The cells had to subsist on whatever could penetrate the wall to the plasma membrane, and this limitation to compounds of low molecular weight can be related to certain other properties of plant cells. They have a less well developed digestive system than animal cells—and very reasonably so, considering that they only 'ingest' simple molecules. This applies even to insectivorous plants, which may capture bulky prey, but also break it down to simple compounds prior to 'ingestion'. Relinquishing solid food (including solid organic material) left only two options. The cells could live in environments which provide soluble organic nutrient. Fungi are present day examples. Alternatively, the cells could develop internal sources of organic nutrient and so minimize their dependence on environmental supplies. In fact, cell walls generally occur along with intracellular structures, chloroplasts, capable of using light energy to synthesize organic compounds from atmospheric carbon dioxide. This does not imply that the evolution of photosynthesis was a consequence of the evolution of cell walls; it merely rationalizes the present day close association between the two.

Most of the differences between plant and animal cells can thus be related to the evolution of the cell wall, which seems to have been a fundamental point of divergence for the two kingdoms. Present day organisms can be viewed as the products of evolutionary diversification, from that point on working within limits set by the biological potential of cells and tissues with and without walls. Having considered the cell wall first, we shall on many occasions in subsequent chapters have to recognize its all pervading influence on plants by referring back to it.

4 Vacuole and Tonoplast

4.1 Introduction

When looking at a section of plant material, zoologists are apt to comment first on how helpful it is to have cell walls which neatly delimit every cell, and second on how 'empty' the cells look in comparison with those of animals. More than 90% of the volume of mature parenchyma cells can indeed consist of *vacuole*—a watery compartment surrounded by a membrane, and containing a wide variety of substances, the most familiar and conspicuous being the red, yellow and blue flavonoid pigments of flower petals. Obviously, not all plant cells are so 'empty'. The vacuoles of cells in meristems can be minute (e.g. Plate 2); alternatively, in cells like those of a legume cotyledon the many vacuoles are full of stored protein and are one of the most conspicuous components (Plate 20a, c). A few plant cell types lack vacuoles (e.g. Plate 22).

The existence of a membrane bounding the vacuole was inferred long before it was actually visualized in electron microscope preparations. Microdissection experiments, and observations on the movement of dyes supplied to cells, left little doubt that there is a membrane and also that it is selectively permeable, capable of maintaining certain solutes at a higher concentration in the vacuole than in the cytoplasm.[657, 658] The vacuolar membrane is called the *tonoplast,* although the word was introduced by De Vries in 1884 for a very different purpose: as a name meaning a body under stress or tension, to apply to small bodies which he considered to be precursors of vacuoles. The idea was that the precursors can become hydrated, the water giving rise to the water-filled compartment of the vacuole, and the precursor metamorphosing to become the bounding membrane. Nowadays the word is used only for the membrane, and does not imply acceptance of any particular theory of the origin of vacuoles.[934]

4.2 Ultrastructure

THE TONOPLAST The tonoplast looks remarkably like the plasma membrane. Both have a tripartite dark-light-dark construction, about 8–10 nm thick. Both have cytoplasm in contact with one face and a watery non-cytoplasmic compartment on the other. Closer inspection does, however, reveal certain differences. For instance, the electron-dense reagent phosphotungstic acid stains plasma membranes but not tonoplasts, probably reflecting the presence of glycoproteins (see page 23) on the former.

The tonoplast frequently (Plate 16c) but not always (Plate 3b) displays asymmetry, the dark layer that faces the vacuole being somewhat thicker and more heavily stained than its counterpart on the cytoplasmic face. This could represent nothing more than an artefact such as precipitation of vacuolar substances during fixation or dehydration of the specimen. Alternatively, it may be perfectly genuine, for asymmetry has also been discovered by means of the freeze etching technique, in which the number of damaging manipulations is minimized. The two tonoplasts in Plate 17a illustrate the point. They bound two spherical vacuoles, so that the convex surface is that which faces the cytoplasm, while the concave faces the vacuolar contents. The convex has between two and three times *fewer* particles per unit area than has the concave. There are also slight differences in the sizes of the particles.[216, 218, 219] It must be remembered that if the tonoplast conforms to the behaviour of other membranes when subjected to the fracturing process used in freeze etching, these two images do not represent the outer surfaces of the tonoplast as viewed from the cytoplasm (convex) and vacuole (concave). Since the fracture is held to pass along the *mid line* of the membrane, they may be views of the inner faces of the split tonoplast, showing the bases of particles that lie embedded in the fabric of the two separated parts of the membrane.

SHAPE AND SIZE OF VACUOLES Turning from the tonoplast to the vacuole itself, it is apparent that this is one of the most variable components of plant cells. Vacuoles can be minute or enormous. As their volume approaches the volume of the cell, so they assume the shape of the cell itself. Where they are small they are usually more or less spherical (e.g. Plates 1, 2 and in the companion cell of Plate 7). Forces of surface tension tend to minimize the surface to volume ratio of vacuoles just as in the analogous case of soap bubbles. Spheres more than any other shapes satisfy this condition. Clearly, however, such forces can be overridden. Many examples of other shapes can be found in the micrographs. The shape of vacuoles can change during the life of a cell (quite apart from *growth* in volume). During the life of cells as strikingly different as the unicellular alga *Chlorella*[19] and the meristematic *cambium*[114, 114a] cells of tree stems the vacuoles alter from being few and large to being many and small. Cells can alter the surface to volume ratio of their vacuolar systems. For a given total volume, the more subdivided the vacuoles, the greater the surface area of the tonoplast. This consideration is important in studies of the manner in which plant cells absorb solutes across their plasma membrane and subsequently accumulate them in their vacuoles. The end result of these sequential processes will depend not only upon the permeabilities of the membranes but also upon their surface areas. It is sometimes inferred that because meristematic cells have small vacuoles (Plates 1, 2), the properties of their tonoplasts can be neglected. However, with the high surface to volume ratios found in small vacuoles, there need not be many of them in a cell for the area of the tonoplast to rival that of the plasma membrane.

Some of the micrographs (e.g. Plate 16b) show vacuoles with irregular, even convoluted, tonoplasts. One possible interpretation—that this might be another device for increasing the surface area of the membrane—should be treated with caution. Vacuoles contain a variety of compounds that are released and could induce artefacts when the cells are fixed during the first stages of specimen preparation. The interface between cytoplasm and vacuole will be prone to artificial stretching or shrinkage, indeed sometimes shrinkage of vacuoles can be so severe that the tonoplasts even come to resemble endoplasmic reticulum cisternae in their general morphology (Plate 16c). There is reason to believe that the smooth contours of frozen etched vacuoles are more lifelike (Plate 17a).[215]

VACUOLAR INCLUSIONS A great variety of inclusion bodies is to be found within vacuoles.[217, 856] In our general state of ignorance of such objects not many can be identified and labelled. One exception is shown in Plate 17c—a proteinaceous suspension in the vacuolar fluid.[291, 721] Other suspensions, crystals, and amorphous deposits are common. A category that deserves to be singled out concerns inclusion bodies containing cell membranes. These probably have several origins. One is the extrusion of lipid myelin figures as illustrated in Plate 3b and discussed on page 13. Another is the invagination of the tonoplast into the vacuole, either with or without an enclosed portion of cytoplasmic material. Cytoplasmic membranes passed in this way into the vacuole can change greatly in appearance, often coming to resemble myelin figures (e.g. Plate 17b); this topic is considered again below. A third origin is the plasma membrane. Observations of living cells by phase contrast microscopy and of fixed cells by electron microscopy indicate that it is possible for the plasma membrane to bulge inwards, not only into the cytoplasm, but even into the vacuole, pushing the tonoplast before it. The bulge can apparently be nipped free, the plasma membrane healing behind it. If the freed portion of plasma membrane remains in the cytoplasm it looks very like a vacuole, and the phenomenon is like that of endocytosis, described for legume root nodule cells on page 25. If the portion is freed into the vacuole it presumably remains surrounded by a portion of tonoplast, and looks like a 'vacuole within a vacuole'.[506]

4.3 Functions of Vacuole and Tonoplast

The functions of vacuole and tonoplast can for convenience be considered under the headings of (1) regulation of the water and solute content of the cell—or, in short, *osmoregulation*, (2) storage, and (3) digestion.

4.3.1 Osmoregulation

MAINTENANCE OF TURGIDITY (1) GENERAL The diffusion of water along its potential gradient, and some of the factors that influence the water potential of aqueous compartments, have already been introduced (page 24). Vacuoles contain dissolved solutes, so the vacuolar water potential is lower than that of pure water. It might be thought that if the concentration of solutes is the same in the cytoplasm as it is in the vacuole, there would be a balanced situation with no net movement of water across the tonoplast. However, the water in the cytoplasm is subject to another factor which lowers its potential beyond the reduction due to solutes. Water molecules become bound to the protein and other large molecules found in the cytoplasm. Thus for vacuole and cytoplasm to be in equilibrium, it is necessary that there be slightly more solutes in the vacuole. Only then will the vacuolar water potential (affected by solutes) balance the cytoplasmic water potential (affected by both solutes and binding)[155, 156] The solvent is, in short, more 'available' in the vacuole. Cells are also capable of transporting molecules into their vacuoles. For this, two mechanisms are thought to exist. One involves molecular 'carriers' in the tonoplast, like those functioning in transport across the plasma membrane (page 24).[462a] The other consists of the delivery of membrane bound vesicles to the tonoplast, where the contents of the vesicles are liberated into the vacuole by fusion of the vesicle and the tonoplast membranes (see also page 78).[502a]

The concentration of solutes in vacuoles has been measured, some vacuoles being large enough to be sampled by inserting micro-pipettes. It seems general that the requirement to have a higher concentration than the cytoplasm is more than satisfied,[588, 804] and that solute pumps in fact load the vacuole until its solute concentration is considerably higher than is needed to achieve a balance between cytoplasm and vacuole. Solute concentrations of the order of 0.4–0.6 molar are general. This means that there is normally a strong tendency for water to diffuse into the vacuole, first from the cytoplasm and ultimately from the environment external to the cell. Since the permeability of the tonoplast to water is greater than that of the plasma membrane, and that of the cytoplasm is very high, it is the plasma membrane that is the major barrier in this pathway.[844] As already stated (page 24) water molecules can permeate relatively easily through the plasma membrane. The vacuole therefore tends to swell, but is restrained from doing so simply because the surrounding cytoplasm would have to be displaced, and this cannot come about because it in turn is contained within a more-or-less rigid cell wall. Thus, unless the plant is so short of water that it is wilting, the high solute concentration in the vacuole ensures that the cytoplasm is always pressed tightly against the cell wall—that the cell is *turgid*.

The net water potential of the vacuolar sap in turgid cells is therefore made up of two major components: an osmotic component due to the presence of solutes, acting in the direction of lowering the water potential; and that due to the inwardly directed hydrostatic pressure exerted by the stretched cell wall, acting to raise the water potential. Whether or not there is net movement of water into or out of the cell depends upon the balance at any given instant between the internal water potential and the external, which likewise is contributed to by osmotic and hydrostatic (and in some instances water-binding) components. As explained on page 32, the existence of the cell wall allows plants (in general) to live in a wide range of external conditions by virtue of their ability to load their vacuoles with solutes, thereby maintaining turgidity. Vacuoles and cell walls *together* provide the major skeletal system of soft plant tissues.

It is in the light of this vital role for vacuoles that attention is being focused on the ultrastructure of the tonoplast. What is the molecular nature of the solute pumps? Do the tonoplast particles (Plate 17a) include solute pumps? Do alterations in the surface to volume ratio of the tonoplast-vacuole system reflect changes in the ability of cells to pump solutes into their vacuoles? Most of these questions await full investigation, but it has been found, for instance, that the electron density of the tonoplast changes while, at the same time, the powers of salt-accumulation of the tissue also change.[310]

MAINTENANCE OF TURGIDITY (2): SPECIAL FUNCTIONS

Maintenance of turgidity is a central theme of vacuolar function but there are a number of significant variations on it, some of which are described in the following paragraphs.

The shape and size of cells is partly determined by the fact that any given volume of protoplasm requires to import nutrients in order to support its activities, and also to export wastes, both at particular rates. It follows that the surface area of a cell must be large enough to cope with these transport processes. A very large central vacuole is one device which has enabled giant cells to evolve. The requirements of the cytoplasm can be met because it is spread in a relatively thin film with a large surface area between the vacuole and the wall. If the same total volume of cytoplasm were to be reformed into a non-vacuolate mass, (or if the cell were to lose its turgidity), it would probably have much too low a surface to volume ratio. These giant cells are found especially amongst aquatic algae, where volumes up to several cm^3 can be attained, but the same phenomenon comes into play in higher plant tissues. In leaves, for example, the presence of large cell vacuoles (up to 0.1 mm^3) allows a relatively small bulk of protoplasm (and that largely chloroplasts) to be spread over an enormous surface area (Plates 32a, 34a). One important consequence is enhancement of the efficiency of carbon dioxide absorption in the mesophyll tissue.

In a young cell with primary cell walls, the forces tending to swell the vacuoles, and hence the cytoplasm that contains them, do more than just maintain turgidity: they bring about cell enlargement, physically stretching the wall from within. The continued relaxation of the wall and pumping of solutes, and the consequent inflow of water and enlargement and coalescence of vacuoles, are thus all integral parts of cell growth. The force that it generates is not exerted in any particular direction, but, as indicated in the previous chapter, the cell wall has specifically aligned microfibrils which determine the distribution of stresses and strains and therefore the shape that will be assumed as the cell enlarges. Perfectly uniform extension would produce a spherical cell. Circumferentially placed microfibrils in a cylindrical cell permit more cell elongation than they do increase in girth. Localized areas of low mechanical strength become localized areas of growth, as in the tips of growing pollen tubes or fungal filaments. Existing walls can be softened to create new growth areas, as when a fungal or an algal cell branches.

A very different, and reversible, type of cell wall deformation, not to be regarded as growth, is powered by the vacuoles of the guard cells of stomata in leaves.[547] Here changes in turgidity combined with the distribution of thickened regions of cell wall alter the shape of the cells and hence regulate the aperture of the stomatal pores (Plate 9a, d), in turn regulating the activities of the leaf in transpiration and photosynthesis.

The spectacular leaf movements of the 'sensitive plant', *Mimosa pudica,* brought about by a mere touch, are probably based on a sudden increase in the solute permeability of the tonoplast of certain cells in the 'hinge' position of the leaves and leaflets. Solutes are released from the vacuoles, there is therefore a loss of turgidity, the cells collapse, and the leaf stalk hinges. Given time, the tonoplast and its solute pumps gradually recover, turgidity returns, and the leaves rise to their normal position.[7] Why a touch should have this effect, and how the stimulus is transmitted to and perceived by the tonoplast, are largely unknown. Potassium ions contribute greatly to the vacuolar solutes that operate movements of both *Mimosa* and stomatal guard cells.

Lastly, vacuoles can enlarge in particular parts of

cells, with profound effects upon their further development. Some of the most important types of cell division in plants are asymmetric, in that the parent cell does not divide equally in two, but into one large and one small daughter cell. This behaviour correlates with the position of the vacuole, which can effectively confine the dividing protoplast to one region of the cell. We cannot tell, however, whether the vacuole determines the position of the protoplast or *vice versa*.

4.3.2 Storage

The chemical composition of vacuolar sap varies so much from plant to plant and from cell to cell that it has been stated there is but one compound common to all—water. Dissolved substances include mineral ions (e.g. chloride, sodium, potassium, magnesium, calcium, sulphate, phosphate), carbohydrates (sugars, mucilages), nitrogenous compounds (amino acids, amides, proteins, peptides, alkaloids), various organic acids, and tannins, flavonoid pigments and their glycosides. Crystalline constituents include salts of organic acids (e.g. calcium oxalate), pigments, and amino acids (e.g. asparagine).[856]

Many and varied functions can be postulated for vacuoles containing specific substances. Where the compounds seem to have no obvious function one has in the past had to resort to the vague concept of vacuoles as depositories for by-products of the biochemical machinery of the cell. Loading a vacuole with such compounds could be regarded as a form of excretion, in that they are removed from the living cytoplasm. It is becoming clear, however, that the concept of vacuoles as dumps is inadequate. True, complex biochemical systems may well inevitably produce wastes, but it is now thought that in many cases the *accumulation* of specific substances has biological significance. There is in particular a subtle chemical warfare between plants and animals in their ecosystems, waged using attractants (e.g. vacuolar pigments in petal cells), unpalatable compounds, or poisons.[890] An example of storage of one such product in vacuoles is given in Plate 17c and its caption.

Excretion is an appropriate word to use in connection with the pumping of sodium ions out of the cytoplasm, a process that occurs both outwards across the plasma membrane and 'inwards' across the tonoplast into the vacuoles. The 'inward' excretion has been extensively studied in *Atriplex*, a plant of salty habitats. It pumps ions into the vacuoles of bladder-like cells on the leaf surfaces until the vacuolar concentration reaches very high values, up to twelve times molar.[498] The quantity of organic acid in the vacuole in at least some cases reflects the transport of positively charged cations—the negative, anionic, organic acids are synthesized in the cytoplasm and maintain electrical balance by accompanying the cations into the vacuole.

Other genuine storage functions do occur in vacuoles. Studies of the ultrastructure of developing seeds show that protein and other food reserves such as phytin (calcium and magnesium salts of inositol hexaphosphate) are deposited in vacuoles by as yet mysterious routes, ultimately filling them to produce the *protein bodies* or *aleurone grains* of the mature seed.[83, 688, 805, 806] Protein bodies of a legume seed are shown in Plate 20a and c. The vacuoles of cells in the root systems of plants also have storage functions. Mineral nutrients traverse the cortex of the root in passing from the soil solution to the centrally located vascular tissue. When the supply is sufficient, nutrients enter the vacuoles of the cortical cells. That this represents genuine storage is demonstrated by experiments in which plants are then removed to conditions where nutrients are in poor supply or even zero supply, as in distilled water. The root continues for some time to provide the shoot with minerals, drawing upon the supplies accumulated in the vacuoles.

4.3.3 Digestion

LYSOSOMES, HETEROPHAGY AND AUTOPHAGY With the exception of the nucleic acid molecules that carry the genetic information of the organism, the lifetime of individual molecules in living protoplasm is usually much shorter than the lifetime of the cell that contains them. They are continually being broken down and then re-synthesized in a cyclic process described as *turnover*. When the cell or the tissue becomes senescent the breakdown processes dominate, and we see this not only when organs such as petals or leaves come to the end of their life, but also as part of the programmed life-history of specialized cells like vessel elements and tracheids (pages 19–20). Except where the whole plant is senescent and dying, very little is allowed to go to waste, and the breakdown products are re-utilized elsewhere. Zoologists are more familiar with breakdown processes than botanists, because as well as exhibiting turnover, animal cells commonly feed on solid matter taken in by pinocytosis or endocytosis (page 24), and subsequently digested. It is therefore not surprising that *lysosomes*, subcellular components specializing in digestion, were discovered first in animals. There they deal with ingested material *and* much of the routine turnover and senescence, whereas in plants, where they are concerned only with turnover and senescence, their discovery was comparatively recent. Intake and digestion of foreign material is described as *heterophagy*, and digestion of the cell's own components as *autophagy*.

PLANT LYSOSOMES It now appears that microscopists have been looking at plant lysosomes for more than a century without realizing their digestive function.[538, 538a] More and more evidence points towards vacuoles as containing digestive enzymes capable of breaking down the lipid, nucleic acid, carbohydrate, and protein of cytoplasmic components that come into contact with them. A rational interpretation of the invaginations into vacuoles, and the membranous inclusion bodies mentioned above on page 34, is that at least some autophagic digestion is accomplished in the vacuole following 'ingestion' at the tonoplast. Plate 17b presents an example of what may well be a stage in the process. It was found in a normal, healthy root tip, emphasizing that such processes do occur naturally.[551] They become more prominent under unfavourable conditions, including oxygen starvation,[144] and, in the embryo of seeds, very prolonged dormancy.[854]

Digestive enzymes are, of course, lethal, and no cell can manufacture and release them into the cytoplasm unless it is programmed to *autolyse*, i.e. digest itself. With the benefit of hindsight, it now seems rather obvious that one safe place for plant cells to store their digestive enzymes is in the vacuole, where they are segregated from the cytoplasm. Even here the enzymes are presumably in contact with the inner face of the tonoplast, and one aim of future research must be to find out how this versatile membrane itself resists digestion. Not surprisingly, its breakdown has been observed to coincide with the onset of rapid lysis of the cytoplasm, for example in maturing xylem (Plate 5b), and in rapidly senescing cells, for example in dying flower petals.[540] A clear distinction has to be drawn between controlled autophagy and wholesale autolysis of cells following rupture of the tonoplast.

Many puzzles remain. How is turnover regulated? What factors control the time at which a given cytoplasmic component enters the vacuole, and the nature and quantity of material taken in? How much turnover can be accomplished without resorting to vacuolar ingestion?

The storage vacuoles of seeds provide a slightly less complex picture, in which the protein and other food reserves that accumulate during seed development are broken down and utilized during germination. Thus at germination the material to be digested is already in the vacuole, and the enzymes either enter at the beginning of the process, or else are deposited there in a latent form during development, to be activated at germination.[610, 735, 919]

The source of vacuolar digestive enzymes is obscure. There is some evidence that they are synthesized and packaged in the endoplasmic reticulum and the Golgi apparatus, and then transported to the vacuole, segregated from the cytoplasm in membrane-bound vesicles, sometimes termed primary lysosomes.[145, 146, 644] The membrane of the vesicle must be able to recognize and fuse with the tonoplast. An alternative view, preferred by some authors, is that the enzymes are formed in cisternae of endoplasmic reticulum which then develop into vacuoles.[48, 538a] At any event, it is clear that formation of the digestive enzymes is a part of the genetically controlled programming of cell development. The amount of the enzymes can alter during the life of the cell. No doubt different cells produce them to different extents, indeed it is probable that some vacuoles completely lack them. There are instances in which cells possess more than one type of vacuole, as judged by their morphology or by the appearance of their contents, and there may be heterogeneity in respect of digestive abilities.

Finally, it is necessary to confuse the picture by recognizing that some digestive processes can apparently be accomplished without the aid of pre-existing vacuoles. The endoplasmic reticulum seems to be the component responsible for initiating digestion in some situations. The end result is, however, similar to vacuolar digestion in that remnants of cytoplasmic components come to lie in vacuole-like compartments. Further information on autophagy and heterophagy mediated by the endoplasmic reticulum may be found on page 67, in the chapter devoted to the endoplasmic reticulum.

4.4 The Origin of Vacuoles

Nobody knows for sure how and where vacuoles originate. Some of the ideas that have been put forward[550] are that they develop: (i) from membrane-bound vesicles originating in the Golgi apparatus, (ii) as swellings of the endoplasmic reticulum, (iii) from pre-existing vacuoles, which can enlarge and divide by fission, so that there are always enough to provide every newly-formed cell with its quota, (iv) by a *de novo* process of molecular association starting when water-attracting (hydrophilic) compounds become surrounded by a tiny aqueous shell, and this in turn develops a membranous skin of lipid and protein molecules to give rise to the tonoplast.

There may be no one answer to the problem, and much depends upon how narrowly or widely vacuoles are defined. There could be several origins for 'watery compartments bounded by a membrane', but if a vacuole is regarded as a compartment bounded by a specific type of membrane, namely the tonoplast, then the above four possibilities can be analysed as follows. The profound differences between the membranes of the endoplasmic reticulum and the tonoplast militate against (ii). Parallels found in the reproduction by fission of mitochondria and plastids add a measure of credence to (iii), as does the apparently ubiquitous occurrence of small vacuoles in meristematic cells (Plates 2, 46c, 48a). The growth of a pre-existing tonoplast, as in (iii), is slightly easier to envisage than process (iv), in which this highly specialized membrane would have to be assembled without benefit of any 'template'. The ability of the Golgi apparatus to produce vesicles enclosed by a membrane that is capable of fusing with the tonoplast (see page 78) indicates that (i) is possible.

The evolutionary origin of vacuoles is an equally speculative subject. It can scarcely be doubted that their primary function was osmoregulation, and contractile vacuoles,[432] still to be found in many primitive organisms, may have been the first type to evolve (and they were the first to be recognized, as early as 1776).[934] By expelling water from cells they would have permitted maintenance of higher internal solute concentrations than found in the external environment. In the ancestors of the plant kingdom, contractility of vacuoles would have become redundant with the development of the cell wall. As in present-day plants, the inflow of water down its potential gradient would have been limited to the relatively small volume needed to stretch the wall sufficiently to oppose the forces leading to entry of water. With loss of contractility and of direct contact between the vacuole and the environment, the advent of solute pumps to load the vacuole would have been all that was needed to ensure turgidity over a wide range of external conditions, and the way would then have been open to exploit the mechanical strength of turgid cells in the evolution of multicellular plant bodies.

5 The Nucleus

5.1 Introduction: The Morphological and Chemical Basis of Inheritance

CELL AND NUCLEUS A century and a quarter elapsed between Robert Brown's establishment of the concept that a *nucleated* cell is the major unit of plant structure, and the modern discoveries concerning the molecular biology of the nucleus.[384] Today it is easy for any student to see nuclei in living or fixed material. Brown's work, by contrast, was based on observations made with a simple microscope, without the aid of the improved lenses that were developed for use in compound microscopes in his time.

Others who observed nuclei, both before and after Brown's 1833 publication, were slow to recognize their significance. Thus when Schwann drew attention to the similarities between animal cartilaginous cells and plant parenchyma cells, he emphasized the 'walls' in each case. His neglect of the fact that nuclei represent another similarity between cells in the plant and animal kingdoms was pointed out later by M. J. Schleiden, and the names of these two men, one a botanist and the other a zoologist, became associated with the 'cell theory' that found its beginnings in Schwann's paper in 1839.

Unfortunately, Schleiden had become over-impressed by his work on endosperm tissue in plants. He was reluctant to accept that this tissue is exceptional in its development from an acellular, liquid state to a cellular, solid state. His consequent, and mistaken, view that in general, cells arise by a process in which nuclei produce walls around themselves, prevailed. Nucleoli, detected in nuclei of epithelial cells from an eel skin as early as 1781 but not generally recognized until much later, were duly incorporated into this scheme. They were regarded as structures which give rise to the nuclei, in turn having their own origin by aggregation of 'granules'.

Confusion over the relationship between cell and nucleus persisted for many years and greatly hindered further understanding. Even experienced investigators were led to misinterpret their own evidence. For example Nägeli, in 1844, observed the division of a *Tradescantia* stamen hair cell into two progeny and correctly reported that the daughter nuclei were derived from the parent cell nucleus. However, he thought this to be the exception rather than the rule, and continued to support Schleiden's theories on the origin of nuclei. Not until the early 1850's was the true origin of cells by division of a parent cell to give two daughter cells generalized by Remak and Virchow. Although the new generalization was by no means immediately accepted, it stimulated more objective studies of the nucleus.

CHROMOSOMES AND HEREDITY Emergence of the idea that the nucleus contains the hereditary material of the cell and the organism was gradual, commencing with the three great conceptual advances of the 10 year period starting in 1855: Virchow's conclusion that all cells arise from pre-existing cells; Darwin's formulation of evolution and the origin of species; and Mendel's demonstration that 'traits' are transmitted from one generation to the next by a series of paired 'elements'. Neither Darwin nor Mendel had much understanding of cellular structures, and other workers in the field of biology who might have interpreted Mendel's results in terms of structure did not know of him or his work until its rediscovery in 1900. By then a structural background to which it was possible to relate the genetical data had been uncovered.[898]

Observations on cells fixed while undergoing nuclear division demonstrated that during the division process a number of small oblong or thread-like bodies *(chromosomes)* move with respect to a system of fibres in the cell (the *spindle*). Fuller descriptions of the sequence of events came in 1879, when the whole process, in 1882 termed *mitosis* by Flemming, was found to be closely similar in plant and animal cells. A verbal description of a continuous sequence of events is always difficult, so the process was considered as a number of *phases*. Originally Flemming described as many as nine, but today only four are recognized. The first, *prophase,* involves the condensation of the chromosome threads so that they appear as discrete entities within the cell. Each chromosome consists of two halves *(chromatids)* lying side by side and joined at a particular region (the *centromere*). The *poles* of the division figure are established on opposite sides of the nucleus and a system of fibres which will become the spindle begins to form between them. The end of prophase is usually accompanied by the disappearance of the membrane surrounding the nucleus (the *nuclear envelope*), and of the nucleoli. In the next stage, *metaphase*, the chromosomes move to the equator of the spindle, midway between the poles. The beginning of *anaphase* is marked by the longitudinal division of each chromosome, fully separating the constituent chromatids, which then move to opposite poles. *Telophase* essentially involves a reversal of the events of prophase. The individual chromatids become thin and extended, and the nuclear envelope and nucleoli are re-formed, giving a new nucleus at each pole. This is usually followed by division of the cells into two at the plane of the spindle equator. Usually the progeny then enter a phase of growth and development known as *interphase*. A detailed discussion of these events will be found in Chapter 12, and the above summary is included only as an introduction to the present general consideration of the nucleus.

The knowledge that organisms develop from fertilized eggs was the starting point for a series of researches that gradually elucidated the nature of the carriers of hereditary information. In 1876 Hertwig discovered that fusion of sperm and egg nuclei is the essential element of the fertilization process. Strasburger, in 1884, showed that the same applies to fertilization in flowering plants. In the previous year van Beneden observed that the male and female gamete nuclei each have the same number of chromosomes, a number that is half the number in the parents. It was later realized that the parents possess two sets of chromosomes, and the gametes one, but not until 1905 was the term *meiosis* coined to describe the type of division by which the chromosome sets are separated during formation of gametes. Fertilization re-unites the two sets of chromosomes, and these are passed on from cell to cell by mitosis as the organism develops, prior to its next reproductive cycle. Since Roux in 1883 had concluded that each chromosome splits lengthwise during mitosis, one product being included in each daughter cell, Weismann was able, two years later, to take note of the continuity and duplication of chromosomes during the life of an organism, and state that they carry hereditary information. All this time Mendel's work lay forgotten, and its rediscovery in 1900 confirmed that studies of cells, nuclei and chromosomes are relevant to the behaviour of the whole organism. From then on, the old subject of cytology and the new subject of genetics developed together.[797]

THE MOLECULAR BASIS OF HEREDITY Meanwhile, studies of the chemical nature of nuclear components had begun. In 1869 Meischer showed that one of them is an acidic, phosphorous-rich compound that he called nuclein. He isolated it from pus cells and went on to show that it also occurs in salmon sperm heads, and that it is resistant to protein-digesting enzymes and soluble in alkali. Later, nuclein was shown to contain a five carbon sugar, deoxyribose, and two types of organic bases, purines and pyrimidines, as well as phosphate; we now know it as deoxyribonucleic acid (DNA). That DNA is located in the chromosomes was indicated by their resistance to proteolytic digestion and their solubility in alkali. In 1914, Feulgen developed a specific staining method for DNA and from 1924 greatly extended the number of observations on the DNA of nuclei and chromosomes in a wide range of tissues and organisms.[384] This ultimately led, in the 1930's, to the theory that the units of hereditary information, or *genes*, consist of DNA—a suggestion originally made by Meischer some 60 years before. More direct evidence was provided in 1944 by experiments which showed that DNA can transmit genetic information from one microorganism to another. However, understanding of the role of DNA in the cell awaited the analysis of its complicated molecular structure, eventually achieved by Watson, Crick and Wilkins in 1953.[866, 893]

We need not be concerned here with the chemistry of DNA so much as with the way in which its molecular structure has been exploited in the course of evolution of living organisms. As already mentioned, the constituents of DNA are acidic phosphate, deoxyribose sugar, and purine and pyrimidine bases. DNA is a polymer, but not as simple a polymer as starch or cellulose. It has four different monomers, each one a nucleotide, that is, a unit consisting of phosphate, sugar, and a purine or a pyrimidine base. Two purines, (adenine and guanine) and two pyrimidines (thymine and cytosine) provide the four nucleotide monomers. They are polymerized in a linear sequence, the backbone of which consists of alternating sugar and phosphate, with the bases emerging as lateral branches at right angles to the long axis. The four nucleotides may be in any order and in any proportion in an individual DNA strand, indeed it is possible to synthesize artificial strands containing just one type of nucleotide, or just two different types alternating with one another etc., etc.[865]

The feature that is of such significance in relation to the transmission of hereditary information is that with the aid of appropriate enzymes (DNA polymerases), a single strand can generate and join with another strand that is precisely complementary to it. 'Complementary' in this sense means that for an adenine in the first strand, there will be a thymine in the second, for guanine a cytosine, for thymine an adenine, and for cytosine a guanine. This comes about because the size relationships of bases permit only the above two pairings, and further, because of the ability of members of these pairs to link together by hydrogen bonding. The two types of hydrogen bonded purine-pyrimidine base pairs form inter-strand links, and the two strands take up the famous double helix configuration.

Nuclear DNA normally exists in this form, but the double helix can be opened into its constituent single strands on two types of occasion (described below), both vital to the life and reproduction of the cell and the organism.

The sequence of the nucleotides on a strand varies from DNA to DNA, but it is not random, any more than the arrangement of the letters of the alphabet in words is random. The arrangement of the nucleotides, like that of the letters, is a form of *information*. In DNA there are only four letters (nucleotides), and it is now clear that three-letter words are employed. 64 such words are possible.[865] Each word represents a code for an amino acid, and the sequence of words determines the sequence of amino acids in the proteins, enzymes and otherwise, that carry out the functions of the cell.

One type of occasion, then, on which the double helix is opened, is to allow one of the free strands to form on itself a copy which can be moved from the nucleus to the cytoplasm, there to govern synthesis of the particular protein that is specified by that length of DNA—by that particular unit of genetic information, or gene. The copy is in fact a strand of ribonucleic acid (RNA), similar to DNA, but with sugar ribose instead of deoxyribose and the base uracil instead of thymine. Because of its function, it is known as messenger RNA (mRNA). Other types of RNA exist in the cell, and they too are synthesized directly upon DNA 'templates'. Since the mRNA is essentially a re-written form of the DNA, it is said to be formed by transcribing the DNA, or in short, by *transcription*. Later, the transcribed information in the mRNA is *translated* to generate the appropriate sequence of amino acids in the corresponding protein. As an example, the transcribed version of the word thymine-guanine-adenine on the DNA would be adenine-cytosine-uracil on the mRNA, and this would code for insertion of the amino acid threonine.[865] The role of ribosomes in translation of mRNA will be considered later (page 60).

The second type of occasion on which the double helix is opened is concerned with duplication of the genetic information, prior to its transmission to the progeny at mitosis. Here *both* strands of the original double helix form complementary copies by means of the base-pairing reactions. The end result is two double helices of DNA where formerly there was one. This mechanism of duplication of the genetic information is the ultimate basis of many of the observations so painstakingly amassed during the historical development of knowledge concerning the role of the nucleus and chromosomes in the life of the cell. Thus dividing cells undergo cycles in which their DNA duplicates, usually in interphase, and then separates into two identical moieties at mitosis. Cells usually contain two copies of the total genetic information of the organism, in the form of two sets of chromosomes. Sexual reproduction entails segregation of the two sets during the formation of male and female gametes, which each possess one set only. Fertilization restores the starting condition. Just as the rediscovery of Mendel's work linked the subjects of genetics and cytology, so the discovery of the DNA double helix, its capacity to store and release information, and its ability to self-duplicate, has made molecular biology an integral part of cyto-genetics.[865]

With this background, we can proceed from the chemical and morphological basis of inheritance to examine the nuclear components of plant cells, and, as far as possible, to the functional interpretation of their structure.

5.2 General Features of Nuclei

The 'typical plant cell' is usually imagined to have a single spherical nucleus in its centre. While very many cells conform to this pattern, there are quite numerous exceptions exhibiting variation in the number, shape, size and position of their nuclei.

MULTINUCLEATE CELLS To a large extent the variation in nuclear number depends upon shifts from the usual one-to-one correspondence between nuclear division and cell division. For example, continued duplication of DNA and nuclei prior to the onset of *cell* division is a feature of the endosperm tissue in many plants. It becomes a liquid, multinucleate mass of protoplasm, remaining so until it is partitioned by the late formation of cell walls. While it is still fluid it is an extremely useful material for observations of mitosis (see Chapter 12 and Plate 44), and it is also a good source from which nuclei can be isolated, that of cereals and coconuts providing bulk preparations.

Endosperm is not the only multinucleate tissue in plants. One type of tapetum (page 22) is liquid, flowing around the developing pollen grains in the anther loculus. Multinucleate 'giant cells' are produced in a whole range of plant tissues under the influence of infection by larvae of certain nematode worms, which presumably secrete a mixture of plant hormones to induce the effect. In tissue culture an imbalance between nuclear and cell division can be created by altering the ratio of the growth regulating substances kinetin and indoleacetic acid. Developing xylem elements in the fern *Marsilea* produce 2, 4, 8 and then even 16 nuclei per cell before the protoplasts are autolysed.[491] The central cell of the embryo sac of flowering plants characteristically has two nuclei (the polar nuclei), which when one male nucleus fuses with them, give rise to the nuclei of the endosperm tissue.[398]

The algae provide numerous examples of multinucleate cells, for which the term *coenocyte* is commonly used. Some algae are partial coenocytes, that is, with easily recognized multinucleate cells circumscribed by definite walls, while members of some other groups are wholly coenocytic, the entire plant (sometimes many cm in size) being one multinucleate protoplast bounded by an outer wall. It seems general that if nuclei sharing a common cytoplasm start to divide, they tend to undergo mitosis in synchrony.

POLYPLOIDY AND POLYTENY In other instances there is an imbalance between DNA duplication and *nuclear* division (rather than *cell* division).[577] In such cases nuclei with 2, 4, 8, 16 or even more times the basic quota of DNA can be found. Such nuclei are said to be *polyploid*. For instance, nuclear polyploidy is much more common in developing xylem than is the development of the multinucleate condition mentioned above in respect of *Marsilea*.[491] A considerable number of plant tissues consistently double or quadruple their nuclear DNA as part of their development. In most cases the chromosome number is increased in proportion. Less frequently, the DNA duplicates but the chromosomes do not, so that bulky *polytene* chromosomes (like those of insect salivary glands) develop. The best example of polyteny in plants is provided by the enormous cell of the suspensor of young bean *(Phaseolus)* embryos, where up to 12 *cycles* of duplication occur, giving rise to chromosomes containing 2^{12} times the basic quota of DNA.[576] Not surprisingly, nuclear volume increases with increase in nuclear DNA-content, and often the two remain in a fixed proportion to one another.

THE SHAPE OF NUCLEI Nuclear size can, however, change without change in DNA content, and so can nuclear shape. Both enlargement without much distortion from the spherical, and distortion both with and without marked enlargement, can accompany cell differentiation. It is common to find elongated, spindle-shaped nuclei in elongated cells (e.g. Plate 5a). The nuclear surface may become indented in maturing cells (e.g. compare Plate 18c with Plate 2, the former being from an older region of the same root). Deep fissures, lined with the nuclear envelope, and containing cytoplasmic components such as microtubules, ribosomes, and mitochondria, may penetrate far into the nucleus. As with the formation of lobes protruding outwards from the nucleus, one obvious consequence is that the surface to volume ratio of the nucleus is enhanced, perhaps facilitating transport of material between nucleus and cytoplasm. The occurrence of deep invaginations and lobes is, although quite common, somewhat unpredictable at present, and not simple to relate to cellular activities. Plate 40a presents an example. One extreme case has been seen in Jerusalem artichoke cells grown in tissue culture.[921] There, long tubular evaginations of the nuclear surface develop in association with microtubules. They are less than 0.1 μm in diameter, and are transitory, existing only during a period just prior to mitosis, when the nucleus moves from a peripheral position in the cell to a more central location in a bridge of cytoplasm extending across the large vacuole.

The last example serves to emphasize that nuclei are like many other cell components in being dynamic structures, subject to changes in shape and position. They are usually centrally located in cells which have few or small vacuoles (e.g. Plates 1, 2). Development of large vacuoles displaces them (e.g. Plate 31a, see also page 36). Where cytoplasmic streaming is very rapid, as in giant algal cells like those of *Nitella*, they are swept around with the other cell components. Their much slower movements where streaming is more sluggish, as in *Tradescantia* staminal hairs, are easily observed. An especially important instance of nuclear movement is that leading to fusion of gamete nuclei during fertilization in the egg and central cells of the embryo sac.[398]

5.3 Chromatin

CHROMATIN[454a] As one part of the historical development of the concept that DNA carries the hereditary information of the organism, it was shown that sperm and egg nuclei each contain a set of genes, and that each and every cell in the organism that develops from the fertilized egg receives copies of these sets, condensed into the form of chromosomes at prophase of mitosis, and distributed into the daughter cells at anaphase. It is thought that when chromosomes are condensed their genes are inactive. Gene action (transcription) is instead a feature of the non-dividing phase of cell development. It is then that the cell may embark on one or other of the many possible pathways of differentiation, selecting and expressing appropriate combinations of genes from amongst the total present in its nucleus. Any understanding of gene action and genetic control of the characteristics of cells, and hence of organisms, must therefore include knowledge of the way in which nuclear DNA is organized in interphase, and of how it changes during development and differentiation.

Formation of readily visible chromosomes at the time of mitosis provides an excellent opportunity to study certain aspects of the organization of the DNA. For instance, cytogeneticists have been able to construct 'genetic maps' showing the location of particular genes on particular chromosomes. Unfortunately, the chromosomal material is much less easy to observe when it is dispersed, as in non-dividing cells. Various staining reactions do detect it, down to the limit of resolution of the light microscope, indeed its affinity for stains is implicit in the name given to it—*chromatin*. The same material is visible in conventionally prepared ultra-thin sections, and the electron microscopist can in addition use a more specific electron dense stain, based on the affinity of a compound of bismuth for DNA.[3a]

HETEROCHROMATIN AND EUCHROMATIN A number of different cell types from oat plants *(Avena)* is illustrated in the Plates. Together, these examples demonstrate that the distribution and form of chromatin varies greatly from one cell type to another, even within a species. In Plate 11c two cell types, in the tapetum and the anther wall, are alike in possessing bulky masses of very densely stained chromatin. In the same Plate, the nuclei of the young pollen grains have many fewer and much smaller dense regions. Plate 31, showing young leaf cells, presents an intermediate condition. Several micrographs of other species (Plates 7, 26b, 40, 49) show an especially common form of dense chromatin, located on the inner face of the inner membrane of the nuclear envelope.

Observations on mosses and liverworts led Heitz, in 1928, to distinguish dense aggregates as *heterochromatin*, and dispersed DNA-containing materials as *euchromatin*. The terms are now applied generally to plants and animals. Heterochromatin zones are of widespread, if not universal, occurrence. Special, recently developed, staining procedures have shown that they can persist through interphase in a condensed form, appear during mitosis at distinct regions in the chromosomes, and reappear again in the subsequent interphase when the euchromatic moiety of the chromosomes disperses. Thus some heterochromatin is to be regarded as a permanent feature of the nucleus, whilst other portions may, as pointed out above, arise in relation to processes of cell differentiation within the organism.[924] The distribution of heterochromatin within a nucleus is another variable, and can alter during the life of a cell.

The first stage of gene activation is (in most cases) transcription of DNA to form mRNA molecules, and this is a process that can be localized by making use of the technique of autoradiography (page 5). If a tissue is supplied with a radioactive precursor of RNA—such as tritiated uridine—the transcribed RNA molecules that are formed on the genes will become radioactive. The distribution of silver grains over autoradiographed specimens will then show where this process was taking place at the time of fixation, provided that the experiment was of short enough duration to preclude movement of completed, radioactive RNA away from its site of synthesis.[492]

There is considerable evidence that heterochromatin is inactive in genetical terms, or at any rate is less active than euchromatin. In most autoradiographic experiments the widespread distribution of radioactivity in euchromatin reflects the fact that in the majority of cells many genes act concurrently. With the exception of the genes that govern RNA synthesis (e.g. pages 47–48), and some genes on polytene chromosomes,[638] it has not been possible to locate specific individual genes by means of autoradiography.

THE ULTRASTRUCTURE AND ORGANIZATION OF CHROMATIN FIBRILS It is estimated that, if it could be uncoiled and measured, the chromatin of a chromosome would be from one to many hundred times longer than the coiled, condensed, chromosome itself. A lily nucleus contains more than 30 metres of DNA double helix, packed into the nuclear volume of about four millionths of a cubic millimetre. It is therefore not surprising to find that at the ultrastructural level, chromatin consists of fine fibrils. Because ultra-thin sections are several times thicker than are the fibrils themselves, superimposed images of stacked fibrils are obtained, and it is difficult to resolve individuals (Plate 19h). Estimates of their thickness average just less than 10 nm. They can be examined by picking them up on support films after forces of surface tension have been used to spread them from nuclei floated on to a water surface.[253]. A greater range of dimensions is observed after this procedure, but there is some indication that the thicker fibrils may arise by deposition of foreign material[908] and that the true thickness is about 10 nm, or perhaps slightly less.[922]

No matter how they are prepared for electron microscopy, chromatin fibrils have a tangled appearance. There must nevertheless be considerable order in their arrangement within the nucleus. Thus somehow the fibrils from different chromosomes are kept separate during successive cycles of coiling and uncoiling. Some observations show that chromosomes can be condensed in the same relative positions that they occupied during uncoiling at the previous telophase. It is exceptional for the sequence of genes and heterochromatic regions in a chromosome to alter during interphase, so the coiling and uncoiling must occur without breaking the DNA strand. A possible role for the nuclear envelope in this connection will be considered later.

The organization of chromatin in the interphase nucleus must allow for duplication of the DNA. This does occur at prophase or telophase in some organisms, but mostly it is a feature of interphase, the period of DNA synthesis (the 'S-period') being preceded and followed by gaps when no DNA synthesis occurs (the 'G1-period' and the 'G2-period' respectively). Eukaryote DNA double helices duplicate at about 0.5–1.2 μm per minute, and chromosomes are observed to duplicate their DNA in a few hours.[382] Since the length of DNA in most chromosomes is sufficient to require *many* hours of synthesis if duplication were to commence at one end and work along to the other at the above rate, it follows that duplication must commence at many sites and spread from them over the total length of the dispersed chromosome.[381] Autoradiographic localization of incorporation of radioactive thymidine into duplicating DNA indicates that the chromosomes in a nucleus may not duplicate in strict synchrony, and that duplication of euchromatic regions may precede that of the heterochromatic regions.[37, 450a] Chromosomes may become more heterochromatic after duplication than they were in the G1 and S periods.[450b] The previously mentioned peripheral location of heterochromatin in the nucleus is especially common early in the G1 period (e.g. Plate 30b).

Techniques additional to electron microscopy have to be used to obtain further insight into the ultrastructure of chromatin fibrils. Chemical analyses show that in eukaryotes, as distinct from prokaryotic bacteria and blue-green algae, the DNA is complexed to basic proteins, for which *histone* is the group name.[774] There are in fact very few histones, and at least one of them has a remarkably wide distribution in plants and animals, varying very little in its amino acid composition. Histones are rich in the basic amino acids lysine and arginine, reflecting their property of associating with the acid groups of DNA. DNA-histone complexes have fibril diameters of 3–4 nm, and this has led to some controversy concerning the substructure of the 10 nm wide chromatin fibrils. Are there two or four DNA-histone strands per fibril? Is there only one strand, coiled so as to appear thicker than the strand itself? The degree to which an object scatters X-rays and electrons is a measure of its mass, a phenomenon which has allowed estimation of the mass per unit length of chromatin fibrils. They appear to be quite open structures, less solid than would be expected on a multi-stranded model. It is therefore suspected that the chromatin fibril is a single strand of DNA-histone (about 45% : 55% by weight) that is twisted into a helix of overall diameter about 10 nm.[76]

REPEATED SEQUENCES OF DNA It has always been something of a mystery that higher plants and animals should have so much DNA per set of chromosomes. It can be calculated that there is much more than is, in theory, needed to code for the RNA and the protein molecules of the organism. One interpretation might have been that important genes are present in multiple copies. This does apply (see pages 47–48) to genes for the RNA of ribosomes (rRNA), but further chemical analyses have led to an additional explanation. From 20–80% of eukaryote nuclear DNA is in the form of relatively short sequences that are repeated very many times—even up to one million times—with varying degrees of perfection.[86] Further, it has been shown that at least some of the permanently-present masses of heterochromatin contain this type of DNA.

Repeated short sequences of DNA, whether in heterochromatin or not, cannot have a direct genetic function in that they cannot be genes. A number of roles has, however, been proposed for them. Arguing that such sequences will tend to fold up and aggregate, like with like, in much the same manner as like molecules come together in a crystal, one idea is that they could account for the dense appearance of heterochromatin in light and electron microscope preparations. Further, there are species in which the chromosomes commonly lie in clusters centred on conspicuous heterochromatin regions. It is suggested that this is an example of association of like with like not only *within* a heterochromatic region, but also *between* heterochromatic regions, an association which could contribute to the maintenance of the three dimensional organization of the total nuclear chromatin during interphase.[924] A comparable, but more specific, type of chromosome aggregation is that involved in chromosome pairing prior to meiosis, when the chromosome number is halved at the time of gamete formation.[369]

The above suggestions may be classed as roles for permanently-present heterochromatin. Possible roles for the type of heterochromatin that appears or disappears during the life of the cell include the protection of vital genes, and the covering or uncovering of genes. For example, a gene that becomes buried in a mass of aggregated heterochromatin may not have space around it in which transcription can occur. Conversely, a gene could be activated by unfolding it from an aggregate. As with other models for the control of gene activity, such as suppression by adding histone to the DNA and activation by removing histone, much more work is needed in order to lift these suggestions from the realms of speculation.

It has not been necessary in this brief account of chromatin to distinguish between plants and animals. The biology of chromatin is probably far too fundamental for the evolutionary divergence of the two kingdoms to have had very much effect. The interphase chromatin has a central role in governing the life of the cell, and no fundamental phenomena are known that belong exclusively to plants or to animals.

5.4 The Nucleolus

In generalizing upon the presence of the nucleus in plant cells, Robert Brown also recorded the widespread occurrence of a conspicuous intra-nuclear body. Further descriptions and studies of this, the nucleolus, were prolific. Montgomery compiled the first major review of the subject in 1898, and cited some 700 references. He was able to conclude that nucleoli occur in almost all nuclei (nowadays we would qualify this to almost all *eukaryote* nuclei); that their number varies from one to several hundred per nucleus, with one, two, three, or four being the most common; that their structure is dynamic, changing in relation to the process of mitosis and also during interphase; that they do not stain in the same way as chromosomes; and that they tend to be best developed in cells that are very active in growth or syntheses.[855]

Since then, the discovery that nucleoli are associated with particular chromosomes may be singled out as being of especial significance. First noted in 1912 by Navashin, this phenomenon has subsequently been documented in greater and greater detail. We now know that the nucleolus is the site of transcription of a particular sequence of genes, and that it consists in part of transcribed molecules of RNA in various stages of maturation prior to their departure to the cytoplasm as constituents of ribosomes. The sequence of genes—the *nucleolar organizer* region—is part of a chromosome, and the nucleolus-chromosome relationship is therefore akin to that between gene and chromosome. In fact the relationship between nucleolus and chromosome was described four years before Bridges for the first time assigned a specific gene to a specific chromosome—the X-chromosome of the fruit fly *Drosophila* (oddly enough, a nucleolar organizer is present on the same chromosome). Much more recently (1968)[557, 558] the nucleolus has again led the way by providing the first view of identified genes 'in action'—or more precisely, genes seen in various stages of transcription in electron micrographs that, as usual, are regrettably static.

5.4.1 *Structural and Biochemical Elements of the Nucleolus*

SIZE AND SHAPE Enough nucleoli are included in the figures to convey something of the possible variations in shape and size. More or less spherical forms are seen in Plates 1, 5e, 11c, 19a–f, 31a, 49, an ellipsoidal shape in Plates 2, 20a, 31c, and a comparatively asymmetric example in the prophase nuclei shown in Plates 1 (N–I), 44a–c, and 45a. In the spherical examples the diameters range from 4 μm (Plate 1) to 2 μm (Plate 49). The long axis of the ellipsoidal nucleolus in Plate 20a measures 15 μm—an example of a very large nucleolus in a cell engaged in intensive synthesis of proteins.

STRUCTURAL ELEMENTS Nucleoli are conspicuously dense in phase-contrast microscopy (Plate 19a–f), implying a high refractive index, and in turn, a high concentration of matter. Electron micrographs reinforce the idea that nucleolar constituents are densely packed. Fibrillar and granular zones, just discernible in Plate 2, are shown at high magnification in Plate 19g and h. Except in the 'vacuoles' within the body of the nucleolus (Plates 1, 19a–c, f, g, 20a, 44a–c) the structural elements are crowded together. The word 'vacuole' is widely used when referring to the comparatively empty, *non membrane-bound* spaces often seen in nucleoli, and to avoid possible confusion with the *membrane-bound* vacuoles of plant cell cytoplasm, we shall use quotation marks for nucleolar 'vacuoles'. They contain both fibrils and granules, but in a comparatively dilute suspension compared with the surrounding nucleolus proper (Plate 19g). 'Vacuoles' are not always seen (e.g. Plates 5e, 11c, 49), though it has to be remembered that they could be present, but not included within the thickness of an individual section.

Many attempts have been made to classify, name, and define in structural and chemical terms the components of the nucleolus. An exceedingly complex nomenclature has been rendered even more confusing by historical changes in interpretation of structure and in application of terminology, particularly when electron microscopists began to use light microscopists' terms. At the risk of over-simplification we prefer to avoid the older wording (nucleolonema, pars amorpha, nucleolini, etc.) in favour of the relatively straightforward descriptions: granular regions, fibrillar regions, 'vacuoles' and nucleolar chromatin.

BIOCHEMICAL ELEMENTS Simple staining reactions can be used to detect two of the major biochemical elements of nucleoli. The basic dye toluidine blue is bound by nucleoli, its colour shifting from blue to purple. The same applies to the less intense staining of cytoplasm, and in both cases the staining is abolished if the material is digested beforehand with the enzyme ribonuclease. The presence of RNA is inferred. On the other hand, if an acidic dye such as acid fuchsin is applied, intense staining reveals the presence of concentrations of basic material. A third component, DNA, is less easy to detect, except through the use of highly sensitive procedures. It has been found possible to isolate nucleoli in quantities sufficient for biochemical analysis, and the following example, determined using nucleoli from pea plants, is seen to correlate with the staining reactions:[52]

11% RNA
7% DNA
6% histone-type basic protein
21% non-histone basic protein
55% 'residual' protein

RELATIONS BETWEEN STRUCTURE AND BIOCHEMISTRY The granular and fibrillar zones both contain RNA and protein, indeed the reason why the individual 6–10 nm diameter fibrils (Plate 19h) are difficult to resolve may be that they lie in a matrix of amorphous protein. The granules, each about 15 nm in diameter, are more readily distinguished (Plate 19g and h). Sometimes the zones of granules take the form of coarse threads 0.1 μm or more (i.e. some 6–8 granules) in width, ramifying in different directions through the body of the nucleolus. The granular and fibrillar zones may be mingled intimately (as in many animals) or the granules may surround a central fibrillar zone (as in many plants). More complete than usual segregation of the zones can be induced by a variety of treatments such as low temperatures,[124] X-irradiation,[128] application of inhibitors of RNA synthesis,[712, 751] and lowering the oxygen supply;[124] it can also occur under certain natural conditions,[241] including prophase of meiosis[273] and mitosis (Plate 45a).

Much of the nucleolar DNA is located in fine strands of heterochromatin. They may lie in clearly defined channels (Plate 2), or in less distinctly separated areas (Plate 19g). Some fortunately orientated sections show them passing into the nucleolus from larger masses of heterochromatin in the nearby nucleoplasm (Plate 19g). This identification of DNA is based on susceptibility to digestion and removal from ultra-thin sections by the enzyme deoxyribonuclease.[124, 453] A technique in which nucleoli are squeezed out of cells and swollen in detergent allows the three dimensional shape of the nucleolar heterochromatin to be seen using phase-contrast microscopy:[452] it consists of continuous loops that are so irregularly twisted as to explain why ultra-thin sections usually include only small, apparently isolated, fragments. Evidence that the nucleolus also contains euchromatic, or dispersed, DNA will be presented later.

The structure and composition of the nucleolus may thus be generalized as follows. Strands of DNA-containing chromatin ramify through admixed or zoned fibrillar and granular regions, both containing RNA and protein. The whole may be rendered spongy by the presence of 'vacuoles'.[454a]

5.4.2 *The Nucleolar Organizer*

Although associations of the nucleolus with a particular chromosome had been noted in 1912, few details were available until de Mol and Heitz published on the subject between 1927 and 1931. By examining telophase nuclei and the prophase stage of meiosis, when chromosomes and nucleoli are visible together, they found that the nucleolus is attached to a structurally distinct part of a specific chromosome. The distinctive feature of this, the nucleolar organizer, is seen when chromosomes are stained for DNA. It is an apparent constriction in the nucleolus-bearing chromosome, placed so that a terminal portion looks as if it is attached to the remainder by an invisible or barely visible stalk. Terminal portions, when small, are termed satellites, and are usually heterochromatic, remaining during interphase as condensed masses at the periphery of the nucleolus (Plate 2). It is not clear what happens to the constricted region itself during interphase. The term 'constriction' is something of a misnomer. The light microscopist sees it as a constriction because of its low content of DNA compared with the neighbouring stretches of chromosome. In the electron microscope it is seen as a mass of fine fibrils,[198] and during mitosis, it is fully the diameter of the neighbouring DNA-rich parts of the chromosome (Plate 47a–d).

It was originally thought that the apparent constriction is the part of the chromosome that organizes the nucleolus, indeed statements to this effect are still frequent. However, in maize plants dosed with ionizing radiation to induce fragmentation of chromosomes, the nucleolus develops in association with fragments bearing the *adjacent* chromatin rather than the constriction itself.[542] Destruction of the constriction by a laser beam reduces nucleolus formation to a *lesser* extent than does destruction of the adjacent region.[50] Also the size of the constriction can vary. It is becoming more satisfactory to think in terms of the nucleolar chromatin loops being part of, or derived from, the region of the chromosome *beside* the constriction. The constriction could simply be a space which accommodates part of the bulk of the nucleolus during interphase, though it seems more likely that the fibrillar material in it must have some special and as yet unidentified functions.

It might be expected that one nucleolus would develop for every chromosome in the nucleus that bears a nucleolar organizer. This does happen, but fusion may reduce the number of nucleoli. For example in the plantain *(Plantago)* a single, large nucleolus can be formed from four units originally organized by four separate chromosomes.[386] Serial sections through the composite structure nevertheless detect four nucleolus-associated masses of heterochromatin, indicative of the presence of four nucleolar chromatin loops. In wheat or hyacinth plants with different numbers of chromosome sets, the *maximum* number of nucleoli is always the same as the number of constricted chromosomes.[157] Chance proximity of nucleoli within the nucleus favours fusion, and in several plant and animal tissues the degree of fusion has been observed to increase as the cells age.

Since chromosomes duplicate during interphase it might further be expected that in the G2 period the number of nucleoli would be twice that in G1. Surveys of onion *(Allium)* plants show that this happens in some cells (anther wall cells, pollen) but not in others (root tips).[272] Where it does *not* happen it is likely that the two organizers of the duplicated chromosome (which after all are very close together, one on each chromatid) work together within one composite nucleolus.[272] Phase-contrast microscopy of such nucleoli reveals the presence of two loops of chromatin.

Within an individual nucleus the nucleoli are nearly always the same size. This balance can, however, be upset by tampering with the nucleolar organizers. They can, for example, be fragmented by X-irradiation.[855] Small fragments produce small nucleoli, as compared with nucleoli that are based on normal organizers, even though the small and the normal may inhabit the same nucleus. Naturally occurring examples of this condition have also been found.[799] A chromosome of maize in which a duplicate copy of the organizer region has become inserted alongside the normal single copy develops an abnormally large nucleolus.[643]

These observations show that the nucleolar organizer is not an entity of fixed size. It can be subdivided and supplemented without altering the nature of its function. Quantitative changes do, however, occur. It seems reasonable to conclude that the nucleolar organizer must consist of an unfixed number of very similar, if not identical, functional units. If it is fragmented the reduced number of units can still operate, but to reduced effect; if it is augmented the increased number can operate to greater effect. We will return later to this important deduction.

5.4.3 *Ribosomal RNA Synthesis in the Nucleolus*

The idea that the nucleolus produces cellular RNA stemmed from early observations that it contains this substance in bulk. In due course structural similarity between nucleolar granules and cytoplasmic ribosomes, which also contain RNA, was noted. Autoradiography of incorporated radioactive precursors (e.g. tritium labelled uridine) was then used to demonstrate rapid synthesis of RNA in the nucleolus and, significantly, subsequent transfer of radioactivity to the cytoplasm, from which ribosomes containing radioactive RNA could then be isolated. More recent work has confirmed that the nucleolus manufactures ribosomal RNA (rRNA). To take two examples: destruction of the nucleolus with a carefully aimed beam of intense ultra-violet light prevents synthesis and transport of rRNA, and a mutant organism in which the nucleolar organizer has become deleted has been found to be unable to make its own ribosomes, and dies when those carried over into the egg cell from the parent come to the end of their life.[104, 855]

How then does the ability to manufacture the RNA

of ribosomes relate to the presence in the nucleolus of chromatin strands that contain multiple units of function? Within limits the question can now be answered, but first it is necessary to digress and look more closely at the end product of the synthetic machinery—the ribosomes themselves.

RIBOSOMES Ribosomes[257] are visible in practically all of the Plates. Most cells contain many millions of them. Exceptions in plants include mature xylem elements, which have lost all of their protoplast, and mature sieve elements, which have lost their nuclei and ribosomes. Ribosomes are 18–20 nm in diameter, and high magnification views (particularly of negatively stained preparations) show them to consist of two subunits, one larger than the other.[578] The two units dissociate when the magnesium concentration is very low, and they reassociate when the normal concentration is restored. Both subunits are exceedingly complex aggregates of RNA and protein molecules, and both participate in the vital reactions by which a mRNA molecule is translated to generate the equivalent protein. The larger subunit contains an RNA molecule about 4000 nucleotides in length, i.e., of molecular weight 1.3×10^6. A much smaller strand is also present. The small subunit contains an RNA strand with just over 2000 nucleotides and molecular weight 0.7×10^6. These figures vary greatly between eukaryotes and prokaryotes, and comparatively slightly between groups of eukaryotes. The RNA strands are complexed to a great variety of different proteins—more than fifty types per ribosome, some present as individual molecules and some represented more than once; some with structural functions and some with enzymatic functions.

THE TRANSCRIPTIONAL UNIT AND ITS MATURATION Returning now to the nucleolus, it has been possible in certain cases where the granular and fibrillar zones are discrete to show that, when radioactive uridine is supplied, the fibrillar zones produce the first radioactive RNA.[49, 323] When extracted and examined, however, it is found to be a much larger molecule than rRNA itself.[390] It varies considerably in size, and is much larger in mammals than in cold-blooded animals and plants. Within the latter groups even closely related species vary, and furthermore, a difference of about 600 nucleotides has been found when samples obtained from roots and leaves of a single plant species are compared.[293] If the radioactive uridine is supplied for a longer period, autoradiography shows that radioactive RNA spreads from the fibrillar to the granular zones of the nucleolus.[49, 323] By then (less than 30 minutes) the initial, large RNA molecule has been progressively 'matured'. Terminal portions have been removed to give an intermediate which in turn becomes cleaved into two unequal parts. Further small portions are removed from each part to generate the two major rRNAs of the two ribosomal subunits. The length of RNA that is discarded varies according to the size of the initial precursor.

The significance of the sequential maturation reactions of rRNA is not understood, but the autoradiographic experiments do provide a link between the structure of the nucleolus and the biosynthesis of rRNA. The first precursor RNA is presumably transcribed on nucleolar DNA; it appears in the fibrillar zone; it matures and moves to the granular zone; later still the products move to the cytoplasm (a process that will be returned to later). It is tempting, but not fully justified, to assume that some of the fibrils wind up to create the granules (Plate 19h). It is relatively certain that the granules are precursors of ribosomal subunits.

THE RATE OF SYNTHESIS OF rRNA Experiments using radioactive RNA precursors show that rRNA molecules are formed at a rate of between 10 and 100 nucleotides per second, depending upon the organism and the temperature.[731] Thus if the original precursor RNA is taken (as in mung bean leaf cells[293]) to be 9000 nucleotides long, one precursor molecule could be formed in from 1.5–15 minutes. But there may be millions of ribosomes in a cell, and their number must be doubled in the time taken by the cell to grow and divide. For example, in a meristematic cell of a pea root tip, about 9 million ribosomes are made during the growth and division cycle time of 10 h, i.e. an average production rate of 15 000 per minute.[390] Clearly, the marked discrepancy between the relatively long time needed to make one rRNA molecule, and the very short actual production time, presents a problem. Somehow the productivity is increased.

In this connection it is worth comparing the way in which protein and rRNA molecules are made. Many copies of a particular protein molecule can be produced under the control of a single gene because many mRNA molecules can be transcribed by the gene, and each mRNA can be translated again and again. By contrast the RNA strands transcribed during rRNA synthesis are end products in themselves: translation is not involved, and the rate at which they are produced is governed by the rate of transcription. Two means of increasing their output have been discovered, and will be introduced in turn.

THE NUMBER OF GENES FOR rRNA It will be recalled that transcription involves opening the double helix of DNA, whereupon base-pairing reactions match the nucleotide sequence of the product to the sequence of the DNA. Indeed, if DNA double helices are opened to the single stranded condition by artificial treatments, and mixed with a transcribed product, the latter will recognize and adhere specifically to the sequence of DNA upon which it was originally formed. Exploitation of this recognition, or 'hybridization', property has disclosed a great deal about the sequence coding for rRNA. The experiments utilize radioactive rRNA obtained and purified from plants or animals that have been supplied with radioactive uridine over a prolonged period, during which they incorporate it into all of their RNA.

Radioactive rRNA can be used as a qualitative reagent by applying it to sections that have been treated to open their DNA helices. It adheres to those regions of the total chromatin on which it itself is transcribed. The excess is then rinsed away and autoradiography used to locate the hybridized radioactive rRNA molecules. In full agreement with the biochemical and ultrastructural observations already described, it emerges that in interphase cells the genes for rRNA lie in the nucleolus.[255] Their disposition in the chromosomes of dividing cells is somewhat variable, but mostly they are in a cluster on the nucleolar organizing chromosome(s).[722]

Radioactive rRNA molecules can also be used quantitatively. Since their molecular weight is known, it is simple to assay how much radioactivity there is per molecule. Then, by finding out how much radioactivity adheres to the DNA, the number of bound molecules can be calculated. If all available sites on the DNA are saturated, the number of bound molecules should equal the number of genes.

The figures obtained using this technique reinforce the conclusion reached at the end of the previous section: that nucleolar DNA consists of multiple units of function.[390] The number of genes per nucleus ranges from about 100 in the fungus *Neurospora* to more than 30 000 in the strain of maize described on page 46, in which the nucleolar organizer is duplicated. Maize with a *single* organizer per nucleolar organizing chromosome has *half* as many rRNA genes.[643] Multiplicity of rRNA genes also occurs in animal nucleoli, though somewhat less strikingly than in plants, and small clusters have even been found in prokaryotes which lack organized nucleoli.[779] Clearly, multiplication of this gene is a phenomenon that transcends biological classification schemes.

THE RATE OF TRANSCRIPTION It may be conjectured that the 'ideal' degree of multiplication achieves a balance between the number of ribosomes per cell, their life-span, and the rate of transcription. Such information is hard to obtain, but the quantitative work on cells in pea root tips, with 7800 rRNA genes per G1 nucleus, shows that even if each gene could be transcribed once in each minute (which is unlikely, see page 47), the required rate of production of 15 000 rRNA molecules per minute could still not be achieved.[390] Comparable work on rat liver cells has yielded a figure of 650 ribosomes produced each minute by the 330 rRNA genes in each nucleolus, that is, a transcription rate of about twice a minute for each gene, thereby maintaining the population of six million or so ribosomes in the cell.[674] Much higher rates of ribosome production than this have been found in rapidly growing animal cells[232]—7000 per minute per 'HeLa' cell nucleus, 12 000 per minute in fibroblasts (both of these cells growing and multiplying rapidly in tissue cultures), 12 000 per minute in the large Ciliate *Tetrahymena,* 1 million per minute in toad oocytes (which have several hundred nucleoli). In all of these cases it is to be noted that the observed rates of synthesis are considerably faster than can be accounted for by transcription at 10–100 nucleotides per second (see above), even after making due allowance for multiplication of the genes. An explanation for the remaining discrepancy is found in electron microscope images of the genes in the course of their transcription.

It is seldom that complex genetical or biochemical systems can be directly visualized. However, in 1968, Miller and Beatty discovered that the multiple rRNA genes of amphibian oocyte nucleoli could be spread on a support film and viewed by electron microscopy.[309, 558] Strands of nucleolar DNA were seen, each very long, 10–13 nm in diameter, and identifiable by their susceptibility to digestion by the enzyme deoxyribonuclease. Clusters of RNA fibrils protruding at right angles to the DNA strands were present at regular intervals along the strands, and each cluster occupied a length of DNA equivalent to the calculated size of a rRNA precursor gene (2.5–3.5 μm). There were RNA-free spaces along the DNA, between adjacent genes (clusters). Within each cluster of RNA fibrils a sequence was clearly seen: starting at one end the fibrils were short, increasing to maximum length at the other end. There were about 100 fibrils per cluster, and it was suggested that each fibril is, or contains, an RNA molecule, and is connected to the DNA strand by a molecule of the enzyme RNA-polymerase, about 100 of which are transcribing each gene concurrently. Each of the 100 enzyme molecules travels along the DNA from the beginning of the gene to the end. The sequence of short to long RNA fibrils is merely the picture that is obtained at a particular instant in time by stopping their movement: some have just started transcription, while others have almost completed the process.

Generalizing from what is now known of the organization of rRNA synthesis in amphibian oocyte nucleoli, it would appear that the overall rate of rRNA production is enhanced both by the multiplication of the genes, and by multiple concurrent transcription of each individual gene.

The molecular architecture of the spread out nucleolar gene system, described above, can be related to the ultrastructure of the intact nucleolus. Feeding with radioactive uridine for a period short enough to label the fibrillar zone of the nucleolus is found also to label the fibrils of the clusters of spread preparations. It seems clear that the fibrillar zone of the nucleolus (Plate 19g, h) must contain rRNA precursor molecules in all stages of formation, also the RNA polymerase molecules, at about 100 per gene during active periods of transcription, and 10–30 nm wide strands of DNA. What remains unclear is the relationship between the loops of nucleolar chromatin seen both by phase-contrast microscopy and electron microscopy, and the dispersed 10–30 nm strands of DNA. The latter are euchromatic and the former seem to be heterochromatic.

TRANSPORT AND ASSEMBLY OF RIBOSOMAL COMPONENTS
Having concentrated so far upon the synthesis of the large molecules of ribosomal RNA, it is necessary to remember that this is only a part of the synthesis of ribosomes. Ribosomal RNA has probably attained its final size and is in granular form by the time it leaves the nucleolus. But when and where is the small RNA strand (the genes for which are outside the nucleolus) added to the large ribosomal subunit (page 47)? When and where are ribosomal proteins synthesized and added? Where are the fifty or more genes for the ribosomal proteins located? When and where do the two subunits first come together? Not all of the ribosomes in the cytoplasm have precisely the same collection of proteins associated with the RNA, and the origin and significance of this heterogeneity poses further questions.[51]

In amphibian oocytes the processes of ribosomal RNA and ribosomal protein production appear to be correlated, and if the former is slow, so is the latter.[308] The evidence regarding transport to the cytoplasm suggests that the large and the small rRNA molecules move independently. In various animals, and in sycamore *(Acer)*[116] and parsley *(Petroselinum)*[761] cells in tissue culture, the smaller rRNA reaches the cytoplasm earlier than larger rRNA made at the same time, implying that the two ribosomal subunits do not associate within the nucleolus. It has been suggested that they are kept from associating by the presence of extremely high concentrations of inorganic phosphate.[810] The basis of this idea is that in nucleoli of maize root tip cells the concentration of phosphate is between 0.5 and 0.8 Molar.[487, 810] In the test tube, concentrations of phosphate as low as 0.01 molar dissociate ribosomes into their subunits.[640] Up to half of the cellular phosphate of maize root tip cells can be concentrated in the nucleolus, and in the absence of membranes, it is a considerable mystery how this condition is maintained. Another function for the nucleolar phosphate might be to keep in solution proteins that precipitate when the ionic concentration is low.

5.4.4 Dynamic Aspects of Nucleolar Activity

NUCLEOLAR 'VACUOLES' Plate 19a–f illustrates the most readily observed indication of the dynamic activity of nucleoli—enlargement of 'vacuoles' and discharge of their contents into the nucleoplasm. One (Plate 19a–f) or more (Plates 1, 20a, 44a–c) vacuoles can be present in a nucleolus. The rather sparse fibrils and granules of 'vacuolar' contents are seen in Plate 19g, where it is also very clear that there is no membrane equivalent to a tonoplast. Nucleolar 'vacuoles' have been known for more than a century, and many observations of both plant and animal material suggest that their pulsation is slow or absent when the cells are in a quiescent state, and active during growth or other biosynthetic processes.[36]

The 'vacuoles' of actively growing tobacco cells in tissue culture (Plate 19a–f) serve to introduce some of the properties of this component of nucleoli.[404] They are surrounded by granular zone material, and reach a diameter of up to 6 μm, in which case each pulsation releases a volume of more than 100 μm^3. Enlargement takes 0.5–2 hours, but discharge only one or a few minutes.[189] The presence of 'vacuoles' is related to the physiological condition of the cells, for instance incorporation of radioactive uridine into nucleoli that contain 'vacuoles' is greater than into nucleoli with no 'vacuoles'. An inhibitor of RNA synthesis, actinomycin D, reduces both formation of 'vacuoles' and incorporation of radioactive uridine, and both processes recover if the inhibitor is removed.[403]

All of the above evidence accords with a role for nucleolar 'vacuoles' in releasing products from the nucleolus, but there are many puzzling features. The 'vacuolar' contents appear dilute and the interface between them and the surrounding granular zone is abrupt. If the 'vacuolar' fluid is aqueous, how is the water accumulated and retained? If it is an electron-transparent matrix with structural rigidity, how are the contents expelled? How is it that individual 'vacuoles' in a nucleolus can pulsate independently?[100] It is known that the volume of the non-'vacuolar' components of the nucleolus remains constant while the 'vacuoles' pulsate within them[713] (Plate 19a–f, caption), but we shall not understand the physical basis of 'vacuolar' pulsation until we know more about the structural integrity of the nucleolus itself, and the forces that hold it from becoming dissipated throughout the nucleoplasm.

If the 'vacuoles' really are releasing pre-ribosomal granules from the nucleolus, the frequency of the particles in the 'vacuoles' should bear some relationship to the overall output. At first sight it might appear that the granules in the 'vacuoles' are too sparse to account for an output of, say, 5000 ribosomes per minute per nucleolus. However, some simple arithmetic shows that if a 'vacuole' accumulates this number of granules for an hour, it will, at a maximum volume of 100 μm^3, contain about 150 granules per square micron in an ultra-thin section that is 50 nm thick. Even if packed together in a monolayer these would occupy less than one-twentieth of a given area of the section, and scattered as they are they would appear sparse.

Objections have been raised to this hypothesis of 'vacuolar' function. A detailed study of nucleoli in slices of Jerusalem Artichoke tuber[713] has shown that the onset and the peak period of RNA synthesis, together with transport to the cytoplasm, *precede* 'vacuole' formation. Indeed 'vacuoles' only appear when RNA synthesis is slowing down. Furthermore, the largest 'vacuoles' of all, containing RNA-rich granules and pulsating just like normal ones, are formed on applying 5-fluorouracil, an inhibitor that is claimed to inhibit selectively the production of rRNA in plants. Bearing in mind also the observations that the rRNA of the small ribosomal subunit appears to be transported to the cytoplasm independently of the large subunit (previous section), it must be concluded that the relationship between nucleolar 'vacuoles' and processing of nucleolar RNA is not simple, though one probably does exist.

STRUCTURE, SIZE, AND ACTIVITY It has already been mentioned that 'vacuoles' are not always seen, and certainly there are indications other than 'vacuole' pulsation of the dynamic nature of the nucleolus. Both the ultrastructure and the size of the nucleolus provide clues. In several tissues, for example dormant seeds, and slices of Jerusalem artichoke tissue held in conditions that are unfavourable for RNA synthesis, the nucleoli lack granular zones.[713] Upon activation, they enlarge and granular zones appear in addition to the fibrillar area.[711a] Such observations are consistent with the view that the granules represent an end-product of nucleolar activity. Maintenance of a given size during periods of activity implies that the production line processes are in a dynamic equilibrium, with the rate at which precursors are supplied being matched by the rate at which the end-products leave. However, sheer size, although it may be a general guide, is not as reliable an indicator of nucleolar activity as the rate of incorporation of radioactive uridine. For instance, nucleoli sharing the nucleus of a root tip cell, and to outward appearances identical, do not necessarily utilize uridine at the same rate.[32] This observation indicates that the activity of nucleoli can be regulated, and serves to introduce the next topic for discussion.

REGULATION OF NUCLEOLAR ACTIVITY In some cases the activity of the nucleolus may simply be limited by the supply of raw materials for synthesis of RNA, but in general it would seem likely that the cell possesses more subtle means of adjusting the output of precursors of ribosomes to match the requirements of the cell, whether it be in a static condition and merely coping with turnover of ribosomes, or whether it be growing and dividing, differentiating or senescing. Indeed since ribosomes are an integral part of the biochemical machinery of the cell, vital for the synthesis of proteins, it is possible that by governing the nucleolus, the cell as a whole can be directed into or out of major phases of activity or inactivity.

Two main categories of regulatory processes can be envisaged: controls that govern the number of rRNA genes, and controls that govern the rate at which the genes are transcribed.

AMPLIFICATION AND DELETION OF rRNA GENES There is good evidence from the animal kingdom that the number of rRNA genes can be altered selectively. Thus certain fruit fly *(Drosophila)* mutants can increase the number of rRNA genes on one nucleolar organizer chromosome if its partner chromosome is removed.[811] Amphibian oocytes enormously increase the number of rRNA genes per nucleus by multiplying their nucleolar DNA and consequently their nucleoli.[855] Different tissues of the insect *Rhynchosciara* contain different numbers of rRNA genes per set of chromosomes, indicating that the number can be altered in the course of the successive mitoses and cell differentiation processes that give rise to the insect body.[256] Brain cells lose rRNA genes as they age.[406]

The available evidence from the plant kingdom is comparable in several respects, but less definitive.[20] The number of rRNA genes alters up to threefold during the cycle of growth and cell division in the green alga *Chlamydomonas*.[377] Structures akin to the additional nucleoli of amphibian oocytes occur at the time of meiosis in developing pollen,[354, 894] and in the polytene nuclei of the *Phaseolus* suspensor (page 42): in neither case, however, is it clear that the rRNA genes have been multiplied selectively, that is, independently of the remaining DNA.[390] The most dramatic alteration that has been seen is the complete loss of the nucleolus. This occurs in a few plant cells,[243] including the vegetative cell of pollen tubes,[344, 401] which are destined not to develop further, and at the time of the loss are close to the end of their life. Even here, however, it is not known whether the nucleolar genes have been deleted, or whether their activity has been abolished.

Many cells in both plants and animals become polyploid during their development (page 42). Doubling or quadrupling of the total nuclear DNA includes doubling or quadrupling of the nucleolar genetic material, in other words the number of rRNA genes can be increased non-selectively, as well as altered selectively as described above. As one example, the large cells of bean cotyledons are extensively polyploid during the phase when they synthesize protein in bulk, which may be one of the factors underlying the presence in them of very large nucleoli (Plate 20a).

REGULATION OF TRANSCRIPTION The other category of control system operates at the level of transcription, and is irrespective of the number of genes. There is evidence that both master 'on-off' switches and quantitative rate controls exist. The former regulate the entire set of rRNA genes and the latter the rate at which some or all operate.

The best example of a master switch in action is seen during mitosis, when (in general) the nucleolar activity ceases between prophase and late telophase (see also page 160). Apart from this, situations are known in which nucleolar organizers can be seen to be present in chromosomes, yet the genes in them remain quiescent, leaving all nucleolar activity to other organizers in the set of chromosomes.[445, 446] Regulator genes located outside the nucleolus have been detected in some organisms.[556]

The existence of quantitative rate controls is implicit in several of the observations that have already been described, such as the activation or inactivation of nucleoli in Jerusalem artichoke cells in artificial culture, and the differences seen in the rate of incorporation of radioactive uridine when two nucleoli within one nucleus are compared. The strain of maize that has two nucleolar organizers side by side on the nucleolar organizing chromosome (page 46) develops a nucleolus that is *less* than twice as large as that in normal maize possessing one organizer: obviously the double dose of genes does not automatically lead to twice the activity. Similarly, the broad bean, *Vicia faba,* with 9500 rRNA genes, has nucleoli which are the same size as those in the related species *Vicia narbonensis,* with only two-thirds as many rRNA genes.[504]

Our treatment of the nucleolus started with generalizations presented by Montgomery before the end of the last century. One of them, that nucleoli are best developed in cells that are active in growth or syntheses, is another way of saying that the activity of the nucleolar genes is subject to regulation. As in so many problems in the world of biology, it is difficult to distinguish between the cause and effect. Does activation of the genetic material of the nucleolus bring about growth and synthesis, or *vice versa?* We cannot tell, but this much is clear: that ribosomes are one of the basic tools of the cell. They do not specify the detailed nature of the proteins whose synthesis they mediate, but they are a vital part of the machinery. Production of ribosomes, and hence nucleolar activity, is therefore essential for cell development, even if it may not be causal.[313]

This account should not be concluded without a warning that the nucleolus almost certainly does more than merely synthesize ribosomal RNA. A great deal is known about this particular activity, and it has therefore been emphasized—unfortunately to the detriment of other, less well understood functions. Perhaps the most salutary reminder of our ignorance about the total properties of the nucleolus is that, despite the attention given to DNA and rRNA, these two substances account for *less than one-fifth* of the nucleolar mass.

5.5 The Nuclear Envelope

INTRODUCTION Our discussion of the nucleus now turns to a consideration of the envelope that encloses it. We have already seen that one activity of the nucleus results in the formation of very high molecular weight particles destined for export in bulk to the cytoplasm. We also know from previous general considerations (page 10) that membranes cannot be perforated by holes circumscribed by free edges. Putting these two facts together, we might expect to find that the nuclear envelope membranes are specialized in some way to cope with the outward flow of particles, not to mention its prerequisite, the inward flow of raw materials.

The discovery that the nucleus has a tough, elastic boundary layer was made long before the advent of the electron microscope. Micro-dissection implements were employed to push the nucleus around inside living cells, and to show that although the surface layer is capable of stretching to many times its normal area, puncturing it leads to the death of the cell. The boundary layer is visible in light microscope preparations, even when there is no chromatin adhering to it (Plate 1, N–I), but the reasons for its prominence relative to other cell membranes were not fully understood until 1950. The first electron microscope studies then confirmed a suggestion first made more than 50 years earlier on the basis of light microscopy, that not one, but two, membranes are present.[133] The presence of pores was also detected for the first time, and it was not long before the overall complexity of the structure was given recognition by conferring on it the name 'nuclear envelope', as distinct from the hitherto acceptable 'nuclear membrane'.[424, 904]

Later work carried out on ultra-thin sections showed that the membrane in contact with the nucleoplasm is separated from the membrane in contact with the cytoplasm by a 15–30 nm wide space, now known as the *perinuclear space*. The circumference of each pore in the nuclear envelope is formed by the fusion of the inner and outer membranes, leaving the nucleoplasm in continuity with the cytoplasm through the encircled pore lumen. In other words, the demand for a bounding structure for the nucleus, able to cope with the passage of relatively large particles, is met by the device of perforating a *double* envelope, thereby avoiding the forbidden free edges that would be the inevitable consequence of perforating a single membrane.

The components of the nuclear envelope must now be examined in more detail, taking first the membranes and then the pores.

5.5.1 The Membranes of the Nuclear Envelope

The membranes of the nuclear envelope are each about 5–6 nm thick, as are endoplasmic reticulum membranes. Freeze etching shows an absence of major substructural differences between the inner and outer layers.[420] The outer membrane does, however, have at least two attributes not shared with the inner. Occasional continuity with the endoplasmic reticulum is one (Plates 18a, 21a), and (generally) being adorned on the cytoplasmic face with chains or spirals of ribosomes is the other (Plates 18d, 21c, 43f).

NUCLEAR ENVELOPE AND ENDOPLASMIC RETICULUM It will be seen in the following chapter that the endoplasmic reticulum can show pronounced regional specialization. As is the case in the nuclear envelope, there can be regions possessing polyribosomes on only one face. Many observers, noting this, regard the nuclear envelope as a regularly occurring, highly specialized, cisterna of endoplasmic reticulum, distinguished by its perforations and by its location in the cell, segregating the enclosed nucleoplasm from the external cytoplasm. Note that the nucleoplasm is surrounded by the whole cisterna, and is not *intra*-cisternal. Portions of cisternae, intermediate in character, i.e. with pores and with ribosomes attached to *both* faces, are seen at prophase (Plate 45c, e) and telophase (Plate 47c, d) of mitosis, adding support to the idea that the nuclear envelope and the endoplasmic reticulum membranes are very closely related. Developmental inter-relationships of the two systems are discussed further on page 67. Of the two, the nuclear envelope is clearly the less dispensable. Thus while nucleated red blood cells retain the envelope but not the reticulum, no nucleated eukaryote cell lacks a nuclear envelope during interphase. Indeed possession of a nuclear envelope is a major distinguishing feature of the eukaryotes.

Whereas in the living cells of a higher plant it is quite possible that the envelope and the reticulum repeatedly form and sever their inter-connections, a permanent connection seems to exist in several groups of algae, where cisternae extend outwards from the outer membrane of the nuclear envelope and completely enclose the chloroplasts (see also pages 123 and 129).[263, 265]

Knowing of the presence of pores in the nuclear envelope and their absence from endoplasmic reticulum, it is perhaps not surprising that biochemical comparisons of the two membranes reveal points of difference as well as points of similarity. The types and relative proportions of most of the constituent liquids are similar. The two share a number of proteins, including enzymes, but there are also differences in terms of protein constituents. For example, out of a total of 32 proteins found in the two membranes as isolated from rat liver, 8 were found only in the nuclear envelope, 12 only in the endoplasmic reticulum, and 12 were common to both.[233] Whether the components that are unique to the envelope contribute to the molecular organization of the pores remains to be seen.

Quantitative as well as qualitative differences exist. Some enzymes found in the endoplasmic reticulum are also present, but in considerably reduced amounts, in the nuclear envelope. It is possible that part of the diminution could be due to differentiation of the two membranes of the envelope, the inner being less like the endoplasmic reticulum than the outer.[419, 424]

NUCLEAR ENVELOPE AND NUCLEOPLASM The dual nature of the nuclear envelope allows the outer membrane to interact with the cytoplasm, and the inner to interact with the contents of the nucleus. Just as the outer membrane possesses attached ribosomes, so the inner one has portions of chromatin attached to it (Plates, 1, 2, 5e, 7, 18c, 21a, 26b, 30b, 31c, d, 39e, 40a). It can be seen in all of these ultra-thin sections that many heterochromatin clumps lie appressed to the inner face of the nuclear envelope. Evidence for definite attachments has been obtained by examining whole mounts of chromosomes and portions of broken and spread nuclei.[138] Also, in plant and animal tissues that have been subjected to centrifugal force before fixation, the nucleolus and chromatin are pushed up against one side of the nuclear envelope, save for thin strands of chromatin that remain stretched across the nucleus from various 'attachment points' on the envelope.[42] Two hypotheses that merit further consideration are that these points may represent places where duplication of DNA is occurring, and that they may be concerned with the spatial organization of the chromosomes during interphase.

DNA SYNTHESIS The idea that DNA duplication might occur at or near the nuclear envelope was prompted by the discovery of a membranous site for the process in bacterial cells. Experiments on eukaryotes have been somewhat contradictory, and early conclusions that duplication commences at the envelope and spreads throughout the nucleus seem to be over-simplified. A major difficulty in the work is that DNA molecules are duplicated at a rate of 0.5–1.2 μm per minute (page 43), so if radioactive thymidine is supplied for five minutes, a radioactively labelled part of the DNA and the site at which it was formed could become separated by as much as 6 μm during the experiment. The indications from very short term experiments are that euchromatin duplicates throughout the nucleus,[450a, b] and that the heterochromatin that lies along the inner face of the nuclear envelope duplicates in that position.[203, 383, 906] Nevertheless DNA-membrane complexes[234] capable of carrying out DNA synthesis[389] have been isolated from eukaryotic cells, so the question of the role of the nuclear envelope in duplication of the genetic material remains open.

A different approach to the same problem has been to carry out kinetic studies with nuclei of different sizes, containing different amounts of DNA. The approach originates from a general consideration of the geometry of spheres, where the surface area to volume ratio decreases with increasing size (since surface area is a function of r^2 and volume of r^3). For example, species of *Antirrhinum* possessing either two or four sets of chromosomes have been compared in this context: the rate of DNA synthesis has been shown to relate more closely to the surface area than to the volume of the nuclei (the latter being proportional to the amount of chromatin).[6] A better comparison can be made in liver tissue, where occasional individual cells contain *two* nuclei, each with the usual *two* sets of chromosomes, while some others contain *one* nucleus with *four* sets of chromosomes. These cells differ in the surface area of nuclear envelope but not in the total amount of DNA per cell. In comparing them it was found that the rate of DNA synthesis is higher in the binucleate cells, correlating with their greater area of nuclear envelope.[6]

ORGANIZATION OF CHROMATIN Autoradiographic experiments and observations of heterochromatin indicate that DNA strands can remain in the same position in the nucleus for long periods. The question of the spatial organization of chromatin is therefore raised again (see also page 52). Attachment of fibrils to the nuclear envelope at frequent intervals along their length could introduce order into the apparent chaos of the chromatin, preventing adjacent chromosomes from becoming inextricably tangled and permitting independent condensation at prophase. Very few types of cell have been examined with this possiblity in mind, but there is evidence in favour of the idea, and further, that the pores of the nuclear envelope constitute 'reference points' to which chromatin could become attached (not necessarily by attachments that consist of chromatin). The organization could become established at anaphase and telophase of mitosis, when new nuclear envelope, together with any fragments of envelope that may have retained their attachments since prophase, ensheaths the chromosomes (Plate 47d).[138, 139] Parts of Plate 18d show fine fibrils radiating from nuclear pores towards clumps of chromatin near the inner face of the nuclear envelope. The same micrograph also shows that pores may be grouped around clumps of chromatin. A different, but perhaps functionally similar arrangement is seen in Plate 18c, where a mass of heterochromatin surrounds a pore.

FUSION OF NUCLEI One more vital function for the membranes of the nuclear envelope should be mentioned. It is their part in the recognition and then fusion of nuclei.[398, 843] In flowering plants there are normally three types of nuclear fusion: (1) fusion of egg and sperm nuclei at fertilization in the egg cell of the embryo sac; (2) fusion of the two polar nuclei in the central cell of the embryo sac; and (3) fusion of the remaining sperm nucleus with the fused polar nuclei to give the progenitor of the endosperm tissue nuclei. In all cases the first contact is made by endoplasmic reticulum cisternae extending from the nuclear envelopes of the nuclei that are about to fuse. The nuclei draw together and the two outer membranes become locally appressed to one another. They fuse and so become continuous round the periphery of the appressed region, in which the two *inner* membranes are by then exposed face to face with one another, the outer membranes having withdrawn from this area when their peripheral continuity was established. The two inner membranes merge in their turn, become peripherally continuous, and withdraw, leaving the two nucleoplasms open to one another. Thus parental membranes, as well as parental chromosomes, contribute to the fertilized egg from which the new generation develops.

5.5.2 The Pores of the Nuclear Envelope

The pores of the nuclear envelope are far from being simple holes through the two layers of membrane. They have a specific structure, the basic details of which seem to be constant throughout the plant and animal kingdoms. Plates 18d and e and Fig. 3 illustrate the major features of what has been called the 'pore complex'.

There are two ways to see the symmetry of the component parts of the pore complex: in sections cut tangentially to the nucleus the face view is presented (Plate 18d); in sections cut across a diameter of the nucleus they are seen 'edge-on' in side profile (Plate 18e). In both cases the details of the image depend to a very great extent on the position of the pore relative to the two successive knife-cuts that produce the section. This problem arises because the overall dimensions of the complex are nearly the same as the thickness of the average ultra-thin section. Only occasionally is the pore centred perfectly, and frequently it lies towards either the top or bottom surface of the section, thereby excluding varying proportions of the total complex. Fig. 1 shows how this factor influences the appearance of simple pores in a cisterna. A further complication is that superimposition of the images of the small components of the pores makes fine detail hard to discern.

The following features are illustrated by the numbered pores in Plate 18d, and in Fig. 3. The curved membrane surface that circumscribes the lumen of the pore is called the pore margin. In those tangential sections in which it is not partially obscured by superimposed material it is seen to be octagonal in outline (pores 1, 8, 11). The diameter measured from any given face of the octagon to the one opposite is 65–75

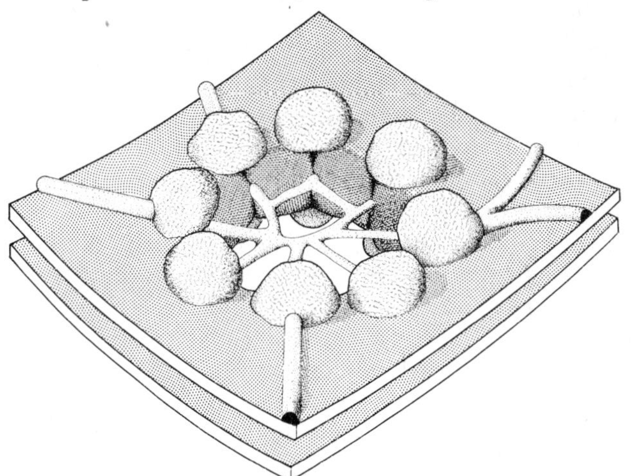

Fig. 3 Diagram of a nuclear envelope pore
A segment of nuclear envelope, consisting of the outer and inner membranes, the intervening perinuclear space, and one pore complex, is viewed as from the nucleoplasm. The pore margin, composed of the joined inner and outer membranes, is octagonal in outline, and is surmounted both inside and outside the envelope by the annulus, of which only the annular granules are included here. The granules may in fact be interconnected both within each ring of 8, and between the two rings. The pore lumen is traversed by fibres, and fibres also pass from annular granules to chromatin (chromatin not shown). The layout of fibrils is purely diagrammatic. No central granule (see text) is shown in the pore complex.

nm. Tangential sections frequently show pores with much less sharp outlines, and with an apparently circular shape. These arise if the section includes portions of another component, called the annulus. The annulus consists largely of granules arranged in two rings, one round the nucleoplasmic face of the pore, the other round the cytoplasmic face. It is difficult to distinguish the individual granules, due to their rather 'fuzzy' and indistinct outlines and to superimposition of images. The scalloped outer edge of the annular rings suggests that the constituent granules are large enough to touch each other (pores 5, 9), but photographic techniques which filter out some of the randomly distributed 'fuzz' usually show them to be discrete. Close proximity to a chain of ribosomes identifies annular rings on the *cytoplasmic* face of the pore (pore 5); corresponding close proximity to chromatin identifies those on the *nucleoplasmic* face. Where pores are tilted appropriately within the section, one side shows part of the octagonal profile and the opposite side some of the granules (pores 4, 5). Such views prove that the granules lie above and below the vertices of the octagon. It is because the granules overhang and obscure the vertices that the pore margin appears rounded rather than octagonal when an annular ring is present in the section.

Occasional fibrils pass over the nucleoplasmic face of the inner membrane of the nuclear envelope, connecting clumps of chromatin to granules of the inner annular ring (arrows on Plate 18d). Tangential sections which show the octagonal pore margin, and which therefore include the central zone of the pore (midway between cytoplasm and nucleoplasm), show an additional component—a network of very fine fibrils and small granules traversing the lumen of the pore (pores 1, 10). Some at least of these fibrils are attached to the pore margin at the vertices of the octagon. Details of this network cannot be resolved, and the representation in Fig. 3 is purely diagrammatic.

One more component is present in some but by no means all pores. It is a relatively large (15 nm diameter) electron-dense granule. Where it is present it is usually close to the centre of the pore (pores, 1, 8, 10). Two such granules are sometimes seen (pore 6). It is not included in Fig. 3.

The profile view of pore complexes adds further information (Plate 18e). The octagonal outline cannot, of course, be seen, and indeed the appearance of the pore margin and the apparent diameter of the pore vary greatly according to how much of the complex is included in the section. The annular material overhanging the pore margin is visible, though the number of granules in the rings cannot be counted. The electron-dense 15 nm particles show up clearly, if present, in the pore lumen. Their position is seen to vary. They may be in the nucleoplasm near the pore, or in the axis of the lumen at the level of the nucleoplasmic annular ring; or midway between nucleoplasm and cytoplasm; or in the cytoplasm just outside the pore. These positions may all be unoccupied, or one granule may be seen in one or other of the positions; alternatively two or three positions may be occupied. Interpretation of this variability is postponed to the following section.

The above description relates particularly to the pores shown in Plate 18d and e, and it must be recognized that certain details may vary between nuclei, tissues, or organisms.[424, 904] Some variable features include the inward and outward extent of the annuli; the degree to which the fibres traversing the lumen are developed; and whether or not the inner annular granules are connected to the outer by fibrous or additional granular material. Features that have by contrast been seen in very many types of material[230] include the annuli with their eight subunits, the octagonal pore margin, and the central particles—which, as in Plate 18d and e, are present in some but not all of the pores of a given nucleus (see also Plate 18c, depicting a different species from that in 18d and e).

It seems preferable to regard the pore complex as a macromolecular aggregate rather than simply as a perforation through two layers of membrane. Thus pore complexes isolated by treating fragments of nuclear envelope by ultrasonic vibration retain their structural integrity. Chemical analyses of isolated nuclear envelopes suggest that RNA is an important component, and that the total molecular weight of the complexes is equivalent to several hundred millions.[730]

PORE DEVELOPMENT AND ARRANGEMENT It is not known how pore complexes develop but the very fact that stages have not been seen and described indicates that development probably is a rapid process. One possible sequence of events is the localized fusion of the outer and inner layers of the envelope to form an initial hole which then enlarges to the correct size, whereupon the normally rather fluid membranes are stabilized by insertion of annulus material, which could bind to the membranes at the edge of the hole and constrain them into the octagonal shape. Alternatively, the constraint could be imposed by attachment of fibres to the inner annulus. Whilst many pore complexes must develop when a new nuclear envelope is laid down at telophase of mitosis (page 162), there is good evidence that they can also arise during interphase.[541]

In some organisms the pores become arranged in regular patterns over the nuclear surface. The club moss *Selaginella* illustrated in Plate 18a has pores placed at the points of intersection of two sets of spirals of different pitch. The final arrangement approaches 'hexagonal close packing', where every pore is equidistant from each of six neighbours, but is distorted here and there by what may be stages of insertion of additional spirals of pores.[817] Spores of another lower plant, the horsetail *Equisetum,* have pores arranged in girdles running around the nucleus.[700] The brown seaweed *Padina* develops alternating bands of pore-rich and pore-free nuclear envelope, the pore-free areas being distinguished by the presence of rows of small particles rendered visible by freeze etching.[583] Images of yeast nuclei showing restriction of pores to one end have been obtained. Where the pore frequency (number of pores per square micrometre) is high the pores may be close-packed in hexagonal or square patterns. Where the frequency is lower the arrangements are usually less symmetrical (Plate 18b). A pattern in which the pores lie in straight, twisted, or branched rows gives way to a more dispersed arrangement as cucumber cotyledon mesophyll cells enlarge during germination.[495] It is not known whether any of these patterns results from interaction of chromatin and nuclear envelope, but the even spacing of pores in arcs around clumps of chromatin, exemplified in Plate 18d, suggests that mutual interactions could be important in governing the distribution of both pores in the envelope and chromatin in the nucleoplasm.

It is obvious that if different cells can have different patterns of pore distribution over the surface of their nuclear envelopes, then the pore frequency too must vary. Values in the range 6–25 pores per μm^2 have been obtained for plant material, the pores in total occupying some 8–20% of the nuclear surface. The captions to Plates 14b and 18b indicate how the frequency can be measured. Some animal cell nuclei have higher pore frequencies, reaching more than twice that recorded so far for plants. There are only a few studies of changes in pore frequency in relation to cell development, and they indicate that the frequency can change, perhaps in response to stimulation of RNA synthesis. An alternative means of increasing the total number of pores is to increase the surface area of the nuclear envelope, with concomitant insertion of new pores. For example, if the radius of a spherical nucleus increases from 5 μm to 6 μm while the pore frequency is maintained at, say, 15 per μm^2, by insertion of new pore complexes, the total number rises by more than 4000 from nearly 9500. Actual data include a more than 30-fold increase to 57×10^3 as newt oocyte nuclei enlarge, and a near doubling to 22×10^3 in *Tetrahymena* macronuclei.[916] The reverse trend is found in animal sperm nuclei, which gradually lose their pore complexes as the cells age and become inert.

There is little doubt that pore frequency and the total pore number are subject to control mechanisms, and that any alterations that occur in them are as much a part of the overall process of differentiation as any other developmental phenomenon in the cell. Specific illustrations are not easily obtained, but comparison of the vegetative nucleus with the generative nucleus of *Tradescantia* pollen provides one example.[457] These two nuclei are derived from the same parent, but have very different functions and destinies. The generative nucleus (which gives rise to the sperm nuclei) not only has a lower pore frequency, but also a smaller surface area. The vegetative nucleus, at a time early in germination of the pollen grain, has about twice as many pores and synthesizes about twice as much RNA as does the generative nucleus.

In most plant cells pore complexes are seen only in the nuclear envelope. However, they have also been observed in layers of endosplasmic reticulum lying in the cytoplasm parallel to the nuclear surface. Such structures are common in animal oocytes and are called 'annulate lamellae'.[424] They do occur in plants,[732] but are not obvious or frequent, and their distribution has not yet been mapped thoroughly. It is not known whether they are formed by casting off portions of nuclear envelope, as has been suggested for oocytes.

5.5.3 The Permeability of the Nuclear Envelope

Previous sections in this chapter have included references to *export* of nucleolar products and other types of RNA to the cytoplasm. Materials that are *imported* into the nucleus include precursors for nuclear RNA and DNA synthesis, many or perhaps all of the enzymes that catalyse these syntheses, the histone proteins that associate with DNA in chromatin and chromosomes, cytoplasmically-produced stimuli which initiate mitosis, and (probably) the protein tubulin (page 142) for producing intra-nuclear spindles in those cells that retain their nuclear envelopes during division (pages 161 and 162). In short, the traffic across the nuclear envelope is two-way, and involves both low and high molecular weight compounds.

PENETRATION OF SOLUTES AND PARTICLES The nuclear envelope restricts movement of small molecules much less than do other cell membranes,[208] though it must be admitted that this generalization is based on very few experiments, none of them involving plant material. Radioactive sucrose, glycerol, and sodium and other ions all diffuse freely into the nucleus of oocytes if they are injected into the cytoplasm.[375] This does not mean that their concentration becomes uniform throughout the cell: in oocytes the water content of nuclei is greater than that of cytoplasm, and the distribution of water-soluble compounds reflects the greater 'availability' of water in the nucleus. Another factor is that nuclei contain large amounts of charged molecules (basic proteins, acidic nucleic acids) that are capable of binding ions of opposite charge. The high concentration of nucleolar phosphate has already been mentioned (page 49). Various small proteins have also been shown to penetrate through nuclear envelope pores.

The free permeability of the nuclear envelope to small molecules is to be expected from the large total surface area of the pores, through which there is *apparently* free contact between nucleoplasm and cytoplasm. Nevertheless, measurement of the electrical resistance of nuclear envelopes has shown that it is possible for them somehow to impede the movement of ions. It is known that this resistance varies during the life of *Drosophila* salivary gland cells, but what is not known is how a system of pores can develop electrical resistances some 1000–10 000 times greater than their apparently open state suggests. Neither can the resistance measurements on *Drosophila* be reconciled with the experiments showing free passage of ions injected into oocytes, except in that oocyte nuclear envelopes do not, it seems, have high electrical resistance.

It is also to be expected that the pore structure will impose an upper limit on the size of particles that can penetrate into or out of the nucleus. Ingenious experiments in which *Amoeba* cytoplasm was injected with electron-dense particles of gold that had been coated to confer uniform surface properties demonstrate permeability to particles up to 9 nm diameter, restricted entry of somewhat larger particles, and exclusion of particles larger than 12–14 nm in diameter. Surprisingly, this exclusion size is about five times smaller than the diameter of the pore lumen, showing that something, perhaps annular or fibrous material, must partially occlude the passageway. Another electron-dense tracer that has been used in this type of experiment is the large iron-containing protein ferritin (page 128). Its molecules are 10 nm in diameter and differ from colloidal gold in many respects, including their surface charges. Organisms vary in the capacity of their nuclear envelope pores to pass ferritin. Further work[208] with gold and *Amoeba* has disclosed that the upper size limit can alter during the life of the cell, and that permeability to these particles is greatest when DNA synthesis is in progress in the nucleus. It is to be noted that the size at which transport is cut off is such that mature ribosomes cannot pass from cytoplasm to nucleoplasm.

EXPORT OF RNA FROM THE NUCLEUS Within this background, the export of nucleolar products to the cytoplasm can be re-examined in relation to the frequency and morphology of pore complexes. Knowing the pore frequency and the rate of RNA synthesis, it has been possible to estimate the 'nuclear pore flow rate' for a few animal cells. In HeLa cells, liver cells, and amphibian oocytes, 1–3 rRNA molecules pass through each pore per minute, while in *Tetrahymena* the flow rate is 20 to 30 times faster.[232, 731]

There is strong evidence that rRNA leaves the nucleolus in the form of pre-ribosomal particles that are about 15 nm in diameter and readily visible in the electron microscope (page 49). Given nuclear pore flow rates of several particles per pore per minute, and many thousands of pores per nucleus, fixation and ultra-thin sectioning of active cells inevitably reveal stages of particle transport. Particles of appropriate size and electron-density can indeed be seen near, in, or just outside pore complexes (Plate 18d, e), and it would seem reasonable to arrange micrographs in a series as in Plate 18e and to interpret them as depicting stages in the dynamic process of transport, separated in time from each other by short intervals. Unfortunately it is an interpretation that has proved to be difficult to substantiate by experiment. The proportion of pores with and without central particles does vary, both from cell to cell and during the life of a cell, but particles are present even in cells with low rates of rRNA synthesis. They do not disappear if rRNA synthesis is inhibited by means of the antibiotic actinomycin D.[184] These and many similar observations are puzzling, for if the particle is a non-migrating component of the pore complex it should never be absent, and if it is a migrating object caught in transit there should be a clearer relationship between its rate of synthesis and the frequency with which it occurs in pores.

Solutions to the puzzle are at present speculative. It will be recalled that RNA is a major component of the pore complex (page 54); in fact the RNA content of one complex is equivalent to about 20 rRNA molecules. While this does not mean that each pore contains 20 rRNA molecules queueing up to leave the nucleoplasm, for some of the pore RNA could be structural (fibrils, annular material, chromatin attachments, etc.), it does suggest that there could be some congestion in the outward traffic from nucleoplasm to cytoplasm. Although the overall traffic may be a few rRNA molecules per pore per minute, an average rRNA molecule might have to wait from several minutes to several tens of minutes to exit.[731] It may even be that some congestion, by building up a critical concentration of rRNA on the nucleoplasmic side of the pores, is needed to effect transport. Also, if there is no transport without congestion, rRNA molecules might become trapped in pores, thereby accounting for the presence of particles in pore lumena in quiescent or inhibited nuclei.[916]

Other factors that must be borne in mind concern the seemingly independent transport of the large and the small form of rRNA (page 49), and the outward passage of other types of RNA, including the very important 'messengers', which cannot unequivocally be recognized in ultra-thin sections.

Underlying all considerations of transport through nuclear envelope pores there is yet another puzzle. It is known that the pore structure has been conserved to a large extent without change during the course of the most profound evolutionary processes, and that throughout the eukaryotes the pore diameter is about 70 nm. Yet particles larger than about 15 nm cannot, it seems, pass through them. If their function is merely to filter out larger particles there is no obvious reason why the pore size should not be matched to the particle size. Do pores, therefore, have functions over and above that of being particle size filters; functions with which the large diameter can be associated? Such functions await identification, but one obvious family of the two prior to the initiation of protein synthesis RNA molecules that are exported from the nucleus. Nuclear pores could well complete the maturation of these molecules, or in some way expedite the association of the two prior to the initiation of protein synthesis in polyribosomes.[549] The structural complexity of pore complexes would seem more reasonable if they have biochemical functions as well as functions in transport.

The nuclear envelope is ideally located to have a role in mediating the control that the nucleus exerts over the cytoplasm and in turn over the life of the cell. Its occurrence throughout the eukaryotes speaks of some such fundamental property. Segregation of the nucleoplasm from the cytoplasm in a manner that does not impair the necessary transport between the two is, presumably, basic, and various additional suggestions have been mooted: roles in organizing the chromatin and its duplication, and roles in completing nuclear products and in mediating nucleocytoplasmic exchanges. As the preceding pages have indicated, enough is now known for these possibilities to be recognized, but a major effort will be needed to assess their contributions to the life of the cell.

6 The Endoplasmic Reticulum

6.1 Introduction

ERGASTOPLASM Nuclei, plastids, mitochondria and vacuoles were all discovered long before the electron microscope became available to biologists—in some cases centuries before. So too was a component of certain specialized animal cells, for which the name *ergastoplasm* was coined in 1897. One of the earliest successes of biological electron microscopy was the recognition that ergastoplasm consists of a network of interconnected membrane-bound compartments lying in the cytoplasm[623] (page 2). In due course it transpired that while a *few* types of cell contain masses of this material so unusually large as to be visible in the light microscope, it is also to be found in diverse forms and variable quantities in virtually *all* cells.[661]

RANGE OF FUNCTIONS All of the gradations in complexity, and the diversity of form, are variations on a common plan—a network extending through the cytoplasm. The internal location and the inter-connections in three dimensions are reflected in its current name, the *endoplasmic reticulum*. Before setting out to see actual examples of its structure and function, it is worth considering the various biologically significant possibilities inherent in the system.

First, the cytoplasm is subdivided into two phases by the membrane of the reticulum. Note that if the reticulum is truly continuous, the singular, membrane, is more apt than the plural, membranes. The enclosed space could, like any other compartment, be a storage facility. Thanks to its three dimensional construction it could also be a channel through which materials are transported around the cell. If portions or derivatives of endoplasmic reticulum do indeed pass in an open form through plasmodesmata (see page 27) it could function in both intra-cellular and intercellular transport.

Turning from the enclosed space to the membrane that encloses it, a second family of possibilities emerges. It is possible, by regarding ultra-thin sections as statistical samples of a tissue,[871a] to estimate the actual surface area of the endoplasmic reticulum cisternae. The massed cisternae which occupy so much of the cytoplasm of, for example, pancreatic cells or the plant gland shown in Plate 21a, provide a surface area of 15–20 square micrometres per cubic micrometre of 'average' cytoplasm. Liver cells, and the tapetal cells[793a] of Plate 22, contain 6–12 $\mu m^2/\mu m^3$. Ordinary root tip cells (Plates 1, 2 and 3) have very much less—1 $\mu m^2/\mu m^3$ or less. These 'surface densities' (i.e. area per unit volume) are rather more impressive if the figures are adjusted to units that are more easily visualized. Since there are 10^{12} square micrometres in one square metre (m^2), and 10^{12} cubic micrometres in one cubic centimetre (cm^3), the above units convert directly to m^2/cm^3. The value of 6–12 m^2/cm^3 for liver cell cytoplasm means that if the endoplasmic reticulum cisternae of an adult human liver, with about 1000 cm^3 of cytoplasm, were opened out and laid flat side-by-side, an area about 100 metres by 100 metres would be covered. Clearly, the development of an endoplasmic reticulum is a highly effective way for a cell to pack large areas of membrane into a small volume of cytoplasm. The surface can be used to expose numerous biochemical reaction systems, whether bound to the surface or incorporated into the membrane, to the cytoplasm. If trans-membrane transport is a significant factor in a particular function of the reticulum, then the greater the area, the greater the efficiency, whether the mode of transport is physical diffusion from one side to the other, or solute pumping by membrane transport proteins packed into the surface.

Thirdly, the reticulum could function in regulating the spatial distribution of events and processes in the cell. In 'positive' functions of this nature, portions of suitably located reticulum endowed with particular enzymes could drive local syntheses. In 'negative' functions, portions of reticulum could act as barriers shielding particular regions from events that are taking place elsewhere in the cell.

RANGE OF FORM The versatility of function envisaged above is matched by the diversity of form actually encountered in different types of cell. It has been claimed that an experienced observer can identify cell types merely by inspecting the endoplasmic reticulum. This is probably correct as long as the observer stays within the animal kingdom. In plants the endoplasmic reticulum is less useful as a diagnostic feature, though extreme and easily recognizable forms do occur (Plates 21a, 22a, 23c, 24c). Also, it is not always possible to distinguish plant from animal, such is the startling parallelism with which the two kingdoms have exploited the possible forms of endoplasmic reticulum.

At one extreme the reticulum consists of interconnected parallel cisternae (e.g. Plate 21), and at the other of tangled tubules (e.g. Plate 23c). The membrane in the former case is rendered especially conspicuous by bearing ribosomes on its outer (cytoplasmic) surface. It is an example of *rough,* or *granular* endoplasmic reticulum. *Smooth,* or *agranular* endoplasmic reticulum lacks attached ribosomes. Various gradations exist between extremely rough and perfectly smooth. The actual density of ribosomes ranges from packed to sparse, and some cisternae, which have been termed either *semi-rough* or *bifacial,* bear ribosomes on one side but not on the other. In both plants and animals the rough endoplasmic reticulum varies more in quantity than in form: whether there is a lot (Plate 21) or a little (Plate 2) in a cell, the flattened cisterna is typical, and is only occasionally varied, as when accumulation of cisternal contents distends it (Plate 3a). The smooth endoplasmic reticulum is rather more variable and appears both as more-or-less flattened cisternae (Plates 22a, 24b, c) and as branched tubules (Plate 23c).

Because plants and animals have in the course of evolution exploited the inherent versatility of the endoplasmic reticulum to a wide variety of ends, *structurally* similar systems found in different cells, tissues, and organisms, can be *functionally* dissimilar. We have therefore considered it best to deal with the structure and function of the endoplasmic reticulum function by function, rather than structure by structure. Unfortunately this mode of presentation leads to the neglect of some fascinating structures and situations, simply because their functions are as yet unknown or conjectural.

Compared with the abundance of descriptive work, elucidation of the function of plant cell endoplasmic reticulum has lagged far behind. One reason is that it is difficult to isolate samples of plant endoplasmic reticulum for biochemical examination, because the forces that are needed to break the cell wall tend also to damage the cytoplasmic contents thereby liberated. Due to this constraint, and also because of the fundamental importance of the subject in medical research, very much more is known about the functions of animal endoplasmic reticulum. The topics considered below have therefore been selected and introduced according to zoological knowledge; in each case we can proceed to see whether the botanical situation is similar: in most it will be clear that the evidence is tantalizingly suggestive but unsatisfactorily incomplete.

6.2 The Endoplasmic Reticulum and Protein Synthesis

With the recent upsurge of interest in molecular biology, much progress has been made in understanding the mechanism of protein synthesis. Parts of the story introduced in Chapter 5 include the genetic information in the DNA of the nucleus, and the origin of ribosomes in the nucleolus. The manner in which the nuclear DNA and the cytoplasmic ribosomes are linked by chemical messages that pass from nucleus to cytoplasm in the form of long, thin, messenger RNA molecules, each carrying a transcribed form of the genetic specification for a particular type of protein, was also mentioned. Ribosomes are part of the biochemical machinery which translates the sequence of nucleotides in the mRNA into a sequence of amino acids in a protein. It is at this stage that the routine procedures of electron microscopy can begin to play a part, for it is usual that mRNA molecules are long enough to accommodate several or even many ribosomes at a time, and the resultant complexes, *polyribosomes,* are visible, and visible indications that protein molecules were being assembled at the moment when the cell was fixed. The size of a polyribosome is a measure of the length of the message and hence the size of the protein (see caption to Plate 20b). A growing protein molecule (itself not visible) extends from each ribosome in the complex, and both ribosomes and protein detach from the end of the mRNA when translation is completed.

In animals, cells which specialize in manufacturing proteins destined for export (e.g. in pancreas, antibody-producing cells, etc.) tend to possess massed cisternae of rough endoplasmic reticulum, with polyribosomes on the membrane surfaces. All cells, of course, have to make protein, and those which do so purely for internal use, as distinct from export, seem to have a higher proportion of their ribosomes and polyribosomes lying free in the cytoplasm. Generalizing, protein synthesis by free polyribosomes culminates in the release of completed molecules into the cytoplasm, while protein synthesis by polyribosomes on the rough endoplasmic reticulum culminates in transfer of the completed proteins into the membrane itself, or through the membrane and into the internal space of the cisternae. They may accumulate or may be transported elsewhere, for example to special derivatives of the cisternae and thence to the Golgi apparatus. General aspects of these processes will be reconsidered after some specific case histories have been described.

There can be little doubt that, as befits as fundamental a process as protein synthesis, the above events are truly biological in their occurrence. For plants, membrane-bound polyribosomes are illustrated in Plate 20b, 21c, and other types found free in the cytoplasm appear in Plate 21c (insert). Accumulation of protein within rough endoplasmic reticulum cisternae is seen in Plate 3a. Only a few plant cells export protein in bulk in the sense that pancreatic or salivary gland cells do, but a number accumulate protein in vacuoles (e.g. storage cells in many seeds, Plate 20a, c) or in cytoplasmic deposits: in such instances a large quantity of flattened cisternae of rough endoplasmic reticulum may be present (Plate 20b). In plants, however, the presence of numerous rough cisternae is not necessarily an indication of bulk protein synthesis. Several cases will be described later, in which as far as can be seen massive systems of rough cisternae are related to functions *other* than protein accumulation (Plate 21a, b).

Investigations of cells that specialise in protein synthesis have yielded the most clear cut information on this role of the endoplasmic reticulum, and it is presumed that less specialized cells differ only quantitatively. Brief case histories of some pieces of research that give an especially graphic visual picture of protein synthesis in plant cells follow. They do not in any way represent a historical sequence, and it should be emphasized that most of the conclusions to be drawn from them are supported by biochemical studies not detailed here.

ACCUMULATION OF PROTEIN WITHIN CISTERNAE The first example is the simplest, and concerns a familiar type of cell and tissue—parenchyma cells in slices of beetroot. If thin slices are washed in aerated water for several days, the cisternae of rough endoplasmic reticulum extend, and eventually proteinaceous crystals develop in the cisternal space.[395] That this is a straightforward illustration of the role of the endoplasmic reticulum in protein synthesis is supported by the fact that the phenomena are markedly reduced if the slices are incubated in water containing inhibitors of protein synthesis by polyribosomes, for example cycloheximide and puromycin. Actinomycin D, a much used inhibitor of DNA-directed RNA (including mRNA) synthesis is without effect, allowing the conclusion that formation of the reticulum and the crystals does not require production of *new* mRNA.[847] Presumably, as in all protein syntheses, some mRNA is needed, so the result implies the presence of long-lived mRNA molecules, surviving from when the cells were in their normal state in the original beetroot. mRNA molecules are known to display wide variation in their life-spans, ranging from a matter of minutes to very prolonged periods—for example some mRNAs are made during embryo development and are retained throughout seed dormancy until translated during germination.[388]

PROTEIN SYNTHESIS AND TRANSFER TO VACUOLES This and the next example of the visual approach to protein synthesis rests largely on the technique of autoradiography. By feeding with radioactive amino acids it becomes possible to find out where in the cell amino acid incorporation produces radioactive protein. If a short feeding period (usually referred to as a 'pulse') is provided, the tissue transferred back to non-radioactive medium, and samples fixed at a range of time intervals ('chase' periods), a picture of the flow of radioactivity through subcellular compartments is obtained.

Storage of protein in vacuoles of cells in seed cotyledons has been mentioned previously (page 36). The development of the cells has been studied in considerable detail, and rightly so, since the protein in them forms a large part of human and animal foodstuffs. To begin with they are normal parenchyma cells, with not many cisternae of rough endoplasmic reticulum, and many more free ribosomes than membrane-bound ribosomes. Later, as the seed begins to mature, the endoplasmic reticulum proliferates commensurate with production of many new ribosomes that bind to it in polyribosome configurations[635] (Plate 20b). The amount of protein synthesis increases and protein appears in the vacuoles,[634] eventually to fill and convert them to 'protein bodies', or 'aleurone grains'[83] (Plate 20a, c). If the length of the polyribosome spirals is measured it is possible to calculate an upper limit for the molecular weight of the proteins that could be synthesized. The longest spirals in Plate 20b are about 500 nm long, corresponding (see caption) to a molecular weight of about 60 000. This accords well with biochemical analyses of the subunits of bean seed proteins.[21, 22]

Autoradiographic experiments employing slices of bean cotyledon supplied with a 'pulse' of a radioactive form of the amino acid leucine indicate that the first site of protein synthesis is the rough endoplasmic reticulum. If a 'pulse' is followed by a suitably long 'chase', the radioactivity is seen to have moved to the vacuoles—clearly, then, the newly made protein is transported from the one location to the other. By fixing the tissue after a range of 'chase' periods, it has been estimated that synthesis and transport of a molecule of storage protein together take no more than about half an hour.[23]

The exact route followed by the protein in moving from endoplasmic reticulum to the vacuole has not been established. Whilst some swelling of the cisternae does occur during the period of protein synthesis, it does not appear that massive accumulations build up within the reticulum itself—not, at any rate, to the extent of producing inflated cisternae with dense contents such as are seen in washed beetroot slices and certain other situations. Gradual transfer of protein to the final destination is to be inferred. This, however, is not what happens in the marine alga *Bryopsis,* which also makes protein and transfers it to vacuoles.[102, 103] There the rough cisternae do accumulate the protein and become distended. It seems that the bulk packages so formed are transferred bodily to the vacuole, where their bounding membrane breaks down. The protein in this case is not a food reserve, but a material with which the alga—a large coenocyte—plugs damaged regions of cell wall (page 20) and produces septa that partition the cell.

PROTEIN SYNTHESIS AND SECRETION IN A GLAND CELL A bewildering variety of glands clothes the above ground surfaces of plants.[746a] Many of them provide spectacular arrays of endoplasmic reticulum, both rough and smooth (e.g. Plates 21a, b and 23). If these glands could be obtained in bulk and in isolation from other tissues, they would provide material for experimentation just as valuable as those specialized animal tissues, such as the pancreas, that have yielded so much information on the functions of the reticulum.

The digestive enzymes of insectivorous plant glands are examples of proteinaceous secretions, and many others, less well understood, exist.[743] In one study, glands found in the leaf stalks of dog's mercury, *Mercurialis annua,* were first of all examined by straightforward electron microscopy and thereby found to offer advantages for experimental work. [210] 'Pulse' and 'pulse-chase' experiments using radioactive amino acid (tritium-labelled glycine) showed that nuclei, mitochrondria, plastids and rough endoplasmic reticulum can all make protein. Some of the protein, possibly made on non-membrane bound polyribosomes, becomes incorporated into proteinaceous masses lying free in the cytoplasm. Additionally, some protein moves from its site of synthesis in the rough endoplasmic reticulum to areas of the cell dominated by dictyosomes (units of the Golgi apparatus) and smooth endoplasmic reticulum. Still later, i.e. about half an hour after the 'pulse', radioactive protein is found in the secretory product that has actually left the cells.[211]

The research on the *Mercurialis* gland hints at the existence of systems within the plant cell that are remarkably like those described in much greater depth for formation, packaging, intracellular transport, and secretion of zymogen by cells of the pancreas. There it has been well established that the protein is first synthesized by rough endoplasmic reticulum polyribosomes, and is then passed into the cisternal cavities. It then migrates to smooth surfaced cisternae or vesicles near the Golgi apparatus. Biochemical modification and concentration occurs here to produce the final secretory zymogen granules (still membrane-bound) that eventually leave the cell.

Whether further botanical work will reveal closer parallels with the situation in the pancreas remains to be seen. There are some obvious differences, apparent in all of the examples so far described. In the beetroot slices the protein apparently stays in the cisternae; in the bean cotyledon storage cells the final destination is the vacuoles; as well as secreting protein the *Mercurialis* gland cytoplasm contains masses of protein lying free without an enclosing membrane.

PROTEIN SYNTHESIS AND SECRETION IN ALEURONE CELLS

The germination of cereal grains is such a familiar process that it is easy to overlook the fact that it encompasses an enormously complex sequence of events, initiated simply by moistening the seed.[84] The rough endoplasmic reticulum plays its part by producing enzymes that convert the insoluble nutrients stored in the endosperm into a soluble form utilizable by the developing embryo. *Aleurone* cells form a secretory tissue surrounding the endosperm, and are stimulated by a hormone, gibberellic acid, thought to originate in the embryo, to secrete amylases (hydrolyses starch to sugar), proteases (protein to amino acids), nucleases (nucleic acid to nucleosides and nucleotides), and several other digestive enzymes. The sequence involves synthesis of rough endoplasmic reticulum, and then synthesis of the enzyme proteins by that rough endoplasmic reticulum, following activation of the necessary genes in the aleurone cell nuclei.[200, 853] In this case it is probable that the enzyme protein molecules enter the cisternal cavities, which can be seen to become distended. There is a later stage at which the enzymes are still in membrane-bound packages within the cells, but once released through the plasma membrane, presumably by reverse pinocytosis, they attack the food reserves in the endosperm tissue (see also page 19).[267, 849]

The overall process is again reminiscent of the synthesis and secretion of pancreatic zymogen. There too hormonal treatments can induce formation of rough endoplasmic reticulum and subsequent protein synthesis. Transfer of protein from rough to smooth cisternae, and involvement of the dictyosomes has not, however, been clearly discerned in the aleurone cells, though smooth surfaced vesicles do occur.

PROTEIN SYNTHESIS: GENERAL ASPECTS

Protein synthesis mediated by the rough endoplasmic reticulum takes several forms and proceeds at different intensities in different cell types and stages of development. Some of the case histories described above belong towards the extremes, but it is quite usual in biology to take advantage of extremes in order to learn something about conditions in more ordinary situations. What, then, are the generalizations that emerge?

To the electron microscopist the visual signal indicative of protein synthesis is the polyribosome. Polyribosomes lie both on the external face of the endoplasmic reticulum and free in the cytoplasm, and protein synthesis occurs at both locations, not to mention in the interior of plastids and mitochondria. It must not be assumed that all of the proteins made at the rough endoplasmic reticulum are destined to be passed into the cisternal cavities, or that all of the proteins that reach the lumen are destined to be secreted. Undoubtedly some, or even most, are, but it is not ruled out that protein can be freed into the cytoplasm, or transferred to other cell components. Transfer to the vacuole has been detected in widely differing materials. The cell membranes themselves are also in constant need of a supply of appropriate types of protein, a need that is thought to be met not only by the activities of bound polyribosomes, but also by synthesis in those lying free in the cytoplasm.[110]

In all instances so far examined the rough endoplasmic reticulum proliferates prior to bulk synthesis of protein, emphasizing that production of a protein in large quantity requires much more than simply activating the gene for that protein. The lipids and proteins of the membrane may have to be made,[405] involving activation or synthesis of other necessary enzymes; ribosome production may have to be promoted; the mRNA must be produced; and the membrane, ribosomes and mRNA assembled into the functional unit recognizable as the rough endoplasmic reticulum, which even then still needs to be supplied with a host of raw materials and essential co-factors.

Cells produce many species of protein, and it is inevitable that many of them have to be synthesized concomitantly, each one specified by its own mRNA. Virtually nothing is known concerning what controls the distribution within the cell of the different syntheses and the rates at which they proceed. Individual cisternae might either process many different mRNA molecules simultaneously, or be restricted to one particular type at a time. As far as is known the ribosomes of the rough endoplasmic reticulum closely resemble and may be identical to those found free in the cytoplasm.[708, 709] The larger of the two subunits of the ribosome (see page 47) binds to the cisternal membrane, and the smaller subunit forms an attachment to the mRNA strand that specifies the nature of the protein to be synthesized. It is more likely to be the mRNA, rather than the ribosomes, that specifies whether a polyribosome will lie free or be attached (perhaps by a special section of the RNA strand) to the membrane.[459]

The fact that proteins which end up in the *lumen* of a cisterna are assembled on the *cytoplasmic* face of the membrane deserves some further comment. Clearly, the protein molecules have to pass through the membrane. It is thought that their synthesis and their transport occur concomitantly, the molecule being in effect extruded from the ribosome, probably through a hole or cleft in the large subunit (which is bound to the membrane) and on through the membrane. A critical stage in the process is reached when the first part of the growing protein has penetrated through the membrane to the lumen of the cisterna. Here it is in an aqueous medium, whereas it was in a comparatively non-aqueous environment when still within the membrane. Hydration of the molecule as it reaches the cisterna might prevent it from returning to the membrane, and may even aid the trans-membrane extrusion.[459]

Polyribosomes, both free and bound, take various forms. The helical type found free in the cytoplasm (Plates 21c, insert, 29a) has in animal material been shown to be based on especially long mRNA strands.[567] More commonly free polyribosomes are mere clusters of ribosomes with no obvious symmetry. The physicochemical basis of the spiral or curled shape of the chains seen on the rough endoplasmic reticulum (e.g. Plate 20b) is not known, nor why in some cases the string of ribosomes loops back on itself to give two parallel chains (Plates 20c, 43f). Other areas of uncertainty concern the movement of mRNA molecules between their site of production in the nucleus and their site of translation in cytoplasmic polyribosomes, in particular how their final destination is specified and recognized.

6.3 Smooth Endoplasmic Reticulum and Lipid Synthesis

Work on animal cells has established that synthesis of lipids is a major function of the endoplasmic reticulum.[663] For example, production of lipid from molecular fragments derived from digested food is a property of smooth cisternae located near the absorptive surface of epithelial cells in the animal gut. Production of a special class of lipid—steroids—is associated with a tubular, smooth form of reticulum seen to best effect (in animals) in cells of the testis and adrenal cortex. Indeed only one component of animal cells, the endoplasmic reticulum, is known to possess all of the enzymes that are required to synthesize the phospholipids that form such an important part of the total lipids of cell membranes.[766] In view of the marked parallels between plants and animals in respect of protein synthesis by the *rough* endoplasmic reticulum, it is pertinent to ask whether parallels also exist in respect of lipid synthesis by the *smooth* endoplasmic reticulum.

In most plant cells the smooth endoplasmic reticulum is inconspicuous, though careful inspection of micrographs will generally reveal a small proportion of smooth membrane amongst the more obvious rough type (Plate 22a–c). Often it is hard to decide whether the smooth portions are biochemically and morphologically specialized in a permanent fashion, or whether they are merely short-lived, perhaps produced by a temporary loss of ribosomes. In other cases, for example in sieve elements, it is clear that smooth cisternae arise by permanent loss of ribosomes from rough precursors (Plate 24).

Tapetal cells of anthers (see page 22) manufacture the lipid cores of orbicules and also lipid precursors (probably in the form of carotenoids) that become polymerized to form sporopollenin on both the orbicules and the pollen grains developing in the anther cavity.[182] Much of the tapetal endoplasmic reticulum is rough (Plate 22a), but smooth portions are clearly seen at terminations of rough cisternae where vesicles are budded off (Plate 22c). The vesicles probably contain lipid, but since much of the reticulum surface is rough it may be that the final product is a lipoprotein rather than a pure lipid.

Smooth cisternae are also seen as sheaths wrapped around tapetal plastids and mitochondria (Plate 22b).[793a] The functions here are conjectural, but an obvious interpretation is that the sheaths could serve in the efficient collection of raw materials and/or the energy-rich compound adenosine triphosphate, produced by plastids and mitochondria, and passed along the cisternae to sites of lipid production. An intriguing parallel is that when a hen synthesizes lipoprotein and protein prior to laying an egg, the distribution of endoplasmic reticulum in its liver cells becomes not unlike that in the tapetum, with sheaths round the mitochondria, and an abundance of other rough cisternae. Again, when excessive demands are made upon the mammalian liver to synthesize plasma proteins, the greater requirement for energy is reflected in a more intimate than normal association between the mitochondria and the endoplasmic reticulum.[661] In liver, as in tapetum, the smooth cisternae are continuous with the rough. Further, the rough form is considered to generate the smooth (see later).

Spectacular arrays of smooth endoplasmic reticulum tubules occur in some plant glands (e.g. Plate 23) and in one of the most highly specialized of all higher plant cells—the suspensor of young embryos.[749] In a few instances, including some sugar-secreting nectaries[201] and sugar-carrying sieve elements,[44] the smooth tubules take up an almost crystalline array, described as a 'coat-of-mail' formation from its appearance in ultrathin section. Like that other semi-crystalline membrane formation, the prolamellar body (Plates 36, 37), it may be a device for storing a large area of membrane in a very compact form.

Well developed smooth endoplasmic reticulum tubules are irregular and tangled (Plate 23). The biochemical link between the various glands where they are found (in plants and animals) is that synthesis of a particular class of lipids known as terpenes is involved. Terpenoid units are incorporated in a vast diversity of biological molecules, including carotenoids (see above), some oils of plant oil-glands, parts of the flavonoid pigments secreted in bulk by the 'farina' (floury) glands of primroses (the subject of Plate 23), the steroid hormones of animals, and cholesterol, a widespread constituent of cell membranes. Oil glands[328, 739, 740, 741, 742, 745, 748] and farina glands[910] are amongst the best sources for smooth endoplasmic reticulum tubules in plants, but biochemical work on much less exotic material suggests that terpene synthesis by the endoplasmic reticulum may be quite general. Thus, preparations of reticulum isolated from bean seedlings that have been fed with radioactive precursors contain radioactive sterol.[435] The reticulum has also been implicated in the synthesis of the terpene and lipid derivatives, resin[591] and cutin.[439] It is not, however, valid to say that all terpene synthesis occurs in the endoplasmic reticulum. There is, for instance, evidence that carotenoids and others can be made in plastids,[282, 327] not to mention in bacteria which do not possess an endoplasmic reticulum. Neither is it valid to say that where terpene synthesis is mediated by the endoplasmic reticulum, it is only the smooth membranes that are active. For example, the gland shown in Plate 21a and b probably secretes terpenoid 'essential oil', and its massed cisternae are obviously rough. What part the ribosomes on the membranes might play in such a system is not clear.

The plant lipids that have been mentioned thus far have all been destined to be secreted. Many others, however, and in particular the lipids of cell membranes, remain within the cells. The role of the endoplasmic reticulum in making phospho-lipids of animal cell membranes was mentioned at the beginning of this section, and evidence that the endoplasmic reticulum plays a comparable part in plant cells is mounting: certainly it synthesizes several key phospholipids which are found not only in the endoplasmic reticulum itself, but also in other cell membranes.[494, 570] If, as is suspected, the endoplasmic reticulum is the primary, or even the exclusive, site of synthesis of membrane phospholipid, the question of how the completed molecules are transported to their final destination is raised. Do they move molecule by molecule, or are whole areas of membrane transferred from the reticulum in the form of vesicles or by momentary establishment of membrane continuities?

6.4 The Endoplasmic Reticulum and Solute Transport

It has already been pointed out that the morphology of the endoplasmic reticulum would suit a role in the intracellular transport of materials. If the cisternae are relatively free of hydrophilic substances that bind molecules of water, the intracisternal water will be more 'available' as a solvent than the water in the surrounding cytoplasm (this concept was applied to vacuoles—page 34 and to nuclei—page 55). Quite apart from any specific solute transport properties of the membranes, there will therefore be a natural tendency for water-soluble solutes to accumulate to a certain extent within the intra-cisternal aqueous environment. With this background, the evidence that the endoplasmic reticulum participates in intracellular transport can be examined.

The mammalian liver provided the first evidence that parts of the endoplasmic reticulum could participate in mobilizing sugars to and from sites of carbohydrate synthesis and breakdown. The liver of a well-fed animal contains stores of glycogen, seen in electron micrographs as rosette shaped particles lying free in the cytoplasm. Significantly, the glycogen is surrounded by a weft of smooth endoplasmic reticulum tubules and vesicles, and when the animal is stimulated to call upon its stored resources, the glycogen disappears with concomitant dilation of the tubules and vesicles. Similar vesicles then appear near the plasma membrane, possibly discharging their contents to the exterior. The observations are consistent with the suggestion that the tubules collect glucose derived by de-polymerization of glycogen, then vesicles are pinched off and transport the glucose through the cell.[661] Confirmatory biochemical work on isolated membranes, and staining reactions on tissue slices, show that the enzyme glucose-6-phosphatase is consistently found associated with the endoplasmic reticulum—not only in liver but many other cells as well.[482]

Glycogen does occur in the cytoplasm of fungi, but in higher plants the major carbohydrate reserve is starch, formed within plastids (except in e.g., the red algae), and hence more remote from the nearest endoplasmic reticulum than in the case of free cytoplasmic glycogen. Nevertheless some types of plant cell contain plastids which are closely enveloped by a cisterna of endoplasmic reticulum. The phenomenon is seen in a not very conspicuous form in Plate 3b. The best examples are found in certain groups of algae,[265] and in the companion cells in the phloem of some, but not all, organs and species.[591] Partly in view of the fact that companion cells function in loading and unloading their neighbouring sieve elements with sugar (see Plate 7), the interpretation has emerged that the 'peri-plastid' endoplasmic reticulum helps in transporting sugar moving to and from the starch in the plastids. A similar idea has already been mentioned (preceding section) in connection with smooth cisternal wrappings of tapetal cell mitochondria and plastids (Plate 22a, b). Endoplasmic reticulum is sometimes seen lying very close and parallel to the tonoplast of vacuoles and once again intermediacy in transport of materials may be suggested (see caption to Plate 27a).[219] Often the wrappings of endoplasmic reticulum are of the bifacial or semirough form, that is, with one face (that nearest the plastid or vacuole) free of ribosomes and the other face rough (Plate 27a).

Glands that secrete sugar abound in the plant kingdom. Floral and extra-floral nectaries are well-known examples.[929] They secrete nectar derived ultimately from sugar reaching them *via* plasmodesmatal connections that lead back to the nearest sieve elements. The concentration, and the types and relative proportions of sugars in the sieve element sap may, however, be changed prior to their emergence in nectar. Also, somewhere along the route between sieve elements and gland cells, a number of compounds, notably amino acids and mineral ions, are taken out of the incoming supplies. For instance, nectar contains much less potassium than sieve tube sap. There is little doubt that these adjustments to the supplies are made by the nectary cells themselves. Clearly, they also transport the nectar to the exterior. It is therefore of interest that the endoplasmic reticulum is one of the most conspicuous elements of nectary cell cytoplasm. It has not been proved that the cisternae contain or transport nectar or its precursors, but where the growth and differentiation of the nectary has been examined, the endoplasmic reticulum has been found to proliferate at the time when nectar secretion commences,[202, 675] In other studies it has been seen to change its form if the secretion of nectar is artificially inhibited.[214]

Once more there is a parallel between plants and animals: delivery of sugar to the exterior *via* derivatives of the endoplasmic reticulum is thought to occur in liver (see above), and it seems very likely that a similar mechanism might apply in nectaries. It must, however, be stressed that difficulties of fixation and interpretation cloud the issue of whether the endoplasmic reticulum cisternae can fuse, even temporarily, with the plasma membrane in order to release cisternal contents to the exterior. Plate 21b and its caption presents a fairly convincing example (though not in a nectary), but if botanists are to make extensive use of the concept, then they will have to seek further evidence. The plasma membrane and the endoplasmic reticulum membrane are usually quite dissimilar, and some local specialization at the molecular level may be needed to permit fusing of the two. The freeze etching technique has in fact revealed exactly this in a variety of membrane-fusion reactions. [729, 850] The small areas where fusion occurs become distinguished by the prior formation of a rosette of particles in the membrane. It will be seen later that a more usual route for transport between endoplasmic cisternae and the experior of the cell passes *via* the Golgi apparatus, and this and the subject of membrane fusion reactions will be raised again in Chapter 7.

The topic of ion transport can be considered very briefly, for the evidence relating to plants is meagre. It is worth including, however, on the by now established grounds that what applies in animals may also apply in plants. The particular form of smooth endoplasmic reticulum that is found in muscle cells is capable of transporting calcium ions in and out of the cisternae, thereby linking the stimulus-to-contract with the molecular events of muscle filament contraction.[663] Again, one of the most extensively developed systems of smooth endoplasmic reticulum is found in the chloride-secreting cells which regulate the salt content of sea birds and fish.

In plants, the evidence so far is little more than suggestive. For example: the endoplasmic reticulum of salt glands (found especially in salt marsh plants) is quite extensive;[359, 765] recent studies on the uptake of chloride ions by certain fresh water algae have revealed an unexpected mode of transport in which the ions are thought to move in packages (rather than as individuals) to the vacuole,[478] with the suggestion that the packages may be vesicles formed from the endoplasmic reticulum;[141] in every type of transfer cell that has been examined (page 30), whether the substances transferred are ions, sugars, amino acids, or unidentified compounds, cisternae of endoplasmic reticulum commonly lie closely juxtaposed to the plasma membrane[633] (Plate 29b). The latter phenomenon is by no means restricted to transfer cells (see, e.g., Plate 10c), but when a cell is thought, as transfer cells are, to be specialized in relation to transport across the plasma membrane, it assumes a greater significance. The suspicion grows that the cisternae may be strategically located to collect incoming, or alternatively to supply outgoing, molecules. Close juxtaposition of plasma membrane and a cisterna of endoplasmic reticulum would reduce the diffusion path for solutes *in transit,* and in addition, it is conceivable that the arrangement might allow for local modification of the molecular constitution and permeability properties of the membranes.

6.5 The Endoplasmic Reticulum and Differentiation

Having considered the capacity of the endoplasmic reticulum to drive various syntheses, we are in a position to examine situations where the reticulum is positioned in the cell in such a way as to modify the chemistry or form of the cell. As envisaged in the introduction, both positive and negative effects are possible. The better documented cases all involve local events in the development of the cell wall.

Comment has already been made upon the precision and delicacy with which callose is deposited in cell walls (page 20). Flattened bifacial cisternae of endoplasmic reticulum generally lie in the cytoplasm adjacent to the callose sites in developing sieve element walls.[591] Knowing that callose is a polymer of glucose, and bearing in mind the likely relationships between sugar transport and the endoplasmic reticulum (above), it has been surmized that the local presence of a cisterna dictates the local supply of sugar, or perhaps the enzymes, necessary for the polymerization and deposition of callose. When, in young sieve plates, the original plasmodesmata enlarge to give rise to sieve plate pores, there is a withdrawal of material from the wall and its replacement by a sleeve of callose (Plate 8e). Again cisternae of the endoplasmic reticulum lie nearby, conceivably controlling callose deposition.[150b]

The above ideas, if correct, are examples of positive actions of the endoplasmic reticulum: that is, events happen at or near the cisternae. To contrast with them there are negative effects: events *not* happening in the vicinity of the cisternae.

One example concerns developing primary xylem elements (page 19), of which it has been suggested that the presence of flattened cisternae lying parallel to and just inside the plasma membrane could 'protect' the underlying wall from secondary thickening. On this view, the bands of secondary thickening with their associated microtubules (Plates 5b) develop only *between* the cisternae, in gaps arranged to generate helical, ring-shaped, or reticular patterns of thickening. The idea is, however, open to criticism—see also page 153.[646]

Another example arises from combined electron microscope and biochemical work on the process of 'budding' by which a yeast cell bulges to produce a daughter.[568] The observations here suggest that just prior to budding, endoplasmic reticulum cisternae extend and come to form a cup-shaped envelope, with an open end where a restricted region of cell wall is exposed to the cytoplasm contained within the cup. Vesicles containing wall softening enzymes (glucanases) are then produced and their contents are released to the wall only at that exposed region. Softening permits bulging of the wall and the gradual outgrowth of the bud. The envelope of endoplasmic reticulum is thus hypothesized to function negatively—by preventing wall softening elsewhere than in its open end. The system may include a positive effect also, for the glucanase-containing vesicles may themselves be derivatives of the endoplasmic reticulum.[539]

Both positive and negative effects are thought to influence wall development in the cell type that develops the most elegant of all walls; pollen. It was mentioned on page 22 and illustrated in Plate 12a that one of the most conspicuous features of lily pollen, the colpus, develops on the part of the wall that lies outermost in the tetrad formed by meiosis. The colpus is a thin-walled region lacking the sporopollenin sculpturing found elsewhere. Examination of pollen development by electron microscopy suggests a mechanism for the prevention of sporopollenin deposition and patterning at the colpus. A sheet of endoplasmic reticulum blankets this part of the wall from within. The idea that this cisterna exerts a negative effect on wall development was tested by centrifuging developing lily pollen hard enough to displace the cisterna. The region that would have become the colpus was then found to develop the polygonal pattern normally characteristic of the *rest* of the grain.[356]

By contrast, positive effects on pollen development are observed in parts of the grain *not* shielded by the blanketing cisterna. In such areas the 'parent lamellae' described on page 22 as being the foundation layers for sporopollenin deposition, are formed external to flattened endoplasmic reticulum cisternae lying just inside the plasma membrane.[165] In other words these cisternae may be positively influencing wall development much as described above for callose synthesis. The chemical nature of the parent lamellae is, of course, very different from that of callose, but being lipoprotein, their constituent molecules do not fall outside the general range of synthetic capabilities postulated for plant cell endoplasmic reticulum.

A very clear example of participation of the endoplasmic reticulum in a morphogenetic process is provided by the alga *Synura petersenii*. This organism covers its cell body with scales that are manufactured within cisternae derived from the Golgi apparatus. Other aspects of the manufacture process are described in the next chapter, but the point that is relevant here is that a cisterna of periplastid endoplasmic reticulum appears to collaborate with the dictyosome cisterna in producing an elaborately shaped mould within which an equivalently-shaped scale develops. Whether the endoplasmic reticulum, which presses into the dictyosome cisterna to form an invaginated pouch, contributes to the biochemical processes of scale production is not known: it does, however, seem clear that it helps to shape the final product.[747]

In these examples, knowledge of the disposition of the endoplasmic reticulum has obviously been helpful in attempts to find the underlying causes of morphogenetic events. Causality is not easy to prove, and even if ideas on the positive and negative effects of the reticulum are, for the moment, accepted, it remains true that the analysis has progressed but one step—in fact just far enough to raise new questions. How do the cisternae come to lie where they do? Are they synthesized there? Do they move to take up the correct position? Are they anchored in position, perhaps by microtubules or membrane-to-membrane cross links? The reticulum seems to emerge as a tool, but we do not know how it is manipulated during the expression of the genetic information that ultimately specifies patterns of cell differentiation.

6.6 The Endoplasmic Reticulum and Digestion

Previous sections have included examples of the synthesis of digestive enzymes by the endoplasmic reticulum: during activation of the aleurone cells of germinating cereal grains, and production of glucanases in budding yeast cells. Extra-cellular digestive enzymes are also produced by the glands of insectivorous plants. That the endoplasmic reticulum may be a source for the digestive enzymes of vacuoles was mentioned on page 37, along with the suggestion that intracellular digestion can apparently be accomplished directly by specialized cisternae. The latter topic can now be elaborated. Descriptions exist relating to tissues as widely divergent as root tips and developing seeds.

HETEROPHAGY The winter aconite, *Eranthis*, affords a particularly clear cut demonstration of the potentialities of the endoplasmic reticulum in bringing about intracellular digestion.[150, 620] During an early stage of seed development not long after fertilization, the cell walls of the innermost cell layers of the ovule wall break down. Their cell contents are absorbed into the endosperm, which at this time is still a liquid tissue, with no cell walls. The absorption of foreign material is accompanied by the appearance of cytoplasmic enclaves surrounded by specialized forms of endoplasmic reticulum, often in concentric layers. The components enclosed in the centre of the whorls of cisternae then break down The descriptions suggest that the digestive enzymes must be released into the central, enveloped, mass, which probably consists largely of foreign ingested cytoplasm. This type of digestion, where enzymes are released from one face (the innermost) of the cisternae, adds to the considerable body of evidence showing that the endoplasmic reticulum can be dorsiventral, displaying different properties, both morphological and biochemical, on the two faces of the cisternae. The phenomenon is also remarkable for a different reason, in that ingestion and then digestion of foreign cytoplasm is highly unusual behaviour for a higher plant, and possible here only because of the absence of cell walls from the endosperm at this particular stage of ovule development. The process is a good example of heterophagy, as distinct from autophagy (as in autophagic vacuoles, page 36), in which the cell's *own* components are digested. We must add heterophagy to pinocytosis as another phenomenon characteristic of animals, to which plant cells can nevertheless revert[836] when freed from their walls.

AUTOPHAGY Given the capacity for mediation of heterophagic digestion by endoplasmic reticulum, there seems in principle to be no reason why plant cells should not accomplish autophagy in a similar fashion. Autolysis of synergid cells in embryo sacs in fact begins with the formation of whorls of endoplasmic reticulum. These surround the nucleus, which breaks down, followed by the rest of the cell components.[758] Convincing micrographs of comparable cytoplasmic enclaves surrounded by whorled cisternae have been obtained for seeds,[854] nucellus cells in ovules[399] and for root tip cells.[33, 144, 551] In the latter, breakdown of the enclosed cytoplasm and metamorphosis of the bounding membranes leads to the development of vacuole-like compartments in the cell, and it is a matter of definition whether these are to be regarded as true vacuoles. It seems reasonable to suggest that there can be more than one type of vacuole in a cell, and that digestive activity in vacuoles can develop either by insertion of appropriate enzymes into pre-existing vacuoles, as described on page 37, or by a *de novo* developmental sequence, as described here.[535]

6.7 Origin, Development, and Structural Modulations of the Endoplasmic Reticulum

Nearly all discussions of the origin of the endoplasmic reticulum emphasize that it is continuous with the outer membrane of the nuclear envelope (page 51). It has been suggested that the nuclear envelope was in evolutionary terms the fundamental system, and that it gave rise to the endoplasmic reticulum. This is impossible to investigate, but it is clear that in present-day organisms the two types of membrane are inter-related developmentally, and that the one can give rise to the other in appropriate circumstances.

The endoplasmic reticulum can grow and develop with considerable autonomy, and it is by no means dependent for such processes upon the nuclear envelope. In the first place, connections between the two systems are probably not as common as at one time thought. They do occur (Plates 18a, 21a), but the earlier impressions of their abundance arose from the use of a fixative, potassium permanganate, which is now largely outmoded because of the danger that it might allow some movement, and fusion, of membranes during fixation or subsequent dehydration. More direct evidence concerning nuclear envelope-endoplasmic reticulum relationships is that if the nucleus is removed from an *Amoeba* cell the reticulum nevertheless can extend and assume new configurations;[222] similarly the characteristic smooth cisternae of sieve elements complete their development (Plate 24) after the loss of the nucleus.[195, 196]

Inspection of the degree to which cisternae are dispersed throughout the cytoplasm of meristematic cells (Plate 2) indicates how unlikely it is that any newly-formed daughter cell (Plate 48a) will entirely lack endoplasmic reticulum, and it seems likely that an initial quota of cisternae received at cell division grows and becomes modified as the cell develops and differentiates. In cells in the core of the root cap this entails a relatively simple process in which the total surface area of the cisternae extends in proportion to the enlargement of the cell, so that surface area per unit volume remains more or less constant.[129] In many other cell types the developmental processes are much more complex, and lead to a wide diversity of quantities and morphologies.

As well as exhibiting changes that are consistent and characteristic features of the development of plant cells, the endoplasmic reticulum can respond, sometimes with dramatic rapidity, to externally applied stimuli. In cells of the cambium it changes in form with the seasons,[114a, 697, 787] as do the vacuoles (page 34). Removal of oxygen, application of high concentrations of carbon dioxide, or of poisons such as chloramphenicol, cyanide, or colchicine, all induce sudden alterations.[162, 163, 214]

One of the more unexpected external stimuli to affect the endoplasmic reticulum is gravity.[415] Analysis of the phenomenon is, however, complicated by the existence of concomitant influences on heavy starch-containing plastids and on dictyosomes. The observations are that when a grass root is displaced from the vertical, the starch-plastids of cells in the centre of the root cap fall, and endoplasmic reticulum cisternae change from being symmetrically distributed around the cell to being accumulated on the uppermost side. If the image of a cell is divided into four sectors by drawing diagonals on micrographs, eightfold disparity in the distribution of endoplasmic reticulum between adjacent sectors is seen after only twenty minutes of gravitational stimulation. This rapid redistribution does not, however, occur in all species, so it may be of only secondary importance in the reactions by which roots grow in the direction of gravity. More drastic experiments involving imposition of artificially high gravitational forces by centrifuging cells or tissues show that the rough endoplasmic reticulum sediments in the direction of the centrifugal force, weighed down by its many ribosomes, and that when the centrifuge is switched off the cisternae rapidly recover and become redistributed.[67] The more the endoplasmic reticulum is studied, the more its mobility and capacity for occupying specific parts of the cell impresses.

MOLECULAR ORGANIZATION So far, alterations in the endoplasmic reticulum, whether of endogenous or exogenous origin, have been viewed at the gross level of structure detectable by electron microscopy. Clearly, it is desirable to enquire into the molecular architecture of the membranes, for it is at this finer level of organization that the ultimate factors governing gross morphology must be sought. Very little botanical evidence exists, but there have been extensive studies of animal cells in which the reticulum is either developing or else is being modified by drug treatment or changes in diet.[766]

It might be expected that a growing membrane would extend by the simultaneous incorporation of all of the molecular components—lipids and proteins. Growing a membrane would on this basis be like spreading a film of paint made to a particular hue by pre-mixing individual colours. The components would always occur in the same relative proportions. However, this procedure seems to be too complex and not versatile enough for the living cell, which would have to regulate many supply lines very precisely in order that all of the components should be continuously available in the right amounts and in the right places. In fact, isolation and biochemical fractionation of growing and changing membranes shows that protein components are added individually. In other words the final appearance of the film of paint, to pursue the analogy further, is obtained by sequentially spraying on new colours, rather than by pre-mixing them. Membrane growth is a 'multi-step' process.[766]

Some complications are that individual species of protein can also be withdrawn from the membrane. Also, every component, lipid and protein, is subject to the process of turnover, in which concomitant breakdown and resyntheses continually renew the population of molecules. Each type turns over at its own characteristic rate. Biochemists and electron microscopists agree that it is difficult to distinguish between 'old' and 'new' parts of growing and changing endoplasmic reticulum systems. It seems that molecules can be withdrawn from and added to *any* part of the surface. There are no specific zones of senescence or growth.

The great diversity of function exhibited by the endoplasmic reticulum implies that its armoury of enzymes must vary correspondingly. As already described, there are variations which look alike to the electron microscopist but have very different functions; likewise there are variations which differ both in structure *and* in function. They are all, however, variations on a common theme, and there appears to be a fundamental collection of molecules that specifies the entity known as the endoplasmic reticulum. The genetic information of the cell specifies this foundation, and can add to it optional components selected from a wide range of possibilities, thus generating a diversity of biochemical attributes. Sometimes the additions confer distinctive structure as well as distinctive function, as when ribosomes and other parts of the machinery of protein synthesis associate with the basic foundation. Sometimes the additions are made without structural change, for example the location of the enzyme glucose-6-phosphatase (referred to on page 64) can be determined by special techniques which show that it enters the growing endoplasmic reticulum of liver cells at random points over the membrane surface. To a microscopist who is merely looking at morphology, the membrane looks the same before and after insertion of the enzyme.[482]

This picture of the development of the endoplasmic reticulum is complicated enough, but when it is remembered that there can be more than one type within a single cell, it seems even more so. It becomes necessary to visualize the total reticulum surface as a mosaic of structurally and biochemically differentiated regions, any of which can be augmented or diminished in response to internal or external stimuli. The foregoing sections have provided several examples of regional differentiation, ranging from the specialized cisterna known as the nuclear envelope (page 51), to continuity of rough and smooth regions (Plate 22b), and insertion of semi-rough regions alongside plastids or vacuoles (Plate 27a). A further, and very important, example relates to the formation of dictyosomes, and will be detailed in the next chapter. One wonders whether the visible, large-scale differentiation seen within the total reticulum is matched by, or founded on, an invisible micro-scale heterogeneity, in which the units are a variety of multi-enzyme complexes (together with membrane lipids) that can associate in varying relative proportions to create the overall biochemical and morphological differentiation.[192]

Many of the concepts introduced in this chapter can be summarized by stating that the endoplasmic reticulum, more than any other cell membrane system, combines to a high degree of sophistication several seemingly disparate features. It is stable in that it is always present, but it is dynamic in that it can change rapidly. It takes many forms and it performs many functions. Nevertheless it is always recognizable as one fundamental entity, the endoplasmic reticulum.

7 The Golgi Apparatus

7.1 Introduction

In these days when so much cell biological research is carried out on just a few favourable and oft-studied tissues—rat liver, root tips, etc.—one wonders what surprises might be found in more exotic tissues, if only they too were to be examined in great detail. For instance, few research workers nowadays would be likely to select as material for a general study an object as unorthodox as a barn owl's cerebellum—yet this peculiar choice illustrates the point. In 1898, Camillo Golgi subjected one, along with other nerve tissues, to an unusual staining procedure called metal impregnation, which involved treating specimens for up to a week, at body temperature or above, in osmium tetroxide and rubidium bichromate. Variations on the method introduce mercury or silver atoms into the cells. Golgi was able to see that a stained network appeared in the cytoplasm, but although other more familiar tissues responded similarly, many of his contemporaries understandably thought that metal impregnation was too damaging a technique to apply to delicate biological material. Numerous additional observations were made, but the suspicion that the network was an artefact persisted, and fifty years were to pass before an equivalent structure was seen in living cells by phase contrast microscopy. Then, six years later, came the first electron micrographs, and it emerged that Golgi had indeed discovered a genuine and widespread component of eukaryotic cells.[41]

The structure has come to be known as the Golgi apparatus, a fortunate choice of name, for Golgi's own description—'appareil réticulaire interne'—is so like 'endoplasmic reticulum' that it would have been necessary to find some other term for the membrane system that now bears the latter title. One reason why 'appareil réticulaire interne' did not survive is that it does not aptly describe many organisms which were found by metal impregnation to possess a population of small particles, rather than a network. The particles were nevertheless thought to be units derived from a net, and so were christened dictyosomes (Greek *dictyon* = net, *soma* = body). The latter term is useful to botanists, because plant cells are amongst those in which the Golgi apparatus is best regarded as a collection of more-or-less independent dictyosomes.

7.2 The Golgi Apparatus: Ultrastructure

7.2.1 The Cisterna, the Dictyosome, and the Golgi Apparatus

THE CISTERNA The structure of the Golgi apparatus is conveniently described by building it up from its smallest component part—a flattened sac, or cisterna, not unlike a small, flattened cisterna of endoplasmic reticulum in that its membrane encloses a thin intracisternal space (Plate 25b). Closer inspection does, however, reveal a number of distinguishing features. First, Golgi cisternae are free of ribosomes. Secondly, the shape and area of their membrane surfaces are much less variable than in endoplasmic reticulum cisternae. A Golgi cisterna is usually a flattened disc, of the order of 1–2 μm in diameter. Thirdly, these discs associate in stacks to form dictyosomes.[574]

Taken together, the above three features allow recognition of dictyosome cisternae in the great majority of plant cells. The wide morphological range exhibited by them in animal cells is not seen in plants, but the basic framework is subject to a certain amount of variation. Thus the cisternae are sometimes relatively simple in shape at the edges of the disc, or they can be perforated by irregular pores; indeed they can be so extensively perforated that the peripheral region becomes a network of branched tubules.[565, 574] The cisternae may be flat, or may be curved, so that the whole dictyosome is concave on one surface and convex on the other. No matter which form is taken, the peripheral region of the cisternae is dilated in all but completely quiescent specimens.

There is generally a population of vesicles nearby (Plates 25a, 27), often with individual vesicles positioned in such a way as to suggest that they have budded off from the swollen cisternal periphery. This is one of the many situations where cell biologists have difficulty in extrapolating from static electron micrographs to the dynamic reality, which in this case almost certainly involves several types of vesicle entering and leaving the dictyosomes.

THE DICTYOSOME Moving up the scale of organizational complexity from cisternae to dictyosomes, it is easy to see that the cisternae associate one with another to form a stack, but less easy to determine how this organization is maintained. The membranes of successive cisternae, although close, do not touch back to back (as in granum stacks of chloroplasts, page 107), due to the presence of a very thin film (about 10 nm) of intervening material (Plate 25c). In some dictyosomes rod-shaped objects have been seen lying side by side along part of the membrane surface within this thin film,[237, 565] but they are only visible in favourable planes of section, and may not always be present. They tend to lie peripheral or adjacent to electron-transparent intercisternal plaques of unknown composition.[566]

It has been suggested that the intercisternal fibres and plaques might be structural elements that link the cisternae to one another. They might also be responsible for the generally flattened shape of the central region of dictyosome cisternae, by preventing them from rounding up to form spherical vesicles. This remains conjectural, but one relevant observation has arisen in the course of experiments aimed at isolating intact dictyosomes from cells. If lysosomal enzymes are also present in the cells, and are released when the cell is ruptured, it is found that the dictyosome cisternae tend to separate, as if a cementing substance was being digested. If lysosomes are absent, the yield of *stacked* cisternae is higher.[571] Also, stacked cisternae can be dissociated by treatment with various chemicals that tend to weaken the mutual attraction that hydrophobic lipid-protein complexes have for one another, so part of the intercisternal 'cement' may consist of a lipoprotein bridge between adjacent membranes.[566]

The number of cisternae per dictyosome varies from plant to plant, from cell to cell within a given plant, and also varies with time within a given cell. The giants of the genre, found in some algae, can have 20–30 (Plate 26b, g and h), but 4–8 is about the norm for most higher plant tissues (Plates 3a, b, 28).

THE GOLGI APPARATUS The highest level of organization concerns the population of dictyosomes, constituting the Golgi apparatus of the cell. There are micro-organisms in which populations, and problems of integration of dictyosomes, do not arise, there being only one in each cell. This is the case in many unicellular algae, where, obviously, the single dictyosome *is* the Golgi apparatus. At the other extreme, the rhizoid cells of the water plant *Chara* are estimated to have about 25 000 dictyosomes.[767] More familiar cells such as those in the root cap have several hundred dictyosomes when mature.[129] Meristematic cells of a shoot apex have been shown to contain 20–40.[12] Changes in the number of dictyosomes occur as part of the process of cell differentiation, for example there is about a tenfold increase during maturation of root cap cells.[129] Again, cells in the outer, secretory layer of an insectivorous plant gland have been found to develop about five times as many dictyosomes as cells in the immediately underlying layer.[743]

Compared with animal tissues such as gut epithelium, where the Golgi region of the cells lies between the nucleus and the external surface of the cell, every cell being the same in this respect, the Golgi apparatus of plant tissues and cells is rarely restricted to a particular intracellular region. This, as usual, is a generalization, and there are exceptions. The most notable is found in the algae, where the dictyosome(s) commonly lie adjacent to the nuclear envelope (Plate 27b). A specific component of the cytoplasm, the endoplasmic reticulum, often lies along one face of each dictyosome, even though the dictyosomes themselves may not be specifically orientated or located within the cells (Plate 27a).

7.2.2 *Dictyosomes as Dorsiventral Structures*

The above section concluded with a hint—the proximity of endoplasmic reticulum to one face—that dictyosomes might be dorsiventral, with distinct tops and bottoms, rather than being symmetrical piles of identical cisternae. Although in higher plant cells the endoplasmic reticulum is not always associated closely with the dictyosomes (e.g., Plates 3a, 5e, 28a), the spatial relationship between the two membrane systems can be as conspicuous and consistent in cells of algae (e.g., Plate 27a) as it is in animals. In some cases the endoplasmic reticulum is replaced by the nuclear envelope (Plate 27b), once again drawing attention to the homology of these two membrane systems (see page 51). Until 'function' has been added to this account of 'structure' let us for the moment call the part of the dictyosome that is closest to the endoplasmic reticulum the 'base'.

Gradations in structure and staining properties provide other indications of the dorsiventrality of dictyosomes. Moving from the 'base' of the stack through the successive cisternae, the dictyosome-attached or -associated vesicles tend to be larger and more abundant (Plates 25a, 27a) and the cisternae to be larger and often more distended (Plates 26g, h, 28a). The cisternal contents may be more conspicuous or elaborate (Plates 27a, 28a). Frequently the substructure of the membranes alters from a 5-6 nm thick form in which a dark-light-dark configuration is not seen, to a thicker form (matching the thickness of the plasma membrane) in which the 'tramline' configuration is visible (Plates 3b, 25c).[296] Freeze-etching shows that the number of particles per unit area of membrane also differs (Plate 25a).[790] Other gradations appear when staining reactions are employed: staining by the metal-impregnation techniques developed by Golgi and other light microscopists;[536, 682] staining to reveal the distribution of enzymes;[159, 905, 925] and staining to reveal the presence of a coating of polysaccharide on the membranes.[681, 905] Finally, to return to the opening point, the endoplasmic reticulum lying at the 'base' of the dictyosome is itself dorsiventral, usually being rough on the face furthest from the dictyosome, and smooth on the face closest to the dictyosome (Plates 25b, 27a).

Several of these points will be examined in more detail in subsequent sections; meanwhile the knowledge that biochemical and morphological gradients exist in the stacked cisternae of dictyosomes allows us to progress towards a functional interpretation of the structure.

7.3 The Golgi Apparatus: Function

The difficulty of interpreting static electron micrographs of fixed, dead cells in terms of the dynamic, living condition has been stressed chapter by chapter in this book. Page 4 focuses attention particularly upon structures which the electron microscopist may not recognize to be dynamic, unless it is possible to correlate structural details with biochemical events by some procedure such as autoradiography. A great deal of evidence now indicates that dictyosomes fall into this category. The hypothesis has been developed that they are dynamic entities which function in the manner of assembly-lines, receiving an input of precursors and generating an output of products. The endoplasmic reticulum is considered to be the route taken by incoming raw materials, and the successive cisternae, starting nearest the endoplasmic reticulum at the 'base' of the dictyosome, represent stages in processing. The finished products leave the assembly line in vesicles or distended cisternae. This hypothesis substitutes terms indicative of the dynamic nature of the dictyosome for those that, until now, have been purely descriptive. The 'base' can now be described as the 'forming face' of the dictyosome, and the opposite face as the 'maturing face'.

The remainder of the chapter develops this hypothesis, starting with case histories of particular types of dictyosome activity which offer direct, visual evidence of an assembly-line type of function.

7.3.1 Case Histories of Golgi Function (1): Scale-Production in Scale-Bearing Flagellates

Single-celled planktonic flagellates might seem as unlikely a source of information about the Golgi apparatus as a barn owl's cerebellum, but they have nevertheless provided cell biologists with a remarkably complete picture, disclosing many features later to be confirmed by study of more conventional experimental materials.

Plate 26 depicts one particularly well studied genus, *Chrysochromulina*. It is the most important of the marine nanoplankton; tiny organisms at the base of marine food chains, capable both of 'producing' by photosynthesis and 'decomposing' by phagocytosis. Like other scaly flagellates, *Chrysochromulina* covers its otherwise naked protoplast with an investment of sculptured scales.[631] In *Chrysochromulina* the scales are predominantly carbohydrate in nature, but in others the carbohydrate serves as a foundation for other materials—calcite, to form the coccoliths of the geologically important Coccolithophorids,[616, 619] or silica, to form the silicified scales of members of the Chrysophyceae.[747]

SCALE PRODUCTION IN DICTYOSOMES The key to the discoveries concerning the Golgi apparatus lies in the fact that the scales are easily recognized. They can be seen in ultra-thin sections, not only in their final destination outside the cells (Plates 26b, e, f), but significantly, in various stages of development within the cells. They lie inside cisternae of the single dictyosome (Plate 26b) of each cell.[520] Since they occur nowhere else it is logical to assume that this must be the site of their manufacture (Plates 26g, h). It is clear that the dictyosomes are dorsiventral, with endoplasmic reticulum at the 'base', and distended cisternae carrying mature scales at the other end of the stack of cisternae. Successively less mature scales are seen by inspecting successive cisternae moving down from the 'top'. Cisternae bearing scales indistinguishable (except in position) from those outside the cell have been observed to have fused in places with the plasma membrane, presumably caught by fixation in the act of liberating their contents to the exterior.

It seems reasonable to suggest that the scales are assembled in the stack of cisternae and are then extruded. There are two possible ways in which this could be done. Cisternal contents (scales) could be passed from one cisterna to the next along the stack, maturing *en route*. No signs of such transfer can be seen, so the second possibility is more likely—that the cisternae *plus* their contents move as units through the stack, starting from the forming face. For the process to be continuous, new cisternae would have to be added at the forming face to compensate for those lost to the plasma membrane from the maturing face of the stack.

INDIVIDUALITY OF CISTERNAE The above observations provide an overall view of the system, which has been embellished by means of other, more detailed, studies. First, scaly flagellates can have more than one type of scale. For example, the *Chrysochromulina* illustrated in Plate 26a produces four types; the other species shown in the same figure, two. The question arises: is there a separate assembly line for each type, or is one type formed and then the assembly line altered in order to produce the next? The answer is neither of these possibilities. What happens is harder to envisage, but is a very important indicator of the functional potentialities of the Golgi apparatus. Each cisterna produces a particular type of scale, so that scanning from maturing to forming face of the single dictyosome in *Chrysochromulina,* etc., one can see different types (Plates 26g, h). Although not illustrated on Plate 26, it has also been found that where very *small* scales are produced, many identical copies can be made within one cisterna.[518, 529] There are even cases[563a] of different scale types cohabiting a single cisterna, and a comparable phenomenon to this has been detected in the cisternae of the Golgi apparatus of a gland in snails.[617] The conclusion is that within a single dictyosome, the successive cisternae can behave as individuals. This is, of course, more complex than most factory assembly lines, where the products are usually identical. Here the products are of more than one type, and because the final investment of scales consistently contains more of one type than of another, it follows that somehow the multi-purpose assembly line is being quantitatively regulated to make its products in the correct numerical proportions.

DORSIVENTRALITY OF CISTERNAE Other relevant features concern the manner of scale development within the cisternae. Most scales are themselves dorsiventral.[520] This is shown clearly by the appearance in section of both the outermost scales (Plates 26b, e, f), and the innermost scales (Plates 26b, e). The shadow-cast preparation (Plate 26c) also illustrates that the two faces of an 'inner' scale are very different in appearance. As well as being dorsiventral, all scales lie in a fixed orientation when in their final destination outside the cell, so that one of the two faces can be labelled the 'outer' and the other the 'inner'. Knowing this, it becomes clear that the cisternae in which the scales are made are, like the dictyosome, also dorsiventral. In the great majority of cases, the scale is specifically oriented within the cisterna so that what will be its inner face is directed towards the forming face. In other words, when a cisterna moves away from the maturing face of the stack, it can deliver its scale to the exterior without any change in orientation, either of itself or of the scale(s) within it (Plate 26b, g).

EXPRESSION OF GENETIC SPECIFICITY Plates 26b and f compare the scale layers of two species of *Chrysochromulina,* and Plates 26b and e compare two forms of a single species that are considered to be genetically different, but perhaps related by a relatively simple mutation.[519, 520] Many other inter-specific and intergeneric comparisons could be made, and they all illustrate that the shape and the patterning of scales are genetically determined traits. Scaly flagellates thus lead us to the next conclusion concerning the Golgi apparatus. It is a structure in which genetic information can be expressed. Further, from what has been learned of the production of different scale types in a given cell, the *individual cisternae* would seem to be endowed with responsibility for expressing individual genetically determined programmes. Plates 26g and h illustrate the point by comparing scale production in dictyosomes of the 'normal' and the 'mutant' form of *Chrysochromulina chiton.*

ASSEMBLY OF SCALES The scales consist partly of elaborately patterned microfibrils, some radial, some circumferential (Plate 26c, d), and some in other specific locations, giving each type of scale its own distinctive character. The development of such complex objects might be governed in at least two ways: there might be a pattern of enzyme molecules built in to the cisternal membranes, acting as a template for the manufacture of the scale; or there might be precise means of regulating the sites and times at which raw materials enter the cisternae, thus generating a specifically patterned end-product. From detailed observations of another scaly alga, *Pleurochrysis,* it appears that both types of mechanism are involved.[91] Both will be returned to later, meanwhile it suffices to recognize that the membranes of the cisternae can possess pattern-generating templates, and also specific recognition sites at which specific types of vesicle can deliver their contents by fusion of the cisternal and vesicle membranes. Plate 26g gives some impression of template activity by showing how the shape of the scale is mirrored by the shape of the cisternal membrane, and Plate 26g and h both show that there are morphologically distinctive regions not only within scale producing dictyosomes, but within individual cisternae.[231]

7.3.2 Case Histories of Golgi Function (2): Production of Mastigonemes

The above case history has introduced various features of the Golgi apparatus, and some remaining points can be added by a brief description of the manufacture of structures called *mastigonemes,* which are very fine hairy appendages found attached to the flagella of very many flagellated cells.[68]

The single-celled alga *Ochromonas* is a good example, and has been studied in detail. Mechanical shocks cause the flagella to fall off, and within minutes the cells start to produce replacements, so that stages in the assembly of the mastigonemes that fringe the flagella can be observed by sequential fixations. Each mastigoneme has a proteinaceous core, and early in the regeneration process structures with a size and appearance appropriate for cores are seen to accumulate inside a cisterna that subtends the forming face of the dictyosome. This cisterna is part of the nuclear envelope, or else is the extension of nuclear envelope that encloses the chloroplasts in so many algae (pages 123 and 129). Somewhat later, developing mastigonemes are seen within dictyosome cisternae. Staining with a specific electron-dense stain indicates that carbohydrates are added to the proteinaceous cores in the dictyosome cisternae.[553] Later still the mastigonemes are extruded.

Whereas in the first case history, the final product (scales) can be seen only in dictyosome cisternae, the second case history has given us a visual indication that materials are transferred to the dictyosome from the underlying cisterna, whether of nuclear envelope or endoplasmic reticulum. Thus scales and mastigonemes between them have illustrated input, processing, and output of Golgi products. Neither, however, can be regarded as a typical plant product, and it is necessary to return to higher plants to see what *their* Golgi systems produce. For convenience the cisternal *membranes* will be examined first and the cisternal *contents* second. It will by then be possible to examine the dynamics of Golgi activity quantitatively, with respect to both contents and membranes.

7.4 The Nature and Processing of the Cisternal Membranes

FORMATION AND LOSS OF CISTERNAE The case histories described above suggest strongly that mature cisternae are lost from dictyosomes at the maturing face. In the scaly flagellates and many other situations their destination is the plasma membrane, with which they fuse. Evidence for another destination—the tonoplast—will also be forthcoming. Obviously this process reduces the number of cisternae that are left in the dictyosomes, and it follows that since dictyosomes very seldom become depleted until nothing is left (this is best shown by those cases where there is always one dictyosome per cell), there must be a corresponding production of new cisternae.

The descriptions given so far indicate that the site of this replenishment is the forming face, where an association with either the nuclear envelope or the endoplasmic reticulum is frequently to be seen. Furthermore, where this association is clear cut, many small vesicles—*transitional vesicles*—lie between the envelope or reticulum (as the case may be) and the forming face (Plates 25a, b, 26g, 27a, b). Now and again some lucky fixations immobilize what appear to be stages in both the production of vesicles and their coalescence at the forming face (Plates 27a, b). The membranes of the Golgi apparatus are therefore pictured as arising (at least in part) in the endoplasmic reticulum or the outer layer of the nuclear envelope, and then being transported in the form of transitional vesicles to the forming face, where they coalesce to generate a cisterna. Generation of new cisternae and removal of mature cisternae are continuous processes. In the course of time a cisterna passes right through to the maturing face. There it either breaks up into vesicles (Plate 27a) or passes in its entirety to its destination (Plates 26g, h, 28a).

Formation of a new cisterna must be accompanied by formation of a corresponding layer of intercisternal 'cement' (page 72). Similarly, the 'cement' presumably breaks down when a mature cisterna leaves the dictyosome, indeed its breakdown could be the process which *causes* departure and allows the cisterna to balloon into one or several more-or-less spherical vesicles.

BIOCHEMICAL CHANGES Biochemical analyses and other electron microscope evidence substantiates the overall picture of formation and maturation of cisternae in dictyosomes. Suitably orientated ultra-thin sections show that the membranes near the forming face are of similar thickness to those of the endoplasmic reticulum, while progressing towards the maturing face, a 7–10 nm tramline configuration appears, resembling that of the tonoplast and plasma membrane.[296] Whereas the electron microscopist can see this transition, which may be gradual (Plate 25c) or abrupt (Plate 3b), the biochemist, who cannot yet isolate and analyse successive individual cisternae, has to be content with an analysis of the total dictyosomal membrane. Nevertheless the outcome is as anticipated. The total membrane contains a population of lipid molecules that is, on average, intermediate in character between those found in isolated endoplasmic reticulum and isolated plasma membrane.[423]

As already pointed out on several occasions, it is the proteins of a membrane that confer on it the greater part of its specificity and individuality. The proteins of the Golgi apparatus have been examined by isolating the membranes and then extracting and separating the constituents. The results are complex. Some widespread proteins occur in the endoplasmic reticulum, the Golgi apparatus, and the plasma membrane. Other proteins are common to the endoplasmic reticulum and the Golgi apparatus. Others are found predominantly in just one of the three systems. The technique for isolating the various membranes is as yet imperfect, but it is very likely that each type of membrane is in some way distinctive in its particular combination of proteins.[923]

Assays of enzyme activities lead to somewhat similar conclusions. The transition from endoplasmic reticulum to the forming face must involve the removal (or else inactivation) of certain enzymes that, like glucose-6-phosphatase, are (page 68) characteristic of the reticulum. Concomitantly, new enzymes enter the membranes (or else are activated). Parallel events take place at the maturing face during the transition between Golgi cisternae or vesicles and the plasma membrane.[925] It has been noted that, in several cell types, a polyribosome lies strategically placed to contribute to synthesis of a protein involved in formation of new cisternae or 'cement'.[237, 240]

RECOGNITION SYSTEMS To these observations on the lipid and protein composition of Golgi membranes can be added something else which is implied, but for which no interpretation in molecular terms is yet available. The concept of the recognition systems of membranes was introduced on page 13, in order to account for the fact that the membranes of the cell do not fuse indiscriminately with one another, there being instead a limited number of permitted types of fusion. The Golgi apparatus provides clear examples. Thus the endoplasmic reticulum gives rise to vesicles which evidently *can* recognize and fuse with one another to generate a Golgi cisterna. By contrast, the starting material (the reticulum) and the end product (the cisterna) differ, and they *do not* fuse with each other. Likewise, successive cisternae in a dictyosome *do not* fuse with one another, but this could be because the intercisternal material acts as a barrier. Provision of specific recognition sites for the entry of specific raw materials has already been mentioned when considering the assembly of scales. By the time they leave the dictyosome, either as whole cisternae or as vesicles, another change has taken place in the membranes. Their particle density is reduced (Plate 25a)[790] and they proceed to fuse with a selected membrane in another part of the cell. Most frequently it is the plasma membrane, and in plants the tonoplast is another option, just as in animals the lysosomes are permitted destinations. The acquisition of a specific recognition pattern by a Golgi cisterna is as much a part of its development as is the processing of its contents, and is in every way as important. In this context the finding that the membranes of vesicles released from the maturing face acquire a carbohydrate component, so that they come to resemble the plasma membrane in chemical composition as well as morphologically, is of especial interest (see page 23).[242, 850, 851, 905] It may be that alterations in the lipid and protein composition of Golgi membranes are insufficient to account for changes in recognition and fusion reactions. Other molecules, perhaps glyco-proteins, in or on the membranes may be responsible for such matters. It is worth noting that many of the vesicles seen in the vicinity of dictyosomes bear a fuzzy outer coat (Plates 3a, 27a). They are called 'coated vesicles', but it is no more than speculation to suggest that the fuzzy coat functions in recognition.

7.5 The Nature and Processing of the Cisternal Contents

The Golgi apparatus has been likened to an assembly line, and it remains to examine the nature of its products. Cell biologists face considerable technical problems in this endeavour, and the situations where a product can unequivocally be specified are few and far between.

SCALES AND CELLULOSE Additional information on the chemical nature of the scales of scale-producing flagellates is now relevant. All that have been analysed are based on polysaccharide. Those of *Platymonas* are very small and are composed of a pectin-like material polymerized largely from galactose, galacturonic acid, and arabinose units.[276, 527] Those of *Chrysochromulina* itself are morphologically and chemically much more complex[287] with fibrillar (Plates 26c, d) and non-microfibrillar components. The scales of an organism called *Pleurochrysis scherffelii* have proved to be so surprising that extremely detailed chemical analyses have been necessary. It now seems clear that in addition to pectinaceous components, they contain fibrillar cellulose to which a proteinaceous material is attached.[348]

Tremors from this discovery have spread far and wide through the botanical world, because cellulose is the most abundant cell wall material in higher plants, and here, in a very lowly organism, is the first indication that it can be made inside the cell. Caution is necessary, for what applies to *Pleurochrysis* need not apply to a maize plant or an oak tree, and it does not follow that higher plant cellulose microfibrils start their life in Golgi cisternae. But an interesting hypothesis emerges from the research on *Pleurochrysis*. Since, in this case at least, dictyosome cisternae can make cellulose, and presumably contain the necessary enzymes, it might be that the cellulose-synthesizing systems of higher plant cells, thought to reside in or on the plasma membrane (page 25), originate in the Golgi apparatus and are carried to the plasma membrane by vesicles or cisternae generated at the maturing face of the dictyosomes. If the enzymes lie in or on the cisternal membranes, here would be an instance where the importance of the *membrane* delivered by the Golgi apparatus exceeds the importance of the *contents* of the vesicles. Alternatively, it is possible that the contents might include a non-fibrillar precursor of cellulose. In that case the main difference between *Pleurochrysis* and higher plants would lie in the timing of the events of cellulose microfibril formation; in the former they are completed before the Golgi vesicle reaches the plasma membrane, in the latter they are delayed and take place outside the plasma membrane.[93]

The idea that the Golgi apparatus gives rise to expanses of plasma membrane, equipped with cellulose synthesizing systems, has been reinforced by work on another alga, *Micrasterias*. This desmid has an elaboratly patterned secondary wall, in parts of which cellulose microfibrils can be observed lying in distinctive arrays.[426] When the secondary wall is being synthesized the Golgi apparatus of the cell generates special flattened vesicles bounded by a thick membrane[425] that incorporates globular particles arranged with the same geometry as the arrays of cell wall microfibrils.[173] It would seem that secondary wall formation in *Micrasterias* proceeds by a sequence involving *template formation* (the specialized Golgi vesicles); *template transfer* and *template incorporation* (movement of vesicles to, and fusion with, the plasma membrane; and finally *template realization* (formation of microfibrils in the same array as the template particles).[426] It is quite likely that higher plants are basically similar, but three variables need to be pointed out: firstly, there need not necessarily be a geometrically symmetrical array of particles—for some syntheses a patterned template would not be needed; secondly, wall materials other than cellulose could be synthesized (see below); thirdly, the timing of the events could vary, some products appearing intra- and others extra-cellularly.

MUCILAGES AND WALL MATRIX POLYSACCHARIDES In situations where the product of the Golgi apparatus cannot be recognized simply by looking at it by electron microscopy, the next best thing is to isolate it. The major practical difficulty is that of sorting out and collecting Golgi-produced vesicles from amongst all the other components of the cell. The approach has, however, been applied successfully to growing pollen tubes. These grow at the tip (see section 3.3) and significantly, the cytoplasm at the tip is richly endowed with vesicles. Further, similar vesicles occur attached to, and nearby, the cisternae at the maturing face of dictyosomes. Electron micrographs, although admittedly static, support the suggestion that the vesicles come from the dictyosomes and deliver their contents to the growing tip by reverse pinocytosis (see section 3.5) at the plasma membrane. When they were isolated, the vesicles were found to contain a non-cellulosic polymer which (like pectic substances, section 3.2) contains a high proportion of galacturonic acid. A very similar material occurs in the cell walls themselves, and the contents of the vesicles are seen by electron microscopy to disappear when samples are treated with the enzyme pectinase, known to digest pectins.[846]

The conclusion is straightforward. It is that the Golgi apparatus generates a cell wall component that is delivered in the lumen of vesicles to the growing region of the wall. The same wording could be used in respect of scale production: it is only the nature of the wall component that differs. The Golgi apparatus of plant cells has thus already been implicated in both wall microfibril and wall matrix production, and further examples continue the story, especially in respect of matrix-type materials.

Other experimental approaches remain, even if the Golgi products can neither be recognized nor isolated. The method of autoradiography has proved to be of especial value. For instance, it has been used to show that five to ten minutes after feeding roots of wheat seedlings with radioactive glucose, radioactivity appears at or near the dictyosomes of the root cap cells.[594] All *soluble* materials of low molecular weight (such as glucose itself) are, of course, removed from the cells during the processing. Chemical extracts made after fifteen minutes in radioactive glucose contain a labelled carbohydrate polymer, and since in these short-term experiments most of the radioactivity is associated with the dictyosomes, these structures probably represent the primary site of sugar interconversions and polymerization. When the 'pulse' feed of radioactivity is followed by a 'chase' period without any radioactive glucose being present, the radioactivity migrates to a layer of mucilaginous material outside the plasma membrane. Very many studies have by now shown that the Golgi apparatus of root cap marginal cells manufactures the slimy polysaccharide mucilage that surrounds a growing root tip.[129, 416, 572] The process illuminates the dynamics of Golgi functioning and will be considered again later. Autoradiographic procedures have also been used to show that addition of sulphate groups to certain specialized mucilages, in algae and animal material, is accomplished in the Golgi apparatus.[198a]

Autoradiography gives information about the activity and contents of *individual* dictyosomes in *individual* cells. By contrast, an *average* picture has been obtained by means of other experiments which commence in much the same way as those based on autoradiography—namely by supplying roots with radioactive glucose. However, instead of processing the tissue for electron microscopy, it is macerated and the subcellular components are separated. Analysis of a fraction rich in dictyosome stacks and cisternae has disclosed the presence of radioactive polymers closely resembling both the pectic substances and the hemicelluloses of cell walls.[314]

Other, somewhat less specific, methods have been applied to many different cell types in attempts to identify the products of the Golgi apparatus. There are 'staining' reactions by which non-cellulosic carbohydrate polymers are rendered dense to electrons, and in general they confirm that Golgi vesicles and cisternae can contain this type of chemical compound. The plant material so studied includes fungal hyphae (not unlike pollen tubes),[324] root cap cells (see Plate 28b),[719] mucilage secreting glands of insectivorous and other plants,[720, 738] and the fascinating case of the reproductive cells of red, green, and brown algae, which after settling on a substratum, secrete a glue that cements them firmly in place.[199] This latter function of the Golgi apparatus costs vast sums of money, for the glue allows seaweeds to become established and grow on the bottoms of ships, appreciably slowing down the world's sea-borne traffic.

Least specific of all of the methods, but paradoxically the one that has been used more than all the others (because it is so easy), is that of relating the appearance of dictyosomes to the timing of events in the life of cells or tissues. Some of the figures illustrate the approach. For instance, the xylem element of Plate 5e was synthesizing bands of secondary thickening at the time when it was fixed, and there are numerous vesicles attached to and close to its dictyosomes. Much the same can be said of Plate 15a, and if the transfer cells in Plate 13 had been making their cell wall ingrowths when they were fixed, their dictyosomes too would have displayed signs of activity. Such correlations are suggestive, but relatively uninformative compared with the analytical procedures; they may also be dangerous, for living cells are rarely so simple as to do only one thing at a time, and the vesicle production may therefore not be related to seemingly obvious events (e.g., wall formation), but to some unsuspected and less conspicuous process. A healthily critical approach must be adopted.[598]

GLYCOPROTEIN Most of the Golgi products described thus far in this section have fallen into the category of polymerized but non-microfibrillar derivatives of sugars, whether extracellular gums and mucilages, or the pectic substances and hemicelluloses of the cell wall matrix. There is a preponderance of pure carbohydrate in this list, whereas amongst products of the Golgi apparatus of animal cells, glyoproteins and mucopolysaccharides predominate.[584] Crudely, glycoproteins consist mostly of protein with some sugar molecules attached, while in mucopolysaccharides the proportion of sugar derivatives is much greater. The glue secreted by the algal spores referred to above approaches the 'typical' animal Golgi product in containing both proteinaceous and polysaccharide moieties.[108a] At any rate, enzymes that attack proteins can render it less effectively adhesive, and so can enzymes that attack polysaccharide.[125] Also, a number of the extracellular enzymes secreted by plants are glycoproteins, including peroxidase (page 19) and various digestive enzymes (pages 62 and 79).[636] Yet other plant glycoproteins[676] and mucopolysaccharides[357] are made in bulk. If what has been discovered about the assembly of mastigonemes (page 74) can be extrapolated, it is to be predicted that the glycoproteins will be found, as in animals,[584] to originate in the rough endoplasmic reticulum, and to receive most, though not necessarily all, of their carbohydrate content after transfer to the Golgi apparatus.

POLYSACCHARIDE-FORMING ENZYMES A start has been made in analysing the enzymatic basis of the production of sugar polymers by the Golgi apparatus. Cisternae or dictyosomes isolated from both plant and animal cells have been found to contain enzymes capable of attaching certain sugars to other molecules. Where the 'other molecules' are themselves based on sugars, the product becomes a sugar polymer; where they are mostly protein, glycoproteins are formed. Several of the enzymes fall under the general heading of 'glycosyl transferases', i.e. sugar-transferring enzymes.[573, 685, 860] It was stated previously that the Golgi apparatus can be endowed with enzymes not found in such high concentrations (if at all) in the endoplasmic reticulum or the plasma membrane, and here now are actual examples. The Golgi apparatus bears no ribosomes, and cannot make proteins for itself, so the enzymes that are specific to it must be synthesized elsewhere, e.g. in the endoplasmic reticulum or at a Golgi-associated polyribosome (page 75), and inserted at the forming face of the dictyosomes, later to be lost. Alternatively, the enzyme protein might be present throughout the different membrane systems, and activated in the Golgi apparatus and inactivated elsewhere. Activation and inactivation could be brought about by altering the enzyme protein itself, or by altering the characteristics of the cisternal membranes in or on which the enzyme molecules reside.[925]

The non-microfibrillar polysaccharides of mucilages, cell wall matrices, and glycoproteins are composed of a range of sugars and sugar derivatives, in each case polymerized in such specific sequences and numbers that it seems impossible that they could be synthesized save by the sequential reactions of enzymes coordinated in their activities through being in multi-enzyme complexes, each enzyme in the complex receiving its substrate from the previous one, adding its particular sugar, and passing the product on to the next enzyme. Complexes of this nature have in fact been isolated from bacterial membranes, where they synthesize the polysaccharide part of the bacterial coat. They have even been broken down into their component transferase enzymes and membrane lipids, which, if added back to one another in the correct order, will reassemble to regenerate functional complexes.[718] It is to be suspected that similar complexes occur in the cisternal membranes of dictyosomes, where they would govern the manufacture of the various Golgi products described above, and of the glycoprotein surface coat of the mature vesicles. They would be the counterparts of the templates invoked when considering the generation of patterns of microfibrils by dictyosome membranes (page 76). It was seen that in higher plant cells, 'realization' of the hypothetical templates for cellulose synthesis is postponed until after 'transfer' and 'incorporation' into the plasma membrane. In the case of non-microfibrillar polysaccharides it is clear, however, that template 'realization' *precedes* template 'incorporation'. As will be seen (below), some reactions continue during template 'transfer'. It is not clear whether 'realization' also continues after incorporation of the dictyosome membranes into the plasma membrane.

The above considerations have centred upon the manufacture of polysaccharides, and it is necessary to clarify one point before leaving the topic. The Golgi apparatus is by no means the sole site of polysaccharide synthesis. It is known from work on glycoprotein production in animal cells that some primary sugar-attachment reactions occur in the endoplasmic reticulum.[584, 733, 818, 926] In plants too, autoradiographic experiments indicate incorporation of radioactive sugars into insoluble compounds in the endoplasmic reticulum.[146, 644] A role for this membrane system in the production and removal of callose is also envisaged (page 66). Finally, one of the most familiar of all polysaccharides, starch, is made in plastids (page 116).

VACUOLAR MATERIALS The tonoplast has been mentioned at several points in this account, as another destination for vesicles arising in the Golgi apparatus. Research on this phenomenon has not progressed as far as in the case of the delivery of material to the plasma membrane, but two plausible, though partially speculative, stories can be put together. The first concerns intracellular digestion and the second osmoregulation.

(1) Evidence that some vacuoles can contain digestive, lysosomal enzymes was discussed in chapter 4 (page 36). To the electron microscopist, one of the easiest of these

enzymes to locate in a cell is acid phosphatase. It releases phosphate from its substrate, and if lead salts are included in the reaction mixture—i.e., tissue containing the enzyme, plus substrate, plus lead salt—the phosphate is trapped as an insoluble precipitate of lead phosphate at the cellular sites where the enzyme is operative. Lead phosphate is electron-dense, and under favourable experimental conditions its deposits (visible in ultra-thin sections) effectively locate the enzyme. The relevant result is that in some circumstances acid phosphatase activity is detectable in both vacuoles and vesicles.[720] The vesicles, although smaller than those carrying the carbohydrate products of the Golgi apparatus, are thought also to arise from dictyosome cisternae, but some may also arise directly from endoplasmic reticulum. Enzyme proteins are presumably synthesized in the endoplasmic reticulum, rather than in the Golgi apparatus, but the suggestion is that at least a proportion of the lysosomal enzymes of vacuoles are processed and packaged in dictyosomes, prior to delivery to the tonoplast by the vesicles known as primary lysosomes (page 37).[145, 146] Primary lysosomes must not only be loaded with digestive enzymes, they must also be equipped with a membrane recognition system that enables them to detect and fuse with the tonoplast.

(2) The osmoregulatory role of vacuoles was stressed in Chapter 4. The Golgi apparatus too can participate in osmoregulation, and at least three types of contribution have been detected, two of them involving vacuoles.

The green alga *Oedogonium* possesses a most unusual mode of cell division, culminating in very rapid cell enlargement driven by turgor pressure that is in turn generated by the low water potential of the vacuolar sap. Just prior to and during this short phase in the life of the cell, the dictyosomes are seen to produce distinctive large vesicles which appear to fuse with the tonoplast. It is suggested that the concentration of soluble, osmotically active material in the vacuolar sap is being increased at this time.[655] The Golgi vesicles might deliver soluble material directly, or (perhaps more likely in view of the known biochemical capabilities of the apparatus) they might deliver polysaccharide which, if depolymerized in the vacuole by enzymes already present in it, would suddenly augment the content of soluble sugars.

In two other algae, *Vacuolaria* and *Glaucocystis,* the Golgi apparatus has the reverse function: that of expelling water from the cells. Golgi vesicles are produced in abundance and appear to deliver into a contractile vacuole which pulsates at intervals, expelling its accumulated fluid content to the exterior.[743, 746a]

The third type of contribution of the Golgi apparatus to osmoregulation may or may not involve vacuoles, depending on how restrictively the word vacuole is employed. 'Water glands' or 'trichome hydathodes' are widespread amongst flowering plants.[498, 737] They secrete a very dilute watery solution, which seems to derive from the content of very numerous Golgi vesicles. Since these are watery compartments they might justifiably be called vacuoles; however, their behaviour is un-vacuole like, for they fuse with the plasma membrane and liberate their sap to the exterior. Several such glands have been examined, and of them none secretes absolutely pure water. Probably there is a gradation between mucilage secreting glands and water glands, with the Golgi system of the former generating a relatively concentrated polysaccharide (as, e.g., in the attracting and trapping glands of insectivorous plants),[743] that of the latter a watery fluid containing mere traces of polysaccharide (e.g. the water glands on *Monarda*),[329] and that of slime glands (e.g. *Rumex* petioles)[738] being intermediate in character.
Many details of the above osmoregulatory roles of the Golgi apparatus remain to be worked out. The water relations of the vesicles involved in expulsion of very dilute solutions are particularly puzzling, and it is also important to find out how widespread such activities are, and to what extent, if any, the carbohydrate content of vacuoles in higher plants is delivered by Golgi vesicles.[598]

The preceding account of the nature and processing of the content of Golgi cisternae has illustrated how the potential of the system has been exploited in the course of evolution. The basic features that are found in animal cells—delivery of material to the exterior and to lysosomes—are also seen in plants. The major specialization of plant cells, the cell wall and vacuole system, is accompanied by matching specialization of the Golgi apparatus. As far as the cell wall is concerned, the Golgi apparatus probably both secretes pre-formed wall material and templates upon which wall material can be fashioned extracellularly. As far as vacuoles are concerned, both the digestive and the osmoregulatory functions are contributed to (though to an unknown extent) by the Golgi apparatus. The abundance of protein in the apoplast of animal tissues is related to the copious secretion of glycoproteins by the animal Golgi apparatus; likewise the preponderance of polysaccharide in the apoplast of plant tissues is related to the polysaccharide synthesizing capability of the plant Golgi apparatus.

7.6 Packaging and Transport

PACKAGING The processing of Golgi products and membranes is not restricted to the dictyosomes themselves. There is, for instance, good evidence that a biochemical change (methylation of pectin-like material) continues after the vesicles have detached from the maturing face of pollen tube dictyosomes.[158, 714] Again, staining reactions performed on root cells show that the vesicle membrane becomes invested with carbohydrate-containing material after detachment from the dictyosome, transforming an initially non-staining membrane to a membrane that stains in the same manner as the plasma membrane (page 23).[851]

The contents of detached Golgi vesicles are often visibly denser than the contents of those still attached. Their staining reactions can also intensify after they leave the dictyosomes.[108a, 324, 846]

It is likely that the latter observations indicate reduction of the water content of the vesicle, and indeed this must be a vital factor at the final, 'packaging', stage in the overall assembly line process. Gums, mucilages and wall-matrix materials are very hydrophilic. In the aqueous environment outside the plasma membrane they tend to hydrate and swell enormously, thus raising the question of how they are prevented from doing so inside the cell, which is also an aqueous environment. Experiments made on an animal system may have provided one answer. There, inhibition of the transport of solutes across membranes caused Golgi vesicles to distend, rather than to follow their normal pattern of condensation.[889] If the vesicle membrane incorporates solute 'pumps' capable of driving the export of solutes from the vesicle to the surrounding cytoplasm, the water potential inside the vesicle could be raised above that of the surroundings, and water would diffuse down its potential gradient, out of the vesicle. The contents would, as is observed, become more and more contrated. However, the forces required to dehydrate hydrophilic gums are very large, and other mechanisms may be necessary to account for the observations.

GUIDANCE Apart from some comments on recognition systems, the ability of vesicles produced by the Golgi apparatus to find their way to their destination has so far been assumed. There is, however, no evidence that they have powers of locomotion and direction finding. They can diffuse, like any other particle, but this is a random process, and does not lead them to any specific part of the cell. Large vesicles diffuse more slowly than small ones, and from a knowledge of the viscosity of the cytoplasm through which they move, it can be estimated that a typical vesicle, of radius 0.1 μm, might diffuse through a distance of the order of its own radius per second.

Cytoplasmic streaming affords a much more rapid delivery system. It could, for instance, carry vesicles towards apical growing points in pollen tubes and fungal hyphae. On the other hand, where it is rapid and random in its direction, as in that classical object for its observation, the *Tradescantia* stamen hair, one has a mental picture of vesicles being swept past their destination. Unless there are specifically directed streams, the time taken to deliver vesicles will depend upon the statistical chance they have of colliding, recognizing, and fusing with the correct membrane, or the correct recognition site on the correct membrane, much as the rate of chemical reactions is determined by the frequency of collisions between suitably oriented molecules diffusing at random in a solution.

Viewed against this background, it will be no cause for surprise to learn that some cells seem to have developed distribution and guidance systems for their vesicles. Time lapse photography shows slow rotation of the protoplast (including the dictyosome) of the algae *Pleurochrysis*, and it is thought that scales can thereby be liberated in succession to different parts of the scaly wall. Microtubules lie near to the dictyosome, but their role, if any, in delivery of scale-containing cisternae is conjectural.[90, 92] In other cells the position of microtubules and endoplasmic reticulum may exert a directing influence upon the movement of Golgi vesicles. This applies particularly to their delivery to specific areas of cell wall, and the role of the endoplasmic reticulum in this has already been discussed on page 66. Microtubules have several times been observed with rows of vesicles aligned alongside, but it is by no means clear whether this is due to chance, or to the vesicles being somehow guided along the microtubules, or to a more specific and functional attachment such that the vesicles can be propelled along the microtubules. The possible role of microtubules and microfilaments will be considered again on page 153. One point that is significant in this connection is that experiments using inhibitors suggest that the final 'extrusion' of Golgi vesicles requires less production of energy on the part of the cell than does the earlier phase in their life, when they are being manufactured.[572]

TRANSIT TIMES Some estimates have been made of the actual transit time of Golgi vesicles travelling from the maturing face of dictyosomes to the plasma membrane. Since the dictyosomes of plant cells are scattered at different distances from the cell surface, it is not sensible to express the estimates in the form of micrometres of journey per unit of time. It is more appropriate to calculate average life-times for vesicles in transit.

One method makes use of the observation that inhibitors such as cyanide, or an atmosphere consisting only of nitrogen, rapidly bring about cessation of synthetic processes in dictyosomes. The vesicles that have already been made are, however, not affected, and continue on their journey. By fixing root cap cells at time intervals following inhibition, it has been found that the vesicles are cleared from the cytoplasm within a period of about half an hour.[572] This therefore represents a maximum estimate of the lifetime of a root cap Golgi vesicle, and the value is in fact in reasonable agreement with that obtained, for the same type of cell, by employing autoradiography to follow the migration of radioactive root cap mucilage in pulse-chase experiments (page 77).

Another method exploits the properties of mucilage-secreting glands on the insectivorous plant *Drosophyllum*. The secreted drops of mucilage are regular in shape and so can be measured. Under optimum conditions they have been found to enlarge by about 1.3 μm^3 per minute for every μm^2 of the secretory gland surface. At the same time, measurements and counts of Golgi vesicles seen in electron micrographs of the gland cell cytoplasm show that the volume of mucilage that is *en route* from dictyosomes to the secretory surface is about 3.4 μm^3 per μm^2 of surface. Since 1.3 μm^3 of this 3.4 μm^3 is discharged every minute (see above), the average life of the Golgi vesicles is 3.4/1.3 = 2.6 minutes.[736, 743] This, however, is probably a minimum estimate, because hydration and swelling of the mucilage outside the plasma membrane is neglected, and the 1.3 μm^3 of *external* mucilage might well be derived from much less than 1.3 μm^3 of *internal,* concentrated mucilage within the vesicles: if this is true, the average life-time would correspondingly be longer than 2.6 minutes. Even so, the speed with which vesicles are delivered is impressively high, and this has further implications which will be considered in the following section.

7.7 The Dynamics of Golgi Activity

INPUT AND OUTPUT Many things are needed in order to keep the Golgi apparatus functioning as an assembly line: a supply of raw materials for the ultimate product, a supply of enzymes to make that product, a supply of membrane for packaging the product, the means of altering the membrane as required, and a continuing supply of chemical energy. A great deal of co-ordination is necessary in order that all of the sub-processes shall integrate into a coherent whole, and it is not surprising that the dynamic flow can be affected by many agencies, influencing both the structure and the function of the system.

Whether assayed in terms of the number of vesicles produced or of the amount of product released from them, Golgi activity is seen to be as sensitive to changes in temperature as are enzymatic reactions in general. Raising the temperature by 10 degrees Centigrade boosts activity 2–3 fold, provided that an optimum temperature is not exceeded.[743] It is at first sight surprising to find that, when the apparatus is working sluggishly at a low temperature, there are more cisternae per dictyosome than there are during rapid functioning. Once again the dynamic nature of the Golgi apparatus provides an explanation. It is that the output of product, and the input of raw materials, can achieve different balances. When output exceeds input, the length of the assembly line (number of cisternae per dictyosome) falls; when output is lowered below input, the length increases. In other words, the organization is versatile enough to accommodate shifts in the dynamic equilibrium.

A measure of independence in the regulation of rates of output and input is implied by the above, and has also been detected by applying substances that indirectly inhibit the Golgi apparatus. The number of cisternae rises when the supply of energy is reduced, as when a root tip is deprived of oxygen.[204] Inhibitors of nucleic acid and protein synthesis also affect the number of cisternae, which may change with time in such a way as to suggest differential effects upon events at the forming and the maturing faces of the dictyosomes.[224, 225, 226] In these and several other observations on alterations of the activity of the Golgi apparatus in plants, it has been found that the dictyosomes react in unison as a synchronous population. They behave as if they are functionally integrated.

MEMBRANE TRANSPORT In making use of the advantages offered by vesicles as transportable compartments, the cell has to cope with the inevitable consequence, that delivery of membrane accompanies delivery of contents. Although this might be considered unimportant, it is a phenomenon that assumes very considerable significance when the rate of membrane output is taken into account.

Scale-producing algae provide a convenient introduction. In *Pleurochrysis scherffelii* the single dictyosome averages, during periods of cell wall scale formation, 32 cisternae. Each cisterna makes a scale, and from the dimensions of individual scales and of the surface area of the cell, it is calculated that 120 cisternae are needed to cover the cell with a single layer of scales. Knowing how long it takes the cell to complete its wall, which averages 22 scales in thickness, the data imply that one cisterna fuses with the plasma membrane every 1–2 minutes; that each cisterna takes from half to one hour to pass through the dictyosome from forming to maturing face and on to the plasma membrane; and that a total of about 40–80 complete renewals of the entire dictyosome are needed in all. The rate of output of membrane can also be estimated. If 120 scales just cover the surface of the cell, and are delivered in the course of 120–240 minutes, it follows that at least twice this surface area of membrane reaches the plasma membrane in the same time, for each scale is completely wrapped, *top and bottom,* by its cisternal membrane. An area of membrane equal to the area of the plasma membrane itself is therefore delivered every 1–2 hours.[93]

Research on other botanical situations testifies that this startlingly rapid rate of membrane flow is by no means an isolated phenomenon. Data already given concerning mucilage glands of the insectivorous plant *Drosophyllum* (see preceding section) are even more striking.[736, 746a] Knowing that 1.3 μm^3 of mucilage is delivered for every 1.0 μm^2 of gland cell outer surface per minute, and that the secretory Golgi vesicles are about 0.15 μm in radius (a very rough approximation), it is possible to calculate the rate of putput of membrane. The volume of each vesicle is 0.014 μm^3, so that nearly 100 are needed to deliver the total 1.3 μm^3 per minute. The surface area of each one is 0.27 μm^2, and so the total area of membrane reaching each 1.0 μm^2 of gland outer surface per minute is of the order of 27 μm^2. In this case a distinction has to be made between 'cell surface' and plasma membrane, because these cells are of the transfer cell type, in which wall ingrowths (page 30) amplify the surface area of the plasma membrane. If each 1.0 μm^2 of cell surface has, say, 10 μm^2 of plasma membrane underlying it, the final estimate for these cells is that when they are actively secreting, an area of membrane equal to the area of the plasma membrane itself is delivered to the plasma membrane in as short a time as 20 seconds.

Another example of very rapid output of membrane is seen in the water glands of *Monarda*[329] (page 79). Possessing about 850 dictyosomes, each with eight cisternae (on average), each gland cell exudes fluid at a rate that implies a throughput of about one cisterna per minute per dictyosome. When calculated in terms of area of membrane, this is equivalent to nearly 2000 μm^2 per minute, that is, an area of membrane equal to the area of the plasma membrane itself is delivered to the plasma membrane in about 30 seconds.

Comparable work on marginal cells of root caps yields a lower, but still impressive, rate of membrane flow. There it is estimated that the dictyosomes (up to 800 or so per cell) contribute 14–26 μm^2 of membrane per minute to the plasma membrane, this in cells with a plasma membrane surface of the order of 1000 μm^2. A new cisterna is added to each dictyosome every 20–40 minutes.[564]

The plasma membrane of growing pollen tubes increases in surface area to keep pace with elongation of the tube, contrasting markedly with the *Pleurochrysis,* the *Drosophyllum* and *Monarda* glands, and the mature root cap cells, none of which display a comparable degree of enlargement while the Golgi apparatus is active. About 2000 Golgi vesicles are produced per minute in a lily pollen tube, and this delivers not only the requisite amount of pectin-type contents to the wall, but also the requisite amount of membrane to account for plasma membrane extension—a beautifully balanced situation.[846]

Although the available information is scanty, it would seem reasonable to generalize that where the Golgi apparatus is active in non-growing or slowly-growing cells, the area of the plasma membrane is augmented far in excess of its requirements. It is only where the growth in surface area of the cell is unusually rapid that this imbalance does not apply.

RECYCLING OF MEMBRANE The most puzzling feature about the above generalization is that nowhere in the literature of botanical electron microscopy can a plasma membrane be found which looks as if it is having its surface area augmented at such incredible rates. It follows that an efficient retrieval system must be operating, taking membrane back into the cytoplasm. From experiments on the rate of 'turnover' of the constituent molecules of animal cell membranes, it is considered that the retrieval system involves withdrawal of molecules or parts of molecules, rather than of fragments of intact plasma membrane.[367] Another suggestion is that coated vesicles function in retrieving portions of membrane (see, e.g., Plate 28a). Since material that leaves the Golgi assembly line must have entered it in some form, it is logical to suggest that the retrieved molecules are re-cycled to the forming face of the dictyosomes. Evidence that the endoplasmic reticulum participates in the events at the forming face by producing transitional vesicles has already been presented, and some quantitative data are now appropriate. The transitional vesicles are about 50 nm in diameter (Plates 25a, b, 26g, 27a, b) and on the basis of their surface area about 200 of them would be required to make one cisterna 1 μm in diameter *if* there is no other source of membrane.

Finally, it appears that there may be at least one 'short cut' in the retrieval and re-cycling of membrane components.[564] It has been noted in maize root cap dictyosomes that formation of vesicles consumes only the peripheral part of each maturing cisterna (Plate 28b, compare the dissimilar situation in *Phleum* root caps—Plate 28a, and the similar situation in an algal dictyosome—Plate 25b). The fate of the central region, which in maize root cap cells is approximately equal in area to the membrane that bounds the Golgi vesicles themselves, is uncertain. It is likely that it too is dissipated into sub-microscopic fragments and re-cycled, without having travelled to the plasma membrane. It should also be remembered that, whatever route is followed, there is much more to recycling than the journey from one place to another in the cell. It will be recalled that the membrane that enters the Golgi apparatus is not the same as the membrane that emerges from it. Recycling must involve adjustments at the biochemical level as well as movements in space.

7.8 Regulation of the Golgi Apparatus

The types of control that are placed upon the Golgi apparatus must be the same as for any other assembly line. They regulate the nature of the finished product, or products, the quantities that are manufactured, the time of commencement, and the speed and duration of assembly.

GENETIC CONTROL The nature of the finished product(s) is specified ultimately by the genetic constitution of the organism. Scale-producing flagellates introduce this concept in a most direct fashion by enabling us to see that highly distinctive structures can be made in Golgi cisternae, characteristic not only of the genus and species, but of closely related forms within the species (Plates 26b and g compared with 26e and h). There are other equally clear examples of the expression of genetic specificity by the Golgi apparatus, but biochemical or other criteria have to be used to detect them. For example, when hemicelluloses are obtained from a range of higher plants, analyses show that the types and relative proportions of sugars present in the molecules, and their three-dimensional construction, to some extent vary, for example, between conifers and flowering plants.[592] Very much better examples of highly specific substances formed in the Golgi apparatus are known for animal cells, where the glycoprotein cell coat, or glycocalyx, that lies on the outer surface of the plasma membrane is a Golgi product, and embodies 'recognition systems' as specific as the antigens of red blood cell surfaces.[886]

Despite these examples of the ability of the Golgi apparatus to produce specific materials, it is not to be regarded as a 'zero-defect' assembly line, of the type aspired to in the space and computer industries, which produces perfectly uniform products. The Golgi apparatus is remote from the ultimate source of precise information in the cell—the genetic information encoded in the form of DNA. Whilst it is possible for nucleic acids directly to control protein synthesis so that all representatives of a particular type of molecule are identical, it is apparently impossible for the Golgi apparatus to make extremely complex macromolecules with comparable precision. The apparatus is composed only of proteins (including enzymes) and a host of other products of enzyme activity. Proteins are *primary gene products:* and, as for example in the Golgi apparatus, they in turn can make *secondary gene products.* Where the latter are complex they tend to show imperfections, that is, to be heterogeneous, each type having an *average* composition about which the individual molecules are distributed. Thus when any one type of flagellate scale is examined, it is seen that the the individuals are not all identical in size, but that they vary about a mean, or average, condition. Similarly, if populations of hemicellulose molecules are taken from one particular source, it is found that the individual molecules vary about an average composition, size, and shape.[592] The same applies to the polysaccharide moiety of animal glycoproteins.[285]

Possibly the fundamental cause of most of the heterogeneity lies in the fact that if presented with a variety of very closely related substrates, an enzyme molecule cannot infallibly reject all but one particular type. In the case of enzymes in the Golgi apparatus, closely related sugars and polymers of sugars must be present during phases of polysaccharide synthesis, so that mistakes in selection of substrate are possible. No doubt there are other contributory causes of heterogeneity of Golgi products; the nature of the products might be changing with time, as during alteration of the matrix composition of a growing cell wall; successive cisternae might through random errors at the forming face of the dictyosomes become equipped with slightly different sets of enzymes.

TEMPORAL CONTROL Amongst cells of plants and animals there are instances where the Golgi apparatus is programmed to synthesize the same product for long periods in the life of the cell. In other cases, and particularly during the development of cells, the apparatus is more versatile, perhaps producing a mixture of compounds, perhaps changing its activities with time. The quantity as well as the quality of the products may alter. Any attempt to picture how such matters might be regulated has to take account of the dynamic processes involved in the synthesis of new cisternae, their progress through the dictyosome to their destination, and the re-cycling of components back to the forming face.

It is not known whether the enzymes that specify and direct the synthesis of the products in the cisternae survive the re-cycling process or whether they have to be made anew after their short active life of about an hour or so during passage through the dictyosome. Probably at least some re-synthesis is necessary, for in the presence of inhibitors of protein synthesis the number of cisternae per dictyosome, and possibly the number of dictyosomes per cell, decreases.[224, 225] Also, if new products are to be made, new enzymes will have to be injected into the system, not just at random, but so that the new cisternae are provided with qualitatively and quantitatively appropriate mixtures. For example, it has been suggested that alteration of the amount or activity of enzymes called epimerases could direct a switch from pectin production to hemicellulose production by the dictyosomes of developing plant cells.[592]

A switch from pectin production to hemicellulose production would involve considerable re-programming of the system that assembles Golgi cisternae. It is not, however, so spectacular a switch as is seen in the scaly flagellates, where *successive cisternae* may possess different templates (page 73). Neither is switching from one scale template to another as remarkable as switching from production of lysosomal enzymes to production of cell wall polysaccharides, yet there is some evidence that not only can one cell be producing both of these very dissimilar materials at the same time, but that the capability for producing both may reside within a single dictyosome.[145, 146] For instance, the development of xylem elements involves production of digestive enzymes which are released when the tonoplast is ruptured during cell maturation, leading to breakdown of the protoplasm and unprotected areas of primary cell wall (page 19). It is most probable that some wall materials are synthesized at the same time as are the enzymes. Straightforward examination of developing cells shows that individual dictyosomes can be producing two morphologically distinguishable types of vesicle, one of which appears to recognize the tonoplast, and the other (presumably) the plasma membrane.[644]

The existence of differences between successive cisternae, leading to the formation of individual products, implies that the mechanisms controlling the Golgi apparatus must be able to respond rapidly enough to endow each new cisterna with specific properties. It is not known whether the specificity is somehow inserted during the short period required to make a cisterna, or whether it arises at a later stage in their one or two hour lifetime. Individuality of cisternae must be especially significant where there is only one dictyosome in the cell, but where there are many, as in higher plant cells, the possibility exists that different dictyosomes, rather than different cisternae, perform different functions. The evidence that higher plant dictyosomes can produce more than one type of vesicle suggests that the individuality of cisternae can be retained despite multiplication of dictyosomes, but on the other hand, localized structural specialization of entire dictyosomes within single cells has been seen.[653] Perhaps the Golgi apparatus should be regarded as having capabilities similar to those of the endoplasmic reticulum—able to become regionally differentiated within a cell, to the extent that either single dictyosomes or even single cisternae become the ultimate units of function, and in addition able to change with time, either slowly (so that entire dictyosomes alter their productivity) or rapidly (so that successive cisternae differ).

No matter whether one or several types of product are being made, by one or many dictyosomes, the cell can regulate the rate and duration of production. The time squence of Golgi activity may be part of an internal programme governing cell differentiation, as in marginal cells of the root cap, or it may be geared to external stimuli, as when glands of insectivorous plants respond to the arrival of their prey.

It should be clear that research on the subject of regulation of the Golgi apparatus has reached the exciting stage in which some of the outstanding problems can be formulated, but not many solutions proposed.

7.9 Origin of the Golgi Apparatus

EVOLUTION Like other internal membrane systems, the Golgi apparatus is absent from prokaryotic cells. Obviously there is no such thing as a fossil record to show when and where in the evolution of eukaryotes it made its first appearance. All that can be done in respect of the question of evolutionary origin is to examine present day organisms to see if they contain any clues to the past history.

In fact, the gleanings are meagre. No eukaryote has been proved to lack a Golgi apparatus, though in some resting stages (cells in dry seeds) or generative cells (animal oocytes) it may temporarily be reduced to a small mass of vesicles and fibrillar material.[574] There is a yeast in which it is normally absent, but appears if isolated protoplasts are prepared and induced to manufacture a cell wall.[322, 773] In some fungi the apparatus takes a simple form that could conceivably be regarded as primitive. It is more realistic, however, merely to describe it as a 'minimum function unit', in which the phenomenon of stacking is not seen, *single* cisternae apparently being capable of processing and packaging products.[574] The fact that cisternae *can* operate on their own poses the question of what special virtues might have been conferred by the evolutionary development of intercisternal 'cement' and the adoption of the stacked condition that is so widespread in the eukaryotes—greater concentration of particular processes in particular parts of the cell?—provision for intercisternal exchanges and re-cycling of membrane components?

In general terms the evolution of the Golgi apparatus may have been an essential part of the development of cells possessing internal compartments. If different compartments are to survive, their membranes must not fuse with one another. Nevertheless, communication and the movement of supplies between compartments is necessary, and it is here that the Golgi apparatus can play a part. Generalizing, it carries and processes materials in membrane-bound packages; operates in one direction only by converting an essentially internal type of membrane (that of the endoplasmic reticulum) to essentially external types (plasma membrane, lysosome, or tonoplast): in short, it is an assembly and communication system that preserves materials from being dissipated into the general cytoplasm, and carries them between compartments that do not normally establish direct continuity.[574]

DEVELOPMENT Turning from the evolutionary origin of the Golgi apparatus to its developmental origin within the life-span of cells or organisms, the three possibilities that might in theory apply to any membranous component of the cell have in fact all been invoked at various times by various authors. Some consider that the apparatus arises *de novo*, independent of any pre-existing supra-molecular structure; some that it is always present, perpetuated by multiplication of dictyosomes or cisternae so that every new cell receives its initial quota; and some that it is derived from other membrane systems in the cell. The first of these three receives the least support from modern research, and the last, the most.

In unicellular organisms with one dictyosome per cell, and in meristematic plant cells, the multiplication of dictyosomes keeps pace with and probably occurs at about the same time as cell division. Thus the unicells always have one per cell, and the number per meristematic cell is likewise much the same before and after division.[12] Contrasting markedly with this constancy, the number of dictyosomes may alter during cell differentiation. The previously mentioned ten-fold increase seen in root cap cells is a good example. It must be remembered that (as far as is known) a supply of transitional vesicles from the endoplasmic reticulum or the nuclear envelope is a pre-requisite for formation of new cisternae. In at least some cases it can be seen that transitional vesicles arise from discrete areas of the reticulum (Plates 27a, b), and thus any increase in the number of dictyosomes is likely to be a result of an increase in the number of these formative areas. Regions specializing in production of transitional vesicles could differentiate from previously unspecialized zones of endoplasmic reticulum or nuclear envelope. Alternatively, the zones could multiply by a process in which existing formative zones extend until a critical size is reached, whereupon two cisternae are generated side by side where formerly there was only one, eventually giving rise to two dictyosomes instead of one. Intermediate stages in this postulated time sequence have been seen,[574, 790] in which the mature part of the dictyosome consists of a *single* stack, surmounting *two* less mature regions (Plate 25a, 27b).

Unfortunately, since many higher plant dictyosomes do not clearly display a close relationship between the endoplasmic reticulum and the forming face, it cannot be assumed that formative zones on the reticulum are always discrete. It is possible, though evidence is lacking, that production of transitional vesicles might in some cases occur diffusely over large expanses, with the forming faces of the dictyosomes being mere foci for coalescence to form cisternae, the number of such foci being determined by some unknown mechanism or perhaps simply by chance.

The control of dictyosome formation can, however, be traced back to a more fundamental region of the cell than the endoplasmic reticulum. Inhibitors of protein synthesis inhibit the formation of Golgi cisternae, and so do substances which inhibit the translation of genetic information.[224] The interpretation is that to make cisternal membranes, proteins are needed,[573a, 744, 880] and gene activation is required to make the proteins. Elegantly direct experiments on this topic have been performed using *Amoeba* and the alga *Acetabularia*, where the cells are large enough for micro-surgery to be possible. If the nucleus of an *Amoeba* is removed, the Golgi apparatus degenerates and becomes unrecognizable after two days.[222, 905] Even after as long as five days without a nucleus, there is a dramatic response to the insertion of a replacement nucleus.[223] Golgi cisternae appear within 30 minutes, and the number of dictyosomes increases during the hours and days thereafter. Comparable experiments with inhibitors have been performed on plant cells, and it is to be presumed that Golgi apparatus of plants is as much under the ultimate control of the nucleus as is that of *Amoeba*. For example, degeneration of the apparatus in enucleate portions of plant cell cytoplasm could well account for the observation (page 17) that such portions fail to make cell wall material.

As in the case of the endoplasmic reticulum, it is necessary before closing this chapter to emphasize that the Golgi apparatus cannot adequately be considered in isolation. Its relationships to other components of the cell will be examined again in chapter 13, where the cell is viewed as an integrated system.

8 Mitochondria

8.1 Introduction

Living organisms need energy for their growth and other synthetic processes; for active uptake or secretion by solute pumps; for movements within cells; for movements of tissues, organs and the whole body; and for maintenance of their structure, which in the absence of a usable energy supply would break down into a disordered and non-living state. In short, the processes of life require a supply of energy—a necessity reflected in the attempt made earlier (page 14) to define the cellular state.

The ultimate source of energy for both plants and animals is light, absorbed by plant pigments, and converted by photosynthesis from electromagnetic form to chemical forms that are transported within organisms and along food chains throughout the biological world.

Three broad categories of biochemical 'energy store' exist. Adenosine triphosphate (ATP) is an example of a type of molecule which is accepted by numerous enzymes, each capable of transferring a portion of its energy-content to acceptor molecules, thereby driving energy-requiring syntheses. Relatively unstable compounds like ATP may be classed as *primary energy stores* because their energy is directly available, and because they are not accumulated. Rather they are produced and utilized as required, a fact highlighted by the finding that the amount of ATP normally present in a growing bacterium is sufficient to maintain the cell's vital activities for only a fraction of a second. Continued growth requires continued production of primary energy from *secondary energy stores,* which consist of compounds that are stable and soluble enough to be moved around the organism. Translocation of energy in the form of sucrose in the phloem of a plant is an example. The third category is the *tertiary energy store,* comprising stable compounds accumulated, stored, and utilized over long time periods. They are usually very stable, macromolecular, and relatively insoluble (e.g., starch, glycogen, fats), their biological advantage being that bulk long-term storage is feasible, since their insolubility means that they do not drastically affect the water potential of the cell contents.

The metabolic systems of the organism are based on the interconversions of the three types of store. *Anabolic* reactions transfer the energy into more and more stable molecules, and *catabolic* reactions progressively release energy. Whereas utilization of primary energy proceeds in most or all parts of the cell, the bulk of its production is entrusted to specific subcellular components. The major source of primary energy is the stepwise oxidation of a secondary store such as glucose or a fatty acid, achieved by sequential enzyme-controlled reactions, some of which release enough energy to drive the synthesis of ATP from two precursors of lower energy content, namely adenosine diphosphate (ADP) and inorganic phosphate. The enzyme systems which in this way *couple* the oxidation of a substrate to the conservation of energy by *phosphorylation* of ADP are associated with cell membranes. In prokaryotes the plasma membrane, or invaginated regions of it, is involved, and in eukaryotes the equivalent 'powerhouse' for the cell is the population of mitochondria.[476]

The name mitochondrion, which indicates that the structures can be thread-like or granular, was coined by Benda in 1898, but the objects themselves had been seen much earlier. Kolliker had released them from muscle cells and demonstrated that they possess a bounding semi-permeable membrane by showing that they swell when placed in water. Staining procedures which enhanced their visibility in light microscope preparations had been developed by Altmann, who speculated that they are some sort of elementary living particle (an idea which will be taken up in chapter 13). That they participate in oxidation-reduction reactions was indicated in the same year that their name was coined, when Michaelis found that mitochondria in living cells change the colour of the dye Janus green B.[467] From this beginning it was 14–15 years before Kingsbury actually speculated upon oxidative functions for mitochondria, and Warburg obtained particulate material from liver cells that was active in some of the oxidative reactions of respiration.

The first electron microscope picture of plant mitochondria was obtained in 1947,[95] and the modern era of collaboration between biochemists and electron microscopists opened with the development of techniques for the isolation of mitochondria. It then became possible to prove that they contain not only oxidative systems, but means of coupling oxidative energy release to phosphorylation, i.e. that they carry out *oxidative phosphorylation.* So much progress has been made in elucidating the structural basis of these reactions that it may be predicted that the mitochondrion will in the future become the first major cell component to be defined in terms of not just ultrastructure and enzymology, but also molecular architecture.[609a]

8.2 Gross Features of Mitochondria

8.2.1 *Morphology*

Observation of living cells establishes both the diversity of form and the dynamic characteristics of mitochondria. The plant hair shown in Plate 43g–i illustrates a rod-shaped form, about 0.5 μm in diameter and up to 6 μm in length. Using phase-contrast optics they can be seen to be swept around the cell in the streaming cytoplasm, changing their shape, bending, becoming globular, thread-like, or branched, sometimes fusing with one another or splitting up into portions.

In cells of the flagellate *Euglena*[109, 611] and the alga *Chlamydomonas*[612] the form of the mitochondria changes greatly if the conditions of growth are altered, and also during the progress of the cycle of growth and cell division, even under uniform conditions. Relatively short, branched or unbranched rods come together at times to generate a mitochondrial reticulum, in which most, if not all, of the previously separate parts become continuous in three dimensions. This type of morphology is difficult to detect in ultra-thin sections, which are too thin to reveal the continuity, and instead create an impression that a population of small and discrete units exists in the cell. Plate 30b and c demonstrate how fallacious this impression can be in the case of the unicellular alga *Chlorella,* where three-dimensional reconstructions are required in order to visualize the shape of what turns out to be a single reticular mitochondrion in the cell.[19] There seems to be no reason why this morphology, which has also been found in yeast,[366] should not be common amongst small cells with sluggish cytoplasmic streaming.

In higher plant cells the most common shape assumed by mitochondria is that of a short rod with hemispherical ends. Plate 29a shows many profiles of mitochondria in a nectary cell, and assuming that the individuals were randomly oriented with respect to the plane of the section, it can be surmized that they are short rods, ellipsoids, or spheres. Longer rods are seen in Plate 29b. Very few studies employing serial sectioning have been undertaken, but the fact that branched mitochondria are only occasionally detected in individual sections (e.g. Plate 29a) suggests that formation of extensive systems interconnected in three dimensions is rare. The fixative glutaraldehyde, which is in general use, does have the disadvantage that plant mitochondria tend to swell and become more rounded than in life. This is probably the case in Plate 13b. It is thus necessary to be aware of the danger that elongated or branched mitochondria may round-off or even fragment during fixation.

Other shapes are not uncommon. A curved bowl such as would be obtained if a thumb were to be pressed deep into a 3 cm diameter sphere of modelling clay has been described for many types of plant cell. Obviously, the appearance of this configuration depends upon the orientation of the section. Plate 30d shows a section that probably has cut through the complete rim of a bowl. Contrary to first impressions there is no need to suggest that the central area of cytoplasm is 'inside' a hollow mitochondrion. Plate 30a illustrates a much less common morphology, where the mitochondrion appears to be a very extensive plate undulating in and out of the plane of the section.

8.2.2 *Numbers and Volume*

The tiny flagellate *Micromonas* (= *Chromulina*) has only one mitochondrion per cell,[515, 528] and as already mentioned, the same applies to the complex mitochondrial reticulum of the alga *Chlorella* and the yeast *Saccharomyces.* Another unicellular alga, *Chlamydomonas,* has been shown by serial sectioning techniques to have a small number (10–15) of highly lobed or branched mitochondria.[14, 754] The number of mitochondria per cell in higher plants is more likely to be in the hundreds or even thousands, depending on the size and type of cell. Changes in mitochondrial number often accompany cell differentiation. A well documented example is that of the central cells of the maize *(Zea)* root cap, which have about 200 when young, and between 2000 and 3000 when enlarged and mature. The number per unit volume of cytoplasm does not, however, alter greatly.[129] Generalizing, the mitochondria in very active cells like secretory cells in nectaries (Plate 29a), transfer cells (Plate 13b) and companion cells in phloem (Plate 7), occupy a large volume of the cytoplasm. In this respect transfer cells can match animal cells such as those in liver, up to 20% of the cytoplasm being mitochondrial material. In one very rapidly growing tissue, the anther stalks of grasses, the proportion may be even higher.[473]

8.2.3 *Location in the Cell*

Cytoplasmic streaming in large cells distributes mitochondria randomly throughout the cytoplasm, and seldom, if ever, are mitochondria able to establish a specific spatial relationship with any other cell component in such cells. It is to be presumed that streaming enhances the rate at which raw materials can reach the mitochondria, and that rate at which products (e.g. ATP) leave.

Observations of animal cells show that pronounced regional specialization is common, and that one aspect of it concerns the deployment of the mitochondria, apparently in relation to localized energy requirements.[207] They may lie closely juxtaposed to the plasma membrane, as for instance in kidney cells, where a massive energy requirement arises in connection with the transport of solutes at the cell surface. The closest analogy to this in plants is the transfer cell (Plate 13b), whose cytoplasm is richly endowed with mitochondria, though it is doubtful whether any long-lasting relationship exists between them and the plasma membrane. Another analogy is seen in the flagellate *Furcilla,* which develops a zone of mitochondria just inside its plasma membrane when it is forced to live by absorbing organic nutrients rather than by photosynthesis.[44a] A second close association is that between mitochondria and the contractile protein filaments of muscle cells, and for this certain flagellates again provide an anology, mitochondria in some cases being positioned alongside the rootlets of the flagella.

A few other associations have been described, but still await interpretation. Mitochondria often lie extremely close to chloroplasts, the outer membranes of the two structures being separated by as little as a few nanometres. This again seems to be an impermanent juxtaposition, and cinematography has been used to show mitochondria moving to and peeling off the outer surface of chloroplasts in leaf cells.[892a] There is no compelling evidence to suggest that the two fuse: they merely lie so close as to imply that a functional relationship is likely, perhaps in the exchange of solutes from the one to the other. Comparable relationships between smooth endoplasmic reticulum and mitochondria were mentioned on page 65, and illustrated in Plate 22a and b.

Interpreting the above observations in general terms, it is likely that the main feature influenced by the location of mitochondria in the cell is the time taken by raw materials to reach them, and conversely the time taken for them to supply primary energy to sites of energy-utilization. Mitochondria may be moved around the cell, thus favouring all regions equally. In more static cells branches of mitochondrial reticulum may penetrate all regions of the cytoplasm (Plate 30c), or spatial associations may develop in which solute transfer is facilitated by minimizing the separation of, and maximizing the area of contact between, mitochondria and donor or receptor structures.

8.3 Ultrastructure of Mitochondria

Early observations of mitochondria in ultra-thin sections revealed the two membranes and two membrane-bounded compartments that subsequently have been found to be of general occurrence (e.g., Plate 29c). An outer membrane surrounds and completely encloses an inner membrane. The latter is invaginated into the matrix and encloses a central compartment containing a variety of mitochondrial matrix components. The invaginations were originally called 'crests', and though this is not always appropriate for their shape, the term 'cristae' is now in general use, having survived from the latin 'cristae mitochondriales', used the first time they were described in detail.[622] There is a narrow space between the inner and outer membranes, and though the whole of this inter-membrane compartment is a continuum, the part that lies within the invaginations of the inner membrane is sometimes given special attention, and called the intra-cristal (or cristael) space.

8.3.1 Conformation of the Mitochondrial Membranes

Conventional staining procedures show the outer membrane as having a smooth surface, though a particulate outer coat, which may or may not be of general occurrence, has been seen in fungal material by treating sections with barium permanganate.[72] Unless phenolic compounds are released and deposited on it during fixation (as in Plate 29c), the outer membrane is generally only weakly stained—a feature that is sometimes useful in distinguishing between mitochondria and very simple plastids (Plates 2, 31a). Outwardly directed vesicles or even small cisternae have been seen on the outer membrane of both plant and animal mitochondria,[73, 236] and it is not yet clear whether these represent connections with smooth elements of the endoplasmic reticulum, perhaps established during brief time periods only, or whether the mitochondrion is delivering membrane to the cytoplasm. It is known that in test tube experiments mitochondria and endoplasmic reticulum exchange several of the phospholipid components of their membranes.[907]

The inner membrane is much more variable in its morphology, though one consistent feature of plant material is that the cristae are quite unlike the 'crests' of the original description, where they were regarded as 'baffles' or partial septa, sometimes extending right across the matrix. Plates 29a–c show that despite the fact that cristae may be abundant, it is a comparative rarity to see a clear connection between the intra-cristal space and the rest of the inter-membrane compartment. The inference is that the invaginations must join to the rest of the inner membrane only at narrow necks. Three-dimensional reconstructions based on serial sections have confirmed this to be the case, and also that the membrane surface opens out from the neck region to form cristae ranging in shape from simple sacs to complex infolded pitchers.[169] An extensive crista is seen in Plate 30d, passing around the circular profile of the mitochondrion, concentric to the outer membrane. Several other bizarre configurations have been described. Simple tubular cristae, found in several groups of flagellates (Plate 26g, h), steroid-secreting cells in animals, and Protozoans, are rare in higher plants.

The ratio of surface area of inner membrane to volume of matrix is the most important variable to be considered when comparing cells of different developmental stages, of different tissues, or of different organisms. It may change greatly even though the overall size of the mitochondria does not, as in cells of the spadix of *Arum* flowers, where the number of cristae per mitochondrion rises as the organ develops and as its respiration rate increases.[768] Quantitative details are available for certain animal tissues, for example liver mitochondria have 30 μm^2 inner membrane per μm^3 of mitochondrial matrix, and heart mitochondria 90 μm^2 per μm^3.[786] In liver the area of the inner membrane is about nine times that of the outer, and the total area of inner mitochondrial membrane per cell is close to that of the endoplasmic reticulum (page 59). Although quantitative data are lacking, it is nevertheless obvious that mitochondria in certain plant cells (e.g., transfer cells, Plate 29b; secretory cells, Plate 29a) have many more cristae than do those in certain others (e.g., young endodermis, Plate 29c; developing pollen, Plate 30e; leaf mesophyll, Plate 41a). It is thought (but see page 94) that these structural differences relate to differences in respiration rate and hence in the capacity for ATP production.

8.3.2 *The Mitochondrial Matrix*

Four components are normally present in the interior of the mitochondrion. Fine granules constitute a ground substance in which fibrillar areas and two sorts of larger particles lie. Both particulate components are more electron-dense than the matrix. One, about 15 nm in diameter, is a form of ribosome, considerably smaller than the ribosomes in the cytoplasm outside the mitochondrion (Plates 29c, 30a (insert), 30e), but like them, containing RNA and protein, and able to mediate protein synthesis.[66]

NUCLEOIDS The fibrils in the fibrillar areas (Plates 29a, b) contain DNA, and the enzyme deoxyribonuclease can be used to digest and so to identify them. They lie in fibrillar *nucleoids* (page 123 describes the origin of this term) which are electron-transparent zones of unknown composition. Their appearance depends greatly on the composition of the fixative used for preparing the material. In the absence of calcium ions they are coarse, but if calcium is present they are very narrow, some 2–3 nm in thickness. The same applies to the fixation of histone-free DNA.[755] Unlike nuclear chromatin, no histone protein is found in association with mitochondrial DNA. Serial sectioning has shown that large mitochondria can have several nucleoid areas,[579] an observation that correlates with biochemical evidence that mitochondria can contain multiple copies of their DNA. Multiple nucleoids are, however, not always present, thus mitochondria in *Beta* leaves regularly contain just one.[443] As in bacteria, the DNA strands are much folded, and are usually in the form of closed circles, a circumference of 5–6 μm being common in a wide range of organisms, though larger circles have been recorded.[580] It is worth recalling Altmann's remarkable speculations that mitochondria are simple units of life (page 87) in relation to these recent discoveries concerning mitochondrial DNA and ribosomes. The subject of mitochondrial genetics and protein synthesis is discussed later (chapter 13).

CALCIUM-CONTAINING GRANULES The remaining component is the second type of particle found in the matrix. It is more dense to electrons than other parts of the mitochondrion (Plate 29a) and varies in size and frequency. Experiments on animal material have proved that the number of dense granules increases if the tissue (or isolated mitochondria) is incubated in a medium containing both ATP and calcium ions.[637] The calcium is absorbed at the expense of energy derived from the ATP, and is deposited in granular form in the mitochondrial matrix. Isolated granules contain both inorganic and organic components, though some of the latter may be due to contamination arising during isolation procedures. The former resembles the minerals hydroxyapatite and whitlockite, with calcium phosphates as major, and magnesium and carbonate as minor, constituents.[879]

Plant mitochondrial granules are less well studied, but probably are like their counterparts in animals. Mitochondria from beans *(Phaseolus)* accumulate strontium (which is acceptable to calcium-transport systems), with concomitant formation of many dense granules.[683] Mitochondria from some plant tissues exhibit the same kinetics of calcium uptake as do animal mitochondria, but the similarity is not universal amongst plant tissues.[118] It is thought[477] that by accumulating calcium, mitochondria in animal cells may influence the activity of various enzymes and transport systems, and the structure of microfilaments and microtubules, all of which are sensitive to the concentration of calcium ions in the cytoplasm. It is not known whether comparable regulatory functions apply to plant cells.

Other structures are occasionally seen in the matrix of plant mitochondria. Large crystals, like those developed in mitochondria of certain animal eggs when yolk proteins are being accumulated, occur sporadically.[470] In one case they have been shown by enzyme digestion to be proteinaceous,[461] but the nature of the protein is unknown.

8.4 Molecular Architecture and Function in Mitochondria

There are of the order of 100 types of protein in mitochondria, some of them non-enzymatic, but the majority with catalytic functions—in oxidizing organic acids derived from carbohydrates and proteins, and fatty acids derived from lipids; in coupling these oxidations to ATP production; in transporting solutes across the mitochondrial membranes; and in the synthesis of lipids and proteins. Many techniques are being used in efforts to elucidate the spatial organization of proteins and lipids in the two compartments and the two membranes of the mitochondrion.

Mitochondria that are suspended in a dilute medium swell as water diffuses into the mitochondrial matrix down its potential gradient. The inner membrane has the greater surface area and so can accommodate more readily to swelling than can the outer membrane, which eventually ruptures. By washing the mitochondria at this stage and centrifuging at different speeds, fragments of outer membrane can be separated from the intact inner membranes. Components released from the intermembrane compartment can also be collected. Matrix materials can in turn be obtained by dissolving or breaking the inner membrane. Components that are incorporated in the membranes or attached to their surfaces can be examined both when in or on the membranes and after isolation and purification.

The above biochemical separations can be related to more detailed aspects of ultrastructure by supplementing the conventional method of ultra-thin sectioning with the freeze etching and negative staining techniques. These give some insight into the substructure of the interior and the surfaces of the mitochondrial membranes, and the results can to some extent be correlated with biochemical observations on the distribution of functional molecules. In intact preparations, the inner membranes are oriented as they are in the cell, that is, with their inner face in contact with the matrix. By contrast, ultrasonic vibration produces vesicles with reversed polarity, for example, cristae that break off and heal at the neck region, and are then released so that the surface previously in contact with the matrix becomes exposed to the medium used to suspend the original preparation. The advantage of being able to obtain suspensions of normal and inside-out membrane bound sacs is that *non-penetrating* reagents (inhibitors of enzymes, antibodies, protein-digesting enzymes, etc.) can be applied to each, thereby giving information about the position in which components lie within the thickness of the membrane.

Some of the information that use of the above (and related) procedures has given will now be summarized.

8.4.1 The Mitochondrial Matrix

Oxidation of the major fuels used by the mitochondrion in primary energy production commences in the matrix. Pyruvic acid derived largely from carbohydrate, fatty acids derived from fats, and amino acids derived from proteins, are all converted to a common denominator, the acetyl residue, with its two carbon atoms. This enters the sequence of reactions known either as the tricarboxylic acid cycle or the Krebs cycle, which effectively separates its carbon, oxygen and hydrogen atoms. While carbon and oxygen are released as carbon dioxide, the hydrogen is transferred to hydrogen-carrying coenzymes, thereby conserving a large proportion of the energy of the original fuel.

With one exception (succinic dehydrogenase) the enzymes of the Krebs cycle are released from mitochondria when the membranes are ruptured. The exception remains attached to the inner membrane, and the rest are dissipated with the matrix. Despite the ease with which they dissolve, there are grounds for suspecting that the Krebs cycle enzymes may be structurally organized in the matrix. When the enzymes extracted from the mitochondrial matrix are assayed as individuals, they can be placed in a series from most active to least active. The significant point that arises from comparisons of different animal tissues is that the order, and the ratio of the enzymic activities, is consistent enough to indicate that the enzymes may be present in fixed relative proportions. One possible explanation is that they take the form of a loose multi-enzyme complex of standard molecular composition.[786] Other multi-enzyme complexes definitely do exist, the advantage of having the component enzymes in a set spatial array probably being that the efficiency with which substrate molecules can be processed (especially if they are present in low concentrations) is greater than would be the case if the enzymes were randomly scattered.

It has been suggested further that any Krebs cycle enzyme complexes that exist might be loosely associated with sites on the inner membrane of the mitochondrion. Certainly, one of them, succinic dehydrogenase, is an integral part of the membrane,[186] and at least one of the others correlates in its activity much better with the area of the cristae than with the volume of the matrix.[786] So far, however, there are no visible indications of ordered arrays of the enzymes, either free in the matrix or loosely membrane-bound. Ultra-thin sectioning and freeze etching do, however, reveal a fibrous network, probably of protein, extending from the cristae into the matrix of active (but not inactive) mitochondria.[815, 915]

8.4.2 The Inner Membrane

The major function of the inner mitochondrial membrane is to release in usable amounts the energy contained in the hydrogen-bearing coenzymes generated by the Krebs cycle, and to couple the energy release to ATP production. A great deal is known about the molecular composition and enzymatic activity of the membrane, but controversy has arisen in attempts to visualize its structural sub-units and deduce their biochemical activity.[625]

MEMBRANE-ASSOCIATED COMPONENTS The dark-light-dark triple layering found in so many cell membranes is difficult to see in ultra-thin sections of mitochondria. The inner membrane usually looks smooth in conventionally stained ultra-thin sections, but in mitochondria of some algal cells 'bristles' extend from its surface into the matrix.[243, 651] The technique of negative staining has yielded the most startling images: in this procedure mitochondria are isolated and broken in a solution of an electron-dense compound, for example phosphotungstic acid. When dried, the preparations appear pale with dark stain penetrating into all crevices and cracks on the membrane surface, and it has been found that inner membranes from both plant and animal mitochondria bear stalked particles when negatively-stained.[625]

The particles are roughly spherical, about 9 nm in diameter, and lie on the side of the membrane that faces

the matrix. Their discovery led to much research and speculation concerning their existence, nature, and possible function, and to controversy when it was later found (a) that their frequency varies greatly according to the precise mode of specimen preparation and negative staining, and (b) that their existence could not be confirmed by freeze-etching.

The observed variation in frequency remains a source of worry, but it is now considered that the particles do have physical reality, and that freeze-etching is not a good technique for visualizing them, in view of the fact that the fracture plane probably travels along the interior of the membrane rather than along the surface that carries the particles. With modification of the staining procedures, it has even proved possible to detect the particles in ultra-thin sections.[427, 814] Finally, the particles have been isolated and examined. In fact, after their removal from inner membranes the isolated particles and the particle-depleted membrane will, if mixed together, re-form particle-bearing membrane. Following biochemical investigations, an early suggestion that the particles are complexes of Krebs cycle enzymes has been discounted. They do, however, possess the capacity to synthesize or break down ATP, depending upon the reaction conditions used, in other words they have adenosine triphosphatase activity. *In vivo* their major function is to couple the synthesis of ATP to the release of energy derived from mitochondrial oxidations, hence the name 'coupling factor' is commonly used when referring to them.[625] It will be seen later (page 107) that they have their counterparts in chloroplasts.

INTERNAL COMPONENTS OF THE INNER MEMBRANE At least four other biochemically active lipid-protein complexes have been obtained by sub-dividing inner mitochondrial membranes.[625] One complex contains the enzyme succinic dehydrogenase, the sole Krebs cycle enzyme that is firmly membrane-bound. The second complex accepts hydrogen atoms carried by the coenzyme nicotinamide adenine dinucleotide (NAD) from Krebs cycle enzymes in the matrix. An intermediate compound, ubiquinone, receives hydrogen atoms from both of these complexes, so that from this point on, all of the energy-containing produce of the Krebs cycle is processed by the same route. Each hydrogen atom is dissociated into a proton and an electron, and the electrons are passed from ubiquinone along a chain of protein-molecules called cytochromes, five of which operate in a fixed sequence. The electrons lose energy *en route,* sufficient at three of the steps to be coupled to the phosphorylation of ADP to make ATP. The enzyme cytochrome oxidase lies at the terminus of the cytochrome *electron transport chain.* Under its influence, the protons and electrons are re-combined, and, together with oxygen that has been taken into the mitochondrion, generate molecules of water.

It is very likely that these components of the respiratory system are built-in to the inner mitochondrial membrane as precisely oriented assemblies of molecules. In addition to the two complexes described above, two others (containing cytochromes) can be isolated. All four complexes contain quantities of lipid and other components, and presumably are subfractions of membrane. Significantly, their enzymic properties alter if the lipid is removed, and it seems that correct functioning only occurs when the constituents are structurally integrated in the intact membrane. The various cytochromes are each represented by approximately the same number of molecules, adding support to the idea that they occur in equi-molar assemblies. Whether the two cytochrome-containing complexes are joined to the other two complexes, or whether ubiquinone acts as a mobile link to integrate spatially separated complexes, is not clear. Calculations show that the inner mitochondrial membrane has ample surface area to accommodate the observed number of molecules and complexes, indeed the data available so far suggests that the assemblies might either be widely scattered or else concentrated in specialized regions, for it appears that a total of only about 10% of the total area of inner membrane is occupied by them.[625]

Comparisons of 'normal' and 'inside-out' vesicles derived from the inner membrane, and some very precise experiments on enzyme localization by means of electron microscopy, have added yet more information on the topography of respiratory components.[625] Succinic dehydrogenase lies at the inner surface, in contact with the matrix, and hence with the other Krebs cycle enzymes. The cytochromes are not in a simple linear sequence in the plane of the membrane. One lies at the outer surface in contact with the intracristal space, while others are in the centre of the membrane. Cytochrome oxidase, at the terminal part of the chain, is in contact with the matrix. The 9 nm spherical molecule that is part of the system for coupling electron transport to phosphorylation is also, as already seen, at the matrix face of the inner membrane. Various arrays of particles have been seen within the inner membrane by the use of the freeze-etching technique, but it is doubtful whether they relate to the hydrogen and electron transport complexes. For instance, inner membranes of mitochondria in a mutant yeast that lacks both coupling factor and a number of cytochrome activities look similar in freeze-etched preparations to inner membranes of normal mitochondria, possessing the full range of biochemical activities.[621]

Of the two mitochondrial membranes, the inner is the main permeability barrier. Thus it retains the acids of the Krebs cycle in the matrix. It can also separate small charged ions so that when the mitochondrion is active, the matrix becomes more alkaline than the inter-membrane space—a factor that may in fact be a means of driving the synthesis of ATP. It contains a variety of specific solute transport systems, amongst them the one that is active in calcium uptake (page 90), others that transport organic acids, amino acids, and phosphate, and a vital one which is geared to the energy production of the mitochondrion in that it exports ATP in precise exchange for import of ADP.[477]

The activities of solute transport systems in the inner membrane can alter the water potential of the matrix and bring about shrinkage or swelling of the mitochondrion. In a number of plants and animals mitochondria that are actively respiring and phosphorylating display an 'energized' configuration in which the matrix has shrunk and the inner membrane has pulled inwards, thus enlarging the inter-membrane space as compared with the 'orthodox' configuration.[463, 822, 902]

8.4.3 The Outer Membrane

Although they are always associated with one another and separated by only a narrow gap, the two membranes of the mitochondrion are strikingly dissimilar. The outer membrane has less protein and more lipid than the inner, indeed it has in certain cells been shown to have a higher lipid content than any other intracellular membrane. The occasionally seen connections between cisternae of smooth endoplasmic reticulum and the outer membrane (page 89) point to a measure of similarity between the two membrane systems, and this receives support from detailed examination of their constituents, both lipid and protein. At least one of the rather heterogeneous collection of enzymes found in the outer membrane has very similar properties to one in the endoplasmic reticulum.[625]

Finally, and as already stated, the permeability properties of the inner and outer membranes are very different. The outer membrane is comparatively freely permeated by solutes of low molecular weight, and it is even permeable to a number of proteins.

8.5 Origin, Development and Division of Mitochondria

Mitochondria contain DNA,[580] a property that brings a new dimension to the topic of development in the cytoplasm, one that has not been part of previous discussions of vacuoles, endoplasmic reticulum, and the Golgi apparatus. What is the role of mitochondrial DNA? Is it a set of genes? Do mitochondria, as Altmann suspected, have genetic autonomy? Or do nuclear and mitochondrial DNAs constitute a cooperative and mutually interdependent genetic system? Similar questions arise in connection with plastids, so to avoid treating them twice, interactions of mitochondrial, plastid, and nuclear DNA will be outlined in brief in chapter 13 as part of an overall view of the cell as an integrated unit. It suffices here to recognize that mitochondrial DNA is functional and essential, and that it participates in most, if not all, of the developmental phenomena now to be considered.[710, 859]

ORIGIN It is a valid generalization that no eukaryotic cell exists without mitochondria. They may become reduced in number, and they may become rudimentary in structure, but they are always represented. Reduction of their numbers from a large population may occur during formation of male gametes;[932] in other cases (described on page 88) there may be few, or in some cases only one, mitochondrion per cell at *all* times in the life cycle. Simplification of their structure can also occur at reproduction (Plate 30e), or can be brought about by environmental effects, for example, if yeasts (but not rice seedlings, see page 94) are deprived of oxygen their mitochondria become little more than vesicles, which however, do retain the essential mitochondrial DNA that allows them to develop again when the oxygen supply is restored.[65]

It has proved possible to 'label' mitochondria. Both their membranes[497] and their DNA[632] have been made radioactive by feeding cells with appropriate precursors. When a population of mitochondria is so labelled, and the fate of the label is determined while the cells containing the mitochondria divide and re-divide through successive cell generations, it becomes clear from the distribution of mitochondrial radioactivity that as cells divide, so too do their mitochondria. The products of mitochondrial division become distributed amongst the products of cell division. The alternative possibility, that the mitochondria of a new cell arise *de novo* while those of the parent remain in the parent, is not supported by the evidence. Just as cells arise from pre-existing cells, their mitochondria arise from pre-existing mitochondria.

Where a cell contains a large population of mitochondria, it is easy to visualize both an increase in their numbers to keep pace with cell division, and the population becoming subdivided on a random basis when the cell divides into daughter protoplasts. Provided that the population is large, the statistical chance that any one of the progeny will not receive an initial quota of mitochondria is very low. For example, in cells undergoing equal divisions in willow herb *(Epilobium)* stem apices, each sister cell formed at telophase receives about 60 mitochondria, the population having increased in size during the preceding interphase.[12] Given a starting population of 120 mitochondria, if each one behaves independently and has the same chance of ending up in one daughter cell as in the other, a mitochondria-free cell will be produced, on average, only once in 7.5×10^{37} divisions. A large tree (let alone a small herb like *Epilobium)* has only about 10^{12}—10^{13} cells, so the possible loss of mitochondria by random statistical chance can be ignored. Unequal distributions are, on the other hand, feasible, and it is to be suspected that there are growth regulating systems which will allow a deficient cell rapidly to build up a normal population, and conversely an over-supplied cell to slow down its mitochondrial growth rate.

Not all cells, however, have large populations of mitochondria. Some, as already indicated, have only one, and hence are in danger of producing mitochondria-free, and therefore non-viable, progeny at cell division. One method of diminishing the risk is to organize the disposition of cytoplasmic components so that an equitable distribution accompanies the equal distribution of daughter nuclei. In very small flagellates (e.g. *Micromonas*), division of the single mitochondrion precedes cell division, and the spatial organization is such that one product becomes a part of each protoplast.[515, 528] It may be conjectured that a second method exists, based on the premise that an extensive mitochondrial reticulum will, if numerous copies of mitochondrial DNA are scattered amongs the various branches, be equivalent in statistical and functional terms to a population: cell division partitions the reticulum at random, and with sufficient copies of the DNA, the statistical chance of producing mitochondrial DNA-free daughter cells would be remote.[19, 366]

DEVELOPMENT Having looked briefly at the means whereby a young cell is provided with its initial supply of mitochondria, there remain the problems posed by their growth in size, complexity, and numbers. The mitochondrial reticulum of *Chlorella* serves to introduce some of these aspects of mitochondrial development.[19] The cells themselves enlarge approximately 6-fold during their growth, and it has been found that throughout this period the mitochondrion occupies a fairly constant 2.5–3% of the volume of the cell. It follows that the total mitochondrion, like the cell, increases in volume about 6-fold. In fact the total length of the reticulum increases from 20 μm in a young cell to nearly 150 μm in a cell that is about to divide. Certain shape changes also take place. Measurements of the rate of respiration, which assess one aspect of the biochemical activity of the mitochondrion, show that as the cell and its mitochondrion enlarge, so the rate of respiration increases in proportion.

A direct relationship between the rate of respiration and the amount of mitochondrial material seems reasonable and straightforward. A complication arises, however, if the activity of individual respiratory enzymes is assayed. For example, the succinic dehydrogenase and cytochrome oxidase of the *Chlorella* mitochondrion do not increase their activity in parallel with the rate of respiration and the size of the mitochondrion.[402] Instead they are synthesized only at a particular period in the life of the cell. The consequence is that at this particular period, the mitochondrial inner membrane must become especially enriched with these two respiratory enzymes. In the subsequent period the amount of enzyme per unit area of membrane falls as the mitochondrion continues to grow while no more enzyme is made. Such fluctuations, it seems, need not affect the rate of respiration, which evidently is controlled by other mechanisms.

Lack of a clear-cut relationship between the area of the inner membrane and its enzyme content tempers the previously cited correlation between the general 'activity' of plant cells and the degree of crowding of the cristae in their mitochondria (page 89). That the abundance of cristae can in fact be a poor guide to respiratory activity is emphasized by observations made on rice seedlings grown either with or without an oxygen supply.[609] The ungerminated embryo has very rudimentary mitochondria, with few cristae. As in other species,[608] mitochondrial development occurs during germination, after which mitochondria in the coleoptile cells possess many cristae *whether or not* oxygen is present. The key observation is that when oxygen is absent, the oxidative activity (assayed as respiration rate and as cytochrome oxidase activity) is greatly reduced. In other words, the mitochondria can form an inner membrane of normal appearance despite the reduced production of at least one of its normal components. Cytochrome oxidase can be inserted later into the 'skeletal' inner membrane, for its activity increases if seedlings that have been grown in the absence of oxygen are given a supply of oxygen.

Comparable studies on other material, as disparate as yeast cells[209] and cells of the mammalian adrenal gland,[918] show that the phenomenon of temporal restriction of mitochondrial enzyme synthesis is general. Components can be added to the inner membrane, or removed from it, irrespective of the growth in area of the membrane. It is *not* the case that every time the area increases by a certain amount, one of each of the respiratory enzymes is inserted. Membrane *growth* is therefore a separate process from membrane *differentiation,* i.e., the acquisition of specific enzymatic properties. In several of these respects the inner membrane of the mitochondrion is similar to the endoplasmic reticulum (page 68).

DIVISION In most cells, division of mitochondria follows their growth and their differentiation. The differences between cells with many mitochondria and cells with just one are probably rather like the differences between a nucleus with the basic quota of DNA, and a polyploid nucleus with several or many copies of the basic quota. The variable in each case is the relationship between multiplication of the DNA and division of the structure that houses the DNA. In most cells the total mitochondrial DNA of the cell is partitioned by mitochondrial division amongst many mitochondria. In yeast (and probably in *Chlorella*), lack of mitochondrial division, with continued DNA synthesis, yields extensively polyploid mitochondria.[897]

The actual process of mitochondrial division has been observed by light microscopy in various living plant cells, but electron microscopists, working with dead material, have to be content with constructing a series of static images which they hope depicts the process.[311] It seems plausible that division is initiated when a critical size or DNA-content is reached, whereupon part of the inner membrane, instead of forming a crista, partitions the previously single inner matrix compartment into two discrete compartments, both enclosed within the intact outer membrane. Plate 29a illustrates this configuration. A constriction forms between the two compartments and eventually separation is achieved, presumably with no more than a transitory (at most) exposure of the intermembrane compartment to the cytoplasm before the outer membrane seals around each of the progeny. As usual, it is easy to describe such events in terms of morphology, but much less easy to envisage what happens at the level of molecular architecture, and what forces dictate the rearrangements of the membranes. It is not known whether there is a specific mechanism ensuring that nucleoid regions become separated by the septum that cleaves the mitochondron, and so that each of the progeny receives some DNA.

It is worth repeating in the context of mitochondrial division, that the rate at which mitochondria grow and divide, and the manner of their differentiation, are geared to the overall differentiation of the cell that houses them. In meristematic cells the divisions maintain the level of the population. In differentiating cells the division rate may alter. The number of mitochondria per unit volume of cytoplasm may be maintained while the cell enlarges, or their frequency relative to the rest of the cyptoplasm may increase or decrease. Finally, superimposed upon the factors that control the size and the density of the population, there are the factors, some internal, some environmental, that control the morphology and the enzymology of the mitochondria in a given cell.

9 Plastids

9.1 Introduction

The previous chapter described how mitochondria, given raw materials, can dispense immediately usable energy in chemical form. The ultimate source of energy in the biological world is not, however, anything to do with mitochondria. It is light, the electromagnetic energy of which is trapped when photons are absorbed by chlorophylls and other pigments in the chloroplasts of green plant cells. Light absorption energizes electrons in the pigment molecules, whereupon the complex biochemical machinery of photosynthesis first extracts a large fraction of the energy, and then utilizes it in organic syntheses. In the first stage (the 'light reactions') electron transport is coupled by systems not unlike those of mitochondria to the formation of primary energy stores, ATP and the reduced coenzyme nicotinamide adenine dinucleotide ($NADPH_2$). In the 'dark reactions, ATP and $NADPH_2$ in their turn provide energy for the synthesis of mobile secondary stores, utilizing carbon dioxide as the source of carbon. The secondary energy stores are predominantly sugars, and they may be utilized directly or transported to other parts of the plant. They may also be converted to temporary or long-term tertiary energy stores—usually in the form of starch grains. It has been estimated that up to the advent of nuclear energy utilization, about 95% of the earth's available energy supply is or was dependent upon these activities of chloroplasts.

In keeping with their fundamental role in the earth's economy, a huge literature[430] on chloroplasts has been amassed since their first description, written by van Leeuwenhoek in letters to the Royal Society just over a decade after Robert Hooke's presentation of the first account of cells to that same body. They were thus the first component to be recognized within the plant cell, but nearly 200 years were to pass before the modern era of physiological and structural investigations of chloroplasts was given a firm foundation. It was in the 1860s that Sachs related the intake of carbon dioxide in the light (discovered, along with oxygen output, during the 18th century studies of Priestley and Ingen-Housz) to the production of starch, grains of which had been observed previously in chloroplasts by von Mohl and Nägeli. Then in the 1880s, Schimper and Mayer published extensive descriptions, not merely of chloroplasts, but also of additional members of what they recognized to be a family of cell components, all related, and now placed within the general category of cell components known as 'plastids'—a word that simply means 'formed bodies', coined in and retained from the days before it was possible to discriminate between plastids (in the modern sense of the word) and other 'formed bodies', such as mitochondria and dictyosomes, which were later to be given their own nomenclature.

There are very few exceptions to the generalization that the cells of plants (other than fungi) contain one or other of the various types of plastid. The family is represented in colourless as well as coloured plants, tissues, and cells. One exception that lacks any plastid is a very specialized cell in a restricted group of green algae.[243] It is of greater significance that the family may die out in male reproductive cells of certain plants.

There is argument about whether they actually disappear from the cytoplasm of the male gamete, but regardless of this point, it does seem that there are cases where plastids from the male gamete do not survive fertilization to become represented amongst the plastids of the egg and embryo.[430] Apart from these exceptions, plastids are present in all cells, and are passed from parent to progeny during cell divisions of all sorts.

The following types of plastid will be emphasized in this chapter: *proplastids*—small precursors of other members of the family, found in meristematic cells; *chloroplasts*—green (in the majority of cases) and functioning in photosynthesis; *chromoplasts*—pigmented, but non-photosynthetic; *amyloplasts*—specializing in the synthesis and storage of starch grains; and *etioplasts*—normally a transitory developmental stage, formed when chloroplast formation from proplastids is interrupted by lack of light.

Some authors distinguish other types, such as lipid- and protein-storing plastids, and several other terms and word usages are common in the botanical literature. The term *leucoplast* has been used to refer to all non-pigmented types of plastid, i.e. proplastids and most amyloplasts. We consider the latter terms to be valuable in that they convey information on function, and will use 'leucoplast' only for those pale or colourless plastids (e.g. in many epidermal cells, or mature cells of colourless parasitic plants) that have not developed much beyond the proplastid stage, yet *under normal circumstances* have lost the precursor function of proplastids. Again, the word 'proplastid' is quite often used wherever there is a precursor function, i.e. for etioplasts as well as for the proplastids of meristematic cells. In our view, proplastids and etioplasts are very different, and therefore their different names are warranted. It should be noted that 'etioplast' is a relatively recent term,[430] hence much of the literature prior to 1967 contains 'proplastid' in its stead.

The features shared by all members of the plastid family are: a double membrane envelope; a system of internal membranes; and a genetic system including DNA, located in *nucleoids,* and ribosomes capable of assembling proteins. Tiny droplets—*plastoglobuli*—are also very widespread. All of these components will be described in detail after the features that characterize the different types of plastid have been considered. The treatment emphasizes plastids of higher plants, but some of the special features displayed in the algae are given a separate section. The final part of the chapter is concerned with developmental inter-relationships within the plastid family.

9.2 Proplastids

It has been known for more than a century that minute bodies with some of the properties of plastids occur in meristematic plant cells. These bodies contain, or can produce, small grains of starch. When the cells differentiate, they may develop into other, more easily recognized, types of plastid. Strugger, in 1950, emphasized their precursor nature by calling them *proplastids.*[801]

Because light microscopists could not readily distinguish them from mitochondria (e.g. Plate 1), there had for long been arguments about the nature of the structures which we now call proplastids. The idea that there is but a single class of particulate entities in plant cell cytoplasm was still being propounded in the early 1950s: it was held that the entity could under some circumstances have mitochondrial properties, and under others could assume obviously plastid characters.[877] Countering this view was the emerging body of knowledge about the distinctive features of plant and animal mitochondria, and significant observations on the inheritance of chloroplasts. Traits shown to belong to the chloroplasts were found to pass from one generation to the next, even though chloroplasts themselves regress to a very simple, mitochondrion-like form in the egg cell. Those who argued that plastids and mitochondria are independent entities with genetic continuity but with no genetic connection ultimately were proved correct. Nevertheless, confusion persisted even into the electron microscope era, and proplastids labelled as 'large mitochondria' crop up in the literature as recently as the late 1950s.

A description written in 1927 by Zirkle is a good introduction to the structure of proplastids.[933] He referred to them as 'primordia', and by studying both living cells and fixed tissues, was able to describe their development into chloroplasts in a range of material. The disastrous effects of fixatives which contain acetic acid were noted (it is unfortunate, and quite unnecessary, that these fixatives are *still* sometimes used for general studies of plant tissues). Various technical hints enabling the minute starch grains of the 'primordia' to be seen were also detailed. An important observation was that 'primordia' in living cells do not accumulate and change the colour of the dye Janus Green B, unlike mitochondria (page 87). Zirkle concluded that 'the primordia are hollow spheroids of essentially the same structure as mature plastids'. A glance at electron micrographs (Plates 2, 3a) substantiates Zirkle's description. The electron microscope adds that the envelope of the 'spheroid' consists of two membranes with a narrow space (approx. 10 nm) between, and various internal components are also distinguished (Plate 31b). Usually there are a few vesicles or small, membranous, flattened sacs, sometimes showing continuity with the inner membrane of the envelope. The stroma contains (in addition to the starch) a small number of plastid ribosomes, and also nucleoid areas with a more transparent appearance than elsewhere, traversed by fine (about 3 nm) fibrils of DNA. A few plastoglobuli may be present.

Proplastids are somewhat larger than most plant mitochondria (Plate 1), and are of no particular shape, ranging from spherical to ellipsoidal to cylindrical to branched forms, with or without bulged or invaginated areas. The early microscopists observed changes in shape, and it is still not clear whether proplastids have an intrinsic power to alter their form, or whether they become pushed or pulled into new shapes by external forces.

In most meristematic cells their numbers are probably in the low tens.[12] There is very strong circumstantial evidence that they divide to keep pace with cell division, and there is no compelling evidence that they ever arise *de novo*. Aspects of proplastid division, and their development to and from other plastids, will be considered in more detail in subsequent sections of this chapter.

Two closely related roles for proplastids are accepted. Firstly, they are precursor structures, and can develop into plastids with specialized biochemical attributes. Secondly, the presence of proplastids or their derivatives in (virtually) all living cells of the plant assures the continuance of the plastid family, not only from generation to generation, but also within one individual. If (for example) a wound stimulates a bud to form in some unlikely and hitherto quiescent tissue, plastids will be represented in the cells of the bud, and ultimately in the cells of the leaves, flowers, and seeds that may develop from the bud. In rather the same fashion that disease organisms are always with members of the human race, so plastids are always with plant cells: given the appropriate conditions, the genetic information deposited in them will be expressed.

These roles being accepted, perhaps the major unanswered question concerning proplastids is whether they have, in the course of evolution, acquired any other functions. In meristems, up to 10% of the cytoplasm may consist of proplastids. Is this very appreciable volume allowed to remain in the form of structures whose usefulness becomes apparent only outside the meristem, or is it made to contribute positively to the life and organization of the meristem, as, for example, by housing in it particular metabolic systems or sites of hormone production? Proplastids can make starch, and those in castor bean seeds certainly are considered to have definite metabolic functions[931]: in developing seeds they synthesize a fatty acid precursor for the main fat store that accumulates as droplets in the cytoplasm, and in germinating seeds, when the fats are broken down (see page 136) some of the products are taken in by proplastids and used to synthesize sugars and starch. It is extraordinary that methods for isolating proplastids were not developed before the 1970s.[820] Now that interest is being focussed on their properties, perhaps more will be learned about this hitherto most neglected of plant cell components.

9.3 Chloroplasts

9.3.1 Introduction

The biochemical and photochemical activities of chloroplasts provide us with the food we eat and the oxygen we breathe. Many textbooks deal with the subject of photosynthesis, and it will be outlined here only as far as is necessary to aid interpretation of the intricate structure of chloroplasts. That photosynthesis consists of two sequential processes, first photochemical 'light reactions', then biochemical 'dark reactions', has already been mentioned. The light reactions generate ATP and $NADPH_2$, utilized in the dark reactions to synthesize sugars following the fixation of carbon dioxide.

Two photochemical systems exist.[306] Each is complex, containing its own characteristic set of light-absorbing pigments together with molecules capable of transferring electrons into which some of the electromagnetic energy of light has been passed. Photosystem I on its own can use the energy to phosphorylate ADP, making ATP. However, photosystems I and II normally act in sequence, their concerted effect being to liberate oxygen from water and to generate the two primary energy stores, ATP and $NADPH_2$, both of which are needed to drive the dark reactions.

The two photosystems may exist as discrete structural entities, and one of the most important quests in current research into relationships between structure and function is that which seeks to marry what is known of the photo- and bio-chemistry of photosystems I and II, with what is known of the ultrastructure of the chloroplast. The great majority of chloroplasts possess both photosystems, but some chloroplasts may be permanently or at transitory stages enriched or depleted in one or other, and the spatial distribution of the two throughout a chloroplast seems not to be uniform.

In the majority of chloroplasts utilization of the products of the light reaction in the fixation of carbon dioxide proceeds by the complicated sequence of enzymatic reactions known as the Calvin cycle. Carbon dioxide is taken in and added to a 5-carbon acceptor molecule, ribulose diphosphate, generating compounds containing 3 carbon atoms. Sugars are drawn off, and the carbon dioxide acceptor molecules are regenerated. The enzyme that attaches carbon dioxide to its acceptor is probably the most abundant protein on earth. It is ribulose diphosphate carboxylase, a molecule large enough to be visible in the electron microscope (Plate 35a), and accounting for more than half of all of the protein in green leaves. It is probable that no green *plant* lacks this enzyme, but recent investigations, particularly of tropical grasses, have revealed that within certain plants, various *tissues*, *cells* and their *chloroplasts* can be without it, possessing instead an alternative and more efficient mechanism for the initial trapping of carbon dioxide. The alternative method initially produces organic acids containing 4 carbon atoms, and the plants that employ it are referred to as 'C-4 plants' (as distinct from 'Calvin cycle', or 'C-3', plants). The ultrastructure of their chloroplasts is considered later.

The early light microscopists established that chloroplasts are two-component systems.[872] Meyer reported in 1883 on the presence in them of minute green 'grains', or *grana* (Plate 32b), which represent one of the components. The second component is the matrix in which the grana lie. It is called the *stroma*, a word that had been used nearly 20 years earlier by Sachs to refer to the residue that remains after chlorophyll has been extracted from chloroplasts, and later by Schimper in a sense more akin to the modern usage, namely 'the plasmatic, non-differentiated part of the plastid'. A fundamental discovery concerning structure and function in the chloroplast was made by separating grana—by then known to consist of chlorophyll-bearing membranes—from the stroma: it was found that the light reactions reside in the membranes, and the dark reactions of carbon dioxide fixation in the stroma. Clearly, since these reactions are sequential, the interface between grana and stroma is of enormous importance. Much of what follows is devoted to a consideration of its architecture.

9.3.2 Chloroplasts: Gross Morphology

The early German microscopists thoroughly documented information on the shape, size, and numbers of chloroplasts in cells of green tissues. A reasonable generalization is that mesophyll cells (Plate 32a) contain 30-500 discoidal or ellipsoidal chloroplasts, each with a major axis of 3-10 μm, and a minor axis about half the length of the major axis. The great diversity of shape and size displayed by algal chloroplasts (page 129) is not featured in higher plants, but there is some variation, consistently to be seen when comparing leaf with stem, young tissue with old, or tissues from plants grown in a range of environments. Rapid shrinkage of chloroplasts occurs when darkened leaves are illuminated, and longer term shape changes have been recorded in films of streaming cytoplasm of leaf cells. The outermost layer of the chloroplast appears to be especially subject to shape changes, and the protrusion, retraction and separation of long 'fingers' can be observed.[892a] Much slower shape changes ensue as a result of accumulation or depletion of starch grains.

Other aspects of the regulation of size and numbers are considered elsewhere (page 132).

9.3.3 The Architecture of the Internal Membrane System of Chloroplasts

Meyer's observations of chloroplast grana were not clearly substantiated for fifty years, and then, in the 1930s, Dutreligne, Heitz, Menke, Weier, and others used a variety of light microscope techniques to examine them.[872] In size they are not far above the resolving power of the light microscope, about 0.2–0.4 μm in diameter. Heitz recognized them to be cylindrical, circular in profile when viewed from 'above' (Plate 32d, e), and rectangular from the 'side' (Plate 32c, f). Sometimes they are not discrete, but are joined to their nearest neighbours (e.g. Plate 41a). The flat circular surfaces of the discs frequently (but not always) lie parallel to the outer surface of the chloroplast (Plates 32f, 34b, 41a). Estimates of their numbers are in the range 40-60 per mature chloroplast, but obviously, small chloroplasts may have many fewer, and very large ones many more.

The earliest electron micrographs of chloroplasts, published in Germany in 1940, [421] included images of grana, and further, suggested that grana contain very thin internal layers. The first post-war studies confirmed the layering, but there were suspicions that the procedure then available for specimen preparation might have created this morphology. Results obtained by examining grana using polarized light, and the appearance of the first ultra-thin sections of chloroplasts, allayed such fears. The layered substructure of grana was seen to be based on membranes of approximately the same thickness (7-9 nm) as other cell membranes. It also became clear that not only are the grana membranous, but that they are interconnected by membranes that traverse the intervening stroma.[486]

Many investigators have been interested in the three-dimensional arrangement of the internal membranes of chloroplasts, and several have coined terms which were convenient for describing the structures seen in their micrographs. As understanding of the structures grew (as in the sequence of Text Figures 4 to 8), so the nomenclatures, and the way in which they were applied, had on occasions to be modified. Our concern here is to select the simplest terminology consistent with modern views, and since one of the most important developments has been the realization that the membrane system is a continuum, we prefer to avoid words suggestive of the existence and aggregation of discrete membranes in favour of words that describe specialized regions of a single, highly complex, membrane surface. The terminology used, which is due to Weier and his associates,[878] will be introduced as it is required in the following account. There is one other useful word, however, which has come into general use since it was introduced by Menke in 1960.[548] It is *thylakoid*, and will be employed from time to time to refer to closed, flattened sac-like *regions* of the membrane continuum.

STRUCTURE OF GRANA The best way of introducing the exceedingly complex arrangement of the chloroplast membranes is to trace the development of ideas up to 1970, when the currently accepted interpretation was published.

One of the earliest reconstructions (Fig. 4)[191] represented grana as stacks of hollow discs, each disc being in the form of a flattened circular cisterna, not unlike a dictyosome cisterna. The discs are, however, more symmetrical, and are quite precisely aligned one on top of the other to give the cylindrical shape of the grana, previously noted by Heitz. This reconstruction views the grana in a chloroplast as being interconnected by way of large flat discs (thylakoids) intercalated amongst the granum discs, and extending from granum to granum across the whole chloroplast. The model is correct insofar as it includes granum discs and intergranum membranes, and also in that the membrane surfaces separate the chloroplast stroma from the spaces enclosed by the various membranes. The major defect is that the enclosed spaces (within the discs) are not interconnected.

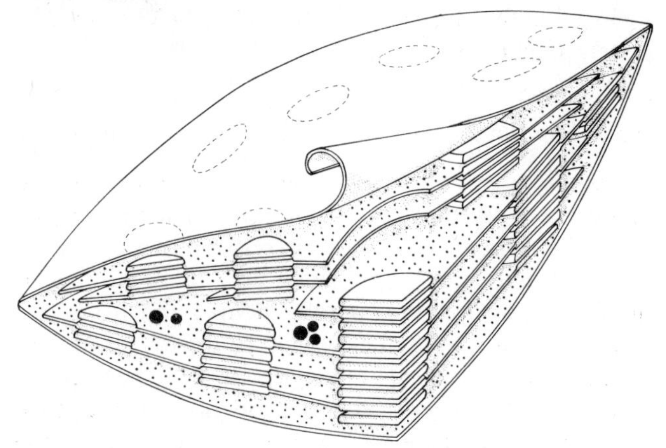

Fig. 4
An early representation of the three-dimensional structure of a chloroplast, cut open to display grana and inter-grana membranes. From Eriksson, Kahn, Walles, and von Wettstein, *Ber. deut. bot. Ges.* **74**, 221–232, 1961.

In due course it was found that the membranes interconnecting the grana are more complex than simple discs of large lateral extent. One of the complications is that they are perforated—just as cisternae of endoplasmic reticulum may be perforated. The degree of perforation seems to vary. In some species it is slight, while in others the perforations are so large that the membrane surface is reduced to narrow membrane-bound channels passing from granum to granum. The inter-granum membranes, instead of being flat expanses, are thus more or less dissected, intricate, and ornate, which is why the name for ornamental carvings or patterns—*fret*—has been applied to them. Frets are membrane-bound connections, narrow or broad, between grana. Figure 5[349] differs from the previous figure by incorporating perforated frets, and by eliminating or greatly reducing the number of granum discs that are isolated from one another through lack of fret connections.

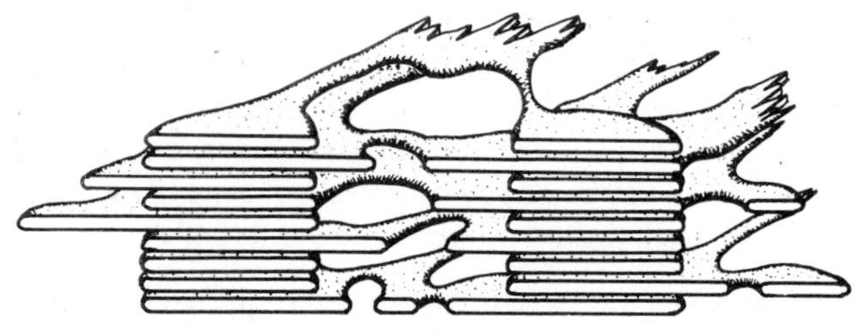

Fig. 5
This reconstruction differs from that of Fig. 4 in the greater complexity of the inter-grana system of membranes (frets). Also fewer granum discs are left isolated, with no fret connections. From Heslop-Harrison, *Planta,* **58**, 237–256, 1962.

The next advance in interpretation has great functional significance. It is that the frets can, by branching, interconnect discs within an individual granum. Figure 6[878] shows a somewhat irregularly branched fret system of this type. In it, the internal space, or *loculus*, of one disc is connected *via fret channels* to others. There may still, however, be isolated granum discs in the model of Fig. 6, and it is now thought that there is a greater degree of organization in the fret system that is displayed there.

Fig. 6
This diagram shows three grana and their associated frets. The fret system is highly dissected and interconnects not only the grana, but also different discs within a granum. From Weier, Stocking, Thomson and Drever, *J. Ultrastruct. Res.*, **8**, 122–143, 1963.

It has proved to be extremely difficult to find out precisely how granum discs are interconnected.[351] The main practical difficulty is that ultra-thin sections are much thinner than grana, so that in order to view all of the connections, it is necessary to cut and photograph the 5-15 adjacent sections that are needed to pass right through a granum and enable three-dimensional reconstructions to be made.

A preliminary interpretation of the mode of interconnection of discs is shown in Fig. 7.[350] A fret ascends at an angle to the plane of the granum discs, and by branching, connects to each disc in turn. Unlike previous figures, it is clear that none of the discs is isolated. The whole membrane surface is confluent, and the space enclosed by it is continuous. Only six discs are included in the drawing, but it is worth noting that any number could be accommodated by extending the ascending fret and adjusting its angle of ascent. As reproduced here, the drawing shows how the points at which the fret connections enter the successive discs describe a right handed helix ascending the side of the cylindrical granum: another six discs would need to be added to Fig. 7 in order to complete one turn of the helix.

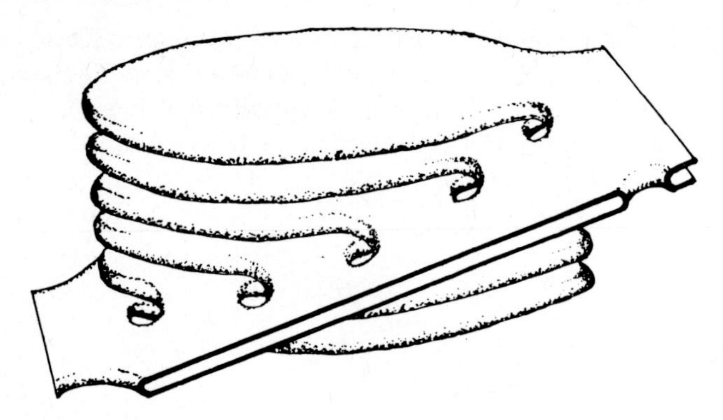

Fig. 7
The 6 discs of this granum are depicted as all being connected to the same fret, which ascends the granum. The points of entry of the fret connections form part of a right handed helix. Mirror image of diagram in Heslop-Harrison, *Planta,* **60**, 243–260, 1963.

The remaining feature is at once the most difficult to detect and to visualize.[626] Figure 7 depicts *one* connection between *each* granum disc and a *single* fret. However, close examination of electron micrographs reveals inadequacies in this interpretation. Plate 32e shows an ultra-thin section which was cut in the plane of the granum discs, and in places *many* frets can be seen passing towards the discs (probably 2 or 3 in number) that lie within the thickness of the section. In short, *each* granum disc must establish *multiple* connections to the system of frets. Figure 8 shows part of a granum and its associated frets, and is based on studies of serial sections. It differs from the preceding figure in that not just one, but many, frets are accommodated. Every fret ascends the granum in a right handed helical path, and connects with every disc.

An important point that is not included in Fig. 8 is that the frets, as in earlier models, interconnect adjacent grana, thus integrating the entire internal membrane system of the chloroplast.[626] As stated before, the internal membrane is a continuous surface, differentiated into granum discs and fret connections, and separating the chloroplast stroma from a continuous compartment lying inside the discs and frets.

Figure 8 may indeed be an accurate representation of some mature grana, but it is emphasized that for many it is an over-symmetrical idealization. Comparison of Plate 32c and Plate 32f shows that there can be wide variation in the number of discs in a granum. Obviously the frets cannot complete more than a small part of one turn of their helical path if only a few discs are present—and the smallest possible granum consists of a mere two superimposed discs. Another important variable is the ratio of frets to discs, shown as 1:2 in Fig. 8. Not many measurements of this ratio have been made, but from the available data it appears that a condition in which there are fewer frets than in the Figure may be common. The ratio influences the number of fret connections per granum disc, and the angle at which the frets ascend the stacked discs.[626] The grana themselves may not be perfect cylinders, and another factor that governs the overall shape of the membrane system is that adjacent grana may not lie parallel to one another (especially near the periphery of the chloroplast—Plate 32c, d). The spacing of the grana within a chloroplast can vary, and stages of development may not show all of the features included in the idealized granum of Fig. 8. The mode of development is examined later (page 110).

9.3.4 Structure and Function of Thylakoids: Macrostructure

Two considerations would seem to be of especial significance in thinking about the structure of the chloroplast membrane system in relation to its function. The first is that a large surface area of membrane is advantageous, for since photosystems I and II are displayed on (or in) the membrane, the greater the extent of the surface the more light energy that can be trapped. The second concerns the dependence of the dark reactions in the stroma upon the light reactions in the membranes. The membranes produce ATP and $NADPH_2$, and these substances have to be moved to their sites of utilization in the stroma. Also their precursors, ADP, inorganic phosphate, and NADP (the oxidized form of $NADPH_2$), have to be moved from the stroma to the sites of the light reactions.

TOTAL AREA It is self-evident that chloroplasts contain a large area of membrane within a relatively small volume. A chloroplast containing 50 grana, each with 10 discs 0.25 μm in radius, possesses about 200 μm^2 of membrane in the grana alone. Although chloroplasts occupy a small fraction of the total volume of a leaf, their combined membrane surface areas therefore exceed by several to many hundred times the area of the leaf. By assaying the amount of chlorophyll and measuring the area of membrane per chloroplast, it is estimated that the area of membrane available to each chlorophyll molecule is between one half and two times the area of the porphyrin ring of the chlorophyll molecule.[819] In fact the molecules may not be spread out evenly, but may be locally aggregated in more crowded areas.

Measurements indicate the extent to which the requirement for large surface areas is met, but do not provide an explanation for the form taken by the membranes. There are other ways of displaying large surface areas (e.g. as in the prolamellar bodies of etioplasts, see page 111 and Plates 36, 37), and indeed some other forms might be more efficient than grana and frets in that there might be less internal shading.

COMMUNICATION WITH THE STROMA It may be that the architecture of the chloroplast membrane system should be viewed as an acceptable compromise between the ideal of maximum efficiency of photochemical activity, and the necessity for providing good communications with the stroma. Without going into the biochemistry of ATP and $NADPH_2$ production,[306, 835] it is easy to envisage that these two substances could arise in a variety of ways. Their precursors (ADP and NADP) must either be transported to the sites where the photochemical energy and reducing power are

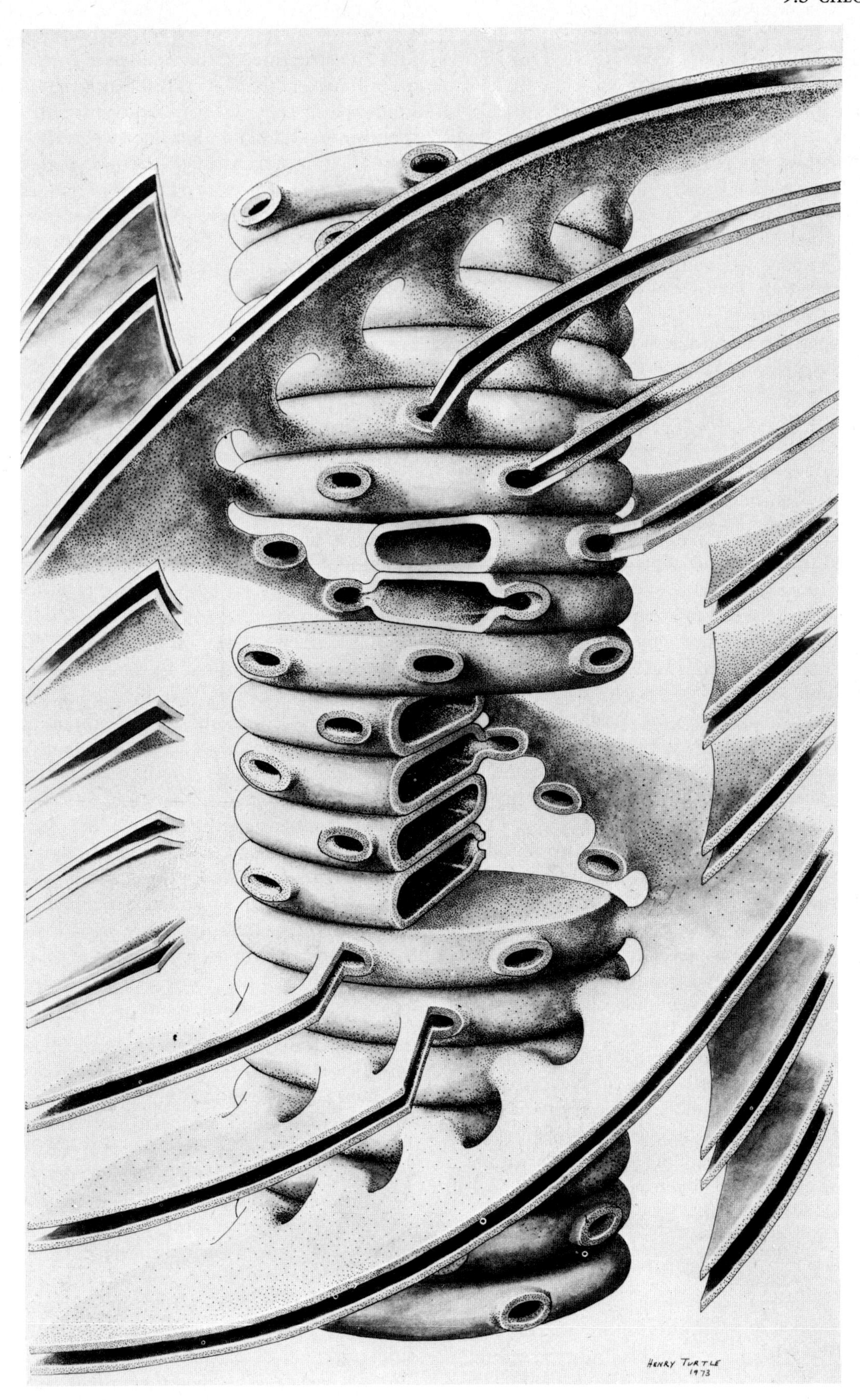

Fig. 8
This perspective drawing of a granum shows the multiple fret connections to each disc. Only one fret, ascending in a right handed helix, is shown in full, the rest being cut away, along with parts of some of the discs.

produced, or mobile molecules must deliver the energy to sites where the precursors are available. On the one hand ADP and/or NADP could traverse the thylakoid membrane and diffuse within the intra-thylakoid space to the appropriate sites. Alternatively the ADP and NADP might stay in the stroma, and the ATP and $NADPH_2$ be formed at the thylakoid-stroma interface. One form in which energy could be delivered to this interface consists of protons, delivered into the intra-thylakoid compartment during the light reactions. After diffusion along the fret channels, there would be a proton concentration gradient between stroma and internal compartment, and this, it is suggested,[306] could drive phosphorylation. All such processes are presumably limited by the fact that only a fraction of the total membrane surface is in contact with stroma (unlike the prolamellar bodies of etioplasts, where *all* of the membrane is bathed by stroma). Clearly, the high degree of communication provided by the fret system of chloroplasts is of very great importance.

PARTITIONS IN GRANA In all of these considerations, the structural feature that is the most difficult to interpret is the close contact between adjacent granum discs. With some methods of preparation a narrow gap can be seen where the disc membranes lie back-to-back (Plate 33a), but with some stains,[586] or with procedures that preserve the lipid material in the specimen,[875] the gap tends to be filled with electron-dense material. The composite structure, i.e. two disc membranes and intervening material, is called a *partition*[878] between adjacent disc loculi.

Adhesion of discs to produce grana would at first sight seem to have disadvantages—the area of membrane in the partitions is bound to be shaded by neighbouring discs, and it is also out of direct contact with the stroma. Yet partitions are found in the vast majority of photosynthetic plants (with exceptions in some algae (page 129) and in the agranal bundle sheath cells of C-4 plants (page 108 and Plate 34)). It would perhaps not be justifiable to argue that because partitions are present, they *must* have some special function. Nevertheless, it seems clear that a more detailed examination of this part of the chloroplast is warranted, in search of functional attributes which might account for its conservation in the course of evolution. This topic will be dealt with further in the next section.

9.3.5 Substructure of Thylakoids

Numerous attempts have been made to present generalized models of the molecular architecture of cell membranes.[79, 769] In most cases, chloroplast membranes have had to be excluded from the schemes: their peculiarities of composition and substructure set them apart.

The constituents which most obviously distinguish the membranes of chloroplasts are the photosynthetic pigments. The phytol chain attached to the porphyrin ring of each chlorophyll molecule and the carotenoids are hydrocarbons. These hydrocarbons probably enter into the structure of the lipid part of the chloroplast membranes, perhaps even substituting in part for some lipids which are present in large amounts in other cell membranes, but are in lower concentrations in chloroplast membranes. Most of the non-pigmented membrane lipids are extremely polar molecules, with specialized chemical groupings, such as one or two units of the sugar galactose (galactolipids), or sulphonic acid groups (sulpholipids), attached at the water-soluble, or hydrophilic, extremity.

ULTRA-THIN SECTIONS Unless potassium permanganate is used as a fixative or as a stain, it is difficult to obtain images showing a dark-light-dark tripartite structure in chloroplast membranes.[351] Indeed it has been claimed that ultra-thin sections reveal a quite different type of architecture, in which the chloroplast membranes are interpreted as planar sheets composed of more-or-less globular subunits, each approximately the same diameter as the thickness of the membrane itself, i.e., 7–9 nm.[873] This model has been criticized on the grounds that units as small as this should not be visible, considering that the sections are up to ten times thicker than the globule diameter. The units should therefore merge into a featureless background of superimposed images.[428] Nevertheless, globules *have* been seen, and suspicions that they arise from some artefact of staining or damage in the electron beam are to some extent countered by the fact that particles also appear when alternative methods of specimen preparation are used.

One alternative method, which like the conventional one relies on ultra-thin sectioning, attempts to minimize damage that might arise during fixation and dehydration. When these processes are much abbreviated, unusually thick membranes are seen. Prolongation of fixation and dehydration treatments 'thins' the membranes to the familiar dimensions. Subunits of various sizes appear in the thick, gently-treated membranes.[444]

FREEZE-ETCHING AND DEEP-ETCHING It is therefore difficult to reconcile the various images that are obtained when variations on the method of ultra-thin sectioning are used. It is even more difficult to relate the appearance of sections to that of freeze-etched specimens, where fixation and dehydration normally are omitted completely. Views of freeze-etched preparations show that the chloroplast membrane is much more complex than a simple expanse of side-by-side globular subunits.[80] Firstly, particles are seen to lie in a smooth, nonparticulate matrix, probably consisting of membrane lipids. Secondly, there are different types of particle, or at any rate, particles of different sizes. Thirdly, the membrane surface is regionally differentiated, with different populations of particles in different areas. Fourthly, it is becoming clear that there can be considerable variation in thylakoid substructure from organism to organism.[555, 639]

Some features of freeze-etched thylakoids are shown in Fig. 9. It must be emphasized that certain aspects are still highly conjectural, and that great controversy has surrounded the interpretation of images of freeze etched chloroplast membranes. Before there could be any reconciliation of the opposing viewpoints, it was necessary to prove that in most cases membranes fracture along their interior, thereby exposing to view their inner constituents rather than their outer surfaces (page 13).[79] The outer surfaces can, however, be viewed if the procedure of 'deep etching', an adjunct of freeze etching, is utilized.[628]

Figure 9 illustrates both freeze etching and deep etching, and gives a nomenclature which has been widely adopted for the designation of the various views.[630] Conventional freeze etching exposes internal faces by splitting membranes. Because chloroplast membranes are asymmetrical, the internal faces are not identical. They are designated the B- and C-faces. The B-face is the exposed surface of that part of the membrane that backs on to the internal compartment, whether disc loculus or fret channel. Underlying the C-face is the inter-disc space in the centre of the partition between adjacent discs. Deep etching involves removal of comparatively thick layers of ice from the surface of the fractured specimen, thereby exposing the *external* faces of any membranes that happened to be just below the fracture, covered by only a thin ice layer. The two external faces are labelled A and D. The D-face bounds the internal compartment, that is, disc loculi and fret channels. The A-face is on the opposite side of the membrane, and two areas are distinguished. The easiest to find is that which is in contact with the chloroplast stroma, for example on the top and bottom extremities of a granum, or on a fret; these are labelled A^1. Less accessible is the corresponding face (labelled A) within grana, that is, the faces that lie back-to-back in the partitions. A-faces are harder to find because the fracture plane tends to jump directly from a C-face down to the next B-face (as in Plate 33b, c).

The appearance of the various faces is described below, followed by attempts to relate the observed structures to known functions.

THE B- AND D-FACES No other cell membrane fractures to show particles as large as those in the B-face of chloroplast membranes (Plate 33b, c). Estimates of their size vary, but 13-17 nm in diameter, by 8-9 nm in height, covers the range of values quoted by workers from several different laboratories. The wide variation in size, even within a single expanse of membrane, suggests that the population may be heterogeneous.[281] It is also possible that the small particles are fragments of the large. It is suggested that the presence of such large particles in the 'locular' side of the membrane causes the membrane to bulge, the bulges being visible when the D-face is exposed by deep etching.[630, 790a] Bulges were in fact observed by means of the shadow casting technique some time before the deep etching process was developed, and both methods have indicated that the large particles sometimes lie in a semi-crystaline array in the membrane.[627] They can also be scattered, as in Plate 33b, c, with up to 2000 particles per μm^2 of membrane area. The dramatic diminution in their numbers on passing from granum discs to frets is also illustrated.

THE A^1-, A-, AND C-FACES The C-face carries smaller particles than the B-face (Plate 33b, c). They are usually

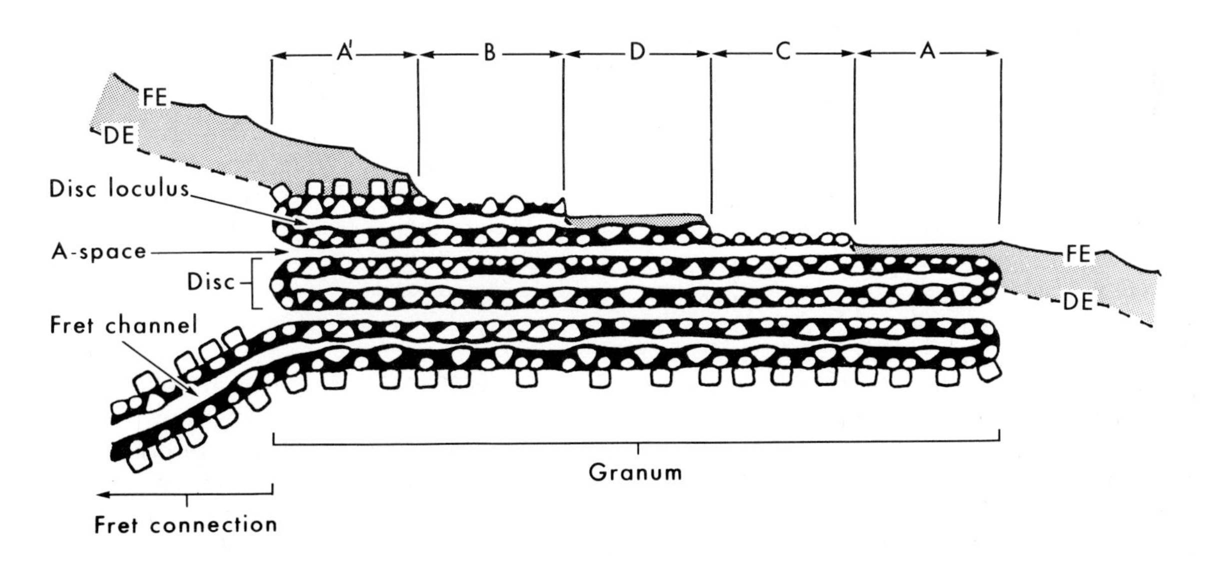

Fig. 9
This diagram of disc and fret membranes is based on observations obtained by freeze- and deep-etching, and by negative staining. Discs, partitions, loculi, frets and fret channels are labelled at the left hand side. The top surface is imagined to be exposed as in a freeze-etched preparation, yielding images of faces labelled B and C. If additional ice (shaded layer) is removed by deep-etching, external faces of membranes, labelled A^1, D, and A, are also exposed to view. The large B-face particles and the smaller C-face particles are shown embedded in the membranes. Molecules of coupling factor for phosphorylation are drawn as squares on the A^1 surface. The solid black representation of the membrane matrix ignores any spatial separation of lipid and protein components that may exist in the non-particulate part of the membrane.

described as having a diameter of 9-12 nm, but in some material there is more variation than this. Up to about 4000 may be present per μm^2. Being small, they cause no bulging of the membrane outer surface (the A^1 and A faces) which is therefore relatively smooth.

Other particles are attached to the exterior of the membranes, and are revealed by deep etching down to the A^1 faces. The attached particles are 10-11 nm in size, and can be removed from the membrane surface by washing it with solutions of the chelating agent, ethylene diamine tetra acetic acid.[16, 629] They were at one time confused with the B-face particles, but it was then shown that the latter are still present within washed membranes.

PHOTOSYNTHETIC UNITS Several independent lines of investigation are converging towards what is hoped will be the eventual elucidation of the relationships between substructure and function in thylakoids.

The researches in fact began long before methods were available for the visualization of ultrastructure. Emerson and Arnold showed in 1932 that 'photosynthetic units' must exist. Their work was based on measurements of the amount of carbon dioxide fixation and oxygen production brought about by flashes of light of varying duration, intensity, and intervals. It was shown that utilization of light energy was more efficient, that is, less light energy was wasted, in some conditions than in others. At *maximum* efficiency, each molecule of carbon dioxide required for its reduction at least 8 quanta of light energy. When the total number of chlorophyll molecules present was divided by the number of molecules of carbon dioxide fixed (at optimum efficiency), it appeared that about 2000 chlorophyll molecules are needed to constitute a functional unit. Working in concert, the chlorophyll molecules in the unit absorb 8 quanta of electromagnetic energy, and pass the energy through a chain of reaction intermediates, eventually providing the total energy needed to bring about one fixation act. If the light reactions of photosynthesis are considered separately from the subsequent dark reactions, the 'unit' is not the set of molecules needed to reduce one carbon dioxide molecule, but the smaller set needed to use light energy to energize one electron, that is 2000 divided by 8, or 250 chlorophyll molecules. These represent a light-harvesting and storing unit associated with one reaction centre, capable of delivering one electron to the electron transport molecules associated with ATP and $NADPH_2$ production.

When, in 1961, shadow casting revealed the pattern of bulges on what is now called the D-face of granum discs, the structural entities represented by the units of the pattern were immediately related to the photosynthetic units of Emerson and Arnold and their followers. Most strikingly, when the area of a bulge was measured and divided into the total area of membrane (for which the total amount of chlorophyll was known) it was estimated that each structural unit contains about 230 molecules of chlorophyll—a value which the authors, in something of an understatement, described as 'surprisingly close to the number of chlorophyll molecules contained in a photosynthetic unit.'[627] Given this correspondence, it was not in the least surprising that the structural units were equated with units of function. They were christened *quantasomes*.

Following nearly a decade of investigation by freeze etching, the position is now somewhat more complex. Each bulge on the D-face is suspected to overlie one of the large particles of the B-face, and such particles have therefore been described as quantasome 'cores'.[630] One of the problems is that the large, B-face particles do not occur throughout the chloroplast membrane system. Their frequency is low in the fret system (Plate 33b, c, Fig. 9), yet frets are known to carry chlorophyll and to be photochemically active. This implies that while there may be functional equivalents of quantasomes in the frets, large quantasome cores are not required.

DISTRIBUTION OF PHOTOSYSTEMS I AND II Once again attention has to be focused on the regional differentiation of the chloroplast membrane. Are frets functionally different from discs? Do the partitions of grana have special properties? A possible rationalization of the available data has emerged from comparisons of frets and grana, and, more generally, of 'stacked' and 'unstacked' thylakoid systems in both algal and higher plant chloroplasts. Frets and grana can be isolated from chloroplasts with reasonable success, and used for the former comparisons.[258] The latter comparisons utilize material which permanently or temporarily lacks grana[630]: examples are the bundle sheath chloroplasts of C-4 plants (Plate 34a-c), plastids at an early stage of conversion from the etioplast condition to the chloroplast condition (Plate 38a, b), and chloroplasts in a range of mutant plants, from algae to tobacco plants, which through genetical deficiencies do not develop grana, though they do have large areas of functional, unstacked membrane.

The comparisons at first seemed to give a clear-cut answer: frets and unstacked membranes possess photosystem I activity, while grana possess both photosystems I and II. Correlating this functional differentiation with structural observations, the large B-face particles ('quantasome cores') appear to be *indicators* of photosystem II activity, while the smaller C-face particles, which occur in both grana and frets, indicate photosystem I activity.[16] It should be noted that the correlation cannot be taken to mean that the particles actually *are* photosystem reaction centres. Later work on plastids developing from etioplasts, and on C-4 plants (see page 109), has shown that the absence of photosystem II from unstacked thylakoids is not general, though the ratio of photosystem I to photosystem II activity may indeed be higher in the stacked than in the unstacked condition.

FORMATION OF ATP Both photosystems generate ATP, as does the electron transport system of the mitochondrial inner membrane. In mitochondria a visible structural entity has been identified as being responsible for the utilization of energy (in the form of electrons moving along the electron transport chain) to phosphorylate ADP and so to produce ATP. The entity is a particle that is attached to the inner membrane, in contact with the mitochondrial matrix, and is visible in negatively stained preparations and in suitably treated ultra-thin sections (page 91).

ATP formation in chloroplasts is, in biochemical terms, reasonably similar to that in mitochondria. The necessary energy is derived from electron transport, and is harnessed, or 'coupled' to ATP synthesis by the 10–11 nm particles referred to above as being attached to the A^1 surfaces.[554, 555a] They are drawn as square particles in Fig. 9, and are described variously as containing the coupling factor for phosphorylation, or the photophosphorylation enzyme, or, most correctly, the calcium-dependent adenosine triphosphatase. In gross appearance they are similar to molecules of the carbon dioxide fixing enzyme, ribulose diphosphate carboxylase (seen negatively stained in Plate 35a), which may also lie on or close to the A^1 faces, being present in very large numbers in the stroma in contact with the A^1 faces.[16]

ARCHITECTURAL AND PHOTOSYNTHETIC PROPERTIES OF PARTITIONS It has already been argued that since stacked discs are so very widespread in chloroplasts, from the algae to the flowering plants, the stacking, or structure of the partitions, may have special functions.

Methods for obtaining ultra-thin sections of cellular material which has retained the bulk of its lipid (unlike most embedded specimens), show that the partition does not simply consist of two appressed disc membranes. There is an appreciable space, known as the *A-space*, between the A-faces (Plate 33a). 'Space' is not an apt word, for stainable material is present. Further, the 'space' becomes stained if chloroplasts are presented with a reagent which gives an electron-dense precipitate as a result of photosystem II activity.[586]

Partitions survive when chloroplasts are suspended in dilute media that cause the discs and frets to become grossly swollen by osmotic forces. The A-faces and the A-space must therefore possess adhesive properties. Some remarkable experimental results have given insight into the nature of the 'glue' that, by holding the granum together, plays such an important part in controlling the overall morphology of the chloroplast membrane system. If chloroplasts are isolated and suspended in an organic buffer solution, the membranes unfold, and integrated systems of frets and grana metamorphose into extended expanses with few grana remaining. If inorganic salts are added to these unstacked preparations, an extraordinary event takes place: the membranes re-assemble frets and grana.[393, 575a]

It appears that the unfolding process is initiated when inorganic ions are absent, but the membrane remains capable of stacking despite the unfolding. The experiment highlights the suggestion made earlier on the basis of observations of the gross morphology of chloroplast membranes, namely that the membrane surface is one highly integrated expanse, now shown to be able to bring about its own morphogenesis. The subject of the development of grana is considered later (page 110), but it may be noted in passing that the unstacking/re-stacking experiments are performed in the test tube, and that granum formation during re-stacking presumably takes place without any growth in area of the membrane. The primary event seems to be formation, recognition, and adhesion of areas of membranes that are able to form partitions. Re-aggregation of the large B-face particles, previously dispersed during unstacking, is part of the overall process.[790a]

The effect of inorganic ions on adhesion of discs to create partitions, and hence grana, indicates that electrical forces are involved. Perhaps divalent ions (calcium or magnesium) form bridges between negatively charged molecules on the back-to-back A-faces, rather as calcium does in the middle lamella of cell walls. Almost certainly, however, other forces participate.[575a] Investigation of interactions of this sort may provide explanations for the occurrence of unstacked membranes in, for example, the bundle sheath chloroplasts of C-4 plants (Plate 34). These have grana when they are young, but later show predominantly unstacked membranes. Since inorganic ions are almost certainly present in them, the factor that brings about their unstacking is likely to be a change in the molecular composition of the A-faces. It is known, however, that mutation of nuclear genes can lead to the formation of chloroplasts with unstacked membranes. This has been found in organisms as far apart as the alga *Chlamydomonas*[279] and tobacco plants,[734] and again, the implication is that the molecular composition of the chloroplast membrane can be modified in such a way that adhesive A-faces are not produced.

Further insight has been obtained by means of more drastic dissection of the membranes than in the above-described unstacking/re-stacking experiments. If the membranes are broken down into their constituent non-lipid material and lipids, and fractions mixed together in various proportions and sequences, it can be demonstrated that lipid is essential for the formation of a membranous organization.[380] One particular class of thylakoid lipids does, it seems, have special properties: sulpholipids (referred to on page 104) alone permit stacking of the membranes that form in the re-association experiments.[130] Other lines of evidence suggest that particular proteins are also needed for stacking, i.e. for formation of partitions.[484]

The primary function of partitions is likely to be something to do with photosynthesis. If their structure somehow confers advantages, then their widespread occurrence, and their abundance in chloroplasts, can be rationalized. What advantages, then, might they confer?

First, there is evidence that formation of partitions *does* create special conditions in the disc membranes. There is a strain of the alga *Chlamydomonas* (strain ac-31) which has been shown to have stacked membranes with particle distributions just as in higher plants.[281] Significantly, if the membranes are unstacked by removal of inorganic salts, the large B-face particles disappear or disperse. In other words, these particles, already suggested to be indicators of photosystem II, aggregate as a *consequence* of stacking. The molecular environment of the partitions leads to their appearance. In reassociation experiments, large particles appear in reconstituted membranes only if thylakoid proteins, lipids including sulpholipid, and chlorophyll, are all mixed together.

An immediate possibility in the search for special photosynthetic functions in partitions is that they permit, or favour, or even generate, photosystem II activity. At present this idea has to be rejected, because experimentally unstacked spinach chloroplast membranes retain their ability to perform the reactions of photosystem II.[393] It remains possible that partitions provide an especially suitable environment for linking photosystems I and II. Both are known to occur in grana with partitions, and they must act in sequence if both ATP and $NADPH_2$ are to be produced. Yet other suggestions have been put forward.[279] One is that molecules of the photosynthetic pigments are highly concentrated in the A-space between the back-to-back membranes of partitions.[873] This could have at least two advantages: one that transfer of energy from molecule to molecule might be favoured; the other that a known property of the carotenoid pigments, that of protecting chlorophyll from becoming bleached by high intensity light, might be exploited to maximum advantage.

Clearly, more work on the biochemical and photochemical functions of partitions, and on their molecular architecture, is required. A great deal of what is already known hints that they will be found to hold many of the secrets of photosynthesis.

9.3.6 Chloroplasts of C-4 Plants

The discovery of C-4 plants arose from the finding that photosynthesis in sugar cane leaves deviates from the pattern established by the researches of Calvin and his associates, which, until then, had been thought to apply to all plants.[57] Survey work then showed that certain other species, mainly of tropical or arid conditions, and including the important crops sorghum and maize, behave like sugar cane. The new biochemical observations were soon correlated with a particular type of leaf anatomy[58] that had been of taxonomic interest 30 years previously (though recognized much earlier), and with peculiarities of chloroplast ultrastructure that had been known for more than a decade.[364]

In Plate 34 *Zea mays* is used to illustrate anatomical and ultrastructural features that are found in many of the known C-4 plants. Chloroplasts occur in both the mesophyll and the bundle sheath cells (compare Plate 34a with Plate 32a, the latter showing a chloroplast-*free* bundle sheath in a C-3, or Calvin cycle, plant). Even in the light microscope it can be seen that grana, but not starch grains, are present in mesophyll chloroplasts while grana are small or absent but starch is abundant in bundle sheath chloroplasts. In some other C-4 plants the bundle sheath chloroplasts do have grana,[453a] and this point will be returned to later. In *Zea,* as in other C-4 plants, the 'agranal' condition typifies only mature leaves, and develops from more normal looking young chloroplasts by a process involving not just unstacking of grana, but destruction of membrane that is present in the grana of the immature state.[77]

The stroma, as well as the membrane system, varies across the leaf. Chloroplasts in the mesophyll lack the carbon dioxide fixing enzyme ribulose diphosphate carboxylase, while those in the bundle sheath contain it. This difference has not been easy to establish, because it is difficult to isolate the two classes of chloroplast from leaves. Recent methods for achieving good separation take advantage of an anatomical feature illustrated in Plate 34b, c—the presence of a layer of suberin in the walls of the bundle sheath cells. The suberized layer resists most cell wall degrading enzymes, which can therefore be used to digest the leaf and so to release mesophyll cells or protoplasts. These in turn yield mesophyll-type chloroplasts. If the enzyme-resistant residues are then mechanically broken to release the agranal chloroplasts, the two types can then be compared in test tube experiments.[122, 418]

FIXATION AND PROCESSING OF CARBON The mesophyll cells, though they lack ribulose diphosphate carboxylase, possess alternative enzyme systems for carbon dioxide fixation.[57] The enzyme phosphoenol pyruvate carboxylase is present, and it is thought that its role is to provide especially efficient scavenging of carbon dioxide molecules, as its affinity for carbon dioxide is much greater than is that of ribulose diphosphate carboxylase. Carbon dioxide fixation by phosphoenol pyruvate carboxylase produces organic acids containing four carbon atoms (hence the name C-4 plants). Many succulent plants take in carbon dioxide in this way, but the C-4 plants have the added ability to transport the organic acids from the mesophyll to the bundle sheath for further metabolism.[614] Because of the presence of the impervious suberized layer in the bundle sheath walls, this is assumed to be by symplastic transport (page 31), and as shown in Plate 34b and c, the necessary plasmodesmata are present. Having reached the agranal, ribulose diphosphate carboxylase-containing chloroplasts of the bundle sheath, the carbon dioxide is released from the organic acids and is finally fixed and processed by the conventional Calvin (C-3) pathway, producing starch and sugars.

The efficiency of C-4 plants does not depend solely on their superiority in collecting atmospheric carbon dioxide. There is an additional, and perhaps dominant, factor. C-3 plants are unable to utilize all of the carbon dioxide that is available to them, because of the phenomenon of *photorespiration*, in which light stimulates the *loss* of anything up to 50% of the photosynthetically-fixed carbon. The biochemical processes involved are not completely understood, but an important factor is that the enzyme ribulose diphosphate carboxylase does not always produce 3-carbon precursors of sugars. When the oxygen concentration is high and the carbon dioxide concentration is low, it yields quantities of a 2-carbon derivative of glycolic acid. In this form the carbon leaves the chloroplast and is respired, that is, oxygen is taken in and carbon is released as carbon dioxide. The role of microbodies in this process of photorespiration is considered on page 137, and it is sufficient here to point out that in C-3 plants the photorespired carbon dioxide is likely to be lost from the leaf, while in C-4 plants the more efficient CO_2-trapping phosphoenol pyruvate carboxylase of the mesophyll is likely to retrieve it for re-utilization. In both types of plant, glycolate production in the light seems to be unavoidably associated with photosynthesis. The structural and biochemical machinery of C-4 plants may be regarded as an adaptation which reduces the adverse effects of photorespiration on the rate of acquisition of carbon, and hence on the potential rate of growth.

PHOTOSYSTEMS I AND II The peculiarities of chloroplast ultrastructure in C-4 plants deserves further interpretation. The mesophyll chloroplast membranes look like their counterparts in C-3 plants, and are capable of linking photosystems I and II to produce both ATP and $NADPH_2$. Since their stroma lacks ribulose diphosphate carboxylase, these products of the light reactions are thought to be utilized in the alternative carbon fixation process, based on phosphoenol pyruvate carboxylase.

There are, however, *at least* two sorts of C-4 plants, in which differences in the processing of the carbon that has been fixed in the mesophyll correlate with differences in the ultrastructure of the bundle sheath chloroplasts.[57, 178] The plant used in the illustrations, *Zea*, has been investigated most fully. Its agranal bundle sheath chloroplasts were at one time thought to be devoid of photosystem II activity, and this observation was fitted into a general scheme purporting to show that partitions in grana are required for photosystem II. As explained in the previous section, this scheme has had to be modified, and more recent work suggests that damage of agranal chloroplasts during isolation led to the loss of key molecules and hence to the failure to detect photosystem II: it is still agreed, however, that in agranal bundle sheath chloroplasts of *Zea*, photosystem II activity is low relative to that of photosystem I.[40] This implies that these chloroplasts are likely to be able to make ATP, but to be relatively inefficient at making $NADPH_2$. The other sort of C-4 plant has grana in its bundle sheath chloroplasts, and it is suggested that they can make both ATP and $NADPH_2$ efficiently.

The correlation that is suggested to exist between the above structural and photochemical differences and the two patterns of carbon processing is expressed as follows. *All* of the C-4 plants use ATP produced by the mesophyll chloroplasts to regenerate their carbon dioxide acceptor (phosphoenolpyruvate). *One* group (including *Zea*) then either directly or indirectly uses $NADPH_2$ (also produced in the mesophyll) to reduce the product of fixation (oxaloacetate) to malate. This four carbon acid moves to the bundle sheath, where it is a source not only of carbon for the Calvin cycle of the sheath chloroplasts, but also of reducing power, needed to drive the Calvin cycle. All that must be added is ATP, and this the photosystem I activity of the agranal chloroplasts can provide. The *other* group of C-4 plants does *not* reduce the product of carbon dioxide fixation (though they do introduce an amino group, making aspartic acid), and so a source of carbon, but not of reducing power, moves to the bundle sheath. The bundle sheath chloroplasts in this group, however, possess grana, and can make the necessary reducing power ($NADPH_2$) as well as the ATP.

Whether other variations will be found remains to be seen. Meanwhile C-4 plants are obviously providing much that is of interest in relating structure to function. The main reason, however, for the intensive studies their discovery has provoked lies in their economic value as efficient collectors of carbon dioxide. The crop growth rates for sorghum and maize silage and sugar cane are at least double those for C-3 plants such as tobacco, spinach, and hay grasses.[927] Their potential in agriculture as a source of new crop plants is being investigated, and in such a programme, the examination and interpretation of chloroplast ultrastructure plays its part.

9.3.7 Development of Chloroplasts

Chloroplasts are derived ultimately from proplastids. Both multiplication of numbers and growth in size are involved. The former is considered later in a general discussion of plastid division, and aspects of the latter are described below.

GROWTH IN MASS It has been known since the last century that the familiar ribbon-shaped chloroplast of *Spirogyra* does not have restricted growing points analogous to the tip of a root or a stem. Rather it grows by incorporation all along its length, though the tips do show more rapid increases than the central region. More recent studies by time-lapse photography of individual growing chloroplasts of *Nitella* over long time periods confirm the conclusion that growth is evenly distributed, except in that strains originating outside the chloroplasts themselves do bring about slight elongation in the direction of the long axis of the cell.[289] It was concluded that new material was added to chloroplast thylakoids all over their surface, in units too small in size to be seen. Still more recent experiments, on incorporation of radioactive precursor into growing *Chlamydomonas* thylakoids, have confirmed that stacked and unstacked thylakoids, and the specialized thylakoids that traverse the pyrenoid (page 130), all grow.[275] The three regions of the total thylakoid surface were found to have slightly different rates of incorporation, but it is difficult to know how much credence to place upon the differences, in view of the known ability of lipids to diffuse within the plane of at least some membranes (page 13).

As with both endoplasmic reticulum (page 68) and the inner mitochondrial membrane (page 94), there is evidence that the molecular components of thylakoids enter the growing membrane as individuals.[428, 603] That this is the case could not be detected in, for instance, a growing leaf, where all stages of development co-exist, but, fortunately, situations more amenable to the analysis of thylakoid development are available. Two have proved particularly useful: the more or less synchronous development of thylakoids that is initiated when dark-grown leaves containing etioplasts are illuminated (page 130 and Plate 38), and comparable synchronous development seen when populations of mutant strains of *Chlamydomonas* are given an appropriate stimulus, e.g. illumination or an alteration of the nutritional conditions. During thylakoid growth in these situations the ratios of individual lipids, pigments, and proteins in the membrane alter as a result of each one having its own characteristic time course of synthesis and insertion. It is inferred that in the more normal conditions in a growing leaf, thylakoid development also proceeds by multi-step insertion of individual molecular components.

We can therefore begin to build up a picture of how thylakoids grow. The components are made individually, and inserted, probably at random, anywhere on the membrane surface. There is no evidence for modes of growth involving derivation of membrane by fusion of pre-formed vesicles, or by progressive invagination from the inner membrane of the chloroplast envelope (two mechanisms which have in the past been favoured; the envelope does, however, seem to have the ability to synthesize galactolipid—a thylakoid constituent.[177a]) No specific growth zones can be distinguished by light microscopy, but the architecture of grana and frets, as seen in the electron microscope, and idealized in Fig. 8, is so complex that growth processes that are localized on a micro-scale have to be invoked in order to account for it.

THE GEOMETRY OF GROWTH Grana are not visible during the earliest stages of the formation of the thylakoid system. Proplastids contain only rudimentary thylakoids—mere sacs, sometimes not even flattened (Plate 31a, b). When development commences, the sacs extend, flatten, and tend to become evenly spaced out. The first signs of grana appear early in the process (Plate 31c), unlike the situation in greening etioplasts (Plate 38), where clearly distinguishable flattened sacs, termed *primary thylakoids*, exist for a considerable period prior to the formation of grana. Nevertheless the mode of development of grana probably is the same during direct development of chloroplasts from proplastids as it is during greening of etioplasts to form chloroplasts. It is at this stage that the growing membrane surface begins to extend in specifically localized regions, though it is emphasized that this does not necessarily mean that incorporation of the molecular components of the membrane becomes restricted to the same regions. The shaping of the granum discs is perhaps more likely to be by a locally directed flow of membrane, the raw material for which is taken in all over the membrane surface.

The profile and face views of Plate 38c and d illustrate stages in formation of the first partition of each granum. Attachment of ribosomes to the thylakoids becomes conspicuous. A pouch of membrane protrudes from one or other face of the primary thylakoid. It does not grow out at an angle, but turns and slides over the surface of the parent thylakoid, extending until it becomes a circular disc, connected by a relatively narrow neck through which the intra-disc compartment (the disc loculus) is continuous with the intra-thylakoid space. At this stage the adhesive properties of the partition must presumably be established, involving the presence of sulpholipids and specific proteins (page 107), and the appearance of the large B-face particles that are so conspicuous a feature of grana (page 105).

It is not known what factors determine where the first granum discs form on the primary thylakoids. The first discs, however, seem to determine the sites of future membrane growth. A second granum disc begins to protrude from the parent thylakoid and grow over the first disc. It appears not just anywhere on the parent thylakoid, but in a position such that when it is mature, its connecting neck, together with that of the first disc, form the beginning of the right-handed helix of fret connections shown in Fig. 8. Membrane growth is considered to continue in this fashion, proceeding in cycles, each of which produces one disc, connected at the next position on the helix. The process has been termed spiral-cyclical, or spirocyclic growth.[351, 430, 867] The extension of each disc over the surface of the previous one must involve the same adhesion phenomena as did that of the first to form, and additionally, there must be a mechanism by which growth usually ceases when the new disc matches the area of the previous one, thus producing a stack of discs all of the same diameter, and hence a cylindrical granum. Eventually, a granum like that of Fig. 7 is produced.

As discussed earlier (page 102), the major differences between the granum of Fig. 7 and the granum of Fig. 8 lie in the multiplicity of frets and their fret connections to the granum discs. The process of spirocyclic growth does not account for the great increase in area of the fret membranes that must accompany the formation of grana. No doubt grana and frets grow at the same time, but whether there are any regularly occurring sequences of events would be extremely hard to discover, so enormously complex is the overall structure. There seems little doubt that membrane fusion, which does not participate in spirocyclic growth, must be important in establishing the many connections between discs and frets, and between neighbouring grana.[626]

It may also be that the 'ideal' granum of Fig. 8 is not achieved directly, indeed it may never be achieved in many chloroplasts. Its regularity and symmetry may be the gradual outcome of many adjustments which, obviously, are not designed purposively to produce symmetry, but which occur at random, some being stabilized if, through their presence, the information needed to sustain the whole system is reduced. The important experiments described on page 107 on the re-stacking of unstacked thylakoid systems are relevant here. They show that, given suitable ionic conditions, a great deal of the assembly of the three-dimensional architecture of grana and frets from previously unstacked membranes can be achieved in the test tube in the absence of an input of energy.

The above account of the development of the internal membrane system is presented in very general terms, and ignores the numerous factors, such as nutrition,[672] light quality,[716] intensity,[500] and periodicity,[362] that can modify the appearance of chloroplasts. The discussions will be reopened in chapter 13 in connection with the interaction in development of the plastid and nuclear genetic systems.

9.4 Etioplasts

9.4.1 *Introduction*

Flowering plants *etiolate* when grown in darkness. Etiolation consists of a set of symptoms ranging from increased stem elongation to production of leaves that are yellow due to biochemical and ultrastructural changes in plastid development. The normal processes which in illuminated plants lead to the development of green, photosynthetic chloroplasts (Plate 31c), are diverted, and proplastids instead develop into a type of plastid called the etioplast (Plate 31d).

Etioplasts retain the general features of plastids—a double envelope, a population of plastid ribosomes, and nucleoid areas containing fine fibrils of DNA (Plate 35a). Their stroma is dilute compared with chloroplasts, and plastoglobuli and occasional starch grains may be present in it.

9.4.2 *The Prolamellar Body*

The outstanding feature of the etioplast is the extraordinary architecture of its internal membrane system, a semi-crystalline lattice known as the *prolamellar body*. It is so named because when etiolated plants are illuminated, it is transformed into lamellae which gradually develop into typical chloroplast thylakoids (Plate 38a-f). The prolamellar body is thus a temporary store of membranes, produced during the limited development shown by plastids in dark-grown leaves. It represents precursor material in another sense also, in that it contains a yellowish pigment, protochlorophyllide, which gives rise to chlorophyll *via* a series of reaction intermediates upon illumination.

We have seen that mitochondrial cristae and endoplasmic reticulum can both exist in tubular form or as flattened sacs. So too can membranes of plastids: in chloroplasts they are sacs (thylakoids), while in etioplasts they are tubular. The tubes are 18-20 nm in outside diameter (lumen about 8 nm diameter), and they branch and interconnect in a variety of regular patterns, generating the extensive lattice of the prolamellar body. In leaf etioplasts the lattice is usually one to several micrometres across, and can easily be seen in the light microscope. In the fluorescence microscope the fluorescence emitted by the pigment of the prolamellar body gives an especially clear view of its regular outlines. The etioplast stroma penetrates into the lattice *between* the tubes, and *within* the tubes is the internal compartment equivalent to the intra-thylakoid space of chloroplasts.

Like so many biological (and non-biological) structures, the prolamellar body is composed of 'repeating units', or 'building blocks'. Whether it grows by stepwise insertion of pre-formed units is doubtful, but at any event a description of their structure is an aid to understanding the overall architecture of the prolamellar body.

THE CUBIC LATTICE Two types of building blocks have been found. One is rare, but the lattice formed from it

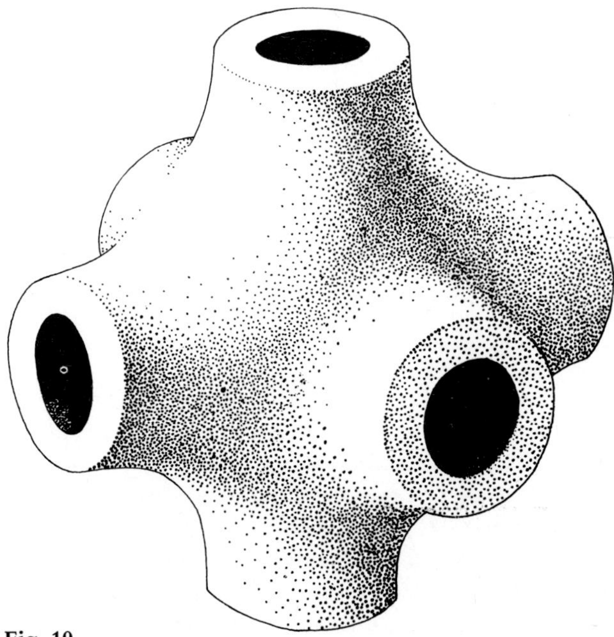

Fig. 10
The 6-armed branched tubular 'unit' of prolamellar bodies possessing the uncommon cubic lattice.

is simple, and so it will be considered first. In it the membrane tubes branch at right angles so as to point outwards in six directions (Fig. 10). It could be enclosed in a cube, with the 6 ends of the branches in the middles of the six faces. Just as cubes can be stacked together, so can these building blocks. In the actual prolamellar body, the units are joined to their neighbours by smoothly confluent membrane surfaces, and no joints can be seen in the tubes. Plate 36b shows a section through such a lattice; Plate 36c a drawing representing a portion of a three-dimensional reconstruction; and Plate 36d a photograph of a model which mimics the position of the tubes (but *not* their dimensions). The model is easily constructed using kits supplied for building models of molecules or crystals out of their constituent atoms; in this case the 6-valent hexahedron of iron is an appropriately shaped atom.

TETRAHEDRAL BUILDING BLOCKS Hexahedrally branched tubes generate a cubic lattice.[300] By far the most common building block is, however, composed of tetrahedrally branched tubes[871, 874] (Fig. 11; appropriate models can be made using models of 4-valent carbon atoms). They can be interconnected in many ways.

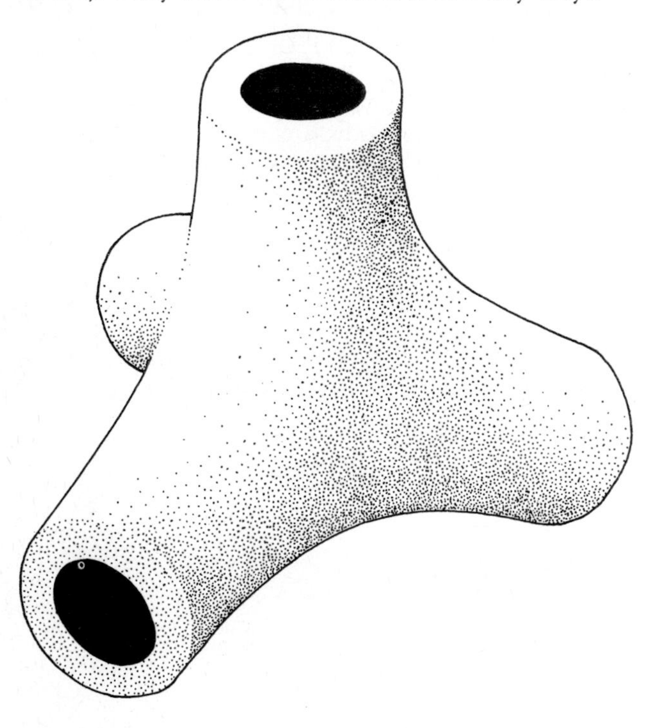

Fig. 11
The 4-armed tubular 'unit' of prolamellar bodies possessing the common tetrahedrally-based lattice types.

THE 'ZINC SULPHIDE' LATTICES Two common types of lattice correspond in their symmetry to the two crystal lattices assumed in mineral deposits by the compound zinc sulphide.[869, 870] In both types the tetrahedrally branched units join together to make three-dimensional networks of 6-sided 'rings'—seen as hexagons in the models of Plate 36f, h and the sections of prolamellar bodies of Plate 36e, g, and Plate 37g. In one of the crystal forms (wurtzite) successive planes each composed of 6-sided rings are connected in register above and below each other: viewed from above (or below) the whole crystal has a hexagonal outline (Plate 36f). The prolamellar body shown in Plate 36e corresponds to this lattice. It has been sectioned at a slight angle to the planes, so that alternating bands appear, first of hexagons belonging to one plane, then the tubes that connect this plane to the next one, then the hexagons of the next plane . . . .

Plate 36g shows a section cut at a slightly steeper angle relative to the crystal planes, so that the alternating bands are narrower, and there are more of them. The lattice in Plate 36g is of the second zinc sulphide type, corresponding to the mineral zinc-blend or sphalerite (or to a diamond lattice). When perfectly grown, it forms an octahedral crystal (Plate 36h) rather than a hexagonal prism, and the main difference is that the successive planes of hexagonal rings are displaced half a ring with respect to one another. This displacement is seen both in the models (Plate 36h) and the prolamellar body (Plate 36g) where the insert demonstrates that the hexagons of two successive planes overlap by half a ring. This does not apply to the 'wurtzite' lattice in Plate 36e. Various combinations of the 'wurtzite' and 'zincblend' types of lattice are possible.

PENTAGONAL DODECAHEDRA Several other prolamellar body configurations are based on the tetrahedral building unit. Fundamental to all of them is a body made up of 5-sided rings—pentagons—rather than the hexagons of the wurtzite and zincblend type lattices. Twenty tetrahedral units join together to give a pentagonal dodecahedron (Fig. 12), that is, a regular polyhedron with 20 vertices (the 20 tetrahedrally branched tubular units) joined together by 30 edges

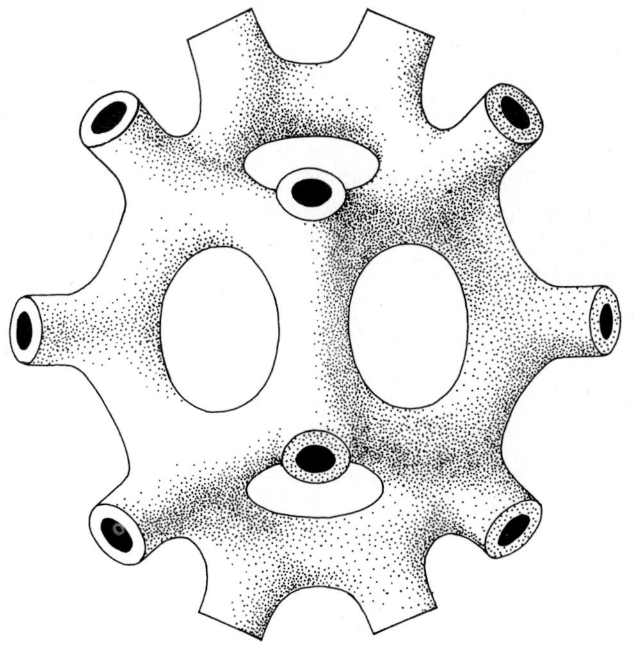

Fig. 12
A pentagonal dodecahedron formed from 20 of the 'units' shown in Fig. 11, viewed along one of the axes of 2-fold symmetry (left and right hand sides identical).

(joined tubes) circumscribing 12 (hence the prefix *dodeca*) pentagonal faces. This structure represents another building block, but at a larger scale than the small tetrahedral branches of which it itself is made. It possesses outwardly directed tubes (one at each vertex) which can join on to further units and so build yet larger structures, as described below.

CENTRIC PROLAMELLAR BODIES In the 'centric' type of prolamellar body a pentagonal dodecahedron lies at the centre of a radiating lattice of interconnected branched tubes (Plate 37b–e).[868] The radiating lattice spreads outwards from *each* of the 20 outwardly directed tubes of the central pentagonal dodecahedron. Most of it consists of the same configuration as in the zincblend type, but special arrangements are needed where the 20 outwardly radiating portions of the lattice meet and join one another. One such portion is shown on its own in model form in Plate 37a. The pentagonal dodecahedron, from one vertex of which it radiates, is at the top. If the outward radiation is perfectly symmetrical, the whole portion takes the form of a large-scale tetrahedron, as in the model. The special connecting rings are seen along the edges, and the central mass is of the zincblend-type lattice already described and illustrated in Plate 36g and h. Such large-scale tetrahedra represent building blocks of a large order of magnitude. It seems that they are put together in two main ways.

The first is illustrated in Plate 37b–e, which are micrographs and models of a prolamellar body centred on a single pentagonal dodecahedron. Twenty large-scale tetrahedral units (each as in Plate 37a) are packed together around the centre. The resulting lattice is not quite spherical. Given perfect symmetry, each of the twenty large-scale tetrahedral units would have a flat, outward-facing equilateral triangle (the triangular base the model in Plate 37a is sitting on). In other words the whole lattice is yet another type of polyhedron, this time with 20 equal triangular faces—an *icosahedron*. In fact prolamellar bodies of this type rarely if ever radiate outwards equally in all directions from their centre (see Plate 37c and compare with the symmetrical model viewed in the solid in Plate 37b and in median section in Plate 37e), and as a result icosahedra are not formed with the geometrical perfection seen in 'spherical' viruses and Buckminster-Fuller domes, which are other objects based on icosahedral symmetry, found in the realms of molecular architecture and man's constructional architecture respectively.

The second way in which large-scale tetrahedral units (as Plate 37a) can be joined to one another is illustrated in Plate 37f and g. The units are symmetrical, and pentagonal dodecahedra can be placed at one, two, three, or all four of their vertices. By so doing, large prolamellar bodies can develop, consisting of large-scale tetrahedral units stacked together and joined by the special connecting rings along their edges and by pentagonal dodecahedra at their vertices. Plate 37g shows a section through one such prolamellar body. A number of pentagonal dodecahedra lie in the plane of the section, and others presumably lay above and below the section. Plate 37f is a layer isolated from a large model so as to mimic part of Plate 37g. The disposition of the fundamental tubes is the same, though the appearance differs somewhat because the section was thinner than the layer of model (allowing for magnification). The overall shape and size of these prolamellar bodies varies according to the number of pentagonal dodecahedra, their spacing, and the extent to which each is surrounded by the twenty possible large-scale tetrahedral space-filling units.[871]

THE 'OPEN' LATTICE Development of a large lattice consisting *solely* of inter-connected pentagonal dodecahedra is impossible because the angle between adjacent pentagonal faces is unsuitable: they cannot 'fill space'. Nevertheless a remaining type of prolamellar body lattice (Plate 37h) is built in large part of pentagonal dodecahedra, arranged and interconnected in such a way that the strains imposed by joining tubes lying at unsuitable angles are acceptably slight. Its structure is based on two features: (a) pentagonal dodecahedra oriented in a particular way can join end-to-end in a straight line, (b) by straining the joints, pentagonal dodecahedra can be joined in clusters of three. A cluster of three is roughly triangular in outline. If the 3 sides of the triangle are extended by adding pentagonal dodecahedra in straight lines, and this process is repeated, a 'honeycomb' network of pentagonal dodecahedra perforated by large holes arranged in a hexagonal pattern is created. Layers made in this fashion can be joined in register one on top of the other by suitably oriented tetrahedrally branched tubes. Plate 37j shows a model viewed from above, with the hexagonally placed holes, and Plate 37h is a micrograph in which the plane of the section is close enough to the plane of a pentagonal dodecahedral network to allow detection of the pattern seen in the model. Plate 37i and k present side views of this most complex lattice. Clearly, the thickness of the section is considerably less than the diameter of the internal space of a pentagonal dodecahedron, and this greatly complicates interpretation of the structure.

OCCURRENCE OF DIFFERENT TYPES OF LATTICE The illustrations of prolamellar bodies in Plates 36 and 37 were selected in order to emphasize one further point. All of the different types shown are from one species of plant. Thus it is *not* the case that a given species, or a given tissue, develops a given type of lattice to the exclusion of others. Indeed, as shown in Plate 36a, different types can not only co-exist within one cell, they can be physically joined within one etioplast. The actual numerical proportions of the different types has been very little studied, and although several authors are agreed that the 'open' type is especially common in barley (*Hordeum*), there are no clues as to how this comes about or to its possible significance.

THE STRUCTURE AND FUNCTION OF THE PROLAMELLAR BODY The visual elegance of ultra-thin sectioned prolamellar bodies and preoccupation with details of their structure should not lead us to neglect general features of their structure and function.

Without doubt the most impressive aspect of prolamellar body structure is the exceptionally high surface area to volume ratio that is shown by the membrane in it. The surface area of the building blocks on which both the cubic and the tetrahedral lattices are based can be calculated, once their dimensions have been obtained from micrographs of known magnification. A 6-armed unit (as Fig. 10) is estimated to have an outer surface area of $7.7 \times 10^{-3}\ \mu m^2$; a 4-armed unit (as Fig. 11), $5.0 \times 10^{-3}\ \mu m^2$. By measuring appropriate dimensions, the number of basic units per cubic micrometre of prolamellar body can be estimated, and in turn so can the surface area of membrane per cubic micrometre of prolamellar body. The infrequently seen cubic lattice contains about 37 μm^2 per μm^3; the abundant 'zinc sulphide' lattices about 54 μm^2 per μm^3. Those prolamellar bodies that are centred on pentagonal dodecahedra contain a slightly less extensive surface per unit of volume than the 'zinc sulphide' type, and the 'open' lattice (Plate 37h) is considerably less efficient.

To help place the above figures in perspective, it may be recalled that the highest observed surface to volume ratios for rough endoplasmic reticulum cisternae (page 59) are about three times *less* than the maximum surface to volume ratio for the branched tubular thylakoids of prolamellar bodies. It may be that the 'coat of mail' configurations of smooth endoplasmic reticulum (page 64) approach prolamellar bodies in the degree to which they pack large surface areas of membrane into small volumes. Such configurations are, however, uncommon.

Nevertheless, the complexity of prolamellar body structure is greater than would seem necessary if their sole function is to store membrane. Myelin figure arrangements (Plate 3b) would do this still more efficiently. What distinguishes the prolamellar body membrane system is (a) that it is a continuous surface, and (b) that it separates two unlike compartments which are also continuous, one being the plastid stroma penetrating *between* the tubules, and the other the space permeating the system *within* the tubules. Although there might be more compact stores for membrane as such, there can be few more compact structures than prolamellar bodies for storing membrane that is at no point out of contact with the compartments that it separates.

It is probable that the observed conformations of prolamellar body membranes are as regular as they are because they represent stable conditions. Comparable geometries could be generated by dipping suitable wire frames in and out of soap solutions. The soap films so formed are subject to forces of surface tension, which cause the films to contract to shapes in which the surface area, and the energy needed to maintain the surface, is minimized. The stability of such surfaces is maximized when their surface area is minimal. In theory, the wire frames could be made to give films that mimic the shape of prolamellar body membranes, and again in theory, the frames could be made of repeating units, and provided that the basic symmetry is maintained, there need be no limit to the overall size, and the number of units in the frame. Mathematicians call such surfaces 'infinite periodic minimal surfaces', and their mathematical properties have been studied since the first description was published in 1865.[752] There is a considerable number of possible infinite periodic minimal surfaces, and their properties have, it seems, been exploited in a variety of situations in Nature, from prolamellar bodies in plants to the single-crystal plates of calcite that sea urchins produce as their exo-skeleton.[176, 587]

Thus, by analogy with soap films, it can be suggested that the form taken by a prolamellar body membrane is determined in part by forces of surface tension. It remains a mystery why there are different types of prolamellar body. Perhaps the type that develops in a growing etioplast is a matter of chance. Once a particular type has been initiated, however, it seems likely that during its growth by insertion of new membrane, the existing configuration would be extended without change. This would not preclude the development of mixed lattices (e.g. Plate 36a), which could arise if two (or more) prolamellar bodies of different type appeared and merged during their growth within a given etioplast. It is notable that when this happens the continuity of the membrane, and its contact with the stroma, is maintained from one type to the other.

The fact that contact between membrane and stroma is maintained throughout all prolamellar bodies (unlike the situation in the partitions of grana) implies that this structural feature may have special functional significance. If components of the membrane are made in the stroma, the arrangement would allow them to be inserted into the membrane at any point on its surface—in which case membrane growth need not be restricted to the periphery of the prolamellar body. The presence of ribosomes within the stroma component of the prolamellar body[301] has a bearing on the nature of the components that could be made and inserted in this way.

One of the most abundant components of the prolamellar body is a large protein associated (in extracts, and probably *in vivo*) with molecules of protochlorophyllide, the precursor of chlorophyll. It is known as protochlorophyllide holochrome, and the light-induced production of chlorophyll, which begins when an etioplast is illuminated, occurs only if the precursor pigment is attached to the protein. The holochrome is estimated to have a particle diameter of about 10 nm, and from assays of the quantity present per leaf and hence per prolamellar body, it can further be estimated that the area of membrane available more-or-less matches the area that would be created if the holochrome molecules were arranged side by side in a single layer.[417]

The subject of the holochrome and the pigment conversions will be returned to in a later section, dealing with the development of chloroplasts from etioplasts (page 131).

9.4.3 Development of Etioplasts

It is a valid generalization that in dark-grown flowering plants, etioplasts with prolamellar bodies develop instead of chloroplasts. The origin—proplastids—is unchanged by the lack of light; only the nature of the developmental process differs. It should, however, be noted that prolamellar bodies can also appear under some conditions of illumination (see Plate 34b), and conversely, that in many algae and other 'lower' plants, light is not required for chloroplast development and so etioplasts and prolamellar bodies are not present, even in darkness.[716]

During early stages of etioplast development the proplastids of the young leaf cells enlarge, starch grains often accumulate in them, [433, 874] and the amount of internal membrane increases. Continuities between the inner membrane of the envelope and the internal system are commonly seen, and it has been suggested that the thylakoids form and extend by continued invagination from the inner envelope. There is, however, no firm evidence for this, and while some synthesis could occur in the manner envisaged, it is extremely likely that thylakoid growth occurs all over the membrane surface, or at any rate, is not restricted to any one small area, such as at a point of invagination.

As to the nature of the etioplast thylakoids, they have been found to contain nearly all of the lipids that are present in mature chloroplast thylakoids, most of the proteins, including some components of the electron transport chains of the photosystems, and carotenoid pigment but no chlorophyll, unless the leaf (etc.) has at some time been exposed to light.[13] As yet none of the differences between etioplast and chloroplast thylakoids, either qualitative or quantitative, can be related to the profoundly different conformation of the membranes in the two types of plastid. This applies even to the conspicuous difference in pigmentation, for prolamellar bodies have been seen under experimentally manipulated conditions in which no protochlorophyllide—normally so characteristic of the structure—could be detected.[831]

In short, it is not known what, in physico-chemical terms, causes prolamellar body development. Even the straightforward description of *how* the structure forms is difficult. The first obvious difference between etioplast and chloroplast development from proplastids is that in the former the developing thylakoids are perforated.[433, 874] The pores are small, about 20-30 nm in diameter, and are remarkably uniformly distributed over the thylakoid surface. As in nuclear envelope pores, the membrane is continuous around the pore margin, and the intra-thylakoid compartment is not in contact with the plastid stroma. It seems to be common (but not obligatory) for these perforated thylakoids to lie wrapped closely around starch grains.

A first step in generating a tubular network from perforated thylakoids is easy to envisage. For instance, pore enlargement would reduce the area of flat membrane between adjacent pores, creating instead a network of membranous tubes. If the original pores were spaced out equidistant from one another, the network would end up with a hexagonal pattern—a more symmetrical version of the network seen at the forming face of a dictyosome in Plate 28a. Formation of the three-dimensional lattice is less easily envisaged, though once formed, its stabilization by forces of surface tension to produce 'crystalline' symmetry and minimization of surface area seems feasible (page 114).

However, perforated thylakoids do not disappear during prolamellar body formation (Plate 36a). Either they are *not* converted to prolamellar bodies, the latter being synthesized independently, or they *are* converted, but concomitantly are synthesized sufficiently rapidly to be always present. Certainly the two forms of membrane are connected at many points. Pertinent data have been obtained for etioplast development in the first leaves of *Phaseolus* seedlings.[74] Starting 6 days after germination, etioplasts in the young leaves produce perforated thylakoids to a maximum area of about 30 μm^2 per etioplast by day 12. Prolamellar body development parallels this growth 3 days later, starting at day 9 and reaching a maximum volume of about 1.5 μm^3 per etioplast by day 15. Using the surface area estimations previously given (page 114), there would be 70-80 μm^2 of membrane in one of these full-sized prolamellar bodies. Only between days 12 and 15 is there any diminution (amounting to about 10 μm^2) in the area of perforated thylakoids. The diminution probably reflects incorporation of perforated thylakoid membrane into the prolamellar body, and it seems reasonable to infer from this that the two forms of membrane *are* in a precursor-product relationship throughout the period of development. It is usual that prolamellar body growth does not consume all of the thylakoid, and ultra-thin sections of mature etioplasts nearly always show the presence of an excess, in which perforations can still sometimes be seen (Plates 36, a, e, g and 37h).

Development stops when the etioplast and its prolamellar body or bodies reaches a critical size, in which the amount of membrane is much less than in a mature chloroplast. What determines the critical size is not known, but it would appear that the supply of key enzymes or precursors runs out, or becomes blocked, and that illumination is a prerequisite for their replenishment. The effects of illumination are considered later (page 131), meanwhile it is worth reiterating two points. The first is that lack of light does not of itself *cause* prolamellar body development. They can appear during chloroplast development in the light.[59, 334, 716, 876] The second is that neither is the presence of protochlorophyllide in prolamellar bodies a causal factor for their development, for they can appear when it is undetectable.[831] Thus to the major problem of determining the mechanism of prolamellar body formation we can add two others: what causes it to grow, and what causes it to stop growing?

9.5 Amyloplasts and Starches

9.5.1 Introduction

The most familiar of the tertiary energy stores of plants is starch.[677] It is a polymer of glucose, but is not a simple, chemically definable compound. In fact starch is usually a mixture of two major types of polyglucoside, amylose (which in the extreme case consists of unbranched polymers, several hundred glucose units in length), and amylopectin (considerably larger branched polymers with branches only 20-30 glucose units in length). Plastids which specialize in the production and storage of starch grains are known as amyloplasts.

Amyloplasts, like other plastids, are bounded by a double membrane envelope. Here and there the inner of the two membranes is continuous with the internal membrane system, which is usually sparse (Plate 39c), but occasionally is well enough developed to show small grana. The term chloro-amyloplast can be used in the latter cases, and if they develop in weak light or darkness they may have a small prolamellar body instead of small grana. The stroma between and around the starch grains contains DNA in nucleoid areas (Plate 39c), and the plastid ribosomes are usually few in number.

The diagnostic feature of amyloplasts is their high content of starch. In the vast majority of higher plants the starch exists as grains large enough to be seen in the light microscope, and easily recognized by their 'maltese cross' appearance in the polarizing microscope, and by their dark blue colour following treatment with iodine. Plate 39 displays a range of examples—though by no means the full range to be found in the plant kingdom. Three variables have to be considered: the chemical composition of the polyglucoside, its site in the cell, and the morphology of the deposits.

9.5.2 Chemistry of Starches

The majority of starches contain two or three times as much amylopectin as amylose, but the relative proportions may alter as the organ or tissue containing the amyloplasts develops.[34] Several instances of increase in the percentage of amylose during growth have been recorded. By contrast the starch grains in sieve element plastids (Plate 39d) come to contain a very highly branched type of amylopectin.[624]

There are also extremes in which one or other of the two major components has been favoured by selection of particular genetic mutations. Sometimes the selection has been natural, and sometimes it has been artificially imposed by plant breeders. For example, maize (*Zea*) breeders have produced varieties[762] in which the starch has an exceptionally high content of amylose (though the grains do not look unusual), others which produce almost entirely amylose-free amylopectin ('waxy' varieties),[363] and others in which the pattern of glucose polymerization generates a substance that is more like glycogen, the storage polyglucoside of animal tissues, than starch ('sugary' varieties of sweet corn). In the latter case the polyglucoside is called phytoglycogen, and there is some evidence that its structure, as well as its chemistry, resembles that of animal glycogen.

Phytoglycogen occurs in another, extraordinary, botanical situation.[691, 692] The flowering plant *Cecropia* and ants have in the course of evolution established an association from which both derive benefit. As part of their 'mutualistic symbiosis' the plant produces numerous small multicellular bodies, up to 1 mm long, that are harvested and eaten by the ants. Although other parts of *Cecropia* trees make normal starch grains, the plastids of the ant food-bodies accumulate phytoglycogen. As in animal cells, the phytoglycogen is in the form of rosettes (55-75 nm diameter) composed of smaller particles (25-40 nm diameter). No other example of a 'glycogen plastid' is known, though the sugary sweet corns deserve to be examined in this regard. Interestingly, there are several records of animals producing a structure that in ultra-thin sections appears very similar to *Cecropia* glycogen plastids; it is, however, a mitochondrion, with glycogen rosettes in its stroma.[809] Also, it is satisfying to find that just as occasional plants can produce the animal product glycogen, so animals can in certain pathological conditions produce the plant product starch.[796]

9.5.3 Cellular Sites of Starch Formation and Storage

Higher plants make starch in the stroma of their plastids. In the case of chloroplasts this is where the carbohydrate products of photosynthesis emerge, and removal of these products (as glucose-containing units) to form insoluble starch in the same compartment is a rapid and economical way of avoiding a harmful build up of high concentrations of soluble, osmotically active material. The starch formed in leaf chloroplasts during the day is usually broken down and transported away during the night. It is referred to as *assimilatory* starch (Plates 32c, d, 34a-c, 35c-g).

Formation of reserve starch (Plate 39a) in amyloplasts in storage tissues also occurs in the stroma, but the raw material is taken into the cells and across the amyloplast envelope, having been translocated from sites of photosynthesis to the site of storage. Amyloplasts frequently lie in tissues that under normal conditions receive no light. Clearly, then, the enzymatic machinery of starch formation is independent of photosynthesis. Indeed starch grains of normal appearance form in plastids of completely non-photosynthetic flowering plants which obtain nutrient by parasitism (e.g., *Cytinus*, *Lathraea*, *Orobanche*) or *via* association with fungi (e.g. *Monotropa*,

Sebaea, Gymnosiphon).[464, 465, 508] It has been claimed however, that one parasite, *Thonningia*, a member of the Balanophoraceae, has lost the ability not only to photosynthesize, but also to make starch grains.[508] It would be of interest to see what would happen if it were to be supplied with an abundance of glucose-containing precursors, for this treatment can induce starch formation in situations where normally none or little is made.

It should not be thought that plastids are the only structures capable of making storage polyglucosides. The green algae behave rather like higher plants in this respect (Plate 35c-g), but in several other algal groups starch grains, or deposits of starch-like material, form outside the plastids.[265] The red algae are particularly noted for their cytoplasmic grains of 'Floridean starch', which is a much-branched amylopectin type polyglucoside.[363] Cytoplasmic deposits of glycogen are common in fungi and bacteria, and, of course, in animals. The slime mould *Dictyostelium* deposits glycogen in its walls during the spore-forming stage of its life cycle. There is also a fungal form of starch. It resembles amylose, and is found in walls of spores, asci, or hyphae of a range of asco- and basidiomycetes. In addition, certain yeasts secrete a starch-like capsular polysaccharide.[546]

9.5.4 Morphology of Starches

GROSS FEATURES Glycogen and phytoglycogen occur as rosettes of small particles, whereas starch accumulates as large grains. Some of the large grains in the mature potato tuber cells shown in Plate 39a and b approach 60 μm in length, but as can be seen in the micrographs, a single cell can contain both small and large grains. Even small starch grains (e.g. in proplastids, Plate 31b) are usually large enough to be visible in light microscope preparations (Plate 1). The very large grains are likely to be formed singly, but it is common to find several to many grains within an individual amyloplast (Plate 39c, e).

The grains themselves may be simple (like those in Plate 39) or compound. Compound grains consist of many sectors that have grown and fused together. Single grains may be round, ellipsoidal, or may develop a variety of shapes that within limits are characteristic of the species. For example,[34] wheat endosperm starch grains are approximately kidney shaped, though their shape, and the extent of a groove that runs around their periphery, changes as they age. In barley endosperm the young grains are somewhat similar to wheat when young, but they do not change shape as they age. Some corn starches are asymmetric, with irregular lobes and branches. Characteristic dumb-bell shaped grains are found in spurge (*Euphorbia*) latex.[505] Such information has been used to compile complex and comprehensive classification schemes.

Generally speaking, characteristic shapes are found only in reserve starches, which are present as long-term stores—in tubers, seeds, etc. Assimilatory starch grains produced during photosynthesis in chloroplasts nearly always assume simple lens shapes, regardless of species (e.g. Plates 32c, 34b).

SUBSTRUCTURE The early light microscopists, starting with van Leeuwenhoek in 1719, found much to interest them in starch grains. Nägeli, in 1858, even envisaged the establishment of a new discipline—the study of the molecular organization of 'bodies',—for which he suggested starch grains would be most suitable objects. Although molecular biology has indeed been established as a new discipline, it has not yet succeeded in showing in detail how starch grains are constructed.

Despite its greater resolution, the electron microscope has not contributed as much as the light microscope to knowledge of the substructure of starch grains. A variety of electron densities is seen in the micrographs, ranging from dense (Plate 39c, d) to almost transparent (Plates 28a, 35d-f). Often the density to electrons is greater at the periphery of the grain than at the centre (Plate 39e), perhaps reflecting incomplete penetration of fixatives, but equally possibly reflecting gradations in composition. Grains are surrounded by electron-transparent haloes in the majority of pictures (e.g. Plate 34b, c), but it is doubtful whether the halo contains a particular substance. Swelling and shrinkage cycles during specimen processing may displace the surrounding stroma and membranes away from the grains.

Many reserve starch grains are built up of concentric layers centred on a 'hilum', which may be in the centre of the grain, but is often placed eccentrically. If the hilum is off-centre (e.g. in potato tuber starch) the concentric layers are ellipsoidal shells, thicker on one side of the hilum than the other. The procedure known as lintnerization (treatment with acids in the cold) may be needed in order to demonstrate the layering, and even where layering is obvious to start with (as in the 2-7 μm thick layers of potato), sub-division into finer layers (50-100 nm in potato) may become evident.[796]

There is general agreement that starch grains grow by depositing new material on to their outer surface, i.e. by *apposition*. The most convincing demonstration of this has been made by autoradiographing grains some time after supplying a pulse feed of radioactive carbon dioxide: the labelled starch (made from labelled products of photosynthesis) accumulates on the surface skin. The stratified layers of reserve starches could therefore arise as a result of periodicity in the apposition process. An obvious suggestion is that one layer might form every 24 hours, and wheat and barley plants have been grown under a range of day-night cycles in order to test this hypothesis. In some experiments it was indeed found that in continuous light the layering was absent (or not obvious). Resumption of alternating day-night cycles led to production of layers, one layer per cycle. Presumably the products of each day's photosynthesis were translocated to the developing seeds, where apposition, interrupted by the periodicity of the photosynthesis, created the layering.[106] It must be emphasized, however, that this is not a general phenomenon. Thus irregular layering has been detected in wheat grown in constant light,[391] and there are several instances, including potato tuber starch, in which layers just the same as those found under natural conditions appear when the plants are kept under continuous light. There must be an in-built periodicity of apposition in these cases.

Another concentrically layered structure—myelin—was described on page 13 and Plate 3b. In myelin figures the layering is much finer than in starch, but the similarity of structure confers similarity in optical properties. Plane-polarized light is transmitted or absorbed according to whether or not the plane of the layers lies in the plane of polarization. Starch grains and myelin both give 'maltese cross' images in the polarizing microscope, and this pattern of birefringence, together with data on the effects of infiltrating solvents between the layers, indicates not only the existence of layers ('form' birefringence), but also that molecular constituents tend to lie at right angles to the layers, that is, radially in the grain as a whole ('intrinsic' birefringence). This applies to both amylose and amylopectin in starch,[796] just as to lipid molecules in myelin figures.

The behaviour of starch grains with respect to polarized light, and the X-ray diffraction patterns given by them, imply the existence of crystalline regions. The proportion of non-crystalline material is, however, much higher than in the case of cellulose—another polymer of glucose known to contain crystalline micelles embedded in amorphous regions. Several different models that aim to depict the manner of packing of amylose and amylopectin in starch grain layers have been proposed,[249] but as yet, none is regarded as perfect or of general applicability.

9.5.5 *Metabolism of Starches*

So many enzymes have been discovered that can bring about the formation and breakdown of starch that there is dispute about which of them actually operate in life. The balance of opinion currently favours for a major role in starch synthesis an enzyme which transfers glucose units from an ADP-glucose complex to the growing molecules of starch. The ADP-glucose complex is synthesized by a separate enzyme, whose postulated role is also vital, for it is activated by 3-phosphoglycerate, a 3-carbon compound that is one of the first products of carbon dioxide fixation in photosynthesis. Thus photosynthesis automatically activates the very enzyme systems that can cope with an influx of photosynthate by storing it as starch.[669]

Forms of both the glucose-transferring enzyme and phosphorylase, which is another enzyme capable of making starch in the test tube or in tissues that are fed with the appropriate substrate, are found tightly bound to starch grains.[509] It seems that they become embedded within their own reaction product. Other enzymes occur, for instance there is one that is capable of converting unbranched amylose to branched phytoglycogen, and another than can introduce branches as in amylopectin molecules. Clearly, each type of amyloplast must have its own characteristic set of enzymes, producing the various chemical forms of starch. It has been suggested that phosphorylase, the ADP-glucose transferring enzyme, and the various branching enzymes, have all evolved by specialization from a primitive multi-functional enzyme.[244]

For an insoluble form of food, such as starch, breakdown is a prerequisite of utilization, and is as important as synthesis. Again several enzymes are known, notably amylase,[410] along with a variety of de-branching enzymes.

In structural terms three patterns of erosion of starch grains have been observed. One is a general erosion which eats into the grain all over its surface.[107] Another occurs where plastid membranes come close to the grain, etching deep holes in it.[361] Both of these patterns are seen in intact plastids, and may reflect the presence of free and membrane-bound starch-degrading enzymes respectively. The third pattern occurs in cells which die during utilization of their food reserves, as in endosperm and some cotyledons. There the amyloplast membranes break down, starch grains are liberated, and corrosion starts at localized points on their surfaces and spreads inwards.[604] The latter pattern is probably due to amylase, but other enzymes, including phosphorylase, may operate in undamaged plastids.

It is of interest that assimilatory starch, which is formed and degraded over a short time period, is more susceptible to breakdown by amylase than is storage starch,[24] which is not usually degraded until long after it has been laid down. Such differences may reflect different degrees of branching in the starch molecules. For instance the grains found in sieve element plastids (Plate 39d) are only attacked by amylase if they are first treated with a special de-branching enzyme, e.g. pullulanase.[624]

9.5.6 *Amyloplasts and the Perception of Gravity*

Roots and stems usually grow in a direction that is related to the direction of gravitational force. Other factors influence the direction of growth, but in darkness and constant temperature and humidity, primary roots grow downwards, main shoots upwards, and various lateral organs may grow horizontally or at other angles fixed with respect to gravity.

The amyloplasts of the central column of cells in the root cap, and of the 'starch sheath' cell layers of shoots, have long been suspected to perceive gravity. They are heavier than most other cell components, and if a root or a shoot is displaced, they fall to the gravitationally 'down' part of the cell. Other responses occur, including sometimes alterations in the position of endoplasmic reticulum (page 68) and dictyosomes, but the movement of the amyloplasts is the most dramatic and consistent.[415] Plate 39c, e show amyloplasts in root caps, where a distinction has to be made between the gravity-sensitive amyloplasts of the central cells, which retain their starch, and the amyloplasts of the peripheral cells, which lose their starch, probably as raw material for dictyosome-directed synthesis of extra-cellular mucilages (page 77 and Plate 28a). The gravity-sensitive types are known as *statoliths*.

Other types of statolith exist. In rhizoids of the alga *Chara* they are round crystalloids, probably derived from endoplasmic reticulum.[756] How any statolith, whether in root cap, starch sheath, or rhizoid, by occupying a particular part of the cell, or falling away from that part, can influence the direction of growth of the whole organ is not understood. Complex interactions with endoplasmic reticulum may be involved, leading to redistribution of growth promoting or growth inhibiting substances being transported to the zones where cell elongation takes place.

9.5.7 *Development of Amyloplasts*

Like other members of the plastid family, amyloplasts develop from proplastids. Proplastids can contain small starch grains (Plate 31b), and the major step in their development towards amyloplasts is the enhancement of their starch-synthesizing capacity, either by activation of pre-existing enzymes or the acquisition of new ones, or both. As already mentioned, the very presence of raw materials is, in at least some cases, all that is needed to induce starch formation.

Setting aside the production of assimilatory starch in chloroplasts, two broad categories of amyloplast can be recognized: transitory amyloplasts and long-term amyloplasts. Chloroplasts, etioplasts, and some chromoplasts and leucoplasts often pass through a temporary amyloplast state during their development from proplastids, and it may be that deposition of starch in these transitory amyloplasts is nothing more than a response to the translocation of food reserves into the tissues where the developments are taking place—part of a reaction system that maintains a favourable osmotic balance in the cells by rendering incoming soluble secondary energy stores insoluble. Later, the starch is consumed. The amount and form of the thylakoid system in transitory amyloplasts varies greatly, depending upon their subsequent fate, and on the conditions of illumination. They can in fact be intermediate in structure between the long-term amyloplasts of storage tissues, which have very little internal membrane, and etioplasts or chloroplasts.

Within both of the broad categories of amyloplast there are pronounced species- and tissue- differences. Variation from plant to plant in the shape of reserve starches has already been mentioned, and within an individual plant it is obvious that specialized cells or tissues can express their own sets of starch-specifying information. For example, root cap starch grains (Plate 39c, e) differ from reserve starch in seeds of the same species. The characteristic starch of sieve element amyloplasts has also been mentioned. The *waxy* character of maize and other cereals is expressed only in pollen and endosperm, and not in maternal or embryo tissue. Perhaps the prime example of tissue-specificity is in *Cecropia* (page 116), where phytoglycogen is produced only in the food bodies harvested by ants, other tissues having starch of conventional chemistry and appearance.

These species- and tissue- specificities presumably stem from the presence of different enzymes or combinations of enzymes, the production of which is in turn controlled by the genetic information of the cell. It is not clear what part, if any, the DNA of the amyloplast itself might play, but it has been observed that deposition of starch markedly reduces the extent of DNA-synthesis in plastids, compared with that seen in starch-free chloroplasts in the same tissues.[345] The available evidence does, however, show that many aspects of amyloplast formation are under the control of the cell nucleus. Indeed one of the characters selected for study by Mendel in his pioneering work on inheritance involves the metabolism and form of starch. He followed the inheritance of 'round' and 'wrinkled' seeds in pea plants and showed that the 'smooth' condition derives from a dominant gene, now known to be nuclear. It is now also known that in the round-seeded variety the starch grains in the seeds are simple, with about 35% amylose when mature; by constrast wrinkled seeds contain compound grains with about twice as much amylose.[34] The numerous varieties of maize described on page 116, with their differing starches, also derive from alterations to the nuclear gene system.

9.6 Chromoplasts

The word *chromoplast* means 'coloured plastid', but it is usually applied in a much narrower sense. Interpreted literally, it would embrace chloroplasts, which are coloured green, various shades of brown, or red, but nevertheless are not regarded as chromoplasts. The distinction is that the bulk of the coloured pigment of chloroplasts functions in photosynthesis, while most (or all) of the pigment in chromoplasts is nonphotosynthetic. In conventional usage, therefore, chromoplasts are non-photosynthetic coloured plastids.

This definition aptly describes those chromoplasts which develop directly from proplastids, for example the chromoplasts of carrot roots, which contain the pigment beta-carotene, and being below ground are obviously non-photosynthetic. In other cases chromoplasts develop from chloroplasts, and they may even revert to a chloroplast state, and here the distinctions between photosynthetic chloroplasts and non-photosynthetic chromoplasts are blurred during the transition periods. The colour of some chromoplasts is due to accumulations of pigments that are also found amongst the photosynthetic pigments of chloroplasts (e.g., beta-carotene). Other chromoplasts manufacture pigments such as lycopene or delta-carotene which are either absent from chloroplasts or present merely as intermediate in pathways of biosynthesis of photosynthetic pigments.

Just as chloroplast or amyloplast development is a part of the differentiation of the cells and tissues in which they are found, so too is chromoplast formation. Chromoplasts can develop in a variety of tissues, but their appearance is associated particularly with reproduction. At the lower end of the evolutionary tree they are common in 'resting' and reproductive spores of algae. In more advanced groups in the plant kingdom they are responsible for the bright red, orange, and yellow colours of flower and fruit tissues in many (but not all) species; as with the more varied hues of vacuolar pigments in petals (etc.) their function here is presumed to lie in the attraction of animal agents of pollination and seed dispersal.

General features of plastids are retained in chromoplasts. The double envelope, plastid ribosomes, and DNA-containing regions in the stroma are all seen (Plate 40b). However, with massive accumulations of specific pigments, and loss or absence of photosynthetic function, it is not surprising that chromoplasts develop characteristic structures. No grana are seen at maturity (Plate 40a), though they may be present during development (Plate 40b). In some species the chromoplast pigments come to lie in the stroma, either in droplets or bound to filamentous structures. In others the pigments crystallize, probably starting their crystallization within the intra-thylakoid spaces, and distorting first the thylakoids and later the shape of the whole chromoplast by the process of crystal growth.

DROPLETS AND FILAMENTS Most chloroplasts, indeed most plastids in general, contain droplets which take up osmium during fixation in osmium tetroxide. They are called osmiophilic droplets, or plastoglobuli (page 127), and they contain plastid quinones, some of which function in electron transport during the light reactions of photosynthesis in thylakoid membranes. In chromoplasts their numbers and dimensions are greater than in chloroplasts, and unlike the situation in chloroplasts, they become repositories for carotenoid and xanthophyll pigments as well as for quinones.[488] *Forsythia*,[473] *Tulipa*,[489] and *Spartium*[595] petal cells provide examples in which chromoplast plastoglobuli are the major, and perhaps the only, site of pigment accumulation. In many other plants, however, they occur along with other pigmented structures (e.g., Plates 40a and b).

Filamentous pigmented bodies are quite common in the chromoplast stroma of fruits (e.g., red peppers, *Capsicum annuum*),[785] and, according to currently available records, less common in petals (e.g., cucumber, *Cucumis sativus*)[775] In cucumber the 'filaments' are in fact narrow tubes, 11-18 nm in external diameter, of uncertain length, lying in bundles, and possibly derived from degenerating thylakoids in the chloroplasts from which the chromoplasts of the petals develop. Neither the composition of the tubes, nor the nature of their association with pigment molecules, has been elucidated.

CRYSTALS Crystalline forms of carotenoids in chromoplasts have been examined in detail in carrot roots (*Daucus*, beta-carotene),[47] narcissus flowers (*Narcissus poeticus*, beta-carotene)[449] and tomato fruits (*Lycopersicon*, lycopene)[46] The techniques of ultra-thin sectioning, electron diffraction, and freeze etching have all been used.

Inspections of sections shows that deposition of what can be inferred to be pigment commences within special thylakoids. In carrot roots these thylakoids arise in proplastid type precursors, while in *Narcissus* and tomatoes they are part of the internal membranes of chloroplast type precursors. Subsequent events are hard to discern, largely because the crystals are seen only as electron-transparent areas with angular outlines, i.e., as 'empty' areas in the sections (Plate 40a, b). An explanation that has been put forward to account for the electron transparency is that the pigment is extracted when the tissue is being dehydrated in organic solvents prior to embedding, but experiments on crystals isolated from carrot roots show that dehydration does not remove all of the pigment.[47] The 'emptiness' may therefore be due to a combination of partial extraction and a lack of chemical reactivity with osmium tetroxide fixative.

Crystallization of pure samples of lycopene and beta-carotene yields crystals with the same outline as the 'empty' angular spaces in ultra-thin sections. Furthermore, freeze etching of chromoplasts (a technique which avoids dehydration) shows that the crystals are genuinely solid structures, composed in fact of sheets of carotenoid molecules. Data obtained by electron diffraction and X-ray diffraction show that the sheets seen in freeze etched material represent planes of the crystal lattice, and has allowed the mode of packing of the carotenoid molecules in these layers to be deduced.[47, 449]

It can be seen from the micrographs (Plate 40a, b) that many membranes undulate through the crystals. Undulating stacks of very electron-dense membranes also occur, especially in early stages of crystal growth. It has been suggested that these expanses of membrane have special functions in the synthesis of carotenoid pigments.[317] The basis of their strange conformation is not understood: some of them, it would appear, are embedded within crystals, as if the overall form of the crystal was generated by the coming together and concerted growth of subcrystals that originated in neighbouring intra-thylakoid spaces, the intervening membranes becoming wavy, perhaps as a result of shrinkage following extraction of crystal material during dehydration.

REGULATION OF CHROMOPLAST DEVELOPMENT Some chromoplasts differ from chloroplasts quantitatively:

they have less internal membranes, less of some pigments (e.g., chlorophyll), and more of others (e.g., carotenes). Other chromoplasts differ qualitatively as well as quantitatively, possessing pigments that characterise the chromoplast state, at any rate when present in bulk. It can thus be anticipated that both qualitative and quantitative controls of the activity of genes are involved in the production of chromoplasts.

Although chromoplasts contain DNA, the genes that so far have been found to regulate carotene synthesis and the development of chromoplasts all reside in the cell nucleus.[430] Despite this, carotenoid pigments appear to be made *within* chromoplasts and chloroplasts (this is, however, not obligatory, for fungi and animals can make carotenoids in the absence of plastids). The most striking demonstration of nuclear control of chromoplast development has come from studies of gene mutations that affect carotene synthesis. In maize (*Zea*), which like other grasses does not normally produce chromoplasts, a mutant known as *lycopenic* has been described.[861] The mutation has been traced to genes in a particular chromosome in the nucleus, and its effect is to block carotene synthesis, lycopene (the chromoplast pigment of, e.g., tomato fruits), instead being accumulated. What would otherwise be chloroplasts become chromoplasts. In other words, mutation of a nuclear gene induces development of a type of plastid that normally is absent from the species. Interestingly, both the bundle sheath (agranal) and mesophyll (granal) chloroplasts of the maize leaves are affected.

It has been suggested that cultivated carrots (*Daucus carota* subspecies *sativus*) arose by a mutation affecting plastid development that has been selected and bred by man. In wild carrots (*Daucus carota* and other species) the plastids of the tap-root do *not* accumulate beta-carotene and are *not* chromoplasts. Similar ideas apply to those varieties of sweet potato (*Ipomoea batatas*) which have orange or yellow chromoplasts in their tuber cells, and are regarded as especially useful sources of vitamin A in the human diet.

Comparisons of cultivated varieties of peppers (*Capsicum*) and tomato (*Lycopersicon*) fruits have shown that nuclear genes dictate the nature of the carotenoid pigments, and hence certain ultrastructural features, of the chromoplasts. Some peppers are coloured by chromoplasts containing filamentous pigmented structures; others by chromoplasts with plastoglobuli.[430] In most varieties of tomato, lycopene is the major carotenoid pigment and crystals of this substance appear in the chromoplasts (Plate 40b).[317] There are, however, varieties with unusually high quantities of beta-carotene (the chromoplasts contain very numerous plastoglobuli, as well as lycopene crystals)[316] or delta-carotene (chromoplasts appearing much as the normal type),[315] and others with an unusually low content of pigments (no crystals, and few plastoglobuli in the chromoplasts): in all this variation it is to be noted that the differences appear only when fruit ripening commences and the chloroplasts (which look the same in the different varieties) start to transform to the chromoplast condition.

Induction of chromoplasts by mutation is, of course, an abnormal occurrence. It cannot account for the normal development of chromoplasts from proplastids or from chloroplasts, processes which do not involve permanent changes to genes. Chromoplast development could involve gene activation and repression, and quantitative regulation of transcription and translation. Since chromoplast formation is one facet of the time course of events of cell differentiation in many fruits and petals, the appropriate genes must be programmed to exert their influences at appropriate times during development. By contrast, there are cases where the control mechanisms are activated by environmental influences, as distinct from being part of in-built developmental programmes. This is particularly evident in the algae, where numerous genera and species react to nitrogen deficiency by turning bright orange, globuli containing carotenoids (usually oxygenated, or 'secondary' carotenoids) accumulating in what under better growing conditions are chloroplasts, and also in the cytoplasm.

9.7 General Features of Plastids

Having examined the major members of the plastid family in respect of their individual features, it is now necessary to insert a general section concerned with the family features shared by all plastids.

9.7.1 The Plastid Envelope

All plastids are bounded by two concentric membranes, separated by a gap 10-20 nm in thickness. Selected illustrations of this double envelope are: Plate 31b (proplastid), Plate 36a (etioplast), Plate 32c (chloroplast), Plate 40b (chromoplast), and Plate 39c (amyloplast). There are few more important boundary structures in our world, for it regulates the inflow of raw materials for photosynthesis, and the outflow of photosynthetic products. Somewhat surprisingly, the envelope membranes resemble cytoplasmic membranes much more than they do the internal membrane system of the chloroplast, having a high content of phosphatidyl choline and a low content of galactolipids,[502] and in their possession of the ability to synthesize large amounts of membrane lipid, in this case galactolipid.[177a]

THE OUTER MEMBRANE The outer membrane is smooth in outline, about 6 nm in thickness, and has internal particles which, when revealed by freeze etching, indicate a molecular architecture which differs from that of the thylakoids.[52a, 555, 582] When chloroplasts with intact envelopes are isolated from cells and suspended in solutions of low-molecular weight compounds, such as sucrose, phosphate, or ATP, all are found to penetrate rapidly into a small fraction of the total chloroplast volume. It is inferred that the outer membrane is unspecifically permeable to these compounds, and permits them to enter the gap between the inner and the outer membranes, this representing the 'small fraction' of the total volume, generally referred to as the 'sucrose permeable space'.[330]

Similar experiments have been carried out using mitochondria (page 92), but there seems to be an important difference between them and chloroplasts. Whereas in mitochondria any solute that traverses the outer membrane into the gap between it and the inner membrane can without obstacle penetrate into the intra-cristal spaces, in chloroplasts the intra-thylakoid space is largely cut off from the gap between the two membranes of the envelope. This is because in mitochondria the cristae are nearly always invaginations of the inner envelope membrane, whereas in chloroplasts the thylakoids are generally separate from the inner envelope (though some continuities are seen, especially during development). Those who write about transport of molecules into and out of mitochondria and chloroplasts therefore have to be careful to specify which region of the structures they are studying: in mitochondria the two possibilities are the inter-membrane space and the matrix, while in chloroplasts the intra-envelope space, the stroma, and the intra-thylakoid space constitute three compartments, each with distinctive properties.[330]

THE INNER MEMBRANE Freeze etching of the inner plastid envelope membrane gives views which in some cases are similar to and in others different from those of the outer membrane.[52a 555] Ultra-thin sectioning, however, nearly always shows morphological differences between the two. The inner membrane is rarely completely smooth, but possesses frequent folds which invaginate into the plastid stroma (Plates 31b, 33a, 34c, 38a, 41a). In some cases neighbouring invaginations anastomose and create a complex membranous labyrinth, known as the 'peripheral reticulum'. This is not, as at one time thought, restricted to chloroplasts in C-4 plants (Plate 34c).[453a] Several C-3 chloroplasts possess it. The peripheral reticulum undoubtedly enhances the surface area of the inner envelope membrane, and it seems likely that herein lies the functional significance of the structure. The inner membrane is known to carry a variety of specific solute transport systems, and it may simply be that the greater its surface area, the greater its capacity to mediate specific import and export to and from the chloroplast stroma. It is, of course, possible that assessment of the significance of the structure solely in terms of surface area is too crude an approach: it could be that the shape of the invaginations, perhaps coupled with specialization at the level of the molecular composition of the membrane, is such as to confer on the invaginations especially efficient conditions for solute transport.

Several specific solute transport systems have been detected in the inner membrane of the chloroplast envelope. As would be expected, they concern both the raw materials for, and the products of, the light and the dark reactions of photosynthesis.

Carbon fixation in chloroplasts involves operation of the complex Calvin cycle in the stroma. The inner envelope membrane maintains high concentrations in the stroma of vital intermediates of this cycle (e.g. those which regenerate the carbon dioxide acceptor molecule ribulose diphosphate), while at the same time allowing products of carbon fixation to leave the chloroplast, either directly or after a temporary period of storage as starch. The major form in which sugar is moved around the plant is sucrose, but this compound does not appear to be formed in large quantity within the chloroplast. Instead phosphorylated forms of 6-carbon sugars (fructose and glucose) are produced, and transported outwards across the envelope, to be linked together in the cytoplasm to make sucrose, a compound which (as already stated), does not leak back into the chloroplast because the inner envelope membrane is impermeable to it.[326]

There is no doubt that energy derived from the light reactions of photosynthesis is utilized in the cytoplasm. For the cytoplasm of cells which are darkened or which contain only non-photosynthetic plastids, the energy is necessarily received by a roundabout route, in which translocated sugar is oxidized by mitochondria to release its contained energy. For cells which themselves contain photosynthesizing chloroplasts other options are available.

It might at first sight be thought that direct export of ATP and $NADPH_2$ (the two energy-containing products of the light reactions) across the envelope to the cytoplasm would occur. However, rates of ATP and $NADPH_2$ production in photosynthesizing chloroplasts are very high (compared, for instance, with rates in the mitochondria in the same cell), and are dependent upon the continual availability of the respective precursors, ADP and NADP. Depletion of the precursors would occur if the products were to be exported, unless there was a matching return flow from the cytoplasm to the chloroplast. It seems that such problems are in fact avoided, by preserving more or less intact the pool of ADP, ATP, NADP, and $NADPH_2$ molecules within the chloroplast. Instead, energy and reducing power are exported to the cytoplasm in the form of phosphorylated 3-carbon sugars[325] and organic acids[800] such as malate (see also page 109). It is thought that 'shuttle' systems exist, in which the exported sugar or acid molecules are stripped of their available energy or reducing power in the cytoplasm, and the residual carbon skeletons are then returned along with inorganic phosphate across the envelope to the chloroplast and stroma for regeneration and re-export.[324a] Some transport of ATP itself can occur, but at a slow rate relative to the carrying capacity of the 'shuttles', indeed it may be that the main direction of ATP movement is from *cytoplasm to plastid,* to cope with energy requirements within chloroplasts during darkness, or within non-photosynthetic plastids.

THE ENVELOPE AND OTHER CELL COMPONENTS Hints that the chloroplast envelope mediates yet other types of solute transport can be obtained simply by looking at the distribution of cell components.[517] For instance the spatial association between microbodies and chloroplasts in leaf cells (Plate 41a) speaks of metabolic interchange across their respective membranes, a topic which is dealt with in the next chapter (page 137).

Other examples are found in the algae, where in many cases the chloroplasts are shrouded in a layer of endoplasmic reticulum lying just external to the chloroplast envelope.[265] This 'chloroplast endoplasmic reticulum' may in turn be continuous with the nuclear envelope (pages 52 and 129). The functional significance of such arrangements is not understood. Chloroplast endoplasmic reticulum does not occur in the red or the green algae, and so no absolutely fundamental role can be assigned to it, but where it *is* present, it is a regular and predictable phenomenon, and has been seen in some instances to persist unaffected even by the breakdown of the nuclear envelope during mitosis. The shroud of reticulum could simply be a pathway which transports chloroplast products to other parts of the cell, or it could be a strategically placed compartment in which chloroplast products can be further metabolized, as well as transported.

Spatial relationships between chloroplasts and other cell components suggest a further degree of complexity, namely that the chloroplast envelope can be regionally differentiated in respect of its transport activity. To a certain extent this is implied in the microbody-chloroplast association already described, but more striking examples are seen in the algae. In some cases a microbody may be found in the cytoplasm just across the chloroplast envelope from the pyrenoid (page 137), and nowhere else; in others, droplets or 'caps' of food reserve accumulate immediately across the chloroplast envelope from the pyrenoid. There seems to be no doubt that the envelope near to the pyrenoid engages in transport processes to an extent not matched in its remaining regions.

RESISTANT CHLOROPLAST ENVELOPES The chloroplast is in general a delicate structure, and skill and expertise are required in order to isolate it from the cell in an intact and functional form. The first sign of damage usually is that the chloroplast envelope ruptures. The stroma can then leak away, and the appearance of the chloroplasts in the light microscope changes from shining refractile bodies to swollen dense masses in which the grana can be seen.

The above changes have become familiar since the first biochemical work on isolated chloroplasts in the 1930s, and it is only recently that an extraordinary exception has come to light. The chloroplasts of certain giant-celled marine green algae can be taken from the cell and will withstand abuse that would shatter higher plant chloroplasts. Their envelopes somehow resist both chemical and mechanical attack.[270a] One much studied example is provided by the alga *Acetabularia*, which yields chloroplasts that remain active for hours in the test tube, photosynthesizing at rates equal to those found in life. In the test tube, however, they tend to retain their products of photosynthesis, whereas in the cell the transport systems of the envelope operate to supply the cytoplasm.[764]

The longevity of chloroplasts isolated from algae with resistant chloroplast envelopes is exploited in a remarkable manner by a number of molluscs which have not only become adapted to feed on these algae, but have specialized in piracy of the algal cell chloroplasts.[833, 834] The algal cells are exceptionally large[575b], and so it is feasible for the animals to suck into their alimentary tract bulk samples of protoplasm. What happened in the course of evolution can only be speculated upon—perhaps initially the resistant envelopes led to excretion of the chloroplasts while the remaining protoplasm was digested. At any event, what happens in the present-day organisms is that the chloroplasts, protected as they are by their envelopes, are taken into the living cells lining the gut. There, because the animals are largely transparent, they can photosynthesize for months, donating fixed carbon to the host mollusc.

Resistant chloroplast envelopes may not be resticted to the large-celled marine green algae. Several higher plants have become adapted to survive in desert conditions through their ability to withstand long periods of desiccation. When moistened, they resurrect and begin to function again, this including a resumption of chloroplast activity. It may be speculated that one mechanism of chloroplast survival in such conditions involves the formation of chloroplast envelopes embodying special qualities not found in more conventional land plants.

9.7.2 *Plastid DNA*

Thanks to the green colour of normal chloroplasts, defective types with altered pigmentation are easily detected. As a result, it has for long been possible to analyse the mode of inheritance of chloroplast defects.[430, 862] Many have been found to be rooted in alterations of genes which, since they conform to the laws of Mendelian genetics during breeding experiments, are known to reside in the cell nucleus. Certain other inheritable defects, however, do not behave as would be expected if Mendelian nuclear genes are responsible for them. The concept that plastids can carry some of their own genetic information therefore emerged, and a search for DNA in plastids began.

NUCLEOID STRUCTURE AND NUMBERS Early (pre-1963) attempts to isolate DNA from plastids were plagued by uncertainties. The minute quantities detected might have come from bacteria or portions of nuclei contaminating the preparations. It was some time until improvements in methodology permitted unambiguous identification and assay of plastid DNA, and allowed investigations of its function to commence.[912] Meanwhile additional evidence was being obtained by light and electron microscopy. Plastids in some favourable materials were shown to contain areas which react positively in the Feulgen procedure (and others) for staining DNA. The plastid stroma was seen to contain matrix-free zones in which fine fibrils, 2-3 nm in thickness, ramified. Pre-treatment of suitably fixed specimens with the enzyme deoxyribonuclease removed the fibrils, thereby identifying them as DNA, or at least, showing that their structure is based on DNA. The marked resemblance of these areas of the plastid stroma to the DNA-containing regions of bacteria (and blue-green algae) led to the application of terminology already in use for prokaryotes, and they became known as nucleoplasm areas or as nucleoids. The latter term has already appeared in this book in connection with mitochondria (page 90), and will now be used for plastids also.

The micrographs of ultra-thin sections presented here show that nucleoids occur in proplastids (Plate 31b), developmental stages intermediate between etioplasts and chloroplasts (Plates 35a, 38a, e), amyloplasts (Plate 39c), and chromoplasts (Plate 40b). With these and many other observations, there is no doubt that they occur throughout the plastid family. Individual sections may not reveal their existence, e.g. Plate 32c-f, either because they are absent from the part of the plastid that fell within the thickness of the section, or because the fibrils may be partially obscured by other components of the stroma. Fortunately, means have been discovered for enhancing the contrast and general visibility of the nucleoids,[346, 441] and these methods, coupled with the use of serial sectioning and three-dimensional reconstructions, has enabled the number, volume, shape, and disposition of plastid nucleoids to be examined.

Generalizing, it appears that the larger the plastid, the greater the chance that it contains more than one nucleoid.[347, 443] For example, in special strains of *Beta vulgaris* that exhibit a wide range of chloroplast sizes, the number of nucleoids was found to range from one in a 0.9 μm^3 chloroplast to at least five in a chloroplast of volume 3.7 μm^3. The nucleoids occupied approximately 10% of the total chloroplast volume, but the volumes of the individual nucleoids varied considerably—they are by no means units of fixed size. Three-dimensional reconstructions showed also that they are not of any fixed shape. Rather they lie in irregular configurations, sometimes flattened, sometimes lobed, between the thylakoids of the grana and fretwork. It is not always possible to be sure that the nucleoids are discrete, because small connecting fibrils could easily escape recognition.

Nucleoids in the chloroplasts of various algae have been studied in some detail. In several cases they have proved to be detectable by light microscopy, thus facilitating the surveying of populations of chloroplasts, as distinct from making detailed observations on just a few chloroplasts by means of the time-consuming procedures of electron microscopy. In the single chloroplast of a *Chlamydomonas* cell there are one or two lens-shaped nucleoids;[693] in various brown[53, 54, 55] and golden-brown[264] algae the nucleoid is single, in the form of a continuous ring passing right around the periphery of the chloroplasts; in the giant cells of *Acetabularia* and *Polyphysa* a minority of chloroplasts have either one or two nucleoids, but the majority (65-80%) have no nucleoids and, judging by the outcome of a range of experimental procedures, no detectable DNA;[911] and in each of the two chloroplasts in a cell of the dinoflagellate *Prorocentrum* there are 80-100 irregularly flattened disc-shaped nucleoids of various dimensions.[442]

NUCLEOID DNA Study of the internal composition of plastid nucleoids has involved both structural and biochemical approaches. The DNA, like that of mitochondria (page 90) and prokaryotes, but unlike that of nuclei, is free of histone protein. The amount present per 'average' chloroplast or etioplast falls within the range 10^{-15}g—10^{-14}g,[428, 783] taking into account data for algae as well as for higher plants. A bacterium such as *Escherichia coli* contains about the same quantity. In terms of length of DNA double helix (page 40), the above range of values is equivalent to 300—3 000 μm.

It is impossible to measure the total length of DNA per plastid by looking at ultra-thin sections. If, however, great care is taken to avoid shearing the molecules, DNA can be induced to flow out of plastids disrupted by detergents and enzymes on to flat surfaces, and picked up on support films on electron microscope grids. By then it is spread in a two-dimensional array and its length can be measured after staining or shadow casting. Interpretation of the values obtained is hazardous. The extent of breakage is uncertain, and in most cases all that can be done is to measure many molecules and present histograms of size classes against frequency of observations. Some investigators have used procedures which in addition allow estimation of the total length per plastid. Without going into detail, it is becoming apparent that DNA strands 40-60 μm in length are of frequent occurrence in chloroplasts from algae to higher plants, but that the total length per chloroplast can be very many times greater than this, for example up to 1000 μm (the alga *Acetabularia*)[911] and up to 5500 μm (*Beta vulgaris*).[443] The 40-60 μm strands may be in the form of circles,[440, 511] just as mitochondrial DNA is often circular. There are also, however, abundant linear strands.

The significance of the frequency with which 40-60 μm strands are found is enhanced by biochemical evidence. Biochemists speak of the 'analytical complexity' and the 'kinetic complexity' of DNA. The former represents the results of gross analyses, such as the above quoted values of 10^{-15}—10^{-14}g DNA per plastid. Within these total quantities there may be different degrees of kinetic complexity. The most complex DNA strand is one in which no two stretches of the DNA have the same base sequence, so that all of the DNA has the potential of representing unique genetic information. Less complex situations are possible. If within the total DNA there are similar or identical units, the total possible genetic information is reduced in inverse proportion to the degree of repetition, or reiteration. The term 'kinetic complexity' arises from the method used to detect reiterated sequences of bases within a total sample. The double helices are opened to form single stranded DNA, and the rate at which the separated strands recognize one another and reassociate to the double helix form is measured. The more repetition, the more rapid the kinetics of reassociation. The kinetic constants can be used to estimate the size of the repeating units.

Determination of the kinetic complexity of chloroplast DNA from both algae and higher plants has revealed the existence of some heterogeneity, but also, the presence of units with a molecular weight of 1-2 $\times$ 10^8, which corresponds to a mass of 1.6-1.3 $\times$ 10^{-16}g, or a double strand length of about 50-100 μm. The total DNA of a lettuce chloroplast would accommodate about 25 such units,[883] and corresponding values are: 10-90 (*Beta*),[443] 25 (tobacco),[816] 25 (*Chlamydomonas*)[884] and 9-72 (*Euglena*).[512]

The structural evidence therefore is that plastid DNA commonly contains strands some 40-60 μm long, that one such strand is in most cases a small proportion of the total DNA of the plastid, and that the total DNA is distributed amongst one to several or even many nucleoids of different sizes found within individual plastids. The biochemical evidence quantifies the DNA content, and shows that a substantial proportion of the total is in the form of similar, if not identical units. Obviously, the biochemically detected units may in fact correspond to the ultrastructurally observable units.

Collating all of the available information, it is apparent that in most cases there are many more units of DNA in a plastid than there are nucleoids—between 4 and 8 times mores in *Beta* chloroplasts,[443] and either 12 or 25 times more in *Chlamydomonas* chloroplasts (depending on whether one or two nucleoids are present in each). Applying the term used for nuclei which have multiple copies of the basic genetic information, the nucleoids must, on average, be polyploid (page 42). It has been pointed out that some nucleoids are larger than others, hence the degree of polyploidy probably varies, even within an individual plastid. Obviously, such variation does not apply where there is only one nucleoid, as in brown algae (see above).

Another indication of variation in DNA content from nucleoid to nucleoid within a plastid comes from autoradiographic studies of DNA synthesis using radioactive precursors. Within a given time interval some presumed nucleoids become more radioactive than others, though the conditions for all are exactly the same.[345] It seems reasonable to conclude that relative to the smaller ones, large nucleoids contain more units of DNA and incorporate more radioactive precursor during periods of synthesis. Autoradiographic experiments also show that synthesis of plastid DNA is not restricted to the same time period as synthesis of nuclear DNA (the S-period, page 43). In some organisms (e.g. *Ochromonas*)[266] plastid DNA appears to be synthesized continously; in others (e.g. *Chlamydomonas*) it is sythesized at a particular period during interphase;[123] in *Euglena* the rate of synthesis is faster than the rate at which the chloroplasts divide, implying that some destruction occurs, as well as synthesis.[510] As with nuclear DNA (page 41), plastid DNA is synthesized by opening the double helix and laying down new single strands upon the pre-existing ones, giving two double helices where formerly there was one.[123]

The existence of sites on membranes of prokaryotes at which DNA synthesis is activated was mentioned when dealing with relationships between chromatin and the nuclear envelope (page 52). In view of their existence, it is of interest that plastid DNA, already seen to be similar in several respects to prokaryotic DNA, appears in at least some algae to be attached to specific points on the surface of thylakoids in the plastid.[56, 442] Associations with thylakoids have also been claimed for higher plant chloroplasts.[783, 913] It remains to be seen how widespread and specific the association is, and whether it functions in the synthesis or the spatial organization of DNA in the plastid.

Additional properties of the plastid DNA system will be considered in sections concerned with the division of plastids (page 132) and the functional integration of the different genetic systems of the cell (page 174).

9.7.3 Plastid Ribosomes

Most of the micrographs of plastids in Plates 31-40 demonstrate the existence of small, electron-dense particles in the stroma. Plate 35a in particular shows several of their more important features. As a result of biochemical and electron microscopical investigations, they have been identified as ribosomes.

Points of similarity between them and cytoplasmic ribosomes are that they stain heavily in ultra-thin sections when uranyl acetate solutions are used; they can assume a range of polyribosome configurations; when isolated they can be induced to dissociate into two subunits; they contain RNA and protein; digestion of suitably fixed specimens with the enzyme ribonuclease removes their RNA and renders them unstainable in ultra-thin sections; and, most importantly, when supplied with the necessary co-factors and raw materials, they will assemble protein molecules.

Points of difference between them and cytoplasmic ribosomes are that they are slightly smaller; the population is usually less crowded; and there are very many biochemical features which show them to resemble the ribosomes of prokaryotes more than they do the ribosomes lying on the cytoplasmic side of the plastid envelope in the same cell.

Several of the above points deserve to be elaborated. The size difference, for instance, between plastid and cytoplasmic ribosomes (e.g. Plate 35a), is directly apparent to the eye, but rather difficult to quantify, for the particles are not perfectly symmetrical. Most 'diameters' as measured in ultra-thin sections of plastids are in the range 17-20 nm. In purified samples that have been negatively stained, long and short axes measure 22 and 17 nm respectively, whereas comparable preparations of cytoplasmic ribosomes are larger, at 26 × 19 (pea) or 20 (bean) nm (all figures with a margin of error of plus or minus 1 nm).[94] Rates of sedimentation during centrifugation imply that the mass of a cytoplasmic ribsome is 1.35 – 1.4 times greater than that of a chloroplast ribosome.[299] A similar difference is seen whatever biochemical parameter is used: plastid ribosomes have smaller subunits, smaller RNA molecules in the subunits, less protein, and fewer types of protein.

The reduced size and complexity of plastid ribosomes (particularly in regard to their proteinaceous components) perhaps indicates that they are less highly evolved systems for assembling proteins than are cytoplasmic ribosomes. Nevertheless the mode of construction of a plastid ribosome is fundamentally similar to that of a cytoplasmic ribosome. In both types of ribosome one subunit is larger than the other, with a larger main RNA moelcule, and containing more species of protein. The RNA molecules of both types mature from large precursors by loss of portions of transcribed RNA[319] (see also page 47). There is some evidence that the genes for the large and the small RNA exist side by side in the plastid DNA, much as in the nucleolar DNA that governs synthesis of cytoplasmic rRNA.[803]

Autoradiographic evidence obtained by feeding the alga *Ochromonas* with radioactive uridine suggests that RNA (probably plastid rRNA) is synthesized at or near the nucleoids of the chloroplast,[264] and then migrates into the general stroma—a remarkable parallel to comparable studies on nucleoli, where the initial incorporation of label also occurs in the vicinity of fibrillar DNA (page 48). Ribosome-like particles can be seen associated with fine fibrils in plastid nucleoids (Plate 35a), perhaps immobilized by fixation at the site of their synthesis. In short term experiments on *Ochromonas*, nearly one third of all of the silver grains indicative of RNA synthesis in the chloroplast were located over the DNA-containing nucleoid.

The extensive multiplication of genes for rRNA that is such a fundamental feature of the nucleolus does not, however, apply to the plastid. There is not enough DNA in one set of genes in the plastid, indeed currently available evidence indicates that each set of genes includes only one, [684] or at most two or three,[803] coding for rRNA. In view of the fact that several sets of genes can be present within one nucleoid, and several nucleoids within one plastid (page 124), there can, however, be considerable numbers of plastid rRNA genes in a plastid and even more in a cell. Thus, in the unicell *Euglena* the 450 nuclear genes for cytoplasmic rRNA are nearly matched by a total of about 350 plastid genes for plastid rRNA.[684] Whether other organisms are so well endowed remains to be seen, but it is not unreasonable to expect them to be, for the total number of plastid ribosomes in a cell can be very large. In the alga *Chlamydomonas* the single chloroplast contains about 50 000, i.e. some 30% of the total number in the cell.[70] In mature green leaves of the broad bean (*Vicia faba*) the number of cytoplasmic ribosomes more or less equals the number of chloroplast ribosomes. Obviously not all types of plastid contain such large populations: etioplasts in *V. faba* possess only about 20% of the total in the cell. Young chloroplasts in cells in a leaf initial about 8%, and proplastids in a root tip cell about 4%.[181] It follows that plastid development involves extensive synthesis of plastid ribosomes.

PLASTID POLYRIBOSOMES Cytoplasmic ribosomes exist either free or bound to the endoplasmic reticulum, and as single particles or as polyribosomes. As described below, plastid ribosomes behave similarly, with thylakoids replacing the endoplasmic reticulum as the membrane surface to which the ribosomes can be bound.

Plate 35a illustrates a variety of plastid polyribosome configurations, including irregular clusters, chains, and what may be a helical arrangement of the type shown for cytoplasmic polyribosomes in Plate 21c (insert). Various other whorls and spirals have been recorded. The distance from the centre of one ribosome to the centre of its neighbour is, on average, slightly smaller than is the case in cytoplasmic material, and occasionally a fine strand, possibly of mRNA, is seen connecting the ribosomes to one another.[205]

Plate 35a also shows the type of polyribosome that lies in contact with thylakoids. Several other examples of association of polyribosomes with thylakoids are illustrated, but nowhere better than in Plate 38e, f, depicting a stage of development in which thylakoid synthesis was proceeding very rapidly at the time of fixation. Plate 38c, showing material fixed at the onset of a period of rapid synthesis of thylakoid, includes a few membrane-associated polyribosomes, but they are not seen in the mature chloroplasts of Plates 32c and 41a. It would appear that the association becomes conspicuous when plastid membranes are being produced, and is less marked at other times.

Biochemical separations have provided additional evidence concerning thylakoids and polyribosomes. The two are, for instance, genuinely attached, and not just loosely juxtaposed. Fractionation into frets and grana suggests that the majority of polyribosomes are attached to the grana,[641] but presumably only to the surfaces that are in contact with the stroma. When treated with the antibiotic puromycin, which causes ribosomes to release the proteins that they are synthesizing, the thylakoid-bound polyribosomes of *Chlamydomonas* detach, indicating that it is the product of polyribosomal activities, namely protein molecules in varying stages of maturation, that link the ribosomes to the membrane.[126, 531a] The detached proteins remain associated with the thylakoids,[121] and the obvious interpretation of both the biochemical and the ultrastructural observations is that at least some of the proteins of plastid membranes are made on the spot, and inserted at sites of membrane growth.

Not all of the polyribosomes of plastids are bound to thylakoids. Plates 32c, 35a, 36a, and 38a-f all show free polyribosomes in the stroma, both during times of membrane synthesis, and at times that would be expected to be relatively quiescent. When these polyribosomes are induced prematurely to release the proteins they are making, the latter remain in free solution,[121] and it is probable that the destination of protein molecules that are matured normally on free plastid polyribosomes is in fact the stroma of the plastid, which is, of course, known to be rich in many proteins.

Differences and similarities of cytoplasmic and plastid ribosomes have already been described, and in concluding this outline of the properties of plastid ribosomes it is worth pointing to the differences and similarities of cytoplasmic and plastid *poly*ribosomes. Free polyribosomes in both situations tend, it is thought, to liberate the proteins they make into the surrounding ground substance. Membrane bound polyribosomes in the cytoplasm can make proteins that remain in the membrane—just as in the plastid, but their other activity, the manufacture of proteins destined to be passed through the membrane into an internal compartment and accumulated or passed to the exterior or to other compartments, has not yet been detected in plastids. Its occurrence is not impossible, however, for some intra-thylakoidal deposits do accumulate (e.g. Plate 39e), and in some cases do seem to be exported from the plastid to other parts of the cell.[15]

9.7.4 Plastoglobuli

Plastoglobuli are, as the word suggests, tiny droplets found in plastids. They lie in the stroma, have no bounding membrane (i.e. they are not vesicles), and are therefore not unlike the small lipid droplets or spherosomes that are often seen free-floating in the cytoplasm. As will be seen, however, plastoglobuli are distinctive not only in their restriction to the plastid stroma, but also in their chemical make up.[488]

They usually take up considerable amounts of osmium during fixation of the tissue, and hence are dense to electrons (Plates 31b, 32e, f, 35b, 38e, f, 40b). Although they are sometimes known as 'osmiophilic globules', their staining pattern does vary, perhaps indicating that their composition is not always the same. Plate 32d shows examples with a dark periphery and a less dense core, and in the prolamellar body remnant of the etioplast shown in Plate 38a there are several examples that are no more dense than the surrounding stroma.

OCCURRENCE The micrographs listed above show that plastoglobuli occur in most types of plastid. They have in fact been found in all, though their abundance and size differ according to the nature and stage of development of the plastid. They are invariably small in proplastids (Plates 3a, 31b). They are frequent in etioplasts, especially near (Plate 37c) or in (Plate 37d, g, h) the prolamellar body. They are not prominent in amyloplasts unless some thylakoid and pigment formation has taken place, but as already seen, they are common in chloroplasts and abundant in some (but not all) types of chromoplast (Plate 40a, b and page 120).

Within limits, the number and size of plastoglobuli reflect what is happening to the thylakoids. Degradation of thylakoid membrane is accompanied by accumulation of plastoglobuli, as seen in senescing chloroplasts, chloroplasts converting to chromoplasts, and possibly during breakdown of prolamellar bodies in etioplasts. By contrast, their numbers do not rise, and sometimes even fall, when thylakoids are being formed, as in growing chloroplasts and during advanced stages of 'greening' of etioplasts to form chloroplasts. It has in several instances been noted that the *end* of the phase of thylakoid development is marked by an increase in the population of plastoglobuli.

CONTENTS Their restricted occurrence and their relationships to thylakoids immediately suggest that plastoglobuli contain material that is characteristic of plastids, and that is also a component of thylakoids. Following their isolation and biochemical analysis it has in fact been found that their major constituents are lipid-soluble plastid quinones, several of which, like ubiquinone in mitochondria (page 92), function in the membrane-located electron-transport reactions that are coupled to phosphorylation of ADP.

Two types of plastid quinone are distinguishable.[490] The synthesis of one type, represented mainly by vitamin K_1, is closely geared to thylakoid formation. Very little excess is produced, and hence not much vitamin K_1 is found in plastoglobuli. Synthesis of the other type (e.g. plastoquinone and α-tocopherol (= vitamin E)-quinone) seems to proceed over and above the requirements of thylakoid formation, and the excess becomes apparent as plastoglobuli. The abundance of some of the precursors (aromatic amino acids) of this second type is enhanced by nitrogen deficiency, and the consequent over-production of the quinones is reflected by development of a large population of plastoglobuli, a phenomenon that is especially conspicuous when certain algae become nitrogen-starved. The abundance of free plastid quinones is augmented when they are liberated as a result of breakdown of thylakoids, thus accounting for the appearance of many plastoglobuli at such times.

The plastid quinones are soluble in lipid because the bulk of their molecules consist of long chain terpene components. Being hydrophobic, they tend to self-associate in droplets—plastoglobuli. If, however, other lipid-soluble materials are present, they too may enter the plastoglobuli. For example, under normal conditions carotenoid pigments are restricted to the thylakoids, but in some categories of chromoplast, and during chloroplast senescence, enhanced synthesis or release of carotene, lycopene, etc., brings these hydrophobic compounds into the non-aqueous environment of the plastoglobuli. Again, mature and ageing chloroplasts of a number of plants produce even longer chain terpene alcohols—polyprenols, and when abundant, these too appear in plastoglobuli as well as in thylakoids.[488]

EYESPOTS IN MOTILE ALGAE Plastoglobuli do not occupy specific regions of the stroma in plastids of higher plants. By contrast, in many motile algae and motile gametes of non-motile algae, a population of plastoglobuli forms a tightly knit group, visible as an orange-red spot in the light microscope, and specifically located in the chloroplast. This, the *eyespot* or *stigma*, has been suspected for nearly a century to function in sensory perception, mediating the orientation of the cells towards or away from a light source, depending upon its intensity. As in chromoplast plastoglobuli, carotenoids are present.[174]

Eyespot plastoglobuli are in many species neatly aligned in flat or curved disc-shaped layers. Within the layer they may be so closely appressed that their individual outlines become hexagonal rather than circular. There may be more than one layer, with stroma and a thylakoid interspersed between them. The size of the plastoglobuli varies from species to species, but a diameter in the range of 100 nm is frequently observed. A layer of plastoglobuli can be several micrometres in diameter.

Not all types of eyespot are restricted to the chloroplast. Some are in the cytoplasm at the periphery of the cell or near the base of a flagellum, and some enter into exceedingly complex spatial associations with other components of the cytoplasm. Of those that do lie within the chloroplast envelope, there seem to be two sub-types, one lying near to the base of a flagellum, and the other at the cell periphery, often in a convex bulge of the chloroplast envelope, separated from the plasma membrane by only a very tenuous film of cytoplasm.[175]

Eyespots pose two especially intriguing problems. How do they function? And how is their structure maintained? As to the first, the major hypothesis at present is that they function by causing a shadow to fall upon some other photoreceptor, probably associated with a flagellum, thereby allowing the direction from which the light is coming to be sensed—rather as a gun is known to point at a light when the shadow of the front sight falls upon the back sight. Actually, rotational movements of the cell mean that the shadowing effect is intermittent, and the photoreceptor structure, whatever it may be, seems to operate by perceiving periodic stimuli.[167]

As to the second problem, it is a considerable mystery how an aggregate of lipid droplets can maintain such precise arrangements, and be present with such regularity in a particular part of the chloroplast. Perhaps the surrounding thylakoids and chloroplast envelope help to prevent the structure from dissipating once it has been formed. It is of interest that microtubules have been seen among the droplets of some cytoplasmic eyespots, and that even in the case of *Chlamydomonas*, where the eyespot *is* in the chloroplast, the drug colchicine, which prevents microtubule formation (page 143), brings about the disruption of the eyespot. When the colchicine is removed, one of the first signs that the cells are recovering is the re-association of granules into a recognizable eyespot.[863]

9.7.5 Phytoferritin

Chloroplasts may contain up to 80% of the total iron found in leaves, and when the supply of iron is sub-optimal, growing leaves fail to turn green, and are unable to photosynthesize. These deficiency symptoms arise because iron is needed not only during synthesis of chlorophyll, but also as part of iron-containing proteins, ferredoxin and cytochromes, which form vital links in the electron transport pathways of photosystems I and II.

Iron tends to be relatively immobile in the plant. For instance, if a solution of an iron salt is painted on to a pale, iron-deficient leaf, the treated area will turn green, but not the neighbouring untreated tissue. It is not readily translocated from old to young leaves. Neither plants nor animals store it in the form of inorganic salts. In both kingdoms it is stored as a non-toxic iron-protein complex, first characterized for animal tissues in 1937, and called ferritin, and later (1962) found in plants,[387] where the name phytoferritin is used (somewhat unnecessarily, as plant ferritin has been shown to be very similar to animal ferritin).

Ferritin molecules (plant or animal) are large proteins, more than 10 nm in diameter, and are visible in the electron microscope (in ultra-thin sections) by virtue of the density to electrons of the iron-containing core of each molecule, where several thousand atoms of iron (as a type of iron hydroxide) lie within the proteinaceous shell. In animals the ferritin molecules are seen to be scattered in the cytoplasm, or in the nucleus, or in mitochondria. In higher plants, by contrast, plastids have become the site of storage of ferritin, though some is occasionally seen in association with lipid droplets in the cytoplasm—which is also where ferritin in found in fungi.[759]

Ferritin has been seen in all of the major types of plastid. It lies in the stroma, where the molecules may be scattered, or, if they are abundant, forming conspicuous aggregates (Plate 35b), sometimes possessing regular substructure with definable crystallographic symmetry.[11] The aggregates are large in storage tissues such as cotyledons of seeds, but they can generally also be found in leaves and most other parts of the plant. If extra iron is supplied to plants (or animals) after a period of iron deficiency, the frequency and size of the ferritin aggregates increases, and the amount of storage may come to exceed that seen in healthy plants grown with a constant supply of iron.[760] Presumably the advent of new supplies after iron-starvation induces the synthesis of unusually large amounts of the protein (apo-ferritin) with which the iron associates to produce ferritin. In view of the fact that plants have virtually restricted their iron-storage to plastids, it would be of interest to know whether the apoferritin is made in the plastids, or whether it is a cytoplasmic product that somehow traverses the plastid envelope membranes, either before or after receiving its quota of iron atoms. This and other problems concerning the interaction of the iron and the apo-ferritin, remain obscure. How is the entry of iron mediated? How is a requirement for iron, e.g. for synthesis of ferredoxin or cytochrome, perceived? How is the iron withdrawn? Is the protein broken down after the iron has been utilized?

9.8 Plastids in the Algae

It has at numerous points in the preceding accounts of plastid types been necessary to refer to members of the algae. Some green algae have chloroplasts which are basically similar to those of higher plants, with their thylakoids arranged in grana and fret systems. In many other green algae, and in other groups of algae, the arrangement of thylakoids is distinctive, sometimes to the extent that its visualization by electron microscopy is an aid to the classification of the organism.[265] Some components of the stroma of algal plastids—particularly ribosomes and nucleoids—are much as they are in higher plants, but others are dissimilar. There are, for instance, few exceptions to the generalization that starch is not stored in the plastid stroma, except in green algae and higher plants; most algal groups develop special, dense regions of plastid stroma called pyrenoids; and many motile algae produce 'eyespots' in their chloroplast stroma (page 128). Except in blue-green algae, where it does not exist, the plastid envelope is like that in higher plants, but as already mentioned (pages 65 and 123), it is in several groups of algae enveloped in its turn by a cisterna of endoplasmic reticulum extending from the nuclear envelope.

The following paragraphs give brief descriptions of the arrangement of algal thylakoids, and of the structure and function of pyrenoids.

9.8.1 Thylakoids in the Algae

Although the 'unit of structure' of the internal membrane system of algal chloroplasts is a closed, flattened sac,[265] i.e. a thylakoid, it is not known whether the visible sacs are mere regions of a continuous membrane system—as is thought to be the case in higher plants—or whether discrete thylakoids cohabit individual chloroplasts. For example, in the prokaryotic blue-green algae single thylakoids lie scattered in the peripheral cytoplasm of the cell (there being no chloroplast envelope), and in the eukaryotic red algae (where chloroplasts *are* segregated from the rest of the cell) the thylakoids are again single, separated from one another by intervening layers of stroma. Much serial sectioning and three-dimensional reconstruction would be needed to establish the separateness or conversely the continuity of these thylakoids.

Unlike other groups of algae, the reds and the blue-greens do not produce any equivalent to the partitions of grana in higher plant chloroplasts. Their thylakoids are in as complete contact with the stroma as are the tubes of the prolamellar body (a structure which can, incidentally, be formed by various algae), [530] or agranal thylakoids in C-4 plants. It may be significant in this connection that the red and blue-green algae possess distinctive photosynthetic pigments (phycobilins) which are tightly bound to proteins of very high molecular weight and large particle size (35 nm). It has been shown by freeze-etching and negative staining that the protein-pigment complexes (phycobilisomes) are present in orderly arrays on the *outer* face of the thylakoids, i.e. the face that is in contact with the stroma.[474] The presence of such large particles in this location may account for the absence of back-to-back association of thylakoids, i.e. of partitions.

Other groups of algae do produce partitions, but do not necessarily restrict them to grana. In other words, thylakoids can be appressed back-to-back over large areas. Some comparisons of membrane substructure in appressed, or 'stacked' areas and 'unstacked' areas, obtained by freeze etching, have already been described (page 105). Both the area over which stacking occurs, and the number of thylakoid layers involved, vary. One extremely widespread configuration consists of two thylakoids sandwiching a third. In 8 major groupings of algae, nearly all of the internal membrane system of the chloroplasts is in this form,[265] i.e. not differentiated into stacked regions and unstacked frets. Layers of stroma lie between adjacent triple layered bands of thylakoids.

Differentiation into unstacked frets and partition-containing stacked regions is seen in only two groups of algae, one being the green algae, thought to be related (in evolutionary terms) to the higher plants. Here, just as in higher plants, there is a great deal of variation in the arrangement of the thylakoids. Examination of mutant strains shows that some of the variation is under the internal control of the genetic system of the organism,[280] and experiments on environmental variables (e.g. carbon dioxide concentration,[260] light quality and intensity, manganese supply) show that the form of the thylakoid system responds to external factors also.

9.8.2 Pyrenoids

Although considered here under the heading of plastids in the algae, it should be recognized that pyrenoids do occur outside the algae (in the Bryophyte *Anthoceros*), and that they are not present in all algae. An example is illustrated in Plate 35c–g. They are electron dense bodies in the chloroplast stroma, sometimes, but not always, traversed by derivatives of the thylakoid system. They are especially conspicuous in the green algae, where the presence of a sheath of refractile starch grains around their periphery causes them to glisten in light microscope preparations. In other groups of algae there is no starch anywhere in the chloroplast stroma, and so the pyrenoids are less easily viewed. Often there is only one per chloroplast, but in large chloroplasts, as in the familiar *Spirogyra*, there are many.

The electron microscope shows the main bulk of the pyrenoid to be a homogeneous deposit of finely granular material. The ribosomes and DNA of the plastid stroma do not penetrate into it. Despite the absence of a bounding membrane, the structural integrity of the pyrenoid is somehow maintained, indeed pyrenoids do not dissipate even when they are isolated by centrifugation from broken chloroplasts. The major component is protein, and the available evidence points to there being one major protein species, the carbon dioxide-fixing enzyme ribulose diphosphate carboxylase. It is estimated that in the green alga *Eremosphaera* about 40% of the total cellular activity of this enzyme resides in the pyrenoid,[368] the remaining 60% presumably being, as is usual for higher plant chloroplasts, free in the stroma or perhaps loosely associated with thylakoids.

Pyrenoid structure varies considerably even within a genus.[265] For example, the strain of *Chlorella* illustrated in Plate 35c–g has no thylakoids in its pyrenoid, yet there are other *Chlorellas* with pyrenoid thylakoids. The penetrating thylakoids may or may not be like the thylakoids in the rest of the chloroplast. In some cases they are flattened sacs outside, and tubular inside, the pyrenoid. In a number of algae the pyrenoid is invaded not by thylakoids, but by narrow, branched invaginations of cytoplasm, lined by the two membranes of the chloroplast envelope. Usually the pyrenoid is amongst the thylakoids of the chloroplast, but in a number of algae it lies in peripheral locations, or even in thylakoid-free stalked bodies extending from the main part of the chloroplast into the cytoplasm. Associations with cytoplasmic structures may be evident in such cases, e.g. membrane bound vesicles or cisternae—'pyrenoid caps'—containing tertiary energy storage material,[517] or a microbody, as described in pages 123 and 137 and illustrated in Plates 30b, c and 35c, d.

Several of the above features are consistent with a role for the pyrenoid as a focal point for the production of tertiary stores of photosynthetic product, or for the dispensing of photosynthetic products across the chloroplast envelope to the cytoplasm. There is the pyrenoid starch sheath of green algae (Plate 35c–f), the proximity of 'food vacuoles' in the cytoplasm just across the chloroplast envelope from the pyrenoid, and the high concentration of ribulose diphosphate carboxylase. However, in the green algae, starch grains can be made, and removed, elsewhere in the chloroplast stroma (Plate 35c–g), so these processes are not the prerogative of the pyrenoid. Again, in the algal groups that store their long-term reserves outside the chloroplast, the pyrenoid can hardly be responsible for the final syntheses, which may be removed from it by several membranes. It seems likely that in acting as a focal point, the pyrenoid merely supplies the raw materials. Not all of the enzymes of the Calvin cycle have yet been found in it, but it is noteworthy that those that *have* are capable of generating 3-phosphoglycerate, a compound known (page 118) to stimulate the activity of an enzyme that participates in starch synthesis.

Multiplication of chloroplasts requires the development of new pyrenoids in most algae, and there are reports of pre-existing pyrenoids dividing to produce two, and other reports of *de novo* production of pyrenoids in young cells formed by cell division. When gametes fuse to produce a zygote in *Chlamydomonas* their chloroplasts (each containing one pyrenoid) also fuse, but one pyrenoid degenerates.[115] Clearly, pyrenoid development is an integral part of cell development in the majority of algae. Equally clearly, we do not yet have sufficient knowledge to understand why this is so. How is it, for instance, that flowering plants, which are held to have basically the same photosynthetic system as green algae, have been able to dispense with pyrenoids in the course of evolution? Just what vital functions are behind their widespread occurrence in the algae?

9.9 Plastid Interconversions

Developmental steps between proplastids and mature chloroplasts, etioplasts, amyloplasts, chromoplasts and leucoplasts have already been described in this chapter. The ultimate common origin of the specialized types illustrates one facet of family relationships within the plastid family, analogous to the differentiation of a range of plant cell types from genetically uniform meristematic cells, or in animals, to the development of several types of blood cell from a common denominator, the stem cell. But there is more to the family relationships of plastids than mere conversion of one basic precursor type into other members: in addition, complex types may revert to the simple condition, or may develop from one another. It is therefore necessary to consider all manner of interconversions of plastids.

9.9.1 From Etioplast to Chloroplast

One of the numerous possible developmental pathways that interconnect specialized plastids has been studied in more detail than others—the 'greening process' leading from etioplast to chloroplast.[62] It is initiated if a dark-grown leaf is illuminated, the first of many events being a very rapid light-induced conversion of the protochlorophyllide, found in the prolamellar bodies, to chlorophyllide. Under most conditions there is then a 'lag period', lasting a few hours, before rapid synthesis of chlorophylls and thylakoids completes the transformation from etioplast to chloroplast.[74]

The lag period can be eliminated by a variety of treatments, including placing the plants in a very humid atmosphere,[4] so it is not a vital stage. Even where it is present, it is by no means a period of complete quiescence, and many processes go on during it. The regularity and symmetry of the prolamellar body is gradually lost[301] (Plate 38a). Chlorophyllide is converted to chlorophyll by the addition of phytol,[830] a terpenoid molecule thought to anchor the pigment molecules by dissolving in the lipid part of the thylakoid (page 104). Other changes in the state of pigment molecules are detected by examining the absorption spectra of leaves.[332] The three-dimensional array of branched tubes in the prolamellar body dissociates to give rise to a set of more-or-less evenly spaced out primary thylakoids (Plate 38a).[454] The area of primary thylakoid produced per plastid probably does not exceed the area of membrane in the prolamellar body, and so this stage of development is interpreted as a re-arrangement of existing membrane, rather than as new synthesis.[74,301,333]

The primary thylakoids formed by dissipation of the prolamellar body are perforated by numerous pores (Plate 38a, b), and they therefore closely resemble the perforated thylakoids that are thought to be precursor material for prolamellar body development (page 115). The pores are *relics* of the lattice spacings of the prolamellar body, whereas during the conversion of proplastids to etioplasts, they may be *progenitors* of the lattice spacings. They soon disappear during greening, giving rise to featureless primary thylakoids upon which grana develop (Plate 38c-f), in much the same manner as when grana appear in chloroplasts that are developing directly from proplastids (page 110). The end of the lag period of greening is marked by the beginnings of grana (Plate 38c, d), and their further development requires the production of new thylakoid membrane, this in turn demanding the synthesis and insertion of all of the component lipids and proteins, and, of course, the photosynthetic pigments.

When a lag period is present during early stages of greening, a multi-step insertion of components into the thylakoids can be detected.[13, 60] Photosystem I activity is present before photosystem II becomes operative, and the latter is predominantly associated with production of grana after the lag period. However, in relation to the previous discussion of the question of whether photosystem II is restricted to partitions (page 106), it is noteworthy that in some algal and higher plant material, photosystem II can be present even though not much stacking has developed. When greening occurs rapidly and uniformly, with the lag period eliminated by providing humid conditions, the order of development of the two photosystems is reversed, thus adding weight to other observations suggesting that the availability of water is important in the conversion of etioplasts to chloroplasts.[5]

The form of the growing membrane system is very sensitive to alterations in the conditions of illumination. If the events of greening are initiated by giving dark-grown plants a flash of light, some will proceed in darkness, and if the light is again turned on, greening will then proceed with no lag. If greening takes place in light of low intensity, prolamellar bodies reappear, and stages of development can be found with small grana, some normal frets, and many areas where the frets are in the form of small prolamellar bodies.[334, 876] In these conditions the accumulation of protochlorophyllide correlates with the appearance of the new prolamellar bodies, but, as mentioned previously (page 115), some prolamellar body areas are even seen during greening in high intensity light, when protochlorophyllide is not detectable.[831] It follows that it is possible to have plastids with some of the structure and functions of chloroplasts and at the same time some of the features of etioplasts. In other words, greening does not necessarily require the destruction of all etioplast characteristics, and then the gradual assumption of the properties of the chloroplast.

Only some of the events of greening are reversible. If only a brief period of illumination is given, the perforated thylakoids can revert to the prolamellar body condition.[333] It seems, however, that once partitions have been established, their stability is sufficient to prevent their reversion if the light is turned off. To take an extreme example, mature chloroplasts in grass leaves in a lawn may turn yellow if they are covered and darkened, but this is not due to transformation of flat thylakoids in grana and frets to tubular prolamellar bodies; it is rather a symptom of breakdown of the thylakoid system and senescence of the chloroplasts. It is only *new* membrane that is being formed that has the option of becoming either flat or tubular. Evidently the presence or absence of light does not completely determine the choice, and it is important to try to discover the nature of the fundamental determining factor or factors, and their mode of operation. Light does, however, trigger the initial conversion of the tubular form to the flat, and subsequent events soon make the change more-or-less permanent.

9.9.2 Other Interconversions

Several plastid interconversions involving amyloplasts have already been mentioned. Amyloplasts are commonly formed as intermediate stages in the development of etioplasts and chloroplasts. Even those amyloplasts that normally are permanent, as in a potato tuber, will produce green thylakoids if illuminated. Some plastids remain for long periods in an intermediate state, sometimes described as the chloro-amyloplast,[725] and comparable combinations of starch storage and pro-lamellar bodies have been seen.[59, 361]

Chromoplasts are very often restricted to tissues, such as fruits, which through being terminal states, give no opportunity to show that formation of this type of plastid is also reversible. Reversion to a green, chloroplast condition, has, however, been observed in oranges[821] and carrots.[295] They can develop directly from proplastids, from proplastids via an amyloplast stage, and from chloroplasts. If the mutant maize which produces chromoplasts instead of chloroplasts (page 121) is grown under low light intensity and then illuminated, there is even a conversion of an etioplast to a chromoplast, with a transitory and rudimentary chloroplast-like intermediate stage.

Leucoplasts too are not often given a chance to express any potential for further development that they possess. Like the other plastids, they can be formed in a variety of ways, especially from proplastids, or by regression from small chloroplasts.

A clear distinction has to be drawn between, on the one hand, interconversions of individual plastids within individual cells, and on the other, interconversions that involve the division of both the plastids and the cells in which they lie. The conversions described above fall into the former category, but there is one more developmental pathway which in most, if not all, cases, is conditional upon cell division. It is the simplification of plastid structure from a specialized state back to the basic proplastid condition.[107a] It is seen when a meristem is produced in a differentiated tissue, when a dormant meristem resumes activity, when differentiated cells are taken into tissue culture conditions and start to divide, and when gametes, particularly male gametes, or other types of reproductive spore, are formed. In all of these cases, whether of natural origin or not, it is of vital importance to the continuity of plastids in the cell lineages that even highly specialized plastids retain their total developmental potential and can express selected facets of it in relation to the developmental state of their host cells. Plant cells are described as being *totipotent*, meaning that they are able, even when highly differentiated, to divide and re-differentiate in a different direction. The versatility of the plastids in plant cells is a part of the overall totipotency, and presumably is achieved via delicate interaction of the nuclear and plastid genetic systems.

9.10 Division of Plastids

POPULATION LEVELS OF PLASTIDS IN CELLS Fundamental features of the division of plastids are very much the same as those described in the preceding chapter for mitochondria. In meristems proplastid multiplication balances cell multiplication, while in other situations cell and plastid division may not be so closely interrelated. Thus in young leaves the population of plastids in each cell continues to rise long after cell division has ceased.[666] Formation of reproductive cells provides examples of the converse, in which cell division without concomitant plastid division reduces the number of plastids per cell from a large number to a small number,[105] or even, in some male gametes, to zero.[113, 400, 473]

It is not, however, sufficient merely to compare the rates and durations of plastid division with rates of nuclear or cell division. Every cell type seems, for any given species, to have a characteristic final number of plastids per cell, and many factors, both internal and external, interact to determine that number. Examples of internal effects are that if the DNA content of the cell nucleus doubles (becomes endopolyploid, page 42), the plastid number increases, on average, by a factor of about 1.7.[105] Interestingly, had the doubling of the nuclear DNA been part of a mitotic cell division in a meristem, the plastid numbers would have *doubled* during the cell cycle, thereby maintaining the size of the population per cell.[12] Again, to have an effect upon plastid numbers, it is not necessary that the *total* nuclear DNA is doubled. Addition of individual chromosomes can have effects: in sugar beet at least four out of nine different chromosomes carry genes that, if present in three copies instead of the usual two, lead to a rise in the number of plastids.[105]

Many external factors influence plastid numbers, particularly light intensity and the wavelength of the light.[849a] Application of the growth regulating substances, cytokinins, induces division of plastids in tobacco leaves,[61] but has only a temporary effect in spinach.[667] A very simple method of observing plastid division is to pull leaves off the moss *Funaria*.[268, 271] Detachment rapidly leads to duplication of nuclear DNA in the leaf cells, and within a few hours the chloroplasts can be observed in dumb-bell shapes, undergoing fission. Leaf detachment obviously does influence chloroplast division in the moss, but it may be that the primary influence is upon the nuclear DNA, which in turn regulates the chloroplasts. The cytokinin effects could similarly be mediated via the nucleus. It has been suggested that the plastids in a cell are always in a state of readiness to divide, but that the size of the population is somehow limited by the amount of nuclear DNA. Thus in meristematic cells nuclear DNA synthesis and mitosis permit a doubling of the population once in every cell division cycle, and a cycle of DNA duplication that gives rise to an endopolyploid nucleus similarly 'releases the brakes', though not sufficiently to allow the population to double.[105]

Correlations between increases in nuclear DNA and increases in plastid numbers do not account for the attainment of a final population that characterizes each type of cell in a given species. In a leaf, for instance, epidermal cells, guard cells of stomata, palisade and spongy mesophyll cells, bundle sheath cells, and the vascular parenchyma cells, each have their own population level (as well as, in many cases, their own type of plastid). It has often been noted that the larger a cell in a leaf, the larger its population of chloroplasts, and that cell enlargement correlates with plastid multiplication. There is one additional, and very suggestive fact: that the proportion of the surface area of mesophyll cells that is occupied by chloroplasts is remarkably constant within a species, irrespective of the age and the size of the cell.[372] The chloroplasts lie mainly in the peripheral cytoplasm (Plate 32a, b) and the area occupied by them ranges from 73% for spinach to 25% for tobacco. How this constant fraction of the total available area is maintained by balancing cell enlargement, chloroplast enlargement and chloroplast multiplication is unknown, and it is doubtful if any such relationship exists in, for instance, meristematic cells, where the proplastids are scattered throughout the cytoplasm.

THE DIVISION PROCESS What types of plastid can divide? Maintenance of populations in meristematic cells show that proplastids can, there being no evidence for any *de novo* production of plastids in such tissue.[12] Chloroplasts certainly can, the actual process having been filmed in growing *Nitella* cells,[289] and described in terms of population sizes for leaf cells in a variety of flowering plants.[372] Data for other plastids are sparse, but there is evidence of increases in the population per cell in discs cut from young spinach leaves and cultured in darkness for 5 days, implying that multiplication occurs during development to the etioplast condition.[665] There is also multiplication during conversion of etioplasts to chloroplasts, and it is thought that in greening maize leaves the developmental processes leading up to and including division occur synchronously throughout the population.[501]

The actual process of chloroplast division has very seldom been watched in higher plant material. The rate of population growth (a 5-fold increase over a 10-day period to a total of about 300 in palisade mesophyll cells of spinach)[666] is such that only about one new chloroplast appears per hour per cell in a growing spinach or tobacco leaf, and the difficulties of observing the process over long periods are compounded by the streaming movements of the cytoplasm. From statistical observations on chloroplasts in growing leaves it is clear that only a very small proportion of the population is derived directly from proplastids. The majority arise from fissions that take place much later in chloroplast development, and it is possible that within the total population in each cell, the fission processes are confined to a quite small sub-population of small chloroplasts. These, but not the very large mature chloroplasts in the cell, can be seen in dumb-bell shapes suggestive of stages in division.[372]

Two sorts of division process have been described for higher plant chloroplasts.[149] One is a gradual constriction which forms a narrower and narrower waist in the mid-region, finally separating the parent into two more-or-less equal progeny. The other begins with the growth of an invagination of the inner envelope membrane right across the chloroplast, partitioning it into two compartments (as described for mitochondria, page 94), but how this type of division is completed is obscure.

As with mitochondria, there is no evidence as yet for the existence of a mechanism for ensuring an equitable distribution of plastid DNA into the progeny of a fission process. However, the presence of several nucleoids, and several copies of DNA per nucleoid, lessens the chance of producing DNA-free progeny.[443] It will be recalled, nevertheless, that most of the chloroplasts of *Acetabularia* have undetectable levels of DNA (page 124), and it is possible that they arose by divisions which failed to give each of the progeny a quota of DNA.

There are profound genetic consequences of a division mechanism that partitions multiple copies of plastid DNA at random amongst the progeny.[429] For instance, if mutated copies of DNA are present, and if they duplicate along with non-mutated copies in a plastid, then the random sorting out process will eventually produce plastids which, by chance, have nothing but mutated DNA. Provided that the non-mutated copies are operative, it is only then that the mutation is expressed, perhaps giving rise to a chloroplast with defective pigmentation. Further divisions will give rise to a sub-population of defective chloroplasts, and if cell divisions are occurring, the same sorting out processes, but on a cellular rather than a sub-cellular scale, may give rise to defective cell lineages, as in certain types of leaf variegation.[430]

FUSION OF PLASTIDS Finally, it should be mentioned that division is not the only way in which plastid numbers in a cell are regulated. The number of chloroplasts per cell *falls* in at least some leaves in autumn.[791] Fusion of chloroplasts has been observed in algae,[115] ferns,[8] and higher plants.[194] One mode of formation of chromoplasts in *Torenia* is by fusion of all of the chloroplasts in a cell, followed by differentiation of chromoplasts from the fused mass.[462]

10 Microbodies and Spherosomes

10.1 Introduction

The advent of adequate preparation techniques for examining cell structure with the electron microscope led to numerous observations on components that are close to or below the limit of resolution of the light microscope. In plants these included a number of roughly spherical particles ranging in size from about 0.5 μm to 2.5 μm. Some were granular, some homogeneous, some with a crystalline core, some bounded by a definite membrane and others not. It is now clear that they fall into at least two classes, which are referred to by rather vague descriptive names, *spherosomes* (spherical bodies) and *microbodies* (small bodies). Notwithstanding their similarity of meaning and derivation, these two words have come to be used in quite different ways.

10.2 Spherosomes

The word spherosome was coined by light microscopists, who, by means of various staining reactions, were aware that different types exist. Now, after electron microscope examination, spherosomes remain as a heterogeneous collection of objects that float around in the cytoplasm,[561, 778] usually with a spherical shape that arises by minimization of their surface area by forces of surface tension.

The major category of spherosome consists of droplets of lipid (Plates 39e, 40a, 41b), which are abundant in some cells and sparse or absent in others. Their numbers and distribution may fluctuate according to the metabolic state of the cell, particularly during development, dehydration, re-hydration, and germination of seeds. The lipid content is usually homogeneous in appearance, although exposure to fixatives such as osmium tetroxide may introduce inhomogeneity. They are not bounded by a membrane, but a surface skin consisting of an outer layer of oriented lipid molecules forms in response to the surrounding aqueous environment of the cytoplasm. An ordinary cell membrane (page 13) is exposed to an aqueous environment on *both* faces, and consequently is based on a bimolecular layer of lipid. In lipid droplets, where the interior is non-aqueous, the skin of lipid molecules is equivalent to just half of the usual triple-layered, dark-light-dark, type of membrane.[920]

10.3 Microbodies

The word microbody belongs to the electron microscope era, and was first applied to small (about 0.5 μm in diameter) but distinctive objects in animal cells.[378] A merely descriptive term was necessary because initially there was no information on the chemical make-up and function of the structure. Meanwhile botanists were also seeing small components of unknown nature and function in plant cells. Some made use of the light microscopists' term, spherosome, and others preferred straightforward descriptions that avoided any possible confusion with other terms then in use: reports of 'crystal containing bodies' were particularly frequent.[150a] In due course the structural homology between the latter and the animal microbodies was recognized, and it was further realized that the crystal is not always present.[248] The term microbody now transcends the classification system of plants and animals, and is applied to small bodies bounded by a single true membrane, containing a granular matrix which may or may not be crystalline in places.

Microbodies have been found in every plant cell and tissue type so far examined in detail,[852a] including leaves (Plate 41a), cotyledons (Plate 41b, c), fruits (Plate 40b), root meristematic cells (Plate 2), root tip outer cap cells (Plate 28a), glands (Plate 23c), developing xylem and phloem and latex tubes,[534] and various reproductive structures. They are also found in unicellular algae (Plate 30b and c), higher algae, fungi, mosses, etc., and are very widespread in animal tissues. They probably occur in all eukaryotic cells. Inflated cisternae of rough endoplasmic reticulum may be distinguished from microbodies by the presence of ribosomes on the outer face of the single bounding membrane (Plate 3a).

The recognition of microbodies as a distinct category of cell component stimulated much speculation about their functions in the cell. Even the early observations indicated that they play some part in cellular metabolism, since they were found in greater numbers in some cells than in others, and were sometimes found closely associated with other cell structures. One of the earliest suggestions was that they are equivalent to animal lysosomes, containing enzymes that could act to break down cellular components during differentiation, or when the cell ultimately died. As we shall see, this view was shown by subsequent studies to be incorrect.

At the same time as the early studies of the ultrastructure of microbodies were being conducted (in the late 1960s) biochemists were investigating particles isolated from cells by the technique of 'density-gradient centrifugation', in which mixtures derived from broken cells are centrifuged through a column of liquid (usually sucrose solution), whose density increases from top to bottom. Particles possessing similar densities come to lie at the same point in the liquid column, and can be collected in bulk. One category of particle, hitherto inseparable from mitochondria, was found to have a characteristic complement of enzymes, notably a high level of catalase (the enzyme catalyzing the breakdown of hydrogen peroxide to water and oxygen). Initially it was not possible to state unequivocally that these biochemically-characterized particles were microbodies. Then isolated material was examined by conventional ultra-thin sectioning methods and shown to contain particles similar to the microbodies on the intact tissues.

Further confirmation of identity with microbodies was provided by a method by which the enzyme catalase can be localized in intact cells. The catalase activity is found to reside in the microbodies (Plate 41a), and in leaf cells (and a variety of others) the microbody crystals seem to show a greater catalase activity than does the matrix surrounding the crystal.[246] Purified catalase is known to form several types of crystal, including some like those observed in microbodies, but despite this, and the staining reactions, the crystalline material of plant microbodies has not, when isolated, been found to be particularly rich in catalase.[379] The presence of catalase in microbodies has nevertheless been demonstrated in a number of plant tissues, including castor bean endosperm,[852] pea cotyledons, root meristems,[668] lacticifers,[534] and various reproductive organs. The enzyme seems to be present in the great majority of microbodies, whether or not crystalline cores are developed. There are, however, a number of algae in which catalase can be detected in biochemical extracts, but not by the specific staining reaction described above; there are also algae where neither the biochemical nor the staining procedure detects the enzyme.[245] Its function is considered later, meanwhile it suffices that, where present, the enzyme is a good biochemical 'marker' that supplements the rather indefinite structural characterization of microbodies.

The biochemical investigations of the past few years have so far revealed three types of microbody, of which two are specialized and have been given special names, *glyoxysomes* and *peroxisomes*. These two names carry biochemical connotations. Each type has a characteristic enzyme content and is restricted to characteristic types of plant tissue. It must be recognized that as methods for isolating microbodies in large quantities from specific plant tissues become available, yet other biochemical varieties may well be discovered. Meanwhile, the third category is merely referred to as unspecialized microbodies. Their complement of enzymes is, it seems, very similar in a wide range of plant and animal tissues. We will first discuss the two specialized types and then proceed to a more general consideration of the development of microbodies and regulation of their activity.

10.3.1 *Glyoxysomes*

The term glyoxysome was coined in 1967,[82] following isolation of a sub-cellular particle containing concentrations of the enzymes that carry out a chain of reactions first elucidated 10 years earlier, and known as the glyoxylate cycle. In structural terms, the particle was seen to be a 'microbody'. Without going into the complex biochemistry of the system, the glyoxylate cycle is capable of utilizing small molecules such as acetate (or its derivatives), with only 2 carbon atoms, in the production of larger molecules, particularly 4 carbon organic acids[832a] and 6 carbon sugars.[43] Through its agency, plant cells that possess it can subsist on acetate, or on food stores that give rise to acetate units upon breakdown. The former ability is found in various algae[262] and fungi, including the *Chlorella*[807] strain pictured in Plate 30b and c, and the latter in fat-storing seeds (Plate 41b and c).

Fats are tertiary energy stores (page 87) and like all such, must be broken down prior to utilization. Participation of microbodies in this is indicated by their frequent close proximity to lipid droplets (Plate 41b and c).[297] The larger droplets have microbodies appressed to their 'half-membranes', while a smaller one may sometimes be seen to be almost engulfed by a single large microbody. Upon isolation, these microbodies have been found to contain enzymes which break down fats to 2-carbon fragments. The enzymes of the glyoxylate cycle, which identify the microbodies as glyoxysomes, then process the two carbon fragments, and produce organic acids. Some of the organic acid is respired by the mitochondria in the cell, but in quantitative terms, most is converted to precursors of sugars.[43] The sugars may then be converted to starch in plastids or be translocated to growing zones such as the meristems of the embryo.

It was stated in Chapter 8 that mitochondria can oxidize fatty acids, yielding two-carbon fragments which are then processed further to release energy and drive phosphorylation of ADP. This certainly applies to mitochondria from animal sources, but as botanists have become more skilled at the difficult task of separating the mitochondria and microbodies of plant tissues, it has become more and more apparent that glyoxysomes are much more active than mitochondria in breaking down fatty acids, at any rate in the relatively few tissues (especially fat-storing seeds) where large quantities of fats are consumed.

The question of why the enzymes that catalyze fatty acid breakdown and the glyoxylate cycle should be housed together in a special compartment of the cell can be posed. Would it not, for instance, be as efficient, or more so, simply to bathe the lipid droplets in cytoplasm containing both sets of enzymes? A partial answer is that the one enzyme system utilizes produce from the other, and it is therefore advantageous that the two be in proximity. The same principle underlies a second and perhaps more telling answer. Breakdown of fatty acids inevitably generates reduced co-enzymes. Whereas in animals these can be processed by the electron transport system of the mitochondria, in the plants under discussion they are re-oxidized by molecular oxygen, yielding as a by-product a compound that has the potential of being exceptionally destructive to the cell—hydrogen peroxide (several herbicides kill by releasing peroxides within the cells of the treated plants). It would seem that a most vital function for the bounding membrane of the glyoxysome is to restrict the possible spread of the hydrogen peroxide generating system, and at the same time to speed the removal of the peroxide by holding within the *same* compartment large quantities of the enzyme catalase, which as already stated, breaks down hydrogen peroxide, and is very widespread in microbodies. Additionally, several of the enzymes in these reactions operate best in slightly alkaline conditions—yet another reason for housing them within the same compartment in the cell, where optimal conditions are, presumably, created.[826]

Fatty-acid oxidation is not the only enzymatic reaction in glyoxysomes to give rise to hydrogen peroxide. There are others which, unlike the glyoxylate cycle enzymes, are found in other microbodies, in tissues as disparate as liver and leaves. Catalase emerges as a protector in all these cases, present in amounts that can reduce the inherent danger of employing peroxidative enzymes. It is not difficult to think of analogies in, for instance, the chemical industry, and it is fascinating to note that some cases exist (e.g. the *Chlorella* of Plate 30b and c) where the microbody lacks hydrogen peroxide-generating enzymes, and significantly, also lacks catalase: here, where there is no danger, there is no protector.

10.3.2 Peroxisomes

In 1954, the microbodies of kidney and liver tissues were named peroxisomes in recognition of their ability to break down formic acid by a peroxidative reaction. Fourteen years later, peroxidative activity was also found in the microbodies of green leaves, and they too have become known as peroxisomes.[827, 828]

Leaf peroxisomes contain some of the enzymes of biochemical pathways which start in the chloroplasts.[827] The biochemical link with chloroplasts is matched by a structural association, for in the intact leaf the peroxisomes are usually found tightly appressed to the outer membrane of the chloroplast envelope (Plate 41a).[246] Spatial associations like this are very puzzling, for no obvious structural attachments between the chloroplasts and the peroxisome have been seen, yet the two adhere closely. It may be that a form of membrane recognition (which in this case does not lead to coalescence of the two classes of membrane) is involved. The association is especially remarkable where, as in certain algae (e.g. Plates 30b, c, 35c, d) a microbody lies close to a specific part of the chloroplast envelope, in the vicinity of the pyrenoid.

The key to the functions of peroxisomes is to be found in their intimate link with chloroplasts and the phenomenon of photo-respiration—a light-induced uptake of oxygen and loss of carbon dioxide.[247] A two-carbon compound (glycolate) is passed from the chloroplast during photosynthesis. Whether the chloroplast envelope is especially permeable in the vicinity of the appressed peroxisomes is not known, but at any event glycolate is oxidised in the peroxisome by molecular oxygen and the enzyme glycolic acid oxidase, with the formation of hydrogen peroxide and glyoxylate. The hydrogen peroxide, as in glyoxysomes, is broken down by catalase. The glyoxylate may be returned to the chloroplast (the 'glycolate-glyoxylate shuttle')[825] where it is reduced to glycolate again, thus completing a cycle in which the net result is the export of reducing power from the chloroplast. Alternatively the glyoxylate can be further metabolized to yield sugars and amino-acids together with the photorespiratory loss of carbon dioxide, the total reaction pathway requiring the co-operative action of enzymes in chloroplasts, peroxisomes, mitochondria, and the free cytoplasmic phase in which they lie, with a massive flow of carbon through all of these compartments of the cell.[689]

The extent of the loss of carbon by photorespiration is regulated partly by the competitive effects of oxygen and carbon dioxide on the carbon dioxide fixing enzyme of the chloroplast stroma, ribulose diphosphate carboxylase (page 99). Under normal atmospheric conditions the relative concentrations and affinities are such that oxygen causes the enzyme to break down some of its 5-carbon substrate (ribulose diphosphate) by an oxygenase activity that yields a precursor of the 2-carbon glycolic acid and a 3-carbon residue. As described above, some of the carbon of the glycolic acid is lost by photorespiration. If the oxygen concentration is increased, the loss is enhanced due to increased production of glycolate: the growth of the plant may even cease, so severe is the effect. By contrast, lowering the oxygen concentration or raising that of the carbon dioxide favours carbon fixation as compared with the oxygenase activity, and growth rates can be as much as doubled.[826] As described elsewhere (page 109), the great advantage possessed by C-4 plants is that they are able to retrieve photorespired carbon dioxide. Their peroxisomes seem to be present mainly in the bundle sheath cells, correlating with the presence in these cells of chloroplasts that contain ribulose diphosphate carboxylase.[247, 360] Most of the carbon dioxide generated in the bundle sheath by photorespiration is re-absorbed by the very active fixation system (page 108) of the mesophyll cells.

What, then, is the function of peroxisomes? Given that glycolate formation is an inevitable accompaniment of fixation of carbon dioxide by the Calvin cycle, what the peroxisome does is to utilize the flow of glycolate advantageously. Elimination of excess reducing power, as described above, is one possible function, perhaps significant in regulating photosynthesis and in helping to protect the photosynthetic pigments from destruction in strong light. Another function is the synthesis of the amino acid glycine, from which mitochrondrial systems can in addition make serine. It is at this step that most of the photorespiratory loss of carbon dioxide occurs. The two amino acids, glycine and serine, can be produced in excess of the amounts needed for protein synthesis in the leaf, and they are essential precursors, not only for proteins, but also for production of chlorophylls. If taken back into the peroxisome, serine can also be converted to glyceric acid, which in turn can re-enter the chloroplast and be fed into the Calvin cycle, thereby boosting the output of sugars.[689, 826]

The net result of all of the above reactions is that carbon flows in the direction of producing a variety of essential precursors, or if not there, in a direction in which carbon is conserved, joining the conventional products of photosynthesis. The sacrifice that is made is the loss by photorespiration of a proportion of the carbon.

10.3.3 Formation of Microbodies and Regulation of their Activity

Superficially, microbodies resemble more closely than most cell components a bag of enzymes, and it might be expected that in the absence of structural complexity, the processes involved in their development might be correspondingly simple. The cell merely places a collection of enzymes in a compartment that is bounded by a single membrane.

Of course nothing in the cell is ever that simple. As seen in the preceding sections, the nature of the enzymes varies with the tissue, the type of microbody, and the developmental stage, and so must be subject to precise regulation. Also, the membrane itself probably has specific permeability, recognition, and enzymatic properties; for instance, at least five of the enzymes of castor bean glyoxysomes are associated, albeit loosely, with the membrane.[379]

From the earliest observations it was suspected that the endoplasmic reticulum might play a part in microbody formation.[248, 283, 852] A cisterna lying alongside the microbody membrane is a familiar feature of ultrathin sections, and an immediate interpretation is that the microbody arises as a swelling in a cisterna of endoplasmic reticulum. It has in fact proved difficult to demonstrate convincingly that the membranes of the cisterna and the microbody are continuous, though continuities have been seen in tissues where microbodies are being formed rapidly. Nevertheless the two membranes have similar dimensions, staining characteristics, lipid composition, and have some enzyme activities in common.[379]

Despite the above similarities, detailed observations on the primary leaves of bean as they grow from the seed do not support the view that large microbodies develop attached to the endoplasmic reticulum.[298] Instead they suggest that in this tissue, microbodies are formed from small, apparently empty, vesicles that are themselves pinched off from smooth elements of the endoplasmic reticulum. The vesicles either grow to form microbodies directly, or fuse together forming larger units that become microbodies.

While accepting that the endoplasmic reticulum may indeed generate the membrane of the future microbody, it is obvious from what is known of the fine structure and enzymology of microbodies that they cannot develop further autonomously. They have, for instance, no ribosomes on which to assemble proteins. Indeed there is good evidence that the main diagnostic enzyme of microbodies, catalase, is made outside the microbody. It appears first as a proteinaceous precursor molecule, a product of cytoplasmic ribosomes. It then appears in the microbodies, where it is matured by the addition of a heme group (also made outside the microbody) and by the association in sets of four of the precursor protein molecules.[468, 469]

The progress of microbody formation and activity has been studied in several tissues.[180a] For example in the bean leaves discussed already[298] the development of microbody (peroxisome) enzymes parallels the development of the chloroplast photosynthetic system. Growth of the peroxisomes from 0.2 μm to 1.0 μm in diameter (a 125-fold increase in volume) is accompanied by a three to ten fold increase in the activities of the peroxisomal enzymes found in extracts of whole leaves. If leaves are made to develop in darkness their etioplasts, being incapable of photosynthesis, do not form glycolate. Nevertheless microbodies develop their enzyme content to quite a high level during etiolation, and soon acquire their normal full peroxisomal activity after exposure of the leaf to light brings about the commencement of photosynthesis. It seems that peroxisome development can proceed, though perhaps not to completion, in the absence of suitable raw materials for the peroxisomal enzymes.

A more striking example of the regulation of the enzymatic function of microbodies has been demonstrated in detailed studies of the cotyledons of germinating sunflower and certain other fat-storing seeds.[297, 832] These, as previously described, develop large numbers of glyoxysomes during early stages of germination (Plate 41b), but by the time the glyoxysomal activity has consumed the stored fat, the cotyledons have turned green and have begun to carry out photosynthesis. *Peroxisomes* are then found closely associated with the chloroplasts (Plate 41c). Thus there is a transition *in the same cells* from glyoxysome-microbodies to peroxisome-microbodies.

Analyses of different stages of the transition process have demonstrated that the enzymes typical of glyoxysomes are still present in the cells when those typical of peroxisomes are forming. What is not clear is whether this involves the formation of a new batch of microbodies (peroxisomes) and the elimination of the glyoxysomes, or whether the existing glyoxysomes move and become refurbished with a different set of enzymes. Although it would appear that synthesis of a new population of microbodies is the most likely possibility, stages of the elimination of old ones have not been observed. One approach to the problem which has met with some success has been to study the catalase, which is present in both types of microbody. Experiments were devized to test whether the *same* enzyme molecules were present in each, and in fact no differences were found.[261] Furthermore, no synthesis of new catalase could be detected during the transition. Judging by this result, the microbodies probably persist intact while their internal enzyme complement is modified by the removal of some units and the insertion of new ones.

Two categories of microbody, peroxisomes and glyoxysomes, have dominated this chapter. The third, the 'unspecialized' microbodies mentioned at the outset, have perforce been virtually ignored through sheer paucity of available information. How long it will be before specializations are discovered in members of the category at present regarded as unspecialized remains to be seen. Knowing that microbodies develop distinctive biochemical properties and functions in two particular types of tissue—photosynthetic cells and fat-storing cells—it seems reasonable to suppose that those in, to take but one example, the highly specialized gland cell of Plate 23 would also, if investigated adequately, be found to be distinctive.

It is tempting to draw an analogy with plastids, a family of structures the members of which are structurally and biochemically distinct, and which, within limits set by the developmental capacity of the cells they occupy, are interconvertible. The developmental change from a storage function to a photosynthetic function in cotyledon cells that, as shown above, seems to bring about the conversion of glyoxysomes to peroxisomes is not, in principle, very dissimilar. There are, however, two major differences between the plastid family and the microbody family. Firstly, microbodies do not display the variation of structure seen in the plastids, and so designation of members of the family must be based on biochemical criteria. Secondly, there is (as yet) no evidence that microbodies possess their own genetic systems or that they show genetic continuity either within the plant from parent cell to daughter cell during cell division, or from one plant to another through successive generations: it is more likely that only the *potential* for their development is inherited.

11 Microtubules and Microfilaments

11.1 Introduction

Much evidence has accumulated that certain tubular and filamentous structures form part of the basic equipment of eukaryote cells. The usual diminutives of the world of electron microscopy have been used in naming them, and they are called *microtubules* and *microfilaments* respectively. Neither is restricted to any particular type of cell or class of organism. Surveys of many cells have disclosed that over the very wide range where they have been found, microtubules show but little variation in morphology, molecular structure, reaction to drugs, and response to changes in physical conditions. It is proposed that their functions are basically similar throughout. There is less information on microfilaments, but those found in plants have both structural and functional features in common with those found in animals.

The discovery of microtubules and microfilaments in cells aroused much interest and many experiments were devized in attempts to determine their respective functions. At the time of the discovery, there were many cellular phenomena which could not be accounted for by the activities of the hitherto known structures (nucleus, plastids, mitochondria, etc.). Amongst the most puzzling were the active movements of cytoplasm that result in either locomotion of the cell (e.g. in amoebae or migration of tissue cells in animals), or changes of cell shape (e.g. elongation or compression of part or all of the cell, or the formation of cilia and flagella), or in streaming of particles around the cell ('cytoplasmic streaming' as found in many higher plant and algal cells), or in the movements of chromosomes during mitosis. Such processes require not only the expenditure of energy, but machinery to mediate the energy release and bring about the observed movements or shape changes. The three dimensional configurations and dynamic activities of many single celled organisms—as may be observed in a drop of pond water, for example—are all the more remarkable when it is remembered that many lack a rigid cell wall, and are contained only within their thin plasma membrane, which can offer little skeletal support to the internal cytoplasm. In short, there would appear to be a need within cells for cytoplasmic systems that are analogous to the skeletons and muscles of animal bodies. The discovery of tubular and filamentous structures in the cytoplasm provided the first clues to the possible nature of such systems.

Microtubules and microfilaments will be discussed in detail individually before considering those systems in which they may act in concert. The present introduction has mentioned animal cells already and the ensuing discussions will continue to emphasize the *biological* approach.

11.2 Microtubules

11.2.1 Structure of Microtubules

INTRODUCTION Microtubules were first discovered in a variety of animal[772] and algal[514, 516] cells in the late 1950s. They were not found in higher plant cells until after the introduction of glutaraldehyde fixation in 1962.[340, 471] Even with this improvement in technique they were not always visible in thin sections of plant material, so that they were regarded as being difficult to preserve. One aspect of the fixation procedure in particular, the use of chilled solutions (1-3°C), actually seemed to have a deleterious effect on their preservation, so this practice was discontinued. Microtubules have now been shown to be present in all plant cells that have been examined adequately, at some stage of their growth and development.[585] In addition to observations on thin sections, similar structures have been found following freeze-etching, confirming that they are present in living cells and are not artefacts of the fixation process.[569]

It is against this background of technical problems that we begin by discussing what is known about microtubule substructure and development, and then what is conjectured to be the function of microtubules. Although microtubules have only recently been visualized in cells, biologists had already accumulated a considerable body of data relevant to their properties. The drug colchicine is obtainable from many species of the genus *Colchicum* (including autumn crocus), and is now known to affect microtubule systems. It has been used by pharmacists from the times of the ancient civilizations of Egypt, Greece and India to the present day. The first observations on cells treated with colchicine were made by Pernice in 1889 in an examination of dividing cells, but the attention of biologists was only drawn to possible uses of the drug in the early 1930s by Dustin's description of its interference with the normal course of mitosis. Thereafter numerous publications appeared covering a wide range of cell types and functions influenced by the drug.[185] Many of these effects are now explicable in terms of an interaction with microtubules.

Three broad categories of microtubule are seen in cells. One, the microtubules found in association with chromosomes during mitosis, is considered in the following chapter. A second, the microtubules found in cilia and flagella, is given only brief coverage here because these structures are not general features of plant cells. The third, the microtubules found in the cytoplasm, will be described here with particular reference to plant cells.[341a] Before any aspects of the disposition and possible functions of microtubules can be considered, however, it is desirable to emphasize that they, like other cell components, change with time: their mode of formation and breakdown will therefore be described before their locations and functions. To understand their developmental dynamics it is in turn necessary to look at their substructure and molecular architecture, so this becomes the first topic in the following treatment of the subject.

GROSS MORPHOLOGY In ultra-thin sections, microtubules appear as unbranched straight or gently curving cylindrical structures (Plate 42b). Marked undulations have been seen, but are regarded as artefacts brought about by shrinkage during specimen processing. Microtubules usually pass out of the plane of section within a few micrometres, hence their true length can only be determined by reconstruction from serial sections. It is, in fact, very difficult to recognize microtubule ends, except when, as sometimes happens, they terminate by abutting on to membranes, e.g. the nuclear envelope or the plasma membrane.

In transverse sections (Plate 42a, c) they are seen to be composed of a uniformly staining wall, with an overall diameter of 24-25 nm and a thickness of 5-6 nm surrounding a non-staining lumen about 12 nm in diameter. The observation that the microtubule walls are sometimes surrounded by a 'clear halo', 5-10 nm wide, has led to the suggestion that a further non-staining outer layer is present.[43a] This is an important point because the surface of the microtubule is believed to be a site of interaction with other microtubules and other cell components. An alternative explanation for the clear halo is that slight shrinkage of the structure might occur during processing, although this appears unlikely in view of the finding that it is present in freeze-etch preparations.[793] In some material electron-dense arms can be seen projecting from one or more points on the circumference of the microtubule, either with free outer ends, or apparently connecting the microtubule to a cell membrane or to another microtubule. These arms and bridges will be considered again later.

SUBSTRUCTURE The microtubule wall normally shows little internal structure in thin sections, but in certain special cases it can be seen to consist of a series of subunits. This image is obtained in plant cells that accumulate or release substances in the cytoplasm that in effect negatively stain the microtubules during specimen preparation, so that the microtubule wall appears in reverse contrast, light against a dark background. The phenomenon was first discovered in certain root tip cells of juniper seedlings and in *Euphorbia* nectary cells,[472] and has been found since in other situations (e.g. *Azolla* roots, Plate 43a). In all cases the microtubule wall appears to be composed of a circumferential ring of roughly spherical subunits approximately 5 nm in diameter. Their exact number around the circumference cannot always be discerned, but image reinforcement methods (see Plate 43b-e and caption) frequently give a value of 13, although other values (11, 12 and 14) have been reported. More recently a similar image has been produced in a range of animal cells by including tannic acid with the glutaraldehyde fixative. In all cases the cross-sectional view of microtubules was found to be composed of 13 circumferentially arranged subunits.[824]

Supporting evidence for a wall structure that is based on side-by-side subunits has been provided by negatively staining microtubules isolated from a range of sources. In these preparations the microtubules lie on their side and the longitudinal rather than the transverse view is seen.[254] The microtubule wall is composed of a series of filaments, each a chain of globular subunits approximately the same size as those seen in ultra-thin sections (4-5 nm). The filaments exhibit a longitudinal periodicity of 8 nm, which results from the association and slight overlapping of globular subunits in pairs.[190] The precise number of filaments in each microtubule cannot easily be determined in the longitudinal view, but again, a value of 13 has been found.[864]

BIOCHEMISTRY OF TUBULIN From the above morphological studies it was concluded that microtubules are made up of a series of subunits linked together. The problem of their chemical composition then attracted much attention. It was already suspected that they might be proteinaceous on the basis of their staining reactions. Isolation of the protein could not, however, be achieved until a method by which it could be recognized in extracts became available.

Progress in this direction commenced in 1967, with an experiment using radioactive colchicine.[63] By then it was known that this drug interferes with microtubules, leading to their disappearance from the cytoplasm.[392] Living tissue culture cells (derived from a human cancer) were treated with radioactive colchicine, homogenized, and the protein separated. It was found that some of the colchicine had become bound to one particular protein. Radioactive colchicine could be recovered from this protein, and it was further demonstrated that one molecule of the protein would bind one molecule of colchicine in a reversible reaction. The experiment has been repeated using various tissues with the same result.[64] Those tissues known to contain a relatively large number of microtubules were found to have high levels of the colchicine-binding protein. Nerves, for example, contain many microtubules and as much as ten per cent of the soluble proteins in brain tissues bind colchicine.[882] The colchicine-binding protein has a molecular weight of 110 000 and is called *tubulin*. It can be dissociated into two subunits, approximately equal in size, but dissimilar in amino acid composition. The colchicine-binding activity is retained by one of the subunits.[618] Similar, but not identical, proteins (they do not always bind colchicine for example[1]) have been isolated from a wide range of animal and plant cells, thus 'tubulins' are a family of related proteins.

Another plant alkaloid, vinblastine (extracted from the periwinkle, *Vinca*) is one of several other compounds that have been found to affect microtubules, in this case precipitating the tubulin as crystals in the cytoplasm.[447] As with colchicine, one molecule of tubulin binds one molecule of vinblastine, but at a different site on the protein, so that both can be bound simultaneously.

MICROTUBULES AS POLYMERIZED STRUCTURES The difficulty of preserving microtubules, the deleterious effects of low temperatures, and the disruptive effects of various drugs, all suggest that microtubules readily break down. Application of high hydrostatic pressures also disrupts microtubules; indeed this treatment is more effective than low temperatures in that the cold-insensitive microtubules of cilia and flagella are broken down as well as the cold-sensitive spindle and cytoplasmic microtubules. In most cases, if the cells are returned to normal conditions the microtubules reappear.[823] Tubulins, it seems, can be present in forms that are either visible or invisible to the microscopist. Microtubules represent the visible, polymerized form, while the invisible form is a population of non-polymerized, i.e. monomeric, tubulin molecules. Virus particles provide a parallel. They too are polymers which can reversibly be dissociated by cold as a consequence of changes in the structure of the water phase around each monomer.[466]

The property of dissociation and reassociation of microtubules is important in relation to their behaviour and functions in living cells. Although there are a few observations claiming to show that microtubules move about the cell as intact structures, it is more usually reported that they 'disappear' from one location and 'appear' at another. An explanation for this observation is now to hand. It is that the cell is able to initiate depolymerization at one site and repolymerization elsewhere.

The work with colchicine provides valuable evidence on how dissociation and reassociation might be managed. It has been established that colchicine does not bind to intact microtubules,[212, 901] yet it leads to their disappearance from the cytoplasm. This suggests that the polymerized form of tubulin is in a dynamic equilibrium with a pool of free tubulin monomers. The effect of adding colchicine is to remove functional monomers by preventing their reassociation. The dynamic equilibrium is thereby shifted and depolymerization of the existing microtubules ensues, this being a response which, in the absence of colchicine, would yield more monomers and so restore the previous equilibrium. In the presence of excess colchicine, what happens is that the tubulin is mopped up as soon as it is released, and eventually all of the polymerized microtubules will dissociate. The system is artificial, but it is significant in that the cell may itself use small molecules similar in their action to colchicine, or in some other fashion alter the affinity of tubulin molecules for one another and thereby lead to a shift in the equilibrium between monomer and polymer. There are several candidates, including calcium ions[881] and the nucleotide guanosine triphosphate,[763] for the role of natural regulator of microtubule assembly.[607]

11.2.2 Development of Microtubules

MECHANISM FOR CONTROLLING DEVELOPMENT Numerous descriptions of the development of microtubule systems in protozoan, plant, and animal cells have been published. Not surprisingly, the details vary almost as much as do the cells themselves, and it would not be appropriate to present a long catalogue of the events and processes that have been seen. Rather, a look at one example will serve to introduce the general principles that seem to underlie the development of many microtubule systems.

The single celled alga *Chlorella* undergoes a simple, purely vegetative life cycle in which the cells grow for 10–15 hours (depending upon the rate of photosynthesis), duplicate their nuclear DNA towards the end of the growth period, undergo one mitosis, followed by a second and maybe a third or even fourth cycle of DNA duplication and nuclear division, and eventually cleave to form uni-nucleate masses of protoplasm which produce their own cell walls and commence their own growth period.[17] For much of the growth period no microtubules can be seen, though free monomer is present. The first microtubules appear in the cytoplasm just prior to mitosis, and the total content of microtubule protein in the cell rises. The microtubules then become restricted to the nucleus as part of the mitotic apparatus, in this case formed within a persistent nuclear envelope. After mitosis they disappear from the nucleus and reappear in the cytoplasm in the plane along which the cytoplasm of the parent cell will later cleave. Their occupation of this site is equally transitory, and before long they are once again seen only in the nuclei during the second mitosis. When it has been completed, they reappear in the cytoplasm, mostly in a plane at right angles to the first cleavage plane. If no further mitoses follow, the parent cell divides in the two successive planes that the microtubules occupied, and once this has been accomplished the microtubules disappear until just prior to the next cycle of reproduction.

The details described above are not, at this point, relevant. Their implications, however, are. The cell evidently possesses at least three types of mechanism by which it could regulate the development of its microtubule system. Firstly, it can control the synthesis of tubulins and so modulate the total amount available for participation in association-dissociation reactions. Secondly, it can control the position of the equilibrium in the association-dissociation reactions in such a way that monomer can be present either on its own or in the presence of formed microtubules. Thirdly, it can control the spatial deployment of microtubules in the cell.

The first and the second control mechanism are now considered together in somewhat more detail, and the third is then examined on its own.

QUANTITATIVE CONTROLS Growth of a microtubule system is now seen to depend partially on the level of monomer within the cell and on the equilibrium conditions that prevail. A given level of monomer should, in theory, lead to the formation of a specified length of polymerized structure. Recent research on the microtubule protein content of cells has therefore been directed to collecting two types of information: one, estimates of the total amount present in a cell; the other, estimates of free tubulin relative to the amount of polymerized tubulin.

The total content of tubulins is rarely as high as in the nerve tissue mentioned previously (page 143), where 1 g of brain contains the equivalent of 3000 kilometres of microtubule! A more usual value for microtubule protein would be a fraction of one per cent of the total protein of the cell, or about one per cent of the soluble protein.[679] Like other components of the cell, tubulins are subject to turnover, and whether the amount present at any given time is being augmented or decreased therefore depends upon the relative rates of synthesis and breakdown. Rates of turnover implying that from 3-13% of the total is renewed every hour have been observed.[679, 895, 896]

The amount of free tubulin can be determined by assaying the total microtubule protein per cell, and then subtracting an amount equivalent to the total observed length of microtubules present. From electron micrographs, 1 μm of microtubule protein contains 1600 molecules of tubulin (molecular weight 110 000). For example, when sea urchin eggs enter their first division, they contain three to four times as much free tubulin as they do tubulin in the form of microtubules in the mitotic apparatus.[135] A similar preponderance of free tubulin is found at later stages of sea urchin embryo development with respect to the formation of cilia.[101, 678, 679] In terms of the monomer-polymer equilibrium there is thus a considerable excess of monomer in these systems.

The existence of a supply of tubulin can in suitable organisms be demonstrated by means of simple, visual experiments. Maltreatment of *Chlamydomonas* cells causes them to shed their flagella.[715] Regrowth, which can be measured by watching the new flagella grow, will then occur. Significantly, some regrowth occurs even in the presence of inhibitors of protein synthesis, showing that tubulin is available. However, for *complete* regrowth, new tubulin does have to be manufactured. In other cases, the mere presence of free tubulin is not enough. The protozoan *Tetrahymena* will not regenerate the cilia of its oral apparatus even though tubulin is available, and it is suspected that assembly of the particular class of microtubule present in them is contingent upon the production of a special 'regulator' protein.[895, 896]

Another protozoan, *Nassula*, provides a clear illustration of the quantitative relationship that can exist between free and polymerized tubulin.[838] The organism contains a large microtubular array known as the 'feeding basket'. After cell division each daughter cell forms a new basket by growth of several hundred microtubules down into the cytoplasm from a series of heavily staining plates lying closely appressed to the plasma membrane. All the microtubules grow to approximately the same length, which determines the depth of the completed basket. If daughter cells are given a mild heat shock just prior to microtubule formation the baskets that form subsequently have fewer microtubules in their walls, but the overall length is now greater. Quantitative analysis shows that there is an inverse correlation between number of microtubules present and their length. There can be relatively few long microtubules, or relatively many shorter microtubules, either state presumably arising from a given quantity of available tubulin.

SPATIAL DEPLOYMENT OF MICROTUBULES Radioactive tubulin, which can be isolated from cells that have been grown in the presence of a radioactive precursor of protein, offers a useful means of studying the mode of growth of microtubules.[161a, 286] When added to them it becomes incorporated at a rate suggesting that the polymerization process is confined to the end or ends of the microtubules. Depolymerization, by contrast, can apparently occur either at the tip or anywhere along the microtubule, in the latter case at several sites simultaneously.[602] An 'end growth' mechanism of this sort[278] implies that the rate at which a population of entirely free tubulin molecules associates, or polymerizes, is low compared to the rate at which tubulin will add on to a *pre-existing* microtubule. Many observations of the growth of microtubule systems in cells in fact suggest that the cell needs 'initiation sites'[838]—sites at which an initial impetus is given to the polymerization process, which, once started, continues automatically by end growth.

The concept of the initiation site can be extended to take into account several aspects of the spatial and temporal deployment of microtubules. If no sites are available, a supply of free monomer could (as already described) exist in the absence of microtubules themselves. Changes in the number and positioning of the sites could bring about corresponding changes in the microtubules.

Unfortunately, most of what is known or conjectured about microtubule initiation sites comes from inference rather than from direct visualization or analysis. Places in the cell from which many microtubules arise usually have a moderately electron-dense, amorphous appearance. They may be associated with particular membranes. For example, the dense plates from which the *Nassula* feeding basket microtubules grow lie closely appressed to the plasma membrane (see above).[838] The nuclear envelope provides other examples. Dense plaques that are seen on it seem to be the origin of intra-nuclear spindle microtubules during mitosis in certain organisms (see page 161). One of the best studied of all microtubule systems—that of the spikes, or axonemes, of Heliozoans and related protozoa—radiates through the cytoplasm from a layer on the outer membrane of the nuclear envelope (see below). Centrioles and the initiation sites on chromosomes—kinetochores—are described in the next chapter. Another type of initiating site in the cytoplasm, seen in various algae, consists of an amorphous zone that is usually in a particular part of the cell, for example in the vicinity of the centriole in organisms where the structure is present, or in a comparable location where there is no centriole. Initiation sites for the cytoplasmic microtubules of higher plant cells have not yet been seen.

The postulate that microtubule initiating sites exist is a logical interpretation of electron micrographs, and a new field of enquiry has now opened with the confirmation by an entirely different type of experiment that such sites do have physical reality. Electron dense objects have been isolated from dividing *Spisula* (surf clam) eggs: when added to crude homogenates obtained from the eggs, they induce the formation of microtubule systems that resemble parts of the mitotic apparatus.[881a]

There seems little doubt but that the cell possesses means of placing initiation sites (*nucleation centres* and *microtubule organizing centres*[647] are other names that have been used) in particular locations. Clearly, there are at least two aspects of this fact that deserve close attention: the macromolecular architecture of the sites and their mode of action, and the mechanisms underlying their specific spatial deployment. Furthermore, it must be asked whether the existence of suitably placed initiation sites is sufficient to account for the geometry of the microtubule systems seen in cells. Do the sites act as a template that restricts microtubule growth into a particular pattern? Are additional orientation systems required? Are alternative systems possible?

The microtubular feeding baskets of *Nassula* seem to be an example of a template-directed system.[838] The heat shock treatments referred to previously produce permanent distortions of the pattern consistent with an alteration of the distribution of initiation sites. The protozoa also provide examples of other mechanisms. The axopods of Helizoans have been studied in great detail. They are long spikes that project from the cell surface. Within the plasma membrane of the spike there is an axial array of several hundred microtubules, the axoneme, extending from the outer extremity back into the main body of the cell and terminating at the central nucleus. Different genera possess different types of array.[35] That found in *Echinosphaerium* is seen in cross sections of axonemes as two interlocking spirals of microtubules. The pattern, however, does not seem to be determined by a template of initiation sites, for during growth the microtubule array, initially disorganized, becomes progressively more and more geometrically perfect until each microtubule occupies a position that is fixed in relation to the positions of its neighbours.[717] Thus, to contrast with 'template-direction', 'self-orientation' evidently is possible.

'Self-orientation' demands that microtubules must be able to interact. A variety of mechanisms may exist, from forces of attraction or repulsion between adjacent microtubules bringing about a uniform spacing, to creation of patterns by the establishment of molecular bridges or links,[545a] whose length and positioning around the tubule axis would determine the pattern generated. In *Echinosphaerium,* for example, the appearance of the two interlocking spirals led to the prediction that two types of intertubule link are involved, one long and one short. Modification of staining procedures eventually confirmed the hypothesis, the links showing up in ultra-thin sections as narrow electron-dense connections of two lengths precisely placed around each tubule.

In operating a self-orientation system, the cell evidently produces and polymerizes a given amount of tubulin, together with the molecules used as links. The progressive sorting out of the components probably depends upon the gradual loss of free energy in favour of more 'order' and stability. The tubules that achieve the greatest number of inter-tubule links would be more stable than the others, which might have to move in order to satisfy their potential for forming links, or might be so relatively unstable as to depolymerize. The dynamic system of dissociation and reassociation of tubulin would always tend to favour the formation of tubules that are, or can become, stabilized through linkage.

Higher plants and animals seldom possess extensive geometrically symmetrical arrays of microtubules. It is not known to what extent template-direction and/or self-orientation operate in them. One phenomenon that is, however, very widespread, is that their microtubules often lie parallel to one another (Plate 42a-c). Two processes that have not been mentioned thus far have been invoked in attempts to account for the observation. One is that a formed microtubule may somehow direct the polymerization of another at a set distance from itself, perhaps *via* establishment of inter-tubule links. The other is that tubules can become linked to cell membranes, thus opening new possibilities for spatial stabilization and organization. Suitably stained sections of higher plant cells have shown that electron dense links, equivalent in length to about one third of the diameter of the microtubule itself, can be present between adjacent microtubules,[900] between plasma membrane and microtubule,[746] and between microtubules and internal cell membranes such as those of vesicles or cisternae of endoplasmic reticulum.[235]

11.2.3 Microtubule Functions: Cell Shape

If the simple polymerized tubulin structure described in the previous section is indeed responsible for all the functions ascribed to it, it must be very versatile. The functions that have been proposed range from determining the shape of cells to defining channels of particle movement in the cytoplasm to the movement of chromosomes at cell division. It is perhaps not surprising to find that microtubules are coming to be regarded as frameworks on which other molecules rest and impart specific functions. Some of these additional molecules can be seen in micrographs, for example the cross-connections described above, and the specific enzymes which form arms projecting from the tubules of cilia and flagella (page 149). It seems reasonable to suppose that others, not yet adequately visualized, might exist.

In describing the functions of microtubules it has been felt that a clear distinction should be made between those involving a comparatively static situation (e.g. cell shape) and those involving movements, either of the whole cell, or of particles within it. While the former functions can be discussed immediately, the latter are held over until after a consideration of microfilaments, since these and microtubules are suspected to act co-operatively in some situations that were formerly thought to involve only microtubule activity. The role of microtubules in chromosome movements will be discussed in the following chapter.

CELLS WITHOUT ENCLOSING WALLS The absence of a rigid enclosing wall is a characteristic of most animal cells and certain protozoa. Some plant cells, especially some unicellular algae and the male gametes of both lower and higher members of the kingdom, also lack an exo-skeleton. Nevertheless, the lack of a supporting wall structure does not prevent cells from maintaining shapes other than spherical, which is what they would look like if forces of surface tension were allowed to minimize their surface area. Observed deviations from the spherical state range from slender spikes protruding from the cell surface, to arm-like extensions of the cell and even to the entire cell adopting a much elongated and slender configuration. Many of these extensions from the main body of the cell contain numerous microtubules lying with their long axes parallel to that of the extension. The conclusion that microtubules might determine the shape of non-spherical, naked, cells has been drawn by many investigators[662] since it was first put forward by Manton in a study of the spermatozoids of a moss (*Sphagnum*) in 1957.[514]

Experimental means of testing the conclusion are now available. Various chemical (e.g. colchicine) or physical (e.g. high pressure) treatments can be tried to see whether disrupting the microtubules changes the shape of the cell. In many cases the treated cells do become spherical, only recovering their original, characteristic shape when they are returned to normal conditions. The following paragraphs describe examples of these studies, selected from the protozoa, algae, and higher plants. The animal kingdom, where similar phenomena are widespread, is neglected here.

One of the protozoa discussed previously, *Echinosphaerium*, has been extensively used for studies of the maintenance of cell shape. As already described, the surface of each cell is covered with long fine radiating axopods, each containing a central axoneme composed of several hundred microtubules lying parallel to its long axis. Depolymerization of the microtubules always leads to the collapse of the axopods back into the cell body, and repolymerization of the microtubules is essential for their reformation. Such retractions and extensions are a normal feature of individual axopods, serving to move the cell across its substratum. They demonstrate convincingly that a fine control can be exerted over the behaviour of microtubules in each axopod separately, and in turn that modification of microtubule systems in a cell is not an all or nothing process; rather, the cell can bring about local adjustments.[717]

The unicellular alga *Ochromonas* lacks a cell wall or pellicle yet maintains a characteristic pear-shape, with the narrow end of the body elongated into a tail and the bulbous end bearing a small projection (the beak) placed to one side of a flattened, rimmed area (the platform).[69, 89] Numerous microtubules are associated with these shapes. The plasma membrane of the main cell body is underlain by a series of curved microtubules which extend into and appear to form a tail, rather as the axoneme microtubules of *Echinosphaerium* extend into the axopods. Others are found associated with the beak and platform regions. Disruption of the microtubules leads to loss of the characteristic shape, the cell becoming spherical. Regrowth of microtubules occurs from two initiation sites which maintain a constant position relative to the other cell organelles, so that the original shape of the cell is accurately regenerated.

Motile male gametes lacking cell walls but containing microtubules are a feature of algae,[562, 563] liverworts,[111, 112] mosses,[179, 514] ferns[516] and cycads. In many cases the body of the gamete is extremely elongated and coiled, and in its interior, the microtubules are massed together forming a long coiled complex extending into the tail of the gamete. Over part of its length the microtubule complex forms a sheath around the much elongated and coiled nucleus.[626a] Application of colchicine during gamete development leads to a breakdown in the normal sequence of morphological changes, for example the nucleus fails to elongate and does not adopt the typical coiled configuration. It has been concluded that microtubules are essential for the normal shape development of these sperm cells.

Within limits, the importance of microtubules for male gamete structure extends into the flowering plants. Pollen grains have a highly developed wall (Plate 11b, c, page 22), but this is formed around the vegetative cell. The male sperm are formed from the *generative* cell, which usually has no wall, and lies within the vegetative cell. Growth of the pollen tube and its wall is accomplished by the vegetative cell while the generative cell adopts a spindle-shape before entering and travelling along the pollen tube. Development of the spindle shape is accompanied by the appearance of microtubules aligned parallel to the long axis of the generative cell.[365] Treatment with appropriate drugs leads to loss of the microtubules and the return of the cell to a spherical shape.[728] The generative cell in turn gives rise to the sperm cells which also lack cell walls. These cells also develop microtubules and become elongated, and again the shape change can be reversed by disrupting the microtubules.[113]

Only a few examples have been given, but the inescapable conclusion from numerous studies on animal and plant cells is that microtubules are in some way concerned with the form of naked cells. However, it is important to realise that changing the form of a cell involves two types of activity, the *generation* of the shape change and the *maintenance* of the altered form. Formerly it was thought that microtubules were responsible for both activities, but more recent evidence (to be discussed below and page 153) suggests that microtubules on their own may be concerned merely with the maintenance of a particular shape.

CELLS ENCLOSED IN WALLS A role for microtubules in stiffening parts of naked cells is relatively easy to envisage. The situation in higher plant cells, where the mechanical properties of the cell wall and the forces of turgor pressure interact to determine cell shape (page 35), is more problematical. Even the earliest study of microtubules in walled cells did, however, offer some suggestion that they could influence shape, not by their own rigidity, but by influencing the deposition of cellulose microfibrils.

The observations were that in cylindrical cells of root tips, the microtubules lie close to the plasma membrane and spaced out along its cytoplasmic face (as in Plates 2 and 42a, b).[471] Along the side walls of the cells, where the cellulose microfibrils are initially placed in a circumferential orientation, the microtubules also lie circumferentially. At the end walls, where the cellulose orientation is random, the microtubules also lie without any preferred orientation. A wide variety of cell types has by now been examined, and the generalization made that where the wall adjacent to the plasma membrane contains ordered arrays of microfibrils, the microtubules are similarly oriented on the cytoplasmic face of the plasma membrane.[117, 259, 530a, 559, 585, 698]

Before analysing the possible functional significance of the above spatial relationship between cellulose microfibrils and cytoplasmic microtubules, it is worth looking briefly at a range of cell types for further details. For example, in *Apium* collenchyma, where the wall consists of successive strata differing in microfibril orientation, the microfibril-microtubule parallelism is usually maintained, despite the temporal and spatial changes from one stratum to the next.[117] Root tip cells manufacture end walls and side walls, with differing microfibril orientations, and as already seen, the parallelism is maintained in the two different situations. Perhaps more striking as a demonstration of the ability of a cell to maintain two sets of microtubule-microfibril associations is provided by the long xylem fibres of *Salix*,[698] which at a certain time in their development lay down microfibrils in two different orientations at once at the middle and the extremities of the cell. Each type of microfibril is associated with a set of similarly oriented microtubules.

Further examples of parallelism between microtubules and microfibrils are seen where cells produce sculptured walls. In developing xylem elements the form of the cell is not only determined by the orientation of the microfibrils of the primary wall, which leads to an overall cylindrical shape, but also by a precise pattern of ingrowths of wall material (thickenings) that form a series of troughs or furrows in the plasma membrane surface (Plates 5 and 6). Microtubules are associated with the growing thickenings (e.g. Plate 5b and e).[335, 340, 645]

A wall with a different arrangement of microfibrils is found around root hairs. These have randomly arranged microfibrils at the growing tip while *longitudinally* arranged fibrils are laid down in the side walls. Microtubules running just under the plasma membrane are also longitudinally orientated but in addition extend closer to the tip region than do the similarly orientated microfibrils. So microtubules extend into the zone of wall formation, where microfibril deposition *subsequently* mirrors their orientation.[585]

The generalization that microtubules mirror microfibrils in their deposition should not, however, be taken as absolute. Two types of exception should be mentioned. One is that microfibrils can be laid down in the complete absence of microtubules—a situation seen under natural conditions in, for example, the unicellular alga *Chlorella*,[17] and under experimental conditions in the presence of colchicine. The other type of exception involves disparity in the relative positions of microtubules and microfibrils. For example in the zoospore of *Cladophora* microtubules disappear before the onset of wall synthesis, which then proceeds with the formation first of a layer of randomly orientated microfibrils and then with an organized layer of transverse fibrils. Microtubules do reappear in the cytoplasm, but are arranged longitudinally, and it is then several days before similarly oriented microfibrils appear in the wall.[701a]

Having recognized that exceptions do exist, what is the significance of the widespread parallelism between microtubules and wall microfibrils? One possibility is that there is a causal relationship in which the cell proper, via its microtubules, controls the orientation in which the cellulose is deposited. On the other hand, it can be argued forcefully that because the two structures are on opposite sides of the plasma membrane, the one cannot influence the other, and that both must be oriented by an independent mechanism, perhaps resident in the plasma membrane.

The main evidence in favour of the view that microtubule disposition governs wall microfibril deposition comes from experiments on the effects of colchicine on growing cells. In cylindrical cells, from the relatively small examples found in the zone of cell extension behind a root apex to the relatively enormous internode cells of the alga *Nitella*, exposure to colchicine leads to the production of swollen cells, sometimes even approaching a spherical shape.[288] Underlying the shape of cells that form in the presence of colchicine is a change in the mechanical properties of the wall, which contains randomly-oriented microfibrils and therefore is ballooned comparatively uniformly by the swelling force of turgor pressure. The normal orientation mechanism is restored if the colchicine is removed, and subsequently deposited microfibrils come to lie in their usual pattern. Again, in developing xylem, where the wall thickenings normally appear outside tracts of plasma membrane distinguished by the presence of microtubules, treatment with colchicine leads to abnormal secondary wall deposition, with microfibrils being laid down randomly.[335, 645, 829]

The obvious inference is that destruction of microtubules by treatment with colchicine has removed the machinery for orienting cellulose microfibril deposition.[585] This is an attractive idea, and it can be postulated that the microtubules influence the direction of microfibril synthesis by stabilizing cellulose-synthesizing enzymes in or on the outer surface of the plasma membrane (page 25). It is to be noted that there is no suggestion that microtubules make microfibrils, nor even that microtubules must be present before microfibrils can be formed. Neither is the existence of other, colchicine-insensitive, orientating systems ruled out. It has, for instance, been observed that in xylem developing in the presence of colchicine, the pattern of wall thickenings is maintained where the developing cell lies alongside another that is already mature.[335] In this situation the mature thickenings of one cell apparently induce the formation of thickenings back-to-back with them in the young cell (Plate 5b). Again, some very interesting results have come from studies of the secretion of cellulose microfibrils by certain bacteria (especially *Acetobacter xylinum*) which form a gelatinous pellicle that enmeshes the bacterial cells. Bacteria, of course, have no microtubules. In young cultures little alignment of microfibrils is apparent, but the alignment is more pronounced in older cultures or in young cultures to which extracts of old cultures have been added. The active component in the extracts has been identified as a terpenoid (tetrahydroxybacteriohopane) but how it brings about alignment of the pellicle microfibrils is unknown.[304a] The possibility that similar compounds may be active in cell wall formation in higher plants is raised.

If colchicine acted only upon microtubules in eukaryote cells, the evidence that microtubules govern cellulose microfibril orientation would be very strong. The facts, however, are otherwise. There are persistent reports in the literature of interaction between colchicine and various cell membranes, and in some cases membrane function has been shown to be altered by the drug.[788] For instance the alveolar sac membranes of the ciliate *Tetrahymena* are characterised by the presence of particles (8.5 nm in diameter) seen on the membrane faces after freeze-etching.[917] The particles are evenly distributed at the normal growth temperature (28°C) but can be induced to aggregate leaving smooth expanses by chilling the cells to 5°C. In the presence of colchicine aggregation occurs only slowly. Reheating normal cells leads to a rapid dispersion of the particles over the membrane face, while in colchicine-treated cells the particles tend to remain clumped. Colchicine is here seen to interfere with the mobility in the plane of the membrane of membrane particles. Such results inevitably raise the possibility that the cellulose-synthesizing enzymes on the surface of plant cell plasma membranes (page 25) may also be affected by colchicine.

There is more to a cell wall than microfibrils. We have already looked briefly at the nature of the cell wall matrix materials (Chapter 3) and their mode of synthesis in the Golgi apparatus and delivery to the growing wall via vesicles (Chapter 7). Microtubules may play two parts in these processes. One will be considered in the next section, and concerns the movement of vesicles through the cytoplasm. The other concerns the sites at which the vesicles are permitted to fuse with the plasma membrane. Unfortunately there is much more conjecture than fact. Thus it has been said of developing xylem that sheets of endoplasmic reticulum lying between the bands of microtubules specify the position of the thickenings by preventing access of vesicles to the plasma membrane, thereby ensuring that wall material is delivered to the actual thickenings. It has conversely been pointed out that the microtubules are so close to one another that it would be difficult or impossible for a vesicle to pass between them to reach the plasma membrane, leading to the counter suggestion that delivery of vesicles must be *between* the thickenings.[282a, 282b] There are comparable wall thickenings in leaf cells of the moss *Sphagnum* and another suggestion stemming from work on their development is that the microtubules may in effect lift the plasma membrane off the inner face of the wall, holding it against the turgor pressure of the cell, so creating a space into which wall material can flow to generate the thickenings.[746]

The latter idea has the merit of being another cytoskeletal function for microtubules, not dissimilar to those found in naked cells. Indeed with such an interpretation, most of the functions postulated so far can be regarded as falling within the general heading of cytoskeletal, others being the maintenance of cell shape and the stabilizing of cellulose synthesizing systems in or on the plasma membrane.

11.2.4 Cilia and Flagella:

As the structure and properties of cilia and flagella are basically similar they will be discussed together, and all references to one will apply to the other unless otherwise stated.[772a] Flagella are the long slender whiplash structures that propel certain unicellular organisms (flagellates) and the motile male gametes (sperm) of animals and lower plants. Cilia are much shorter, and occur in great numbers over the surface of some unicellular organisms (ciliates, *Paramecium* for example has approximately 17 000 cilia) and also on the surface of some epithelial cells forming a ciliated epithelium (lung tissue for example). Cilia beat in a coordinated fashion, either driving the organism through the water, or creating a flow of water which may bring food particles to an organism, or pushing a film of water past the ciliated epithelium of larger animals. Both cilia and flagella are bounded by an extension of the plasma membrane, and originate from a *basal body* located just beneath the cell surface.

Cilia and flagella have for long been studied by microscopists, and more recently by biochemists. A series of papers by Ballowitz (1888) on sperm flagella of various animals include some remarkable observations and provide the first clues to their fine structure. He allowed his preparations to stand for a few days, which led to partial degeneration of the cells (autolysis), before staining with gentian violet. Microscopic examination revealed that some of the flagella had split lengthwise into varying numbers of smaller fibres. The greatest number, found in chaffinch (*Fringilla coelebs*) sperm, was eleven. No further information was available until the early 1950s, when Manton and her colleagues reported their observations, made first by ultra-violet light microscopy[513] and soon afterwards by electron microscopy, on the structure of flagella of certain algae and motile gametes.[521, 522] In these whole mount preparations a series of smaller, longitudinal, fibres, usually eleven, could be seen to originate from a single flagellum by fibrillar disintegration. Even before ultra-thin sectioning methods became available it was deduced that the structure of the flagellum consists of a circle of nine outer filaments around a central pair (referred to as a '9+2' pattern).[521] Sections show that the nine outer filaments lie on a circle and are composed of a pair of microtubules (doublets), one (the 'A' tubule) slightly inside the circle compared with the other (the 'B' tubule). The A tubule of each pair bears two short arms on the side away from the other (B) tubule. All the arms on the doublets point in the same direction around the circle, this is always clockwise when the flagellum (or cilium) is viewed from the distal end looking towards the cell body.[180, 864]

The above description applies to cilia and flagella from unicells to the higher plants and animals. The only exceptions[1a] involve loss of one or both of the central pair, giving (9+1) and (9+0) arrangements, although a case with a (9+7) arrangement has been reported. Loss of the central pair does not seem to impair flagellar motions, so it is thought that the principle structures generating movement are the nine paired tubules and their arms.

Cilia and flagella can be broken from the cell surface and collected by centrifugation. Addition of adenosine triphosphate (ATP) leads to a resumption of their beating movements, demonstrating that this activity is independent of other cell structures such as a basal body. A search for the site of ATP breakdown in isolated cilia was carried out by selectively solubilizing each component and testing to find whether the solution or the residue contained the enzyme adenosine triphosphatase (in cilia and flagella called 'dynein', from its role in converting chemical energy to mechanical forces).[262a] At each stage samples of the residue were embedded and sectioned to determine which components were still present. Treatments which removed the arms of the A tubules also solubilized dynein. Examination of the dynein by negative staining showed that it is a flat rectangular particle of varying lengths due to polymerization of the molecules. The length of the short edge of the particle is the same as that of the arms seen in ultra-thin sections of the A tubules, that is, each arm seen in cross-section is a long slab of enzyme running along the length of the tubule.

The energy released by ATP breakdown at the A tubule is believed to induce a conformational change in the protein of the doublet microtubules. Co-ordinated changes in successive doublets around the circle and along the length of the flagellum or cilium leads to the observed movements, although the exact mechanism is as yet unknown. Several proposals have been put forward based on the above information and on theoretical considerations of the type of movements observed.

The essential homology between the microtubules described in preceding sections and the tubules of cilia and flagella can be seen by comparing their structure. In negatively stained preparations they are both composed of subunits of monomers of about the same size, organized in the same way. In cross-section the microtubules and A tubules are each composed of thirteen monomers forming the wall. The B tubule however contains only ten monomers around the circumference, the remainder of the wall being shared with the A tubule. The central pair, like cytoplasmic microtubules, each have thirteen monomers around the circumference.[824, 864] Further, it has been shown that the isolated flagella tubule protein binds colchicine in the same manner as microtubular protein. However treatment with colchicine does not lead to the destruction of flagella, although it will preclude their regeneration after experimental removal. It seems that the flagella tubules are very stable structures compared with their cytoplasmic counterparts, showing only a very small tendency to undergo reversible dissociation into monomers.

At the base of each flagellum (or cilium) is a basal body which is identical in structure to, and may have originated from the kinetosome (centriole) which lies in a depression of the nuclear envelope of most animal and protist cells. It is a short cylindrical structure, not enclosed by a membrane, whose walls are composed of nine triplets of microtubules organized as are the doublets of the flagellum, but with a further tubule (the C tubule) added to the B tubule so that all three lie in a straight line at a slight angle to the cylinder surface. A short wall joins the A and C tubules of adjacent triplets. The A tubules do not bear arms and there are no tubules in the centre; instead a central rod or filament is present from which radiate a series of curving arms forming a cartwheel structure, though this is often confined to one end of the basal body (the proximal end). The A and B tubules at the distal end grow out to form the doublets of the flagellum.

Centrioles are absent from the cells of higher plants. Among lower plants many algae possess centrioles in the vegetative and reproductive cells, but others (some algae, Bryophytes, Pteridophytes and Cycads) lack centrioles, only forming them during the development of motile gametes.[626a] A discussion of centriole behaviour during mitosis will be found in Chapter 12.

11.3 Microfilaments

ANIMAL SYSTEMS Many different filamentous structures are common in the cytoplasm of animal cells, where they are thought to perform a variety of functions. One of the best understood systems is that of muscle, containing an interacting series of actin fibres. Another system of fibres, distinguished by their small size and hence termed microfilaments, has been shown to be present in a wide range of animal cells and seems to participate in a number of activities connected with movements of and within cells.

Microfilaments appear in thin sections as slightly beaded structures, 5–7 nm in diameter, that are often more lightly stained than other cell components such as ribosomes and membranes. They are at their most conspicuous when they occur associated together in bundles or sheets, the filaments lying roughly parallel to one another and usually separated by a space of about 10 nm. The filaments are often seen to branch and join up with each other, though in some cases these images may only be a result of the overlapping of microfilaments in the depth of the section.

One of the situations where microfilaments are regularly found[885] is in the form of a 'contractile ring' just below the plasma membrane associated with the cleavage furrow of dividing animal cells (the constriction that forms between the two daughter nuclei to form two new cells). The observation that the drug cytochalasin B inhibits animal cell division by preventing and even reversing cleavage furrow formation prompted an investigation of the effects of the drug on microfilaments. It was found that the contractile ring microfilaments disappeared in the presence of cytochalasin B and were reformed if the drug was removed. This discovery has led to studies on the reaction of various cells to treatment with cytochalasin B, which is often, and without justification, assumed to act directly and specifically on the microfilaments. From the examples of animal cells studied so far, it seems that microfilaments are involved in specific movements of the plasma membrane or of cytoplasmic components.[885] Such matters are discussed further in the next section.

PLANT SYSTEMS Electron microscopists have not often observed microfilaments in plant cells (Plate 43j).[341a] The reasons for their elusive quality are not entirely straightforward and various explanations have been offered, such as their possible limitation to certain cell types and the difficulty of detecting them unless they happen to be parallel to the plane of the section. Also there are instances where it is relatively simple to observe presumed microfilament bundles by phase contrast or interference contrast light microscopy, but extremely difficult to obtain comparable images by electron microscopy after fixation and specimen processing. It has been demonstrated that they are seen more frequently if modifications are made to the conventional staining and dehydration procedures in the embedding processes to give improved preservation of the material.[633a] As many plant cells are notoriously difficult to preserve it does seem plausible that the microfilaments observed to date have been chance survivors of inadequate preparation methods. The improved method has revealed microfilaments as a regular feature of elongated cells in a number of higher plant tissues from several different species.[341a] They occur associated together into several discrete bundles or fibres running along the length of each cell. In one instance an elongated mitochondrion was seen apparently attached by one end to a fibre.[633a] It would appear that microfilaments may be associated with cytoplasmic movements in plant cells, as in animals, a topic that is discussed further in the next section on cell movements.

FUNCTION Some investigators have compared microfilaments with the actin filaments seen in animal muscle, suggesting that microfilaments are composed of 'cytoplasmic actin'. Muscle filaments are characterized by the ability to bind molecules of the protein myosin, and to release them in the presence of adenosine triphosphate, and to contract when exposed to calcium ions. Similar properties have been claimed for some microfilament systems. However, experiments designed to test the interaction between cytochalasin B (known to affect microfilaments) and actin, both *in vivo* and *in vitro*, have yielded conflicting results.[228, 673, 784] Whatever conclusions are finally reached it seems unlikely that cytochalasin B acts on microfilaments in a manner analogous to that of colchicine on microtubules. One suggestion put forward is that the drug disrupts cross connections between the plasma membrane and the microfilament system, so inhibiting the characteristic membrane movements.[235]

11.4 Microtubules, Microfilaments and Cell Movements

Cells exhibit a variety of movements, one of which, propulsion by cilia or flagella, has already been discussed briefly (page 149). Broadly, the remainder can be divided into movements involving a bulk flow of cytoplasm, and those involving specific movements of individual cell components. Examples of bulk flow range from amoeboid movement to cytoplasmic streaming, while specific movements concern components as small as dictyosome vesicles or as large as the nucleus. Many of these phenomena are still poorly understood, although cellular structures thought to be associated with the movements have been described—in some cases microfilaments and in others microtubules. However, there is an increasing body of evidence to show that these two structures act in a cooperative manner in many cases.

11.4.1 Amoeboid Movement and Cell Shape Changes

'Amoeboid movement' takes its name from the characteristic locomotory activity of amoebae (e.g. *Amoeba, Chaos*),[813] although it is also found amongst cells of higher animals (white blood cells for example) and animal cells maintained in culture.[782] A localized forward movement of the cell occurs which usually also involves a retraction of the cell from another point. This may be accompanied, especially in cultured cells of higher animals, by a characteristic activity of the plasma membrane variously described by the terms 'ruffling', 'blebbing', 'undulating', and 'micro-spike formation' depending on the cell type involved.

The advancing cytoplasmic front contains a variety of filaments and fibrils, the finest of which resemble microfilaments. Exposing the cells to cytochalasin B inhibits the forward movement of the cytoplasm and also inhibits the characteristic plasma membrane activities. Microfilaments that were previously abundant at the moving front are either lost or replaced by a dense mass of fibrils and granules after drug treatment.[782, 813] Microtubules are not usually a prominent feature of the cytoplasm of these cells and do not appear to be involved in amoeboid movements, although they are found associated with a complex array of microfilaments and other fibres beneath the plasma membrane of cells which are stabilized in contact with a solid surface, and which are carrying out phagocytosis.[686]

CELLS IN TISSUES These activities of microfilaments are not confined to the movements of free-living cells, but are also found in cells of various animal tissues. Morphological changes in tissues, especially of epithelia, can be inhibited by cytochalasin B and may even be reversed by the treatment. In such tissues microfilaments are located at specific places in the cells, at one end for example, and appear to act in a coordinated fashion to bring about the change in conformation. The change can usually be traced to a contraction of the membrane of each cell in the region of the microfilaments. In the simplest cases the epithelium may just roll up to form a tube, and in more complex cases the specific distribution of microfilaments among the sheet of cells can lead to the localized formation of bulges, pockets, etc.[885]

TIP GROWTH By contrast, the growth of part of a cell to form one or more extensions from the main cell body seems to involve both microfilaments and microtubules. For example the growth of a nerve axon from the main cell body depends on microfilaments during the forward extension of the cytoplasmic front ('tip growth') *and* on microtubules to maintain the elongate form so generated.[885] It is possible that many of the cell shape phenomena described previously (page 146) depend on microfilaments (acting as intracellular 'muscles') for their generation and on microtubules (acting as intracellular 'bones') for their stabilization.

11.4.2 Cytoplasmic Streaming

INTRODUCTION Cytoplasmic streaming is a feature of many plant cells and is especially easily observed in giant algal cells such as the internodes of *Nitella*. It has been the subject of extensive studies that commenced long before the introduction of electron microscope techniques. Many of the experiments were marvels of manipulative skill and careful painstaking observations with the light microscope,[417b] and demonstrate that we still have much to learn by examining living cells before subjecting them to the rigours of fixation and embedding. The type of streaming shown by many algal and higher plant cells often involves a continuous movement around the vacuole and is termed 'cyclosis'. The movement of cell components occurs in a definite layer of cytoplasm (the sol) underlying a cortical stationary layer (the gel). The movement is energy dependent, its rate being increased by metabolic promoters and decreased by inhibitors. Agents which block disulphide bridge formations in proteins are in some way involved in the process.

ALGAL CELLS Isolation of cytoplasmic droplets from streaming algal cells allowed observations to be made on the cytoplasm without interference either from the cell wall or from the layer of chloroplasts which often lies immediately beneath the plasma membrane, obscuring the view of what is happening within. Although the droplets did not exhibit cyclosis as such, many movements typical of streaming cells were noted.[395a] The movements were found to be associated with a system of very fine fibrils that were constantly moving in the cytoplasm, aggregating to form transient thicker bundles before again breaking up into fibrils. The bundles exhibited wave motions much like those of a flagellum.

Observations on intact *Nitella* cells, the microfilaments of which have many of the properties of muscle actin,[623a] can be made if a 'window' is opened in the layer of chloroplasts by illuminating with a very bright spot of light for a few minutes (as provided by the usual light microscope lamp and condenser system). After several hours the chloroplasts begin to disappear from the area that was illuminated, thereby exposing the underlying cytoplasm.[417a] Streaming continues across the window, but at a reduced rate. Eventually a system of parallel fibrils grows across the window and streaming returns to its former speed (about 50 μm/second). Presumably the parallel fibrils are normally present in the sub-cortical cytoplasm at the boundary between the stationary and the mobile cytoplasm, but are removed, along with the chloroplasts, by intense illumination. The streaming that is observed in the absence of fibrils must represent material that is 'free-wheeling' across the window.

The window technique has allowed other detailed observations of the streaming cytoplasm to be made. According to one report[7a] the system of fibrils (sub-cortical fibrils) that lies just beneath the chloroplast layer is not responsible for the streaming motion but forms the outer boundary of a 10 μm thick layer of moving cytoplasm. Within the mobile layer some particles move up and down in a wave motion, as if they were attached to very fine fibrils (endoplasmic fibrils) that undulate at about 3.5 cycles per second, with a wavelength of about 25 μm and amplitude of about 5 μm. It is calculated that the observed numbers of endoplasmic fibrils (about 17/μm^2) would produce a sufficient flow past themselves to account for the observed streaming rate. They are assumed to be anchored to the sub-cortical fibrils so that they drive the cytoplasm round the cell, rather than propel themselves through the cytoplasm. So far it has not been possible to preserve this system sufficiently well to allow observation of it in ultra-thin sections.[7a]

HIGHER PLANT CELLS Fibrils have been seen in living cells of higher plants[600a] (Plate 43g, h, i). Mitochondria and spherosomes appear to become 'attached' to the fibrils, whereupon they move along at about two micrometres per second. Such observations are reminiscent of the fine structural work discussed previously which showed a mitochondrion 'attached' to a microfilament bundle. In the vicinity of the fibres the cytoplasm appears to move much faster than the mitochondria and spherosomes, at about 20–50 μm per second.

STREAMING AND MICROFILAMENTS In the earliest reports on microfilaments in plant cells it was suggested that they make up the fibrils that had been connected by light microscopists with cytoplasmic streaming, although at the time there was more support for the view that microtubules are the structures involved in this activity. However, although microtubules may indeed be involved in some algal cells (e.g. *Caulerpa*),[723] in other algal cells and in higher plants the microtubules lie against the plasma membrane and often are orientated at right angles to the normal direction of streaming. Observations on streaming, by contrast, suggest that it occurs in deeper layers of cytoplasm, where the microfilaments may be orientated in the direction of flow.

The discovery that cytochalasin B acted against microfilament systems in animal cells led to it being used in experiments on cytoplasmic streaming in plants. In both algal and higher plant cells application of the drug leads to a rapid cessation of streaming.[75, 885, 914] If the drug is removed, by transferring the cells to fresh medium for example, then streaming soon resumes at the previous rates. Observations with the light microscope on *Nitella* rhizoids during cytochalasin B treatment showed that the fibrils become progressively distorted with the formation of motile looped and linear fibrils. These eventually become stationary, forming thicker looped structures accompanied by formation of polygonal fragments, much as seen in the experiments with isolated cytoplasmic droplets. It was suggested that in these cells cytochalasin B causes detachment of the microfilaments from the cortical cytoplasm.[120]

The sensitivity of plant cells to cytochalasin B has been used to support the view that they contain microfilaments similar to those found in animal cells.[341a] However in the few plant systems examined in detail no evidence has been found for the destruction of the microfilaments by the drug, as reported in some animal cells. This has led some authors to question the assumption that the drug acts specifically on microfilaments. They argue that it could equally act on some other component of the streaming mechanism, such as the energy-releasing mechanism that drives it.

ANIMAL CELLS Streaming of the type found in plant cells, involving a continuous circulation of the cytoplasm, is not a common feature of animal cells. On a smaller scale, however, streaming is quite often found associated with specialised cell extensions, for example the axopods of *Echinosphaerium*, which have already been described. Other protozoan organisms, the suctorians (e.g. *Tokophyra*), pump the cytoplasmic contents out of their prey through long cytoplasmic tubes. Arrays of microtubules are responsible for maintaining the structure of these cell extensions, but it has also been assumed that they are responsible for the streaming of particles along the extensions. In the case of *Echinosphaerium*, the effect of removing the microtubules without at the same time collapsing the overall structure of the axopod, has been tested by passing a fine glass needle through the cell body of an *Echinosphaerium* and pushing it out at the far side, thus forming an artificial axopod consisting of a thin film of cytoplasm supported not by an axoneme of microtubules, but by glass.[184a] Characteristic particle movements were nevertheless observed in the artificial axopod. Furthermore, movements continued in the presence of concentrations of colchicine strong enough to collapse the normal axopods. Fine filaments (microfilaments?) were observed in the artificial axopods and may have been responsible for the streaming activity. It would appear that once again microtubules provide a skeleton but are not wholly responsible for movements.

A further example of a system that was originally thought to involve only the activity of microtubules, but can now be shown to involve both microtubules and microfilaments is seen in the pigment cells (melanophores) in the skin of certain animals. A system of cell processes extends out from the melanophores into the surrounding tissue. Pigment granules (melanin) migrate outwards from the main cell body along the processes, lending a dark colouration to the skin; a reversal of flow leads to a lightening of colour. In the killifish (*Fundulus*) the migration occurs along cytoplasmic channels that are bounded by microtubules which are believed to be involved in the movement of pigment granules.[662] However similar cells in the skin of the frog contain microfilaments as well as microtubules. Exposure to cytochalasin B reverses the outward migration of granules, although it has no effect after outward migration has ceased.[503, 543] Application of colchicine to darkened skins inhibits the lightening process (granule aggregation) but does not inhibit the darkening process when applied to light skins. It would appear that both microtubules and microfilaments are involved in this specialized class of intracellular movements. Note that the movement differs from the bulk streaming processes described previously. A specific cell component moves in a manner described as *saltatory* (from *saltare*, to dance). Other movements of specific components are now considered, some of them, like the migration of melanin granules, being classed as saltatory.

11.4.3 Movements of Selected Cell Components

DICTYOSOME VESICLES It is worth re-opening the subject of the delivery of dictyosome vesicles, already introduced on page 80, in the light of what has been learned concerning microtubules and microfilaments.

Dictyosome vesicles are usually too small to be clearly resolved with the light microscope, so conclusions as to their movements based on electron micrographs are bound to be of a tentative nature and subject to criticism. In some cases moving particles in living cells can be positively identified as vesicles if subsequent examination of ultra-thin sections shows that they are the only component of that size present. This still leaves questions about their movements: are they moving in response to a specific force applied to each one individually, or are they merely carried along in the general flow of cytoplasmic streaming? We shall see that there is evidence that both types of movement occur.

Localized delivery of dictyosome vesicles concerned with wall synthesis can be observed, especially at the tips of pollen tubes (page 77). This tip growth is analogous to the tip growth of animal cells, but occurs by additions to the plasma membrane and cell wall at the growing front, rather than by bulk flow of cytoplasm. Treatment of a variety of cells showing tip growth (pollen tubes of *Clivia* and *Lilium*, root hairs of some *Brassicas*, rhizoids of the giant alga *Caulerpa*) with cytochalasin B leads to a cessation of streaming and growth.[75, 119, 235, 606, 885] In *Tradescantia* pollen tubes the extent of the inhibition was shown to be dependent on the concentration of cytochalasin B.[537] Microfilaments were found in untreated pollen tubes, apparently cross connected by short side arms to cisternae of the endoplasmic reticulum and to the plasma membrane and sometimes to microtubules, but not specifically associated with vesicles. Little change in these relationships was noted when material treated with cytochalasin B was examined.[235]

The movement of dictyosome vesicles to the plasma membrane is a general phenomenon associated with wall growth (page 80). It has been suggested that microtubules might somehow effect this movement, for example during the growth of localized thickenings of xylem elements. These thickenings have microtubules associated with them (Plate 5b, e) which have been supposed to direct the vesicles to the underlying wall. Growth of the intervening wall was assumed to be prevented by the screening effect of cisternae of endoplasmic reticulum lying flat against the plasma membrane. It has been pointed out, however, that the spacing between the microtubules is insufficient to allow objects of the size of dictyosome vesicles to pass between them and so to reach the membrane surface.[282a, 282b] There is no doubt that the disposition of microtubules and cisternae of endoplasmic reticulum around the side walls of developing xylem elements undergoes a regular sequence of changes, but their significance is not understood. As pointed out previously (page 148) the whole system can be disrupted by applying colchicine, leading to the wall becoming irregularly thickened. It is not yet clear whether the colchicine effect, which presumably is mediated via disruption of microtubules, indicates a breakdown of the system that directs the delivery of dictyosome vesicles. The observed effects on wall development could well result from disorganization of a microtubule-stabilized wall-synthesis system located at the plasma membrane (page 148).

Cell plate formation probably involves a movement of dictyosome vesicles from both future daughter cells through cytoplasm that is packed with microtubules (see Plate 48 and page 169). The vesicles may become attached by short links to some of the microtubules. In the presence of colchicine the outward growth of the new wall ceases, as does the movement of vesicles. The conclusion that the microtubules are responsible for moving the vesicles should be viewed with caution in view of the experiments on artificial axopods in *Echinosphaerium* mentioned above. The microtubules might only be playing a structural role which is essential for normal growth, although it seems that microfilaments may not be associated with them in that cytokinesis is not known to be interrupted by cytochalasin B.

In conclusion, there seems to be little evidence for a general mechanism present in plant cells that results in movement of dictyosome vesicles. In some cases they may be passengers in a general flow of cytoplasm, mediated by a cytochalasin B-sensitive system, perhaps based on microfilaments, while in others the integrity of microtubules appears to be necessary, though this should not be taken to mean that microtubules necessarily exert a motive force.

CHLOROPLASTS It has been known since the observations of von Mercklin in 1850 that the position of chloroplasts within leaf cells can change, for example if the intensity of the incident light is altered. In general, high intensity light induces a movement of the chloroplast to a position in the cell where it is shaded, or to a change in the orientation of the chloroplast so that only the smallest possible profile is exposed to illumination. Low intensity light leads to chloroplasts presenting the maximum possible surface area to the incident illumination. In the alga *Mougeotia* it has been shown that phytochrome molecules, thought to be arranged in a screw-like (helical) orientation in the plasma membrane or cortical cytoplasm, provide a sensing system, which is assumed to pass a signal to the cytoplasmic component that is responsible for moving the chloroplast.[858]

The identity of this cytoplasmic component has been the subject of several investigations. For instance, in a study of a plant cell tissue culture it was noted that cross links were sometimes present between microtubules and plastids. It was suggested that chloroplast movements might be initiated if these links somehow travelled along the microtubules. However, there is as yet a lack of similar evidence from other cells, and the postulated mechanism cannot be regarded as proven. Other approaches have been more fruitful. The flat, plate-like chloroplasts of the green alga *Mougeotia* re-orient when the illumination is changed between high and low intensity, and furthermore the movement can be inhibited by treatment with cytochalasin B.[752a, 858] The drug seems to act specifically on the chloroplast movement system and not on the photoreceptor system, strongly suggesting that a microfilament-type system is present in the cell. The way is open for other similar experiments on different plants to establish how widespread the system is throughout the plant kingdom.

NUCLEUS Nuclei, like other cell components, can be carried along in the general flow of cytoplasmic streaming. However, in a few cases there is evidence that specific movements are undertaken by nuclei, independent of the remaining cell contents.[750] For example, nuclei migrate along the hyphae of certain fungi. In *Polystictus versicolor* it has been shown that the two 'polar plaques' (see page 161, dense laminae attached to the nuclear envelope on opposite sides of the nucleus) have microtubules radiating from them into the cytoplasm.[274] The nucleus is slightly elongate, and moves along the hypha with one polar plaque pointing forward and the other pointing back. It has been suggested that interaction of the microtubules with the local cytoplasmic environment and with other components (mitochondria, endoplasmic reticulum) results in a force that moves the nucleus.

Various types of movement, either of whole nuclei or of their envelopes and chromosomes, occur at mitosis. These usually involve microtubules and are discussed in Chapter 12.

LIPID DROPLETS An unusual type of cell movement has been described for lipid droplets in epidermal cells of *Ornithogalum* (Star of Bethlehem) ovaries.[451] It is included here because all the examples described so far have involved a linear movement of particles, while in this case the lipid droplets are observed to spin rapidly on their axes. Examination of thin sections of this material has shown that each droplet is enclosed within a basket of microtubules which are presumably involved in the generation of the spinning movement. There is no evidence to indicate how the movements are generated, although careful observations suggest that the microtubules vary slightly in diameter according to the rate of spinning. The microtubule walls are thought (on the basis of staining reaction with ruthenium red) to be coated with a layer of polysaccharide material, but the significance of this is unknown.

The success of experiments in which cytochalasin B is used as a tool to investigate the activity of presumed microfilaments has diverted attention away from the paucity of information on the structure and mode of functioning of these structures. It is also a fact that little is known about the site and mode of action of the drug. No real gains in our understanding of microfilaments as cell structures will be made until these gaps in our knowledge are filled. The importance of the work with cytochalasin B lies in the gathering body of data that questions some of the long-held views on microtubule functions. In *some* cases it is now possible to remove the dynamic role ascribed to microtubules and leave them as relatively static structures acting as a framework, or cytoskelton, for the cell.

Microfilaments emerge as the 'muscle' of the cell, and like muscles of animals, seem to require anchorage to some substratum. In some cases this anchorage may be provided by the plasma membrane, in others it may be provided by microtubules. Although so little is known about them, microfilaments already add a further dimension of complexity to the functioning of the cytoplasm. Presumably they require a system for determining their production and location in the cell, as well as one for regulating their activity.

12 Cell Division

12.1 Introduction

The dynamic sequence of events that culminates in the formation of two identical nuclei from a single progenitor has long been the subject of study.[898] The work of Remak and Virchow in the 1850's (page 39) established that nuclear division precedes cytoplasmic division, but the full complexities of the overall process of cell division were not appreciated until, starting in 1873, a number of observations were made by the leading cytologists of the day, showing it to be a far more elaborate process than had previously been supposed. Flemming's work included the proposal (1882) that the term *mitosis* should be applied to the process of nuclear division, the word being appropriate in its derivation to the movement of the thread-like bodies, later called chromosomes by Waldeyer, that were seen to participate. Although 'mitosis' aptly describes the movement of chromosomes, it conveys less of the overall process than the word it has largely replaced, namely *karyokinesis*—the division of nuclei. Neither word covers the changes in the cytoplasm that usually accompany nuclear division, and for these Whitman in 1887 used the term *cytokinesis*—the division of the cell.

Two categories of structure were seen to participate in mitosis, and were distinguished on the basis of their affinity for stains. One category, the 'chromatic figure', like its constituent chromatin and chromosomes, was so-named because it could be densely stained. The other, the 'achromatic figure', being relatively unstainable, was, and is, less amenable to light microscopy, but we shall be devoting considerable space to the modern studies of its properties by means of electron microscopy. The achromatic figure is usually a symmetrical object, in which the chromatic components move. One type was described as *amphiastral* by Fol in 1877, in recognition of the symmetry with which two star-shaped bodies (asters) lie, one at each end (*pole*) of the intervening *spindle*. It must be emphasized that amphiastral spindles are typical only of animal cells and certain protozoans and lower plants, and that the spindles of the vast majority of higher plant cells have more diffuse, less pointed poles, and, having no asters, are described as *anastral*.

Whilst the origin of the chromosomes from the chromatin of the nucleus could easily be followed, there was some confusion over the origin of the achromatic figure. Strasburger believed that it originated entirely from the cytoplasm, while Flemming regarded it as of entirely nuclear origin. Fol believed that the spindle arose from the nuclear material and the asters from the cytoplasm. The issues were partially resolved by van Beneden and Boveri in 1887, who independently discovered that the centre of each aster persists from one cell division cycle to the next, lying in the cytoplasm close to the nucleus. Boveri, eight years later, showed that a very small body, the *centriole*, lies in the middle of each aster: it was soon shown to be an autonomous body, arising from a pre-existing centriole.[384] The presence of asters in animal cells and their absence from plants was linked by some investigators to the presence of a rigid cell wall in plants. Thus asters were believed to assist in positioning the spindle in the animal cell and in the formation of the typical dumb-bell shape configuration at the end of the cell division process. However, it now seems more likely that the loss of asters in higher plants is correlated with the loss of cilia and flagella, centrioles being very closely related to the basal bodies of these organs of locomotion. It should be mentioned, however, that certain animal cells can form asters in the absence of centrioles, and that aster-like structures have even been found in certain higher plant cells (e.g. in endosperm of *Crepis*[29]).

The principal features of mitosis that could be detected with the light microscope were therefore established by the beginning of the present century. A summary was presented on page 39, and is amplified here in order to set the scene for the remainder of this chapter. In plants the progressive condensation of the nuclear chromatin at *prophase* leads to the appearance of discrete chromosome structures, each composed of two sister chromatids. At the same time the nucleolus becomes more diffuse and gradually disappears. A region around the nucleus (now called the *clear zone*) becomes devoid of large cytoplasmic components and the nuclear envelope fragments and disappears, marking the end of prophase. Formation of a barrel-shaped fibrous spindle with diffuse poles is then accompanied by the formation of points of attachment between fibres and chromosomes. The attachment points are known as *kinetochores*, or centromeres, and each chromosome usually has its kinetochore at a particular point along its length. This stage has become known as *prometaphase*, and the events and associated chromosome movements are distinguished from those of *metaphase*, during which the chromosomes lie with their kinetochores arranged at the mid plane (or equator) of the spindle. The separation of sister chromatids and their movement to opposite poles at *anaphase* was at first thought to be initiated by the division of the kinetochore region, but it is now known that each chromatid develops a kinetochore, and that the separation involves movements of the already paired kinetochores and the fibres that have grown from them. The kinetochores 'lead the way' towards the poles, giving characteristic V or J shapes if they are non-terminal in the chromatids. At *telophase* the nuclear envelope re-forms and the nucleolus reappears as the chromosomes become more diffuse and revert to the *interphase* condition.

Experimental work continued since these early descriptive studies of mitosis, with the structure and function of the spindle fibres attracting particular attention. Information that was invaluable to cell biologists when experimental work on microtubules was undertaken in the 1960's was obtained. The work of Inoué and his colleagues, starting in the 1950's, was a major landmark.[392] They demonstrated that spindle fibres could be observed in living cells, and, by examining specimens with polarized light, that the fibres possess a regular substructure. The basis of the substructure was discovered later, when improved techniques of electron microscopy became available. Meanwhile it was also shown that a variety of treatments, such as cooling or exposure to the drug colchicine, could reversibly decrease the amount of fibres present in the spindle, while others, such as warming or exposure to heavy water (deuterium oxide, D_2O), could increase the amount present. Loss of the fibres led to a cessation of normal mitosis, proving that they are essential for chromosome movements.

The advent of suitable thin-sectioning techniques for electron microscopy in the early 1960's did not lead to the hoped-for unveiling of the mitotic mechanism. The structures involved were, however, defined to a better level of resolution, one of the most important findings being that spindle fibres are bundles of microtubules.[660]

The reasons for the lack of instant success in elucidation of the mechanism lie in the fact that mitosis is a peculiarly difficult subject to study. The mitotic apparatus is large and may fill nearly the whole cell, yet some of its vital components are so small that they cannot be resolved with the light microscope. The structures are constantly moving relative to one another, thus introducing a time factor into the system. While observations with the light microscope on living material provide an accurate analysis of the dynamic activity of the large components, such as the chromosomes, they cannot effectively be used to study the detailed behaviour of the small components, such as the microtubules. Detailed observations on the latter can be made on fixed material sectioned for the electron microscope, but difficulties remain. Fixation arrests the process at some particular point in the cycle, perhaps after inducing abnormal movements of the structures. The total apparatus is so large relative to the thickness of an ultra-thin section that location of the precise area of the mitotic apparatus included in a micrograph becomes a problem.

Recently a systematic study of cell division using both light microscopy of living cells and electron microscopy of the same cells, kept under observation during fixation and embedding, has been undertaken by Bajer and his colleagues. The cells used are from the endosperm of the blood lily, *Haemanthus*.[29] In early ovule development this tissue is liquid because although individual cells are dividing continuously, cell walls are laid down later (page 41). The cells are carefully removed from the ovary with a pipette, and a drop placed on a cover slip which is inverted over a supporting ring on a microscope slide. Studies of the mitotic process in these living cells are made using time lapse photography, so that the entire sequence of several hours can be compressed into a film lasting a few minutes. Stages of especial interest can then be picked out and the individual frames of that part of the film studied in more detail (Plate 44). Judicious application of fixatives allows selected stages to be examined by electron microscopy. Many of the descriptions in this chapter are taken from work on liquid endosperm preparations.

The two stages of cell division, mitosis and cytokinesis, follow one another in the cell and are treated separately here in their natural sequence. Although mitosis usually is considered as a number of phases these do not provide an optimal format for a discussion of the whole process. Instead, we select facets of the subject that follow on from earlier chapters, describing the behaviour during mitosis of the chromatin, the nucleolus, and the nuclear envelope. The central role of the spindle in mitosis has attracted much attention and a wealth of data has been accumulated and subjected to a variety of interpretations. Here we try to separate the observations, which are mainly incontrovertible, from the interpretations, which may be open to question. A description of the development of the spindle and its interaction with other cell components is followed by a discussion of the mechanisms of spindle formation and of chromosome movement.

12.2 Chromosome Condensation and Structure

Distinctive changes in the appearance of the nuclear chromatin provide the first visible hint that a cell is about to enter mitosis. The chromatin becomes progressively more clumped and, at early prophase, the chromosomes can be resolved as a series of long intertwined diffuse structures packed into the nucleus (Plate 1, N-1). They continue to become shorter, thicker and more intensely stainable as mitosis progresses to metaphase (Plate 44a–e). If division is interrupted, for example by disrupting the spindle, the condensation process continues for longer than normal, resulting in the formation of especially compact chromosomes, valuable to those who investigate the chromosome complement of cells, e.g. in searching for abnormalities, because they can be spread out on a slide and stained for light microscopy.

Many light microscope studies have been devoted to the study of the condensation of chromosomes. Examination of suitably-fixed chromosomes sometimes reveals that each arm (on either side of the kinetochore) is composed of a coiled thread-like structure, the *chromonema,* the visibility of which can be enhanced by various chemical treatments (e.g. treatment with ammonia vapour) so that they may be observed at various stages of mitosis. It has been concluded that each chromosome arm consists of a single thread that continues to coil up upon itself, so that the chromosome appears to get shorter and shorter.

Unfortunately, use of the electron microscope has not yet established any general clear relationship between the chromonema of the light microscopist and the chromatin fibres of the electron microscopist (see chapter 5). Ultra-thin sections (Plates 45–47) show that the chromatin is evenly dispersed across the chromosome. The gyres and spirals of the chromonema can, however, be revealed in 3-dimensional reconstruction.[16a] Ultra-thin sections reveal little about the detailed arrangement of the chromatin fibres. They can, however, be seen with greater clarity in whole mount preparations, where the chromonemata are still not discernible, each chromosome arm appearing as a confused mass of chromatin fibres bunched together. Whether the chemical treatments that are used to visualize chromonema produce images that are entirely artefacts, or instead induce a partial breakdown of some structure of higher order, remains to be determined. The mechanism by which the enormous lengths of chromatin present in each nucleus fold to give a series of discrete chromosomes not entangled with one another is a major problem in biology to-day.[180] One suggested mechanism envisages a series of special areas spaced out along the fibre that can adhere to each other, and so lead to the folding up of the intervening thread. Each folded thread would then be folded again and so on. However, evidence for such a scheme has not yet been found.

As the chromosomes condense, other structures associated with the arms become visible. The most noticeable of these is the *primary constriction* on each chromosome. It stains less intensely than the chromatin, and, to the light microscopist, appears to represent a site where the two sister chromatids join together. It is more usually referred to as the kinetochore or centromere, and is described more fully below. Other constrictions, the 'secondary constrictions', may also become visible, sometimes in specific patterns along the chromosome arms. The most studied of the secondary constrictions are the nucleolar organisers, discussed elsewhere (see Plate 47 and page 45).

KINETOCHORES True kinetochores are absent from some lower forms, such as the amoeba *Pelomyxa*[716a] where each of the numerous short chromosomes is attached to the spindle at several points along its length. Differentiation of specific parts of the chromosomes into kinetochores is found in virtually all higher forms, with a trend from individual chromosomes having multiple kinetochores (as in the wood rush, *Luzula*[81]) to those having a single kinetochore (most higher animals and plants).

Single kinetochores can be found at all possible positions along the length of a chromosome, however the actual position is specific for each chromosome and so forms the basis for an important method of classification. Thus chromosomes may be *metacentric* (kinetochore in the middle), *submetacentric* (off-centre), *acrocentric* (towards one end) and *telocentric* (at the end).

Surveys of kinetochore structure over a wide range of organisms show an increasing degree of complexity from lower forms to higher ones with marked differences between the structures found in higher plants and animals. For example in *Luzula* each of the multiple kinetochores lies in a slight depression on the chromosome surface and is surrounded by a halo of non-staining material.[81] In a number of algae[409, 575] and in many animal cells,[408] the kinetochore consists of a triple-layered flattened plate or disc lying on each sister chromatid. An outer heavily staining layer is separated from a similar inner layer by a non-staining space. The development of this type of kinetochore has been followed:[711] at early prophase the chromosomes bear a spherical or ball-shaped fibrillar mass from which microtubules appear to grow out. As mitosis progresses the mass differentiates into the triple-layered structure with the microtubules terminating in the outer layer.

Observations with the light microscope of higher plant chromosomes reveals little of the structure of kinetochores, other than the presence of a number of faint strands of chromatin crossing the constriction. In the electron microscope most higher plant kinetochores appear as a more or less prominent ball of fibrous material attached to the chromatid (Plate 46d). In some plants (e.g. *Haemanthus*[27] and *Tradescantia*[899]) the ball may be situated in a cup-shaped depression in the chromatid surface. Numerous microtubules run into and terminate within the fibrous material; evidence for the growth of microtubules from the kinetochore will be discussed later (page 165). The composition of kinetochores can at present only be judged from their staining properties; they appear to be proteinaceous in nature.

Some experiments suggest that the DNA associated with the kinetochore region may be especially concerned with the synthesis and positioning of kinetochore protein(s). For example, an experimentally induced chomosome break within the kinetochore region yields two fragments, each with unimpaired kinetochore functions. This could be interpreted to mean that in the kinetochore there are multiple copies of units of function, much as in the nucleolar organiser (page 46). Kinetochores cannot be seen in interphase cells, so either they are present in a diffuse state, like the chromatin, or they are not laid down until early prophase. Clearly the mechanisms that control the formation of kinetochore proteins and their deposition at predetermined sites on each chromosome are of great importance; however, at present they are poorly understood.[499]

KINETOCHORE FUNCTION An essential feature of mitosis is that after the duplication of the nuclear DNA, sister chromatids at first adhere to one another, and then become separated. Kinetochores do not seem to function in adhesion, for they are not single structures spanning both chromatids, but are paired, lying back-to-back on the sister chromatids (Plate 46c), which may even be slightly separated from one another at this region. Their function begins later, at the end of prophase, when kinetochore fibres (i.e. bundles of microtubules connected to the kinetochore) develop. Details are described later in the sections dealing with the spindle, but it is probable that two aspects of function can be distinguished. The first is concerned with the orientation of still-paired chromatids on the metaphase plate (Plate 44d, e, f), and here the back-to-back positioning of the kinetochores, together with the activities of the spindle fibres (including those emanating from the kinetochores themselves), are important. The second function is seen after metaphase, when the kinetochores appear to 'lead the way' as the now-separated chromatids move towards the poles of the spindle (Plate 44g–i). The motive forces that drive the movements of anaphase are transmitted to the chromatids at the kinetochores. The two aspects of function are highlighted by the behaviour of *acentric* chromosome fragments, produced by breakage in such a way that no kinetochore is present: such fragments participate neither in the movements to the metaphase plate nor in the movements of anaphase.[29]

Having mentioned adhesion between sister chromatids, this, and the initial separation process, need further comment. Chromatid arms adjacent to the kinetochore region tend to show a greater degree of condensation in early stages of prophase than do more distal portions, and while the sister chromatids are associated along their entire length during metaphase, the region adjacent to the kinetochore is the last to separate when anaphase starts. Sister chromatid association has been thought to be due to specific areas of late-replicating DNA which maintain links between sister strands. However, there is little evidence for such a system. Sister chromatid separation seems to occur simultaneously for all chromosomes in a cell (with some exceptions), whether they are attached to the spindle or not (acentric fragments). A similar separation occurs even in cells treated with colchine to remove the spindle. Thus the initial separation of sister chromatids appears to be independent of the microtubule system, instead being brought about by some factor which influences all chromosomes throughout the cytoplasm, even those in the neighbouring nuclei of coenocytic organisms.

12.3 The Nucleolar Cycle

The 'textbook' story of the behaviour of the nucleolus at mitosis is that it fragments and disappears towards the end of prophase, re-forming during telophase and early interphase. However, as the range of organisms studied has been extended, it has become apparent that there are several patterns of nucleolar cycle. Many of the deviations from the general cycle are to be found among the lower eukaryotes, though higher plants and animals also provide exceptions.[454a]

Analysis of the different patterns of nucleolar behaviour suggests that there is a very broad evolutionary trend from persistence of nucleoli in lower organisms to fragmentation and re-synthesis in higher organisms.[650] Whilst this is an attractive idea, it has already been pointed out that some higher organisms retain their nucleoli throughout mitosis, and equally, there are well documented cases of nucleoli disappearing during mitosis in lower organisms. A brief survey of some of the patterns found will serve to illustrate the range of behaviour and form the basis of a more detailed examination of the situation as it pertains in most higher plants.

In the rather primitive mitotic mechanism found in the Dinoflagellates[448], the nucleolus persists in the nucleoplasm and is divided between the two daughter nuclei at the end of mitosis by the ingrowing nuclear envelope. *Euglena* also exhibits a persistent nucleolus which is large and of unusual structure. It elongates between the spindle poles at mitosis, eventually splitting into two so that each new nucleus receives a portion. In some algae the nucleolar fragments formed at prophase adhere to, and thickly coat, the chromosomes, completely obscuring them in some cases (e.g. *Spirogyra*). In other algae the nucleoli, much as in higher plants, become more and more indistinct until they disappear at metaphase, re-forming again at telophase (e.g., *Closterium,* which is related to *Spirogyra* despite having a very different nucleolar cycle). Nucleolar loss seems to be a direct result of spindle activity in one species of the filamentous green alga *Oedogonium*, where it persists in the metaphase plate region at anaphase, to be excluded into the cytoplasm by the formation of the daughter nuclei, which then form their own new nucleoli. Elimination of persistent nucleoli from the spindle also occurs in the primitive pteridophyte *Psilotum*, as well as in certain grasses. Another variable is seen within individual species: the timing of the events of the nucleolar cycle can differ from tissue to tissue, e.g. amongst the different cell types in a root tip[161].

In higher plants, nucleolar loss is generally described[455] as a progressive dispersal of the granular and fibrillar components into the matrix of the forming spindle until its identity is lost (Plate 44a-d, Plate 45a-d and Plate 46a). Some interpretations state that the dispersed material coats the chromosomes (Plate 47b, c) and subsequently reaggregates at telophase to contribute to the new nucleolus. On this view the nucleolus is not completely lost at each division, but merely temporarily changes its form in a behaviour pattern that is described as *dispersive*. While this may be true in some cases, other information suggests that new nucleoli arise from the activity of the nucleolar organizer regions on the chromosome arms. The latter are clearly distinguishable in condensed chromosomes as less densely stained regions compared with the surrounding material (Plate 47a-c). At telophase they start to swell and expand, initiating the new nucleoli even before the identity of the rest of the chromosome is lost (Plate 47d and Plate 1, N-2).

The molecular biology of the nucleolar cycle has been investigated in some organisms. Understandably, greatest interest has been centred on the nucleolar ribosomal RNA genes (rDNA). These have been shown to remain unchanged during mitosis in Chinese hamster cell tissue cultures, which have persistent nucleoli.[10, 589] However, in *Chlamydomonas*, which loses its nucleoli in prophase, there seems to be a loss of a substantial part of the rDNA.[377] Restoration of the rDNA to its original level takes place during the early part of interphase (G_1) before the rest of the nuclear DNA is duplicated. This is taken to imply that only a limited number of genes coding for rDNA are present as an integral, nucleolar organizing part of the chromosome. Further copies are formed at telophase, persisting throughout interphase, and being broken down at prophase. It will be interesting to see if a similar loss of rDNA occurs in other organisms that lose their nucleoli during mitosis, although there seems to be no evidence for this in *Physarum*.[928]

In view of the loss of the nucleolus at prophase in some organisms, attention has been directed towards the fate of the ribosomal RNA precursors that constitute such an important part of the nucleolus (page 47). It has been shown that in cultured mammalian cells production of completed ribosomal RNA molecules ceases between prophase and telophase, yet the precursor molecules are still present, apparently in a stabilized form.[206] The appearance of a new nucleolus at telophase can occur in the absence of RNA synthesis during mitosis, but can be blocked by inhibiting RNA synthesis in the period *before* prophase. It would appear that the nucleolar material is conserved throughout mitosis and reaggregates in the new nucleolus during telophase.[642] The conserved material may correspond to the coats seen ensheathing the chromosomes at anaphase and early telophase[650] (Plate 47b, c), a view supported by the finding that under certain conditions ribosomal RNA precursors are found bound to the chromosomes.[206, 626b] Alternatively the conserved material could lie free in the spindle region where it would be indistinguishable (to the microscopist) from the ribosomes which invade the spindle after the nuclear envelope ruptures (Plate 45c, e and Plate 46).

12.4 The Nuclear Envelope in Mitosis

The nuclear envelope displays a considerable variation in behaviour during mitosis according to the organism being studied. In general it is more or less completely disrupted and lost at the end of prophase in higher plant and animal cells, but in lower forms it persists to a greater or lesser extent, so that there appears to be an evolutionary trend from 'closed' spindles (nuclear envelope persists and the spindle develops within it) to 'open' ones (nuclear envelope disperses).[647]

Where the nuclear envelope does persist, it may play an active part in mitosis. In Dinoflagellates, for example, the nucleus is pierced by a number of cytoplasmic channels which contain a system of microtubules apparently attached to the outer membrane of the nuclear envelope.[448] The chromosomes are attached to the inner membrane, and it may be that forces exerted by the microtubules on the envelope move the chromosomes to opposite ends of the nucleus. In organisms with closed spindles (numerous lower fungi, algae and protozoa) the spindle microtubules terminate on opposite sides of the nucleus at dense plate-like structures attached to the envelope (the polar plaques).[647] Formation of two new nuclei in cells with closed spindles is achieved either by a constriction of the nuclear envelope at the equator, or by the poleward separation of the two chromatin masses drawing out and pinching off the envelope in the middle. Other members of the algae and fungi show at least partial fragmentation of the envelope at the poles or more open spindles with little or no nuclear envelope persisting in the equatorial regions. Examples of these are the red alga *Membranoptera* and the basidiomycete *Coprinus*. Bryophytes (e.g., *Marchantia*) and ferns (e.g., *Dryopteris*[97]) have open spindles typical of those found in higher plants.

Knowledge of what happens to the pieces of nuclear envelope that are formed at the end of prophase is incomplete. In seeking to find out whether they survive to contribute to the new nuclear envelopes at telophase, the light microscopist is hampered because, though he can observe living, dividing cells, he cannot with certainty distinguish fragments of nuclear envelope from other cell membranes. The electron microscopist can identify them by the presence of pore complexes, but he is restricted to samples which are dead, and in which the fragments of nuclear envelope are scattered so widely through the cell that their distribution cannot readily be mapped.

A combination of electron microscopy and the light optical technique of interference contrast developed by Nomarski has provided a picture of the behaviour of the nuclear envelope at prophase and metaphase in *Haemanthus*.[27, 28] The first changes follow formation of the clear zone at prophase. At first smooth and spherical, the envelope soon begins to pucker and become irregular (there is some evidence of this in Plates 44a and b and 45a). By then, microtubules are found associated with the envelope in a manner suggestive of pulling and pushing movements. It seems unlikely that the microtubules are joined by their ends to the envelope, but rather pass close to the surface, perhaps interacting with the membrane *via* short arms or links. Shortly before disruption of the nuclear envelope, the faces nearest each pole become somewhat flattened (Plate 44c). The activity of the membrane concentrates in these areas, giving rise to a vigorous boiling appearance when viewed in speeded-up time lapse films, and eventually to the fragmentation of the membrane. Equatorial regions of the envelope persist for a short while (Plate 44d).

The fate of the fragments of nuclear envelope seems to vary from organism to organism, and maybe even within individual cells. Some lose their pore complexes. Fragments which have retained pores but resemble rough endoplasmic reticulum in having ribosomes on *both* membranes are illustrated in Plate 45c, e. Other observations indicate survival of pore complexes throughout mitosis, especially in fragments lying near the equator or in association with chromosomes. Because of this diversity of behaviour, it is not possible to be precise about the events of anaphase and telophase, when new nuclear envelopes form. In organisms with closed spindles the old envelope clearly contributes to the new ones, and this probably applies to open spindles also. Whether all of the original envelope is retrieved and used, or whether some changes into, and remains as, ordinary endoplasmic reticulum is not known. Certainly some of the cisternae which associate with the trailing arms of chromosomes in late anaphase have pores while they are still separate from the chromosomes (Plate 47c), while there is also evidence that cisternae of rough endoplasmic reticulum can attach to the chromosome surfaces, develop pores, and lose the ribosomes from what by then is the nucleoplasmic membrane surface (Plate 47d).

In some cases individual chromosomes are nearly completely sheathed by new nuclear envelope by late anaphase. In others (Plate 47c) the chromosomes are in this condition only at the poles. Coalescence of the chromosomes at telophase gradually leads to the merging of individual pieces to form the intact nuclear envelopes of the daughter nuclei. It is known that even before the new nuclei start to grow, their combined surface area may exceed the surface area of the parent. In such cases the total area of nuclear envelope is augmented during mitosis. Experiments in which inhibitors of protein synthesis have been found *not* to interrupt formation of new envelopes indicate that pre-existing cisternae, rather than newly-synthesized membranes, are employed.[206]

The attachment of chromatin fibres to the inner membrane of the nuclear envelope was described on page 52. The molecular basis of the attachment is not understood. Obviously, the attachments break at prophase. They are not reinstated until late anaphase, and it is noteworthy that by this time the chromosomes may have acquired a specialized surface coat (Plate 47a–c), which could have a role in attracting and holding the membrane, or in mediating formation of pore complexes.

12.5 Formation and Function of the Spindle

12.5.1 Observations on spindle formation

Spindle formation involves the development of an extensive microtubular system which can be detected in living cells as it is birefringent. In higher plants short segments of microtubules appear in the clear zone around the intact nucleus, forming disorganised arrays that only give rise to a rather weak birefringence. Further development of the clear zone leads to an increase in its thickness at the future poles and this is accompanied by an increase in intensity of the birefrigence as the microtubules elongate and come to lie parallel to the long axis of the developing spindle. Not infrequently three, or even more, poles develop (Plate 44b and c). These usually merge together (Plate 44d), giving a normal spindle which may be of intermediate orientation if two of the poles move towards each other from their original positions.[29]

Fragmentation of the nuclear envelope leads to an invasion of the nucleoplasm by microtubules from the clear zone and by cytoplasmic ribosomes (Plate 45c–e). At the same time the microtubules forming the kinetochore fibres start to develop on the chromosomes. The precise timing of these events is the subject of some debate which is centred on the presence or absence of microtubule monomer protein in the nucleus before envelope fragmentation. Clearly if monomer is not present then there can be no formation of kinetochore fibres at this stage. Some observers have claimed that kinetochore fibres are formed before nuclear envelope fragmentation. In practice it is not easy to show unequivocally that the envelope is still intact, since a single section only includes a small fraction of the total nuclear surface. In many lower organisms the envelope remains intact throughout mitosis with the spindle completely enclosed, so there seems good reason to suppose that microtubule monomer is able to pass into the nucleus.

A rapid increase in birefringence occurs after nuclear envelope fragmentation, with the development of kinetochore fibres and continuous fibres. Initially the continuous fibres move laterally from the clear zone to establish the spindle and then their numbers increase.[27] It now seems probable that continuous fibres do not represent continuous microtubules in the sense that each is continuous from one pole to the other. Judging from the few studies available, there seems to be a wide variation in the proportion of truly continuous microtubules present, a fact which has to be taken into account when postulating functions for the continuous fibres (see 12.5.5). There is no numerical relationship between fibres seen in living cells with the light microscope and microtubules in the electron microscope.[26, 397]

Development of the kinetochore fibre system proceeds by both an increase in number of microtubules attached to each kinetochore and an increase in their length. For example, in *Haemanthus* the number per kinetochore rises from prometaphase to metaphase and thereafter declines.[27, 397a] The microtubules appear to be formed directly at the kinetochore, growing out into the spindle. Occasionally observations have been made which suggest that microtubules of the clear zone curve in and attach to the kinetochores directly, but even if this conclusion is correct only a very low proportion of the kinetochore tubules would originate in this way.

In the fully formed spindle, microtubules of the kinetochore and continuous fibres appear indistinguishable from one another. Those of the kinetochore fibres account for 10–40% of tubules in the whole spindle and can be traced for considerable distances towards each pole. Although there is generally no ordered arrangement of the microtubules within the spindle, they do lie close to one another and cross links 10–40 nm long have been seen between them in both plant and animal cells.[339, 900] Such links may be of great importance to the interaction of microtubules during chromosome movements (see page 167).

12.5.2 *Movement of chromosomes and cytoplasmic particles*

A characteristic feature of the spindle is that it induces specific movements of all components that come into contact with it, whether these are chromosomes or cytoplasmic particles that have become trapped in the structure during its formation, or come into contact with the completed structure. Even before nuclear envelope fragmentation, cytoplasmic particles are seen to move in the clear zone, generally towards the poles as these become established. Movement of these particles, as well as of acentric chromosome fragments (those without kinetochores) and remnants of the nucleoli, always occurs towards the nearest pole in the completed spindle. If the cell has more than two poles the same rule applies, movement is always towards the nearest pole. The rate of movement is generally similar to that observed for chromosomes at anaphase, or somewhat faster.[28, 29]

Chromosome movements commence virtually as soon as the spindle is formed, with the kinetochore region apparently the point to which the moving force is applied. There appears to be no regular pattern of chromosome movement culminating in their arrival at the metaphase plate. Although some will move directly to the plate, others may move initially towards one pole before returning to the plate and may even oscillate backwards and forwards. This behaviour of the chromosomes gives a very strong impression that the metaphase configuration results from a balance of the forces acting on individual kinetochores (see page 166). In addition to these kinetochore-directed movements of the chromosomes, their arms may be subject to 'neocentric' movements as a result of microtubule attachment to points other than the kinetochore, resulting sometimes in one arm moving in the opposite direction to the rest of the chromosome. Such movements are usually of short duration in plants. A more general feature of plant mitosis is that the arms appear to be subject to the general transport properties of the spindle. Thus, although the kinetochores are arranged at the equator at metaphase, the attached arms come to lie longitudinally in the spindle pointing towards the poles (Plate 44e and f).[28, 29]

At anaphase the chromatids separate and the kinetochores move to the poles, pulling the chromosome arms behind them. Initiation of anaphase movements seems to occur simultaneously for all chromosomes on a spindle (exceptions are known in which certain chromosomes always move later than others), regardless of whether they have all reached the metaphase plate. Those that are not in this position behave normally, so that one chromatid moves to the further pole past the chromatids going in the opposite direction toward the other pole. Thus the force promoting chromosome movement acts on each independently and is not a general phenomenon resulting, say, in the movement of all material in one half of the spindle in one direction and in the other half in the opposite direction. In coenocytic systems with nuclei dividing synchronously, neighbouring nuclei usually enter anaphase together, although frequently a wave of anaphase initiation can be seen to pass through the tissue as a whole.

The rate of chromosome movement during anaphase is constant for the greater part of the distance from the equator to the pole, ranging from 0.2–5.0 μm per minute.[227] All chromosomes, regardless of their size, move at the same speed in a given spindle. Such information implies that the rate of movement is not governed by the total force available but by the rate at which the force is developed. This argument is supported by calculations which show that the hydrolysis of only about 30 ATP molecules is required to move an average chromosome to the pole, a very small fraction of the normal turnover in the cell.[227] Dicentric bridges (two kinetochores on a single chromatid travelling to opposite poles) do show reduced rates of kinetochore movement, until the bridge is broken, and this phenomenon can be used to estimate the total force available.[172]

Movement of acentric fragments and cytoplasmic particles towards the poles continues at anaphase ahead of the advancing chromosomes. In the interzonal region between the separating chromosomes these movements generally come to a halt and may reverse, transporting particles to the equator (see page 169).

12.5.3 *Mechanisms of Spindle Formation*

In Chapter 11 some observations on microtubule formation were discussed and it was concluded that in many cases these structures originate at specific initiation sites, growing out by addition of monomer protein to the distal, free end by polymerization from a monomer pool. Since the structural elements of the spindle consist of microtubules, the problem of spindle formation is one of microtubule initiation and growth (other spindle components such as microtubule cross links, associated enzymes, etc., being added to this structural matrix subsequently). The presence of two microtubule systems in the spindle, those of the continuous fibres and those of the kinetochore fibres, leads to the proposal that there may be two distinct types of initiation site involved. One of these clearly resides at the kinetochores, the other may reside at the poles in lower organisms and in animal cells, but its location in the anastral spindles of higher plants is less obvious. The view that even in animal cells microtubule initiation sites are scattered throughout the spindle, as well as being present at the kinetochores, has been expressed[171], and the reader is directed to more detailed discussions[29, 499] than will be included in the remainder of this section.

FORMATION OF CONTINUOUS FIBRES Earlier views of biologists working with animal cells, that centrioles are responsible for the microtubule organization at the poles, have not been substantiated.[652] Apart from many exceptions in the animal kingdom itself, the whole of the higher plant kingdom is now known to lack centrioles, and even in lower plants that form motile gametes (e.g., mosses, ferns and liverworts) the occurrence of centrioles is limited to the last few divisions of spermatogenesis; they are not found at all in vegetative cells.[626a] Thus these plants can express genetic information at a given time and form centrioles. In many animal cells centrioles are always present and their positioning at the poles in mitosis has been explained as a mechanism for ensuring that each daughter cell receives one.

The concept that some sort of pole determinant (mitotic centre) must exist in cells to ensure the formation of a bipolar spindle still attracts much attention, especially in animal and many protozoan cells. The initiation of poles in higher plants does not always seem to fit into such a scheme; for example the regular formation of multipolar spindles already referred to in *Haemanthus*, and their subsequent coalescence to form a normal spindle, argues against a rigid system of two pole determinants per cell. However, in a few instances it has been possible to induce animal cells to form spindles with four poles by temporarily interrupting the mitotic cycle. This phenomenon has been interpreted by some observers to mean that the ordered cycle of replication of pole determinants has been upset, two such cycles having occurred between successive mitoses instead of one.[499]

Returning to the origin of microtubules at mitotic poles, it seems clear than in centric spindles the initiation sites are clustered around the centrioles, whereas in anastral spindles they are assumed to be more dispersed at the poles and undetectable with present microscope techniques. An indication that this may be the case is afforded by work on post-mitotic events in the desmid *Closterium*.[652] The microtubules, which persist at the acentric poles after telophase, are gradually drawn together and become focussed on one region in which a rather ill-defined ball of granular material appears. The ball then migrates into the daughter cell, followed by the microtubules and nucleus. The granular material is believed to represent the polar microtubule initiation sites which only become visible when they aggregate. However, the possibility that the structure is only synthesised at telophase has not been excluded. The asymmetric division of certain pollen cells[353] (see page 170) involves the appearance of a small granular body at one pole of an otherwise typical anastral spindle.

In plants there seems to be a choice of two possible modes of continuous fibre formation: either they are formed by the initiation of each tubule by a specific site at the pole or they are merely seeded in a mass of monomer protein, the subsequent growth and orientations of the tubules depending on the properties and interactions of the polymerized structures.

Work on protozoan microtubule systems has shown (in *Nassula*, page 144) that microtubule number is dependent on number of initiation sites, while ultimate microtubule length depends on the number of tubules forming and the monomer pool size. Thus it follows that quantitative work on microtubule number and length, combined with experimental procedures designed to alter the initiation site and microtubule monomer levels in dividing cells, should yield data that would allow judgement of the polar initiation site hypothesis. However, few quantitative studies have been undertaken and most of these cannot be discussed in isolation from the other microtubule system, that originating from the kinetochores. One experimental treatment has, however, been shown to affect only the poles of the spindle, that is exposing cells to isopropyl N-phenylcarbamate, which leads to the formation of multipolar spindles.[140, 338] There is as yet no quantitative data to show whether the original poles split up during treatment (leading to the same total number of microtubules as in the untreated cells) or whether polar initiation sites replicate (leading to greatly increased numbers of microtubules).

MICROTUBULE INITIATION AT KINETOCHORES The origin of kinetochore fibres in plants is less obscure than that of continuous fibres. From each kinetochore microtubules grow out into the spindle, a process which can be observed in living material under both normal and experimental conditions.[29] The most successful experimental approach has been to use a microbeam of ultraviolet light to irradiate selected areas of the spindle, leading to the disruption of treated fibres which can be detected in living cells by a localized loss of birefringence.[227] Kinetochore fibres disrupted in this way can be seen to grow out again, re-establishing lateral contacts with other microtubules in the spindle.

The variation in the number of tubules associated with each kinetochore (page 163), both between cells of the same species and in the same cell during the course of mitosis, suggests that neither the number of initiation sites nor their activity is closely controlled. The extent to which the available microtubule monomer pool governs the number as well as the length of microtubules formed remains to be determined.

QUANTITATIVE STUDIES ON SPINDLE FORMATION Although it is apparent that any hypothesis concerning the formation and function of the spindle must be compatible with the structures and activities found in dividing cells, remarkably little precise information on the subject has been collected. The observations on living *Haemanthus* endosperm cells have greatly improved our knowledge of the dynamic changes involved,[29] but since precise analyses of the structural changes are not yet available it is impossible to draw firm conclusions about the mitotic mechanism in these cells. The problems involved in such an investigation are considerable, since it involves serial sectioning at a known plane through an entire cell at a known stage of mitosis. This then has to be repeated for as many stages of mitosis as possible.

One of the earliest investigations of this type was carried out on the final series of divisions, both mitotic and meiotic, leading to the formation of male gametes in the marine diatom *Lithodesmium undulatum*.[523–526] This material provides a natural and highly predictable variation in the number of kinetochores and chromosomes associated with each of the divisons, and since these follow one another in rapid succession it is possible that the pool of microtubule monomer protein present at the beginning provides the spindle tubules for all the subsequent divisions, the pool being divided equally between daughter cells at each stage. The observations on *Lithodesmium* show that the number of microtubules in the spindle at metaphase I is about 320 compared with about 150 at metaphase II. Since the number of kinetochores present falls by half between these stages, as does the cytoplasmic volume, these counts support the view that the microtubule monomer pool has been halved as well as the number of initiation sites. However, the previous mitotic division, involving the same number of kinetochores but double both the number of chromosomes (since they are not paired) and cytoplasmic volume, involves fewer microtubules than metaphase I (200 compared with 320). Mitosis is somewhat atypical in this organism, the chromosomes lying in a series of layers at the equator and moving sequentially, rather than simultaneously at anaphase, almost as if the microtubules were re-used after each chromosome movement for the next one.

Perhaps the most surprising aspect of the analysis of this diatom is the remarkably low number (about 10%) of tubules that are actually continuous from pole to pole: the majority of microtubules only interdigitate at the equator. The tubules originate from a spindle precursor appressed to the outer membrane of the nuclear envelope. It consists of a rectangular plate several layers thick which splits along a median layer to form two polar plates; these are separated by the growth of a microtubule system between them. Each plate bears a strong functional and morphological similarity to the plates of material that carry the initiation sites of *Nassula* feeding baskets. Initially it appears that all the tubules are continuous from one plate to the other, but as the spindle elongates between the plates it becomes apparent that they are interdigitated at the future equatorial region. Here the lateral interaction between tubules from opposite plates seems to lead to an increase in their stiffness and stability, since longitudinal sections reveal a clear distinction between straight microtubules closely packed at the equator and rather more disorderly and wavy microtubules near the poles. The continued elongation of the spindle tubules is of interest since the growth of the free tips would only increase the zone of interdigitation, which does not occur, so monomer molecules must either be added at the polar plates or by intussusception into the existing structure.

Other quantitative analyses have been reported from animal cells; however, these are more closely concerned with the search for an acceptable hypothesis for chromosome movement than with spindle formation, and so will be considered in the next section.

An indirect method of assessing the microtubule content of the spindle involves using the birefringence observed in polarized light as a measure of the total polymerized material.[794] This method has been successfully used with sea urchin eggs to demonstrate that the spindle structure at metaphase is in equilibrium with a pool of microtubular monomer protein.[795] Eggs were grown at two temperature extremes to produce cells with two different monomer pool sizes. Metaphase birefringence was then measured in cells of these two types maintained at a range of temperatures. The magnitude of the increase in birefringence which is observed at successively higher temperatures, due to a shift of the equilibrium in favour of the polymerizations, corresponds to first order polymerization and dissociation kinetics, regardless of how much monomer is present. Immersion of the cells in a relatively high concentration of heavy water, deuterium oxide (D_2O), led to spindles with a uniformly high birefringence value at all temperatures. This is in agreement with work on other protein polymer systems (e.g., virus coat proteins)[466] where deuterium oxide drives the equilibrium strongly in favour of the polymer so that the reaction goes nearly to completion.[99a] These experiments suggest that the number of microtubules present in the completed spindle depends on the equilibrium between monomer and polymer rather than on a fixed number of initiation sites at the kinetochores and poles.

Nevertheless, when spindles are isolated from cells, they will elongate if suspended in a medium that contains tubulin (even tubulin obtained from another organism), but they are surprisingly stable when the surrounding free tubulin is removed.[392a, 686a] If spindle structure depends in a simple fashion upon monomer-polymer equilibrium, they should depolymerize under such conditions, and there is therefore a discrepancy between the conclusions drawn from work on isolated and intact systems.

12.5.4 Mechanisms of chromosome movements

Chromosome movements during cell division are of two types, those occurring before metaphase, culminating in the equatorial arrangement of *paired* chromatids, and those occurring at anaphase resulting in the separation of *sister* chromatids to the poles. For both types of movement the presence of a normal kinetochore with attached kinetochore fibres on the chromosome is essential. Thus acentric chromosome fragments behave as other cytoplasmic particles that come into contact with the spindle (page 163). Chromosomes are seen to follow passively the movements of kinetochores in most cases, so that the moving force must be exerted through the kinetochore fibres. Since there are two types of chromosome movement it may be suggested that two types of mechanism are involved; however there is little evidence of this, and we shall see that the differences in chromosome behaviour before and after metaphase are due to the association of sister chromatids *before* and their dissociation *after* metaphase.

We will first consider some general conclusions that can be drawn from chromosome movements before considering some proposals that have been made for the origin of the forces involved. In section 12.5.2 it was concluded that each half spindle acts as a transporting system, moving material to the poles. If it is assumed that the kinetochore fibres are subject to the same forces, then each set of fibres will tend to be moved to the nearest pole. Since the back-to-back arrangement of the kinetochores on sister chromatids ensures that their fibres point in opposite directions, then sister kinetochores will tend to move to opposite poles. The implications of this are now considered for each of the mitotic stages:

PROMETAPHASE For those chromosomes that are lying near the equator at spindle formation the force exerted will be equal, but opposite on each kinetochore. Evidence for this has been found in many cells which exhibit stretching of the kinetochore region before metaphase, resulting in the slight separation of sister kinetochores, but not of chromatids.[385] Chromosomes lying nearer one pole than the other at spindle formation develop kinetochore fibres in both directions; however, those to the further pole are longer and this appears to mean that a greater force is exerted on them than on the shorter fibres of the sister kinetochore since the chromosomes move to the equatorial position. There is evidence to support this view from ultra-violet microbeam studies in which kinetochore fibres were dissociated on one side of the chromosome during prometaphase. The chromosome then moved to the pole on the undamaged side of the chromosome as if the equilibrium had been upset.[394] Subsequent regrowth of the fibres led to a return of the chromosomes to the equator. Additional evidence comes from the study of trivalents (certain types of sex chromosome pairing at metaphase I having three functional kinetochores) and suggests that the pulling force is proportional to the length of the kinetochore fibre. In certain orientations two sets of kinetochore fibres are directed towards one pole (the nearer) and the other to the further pole. At equilibrium the additive length of the two sets of short fibres is approximately equal to the length of the third set of fibres.[38]

METAPHASE Metaphase can be regarded as an equilibrium position maintained by equal and opposite forces acting on sister chromatids. In many cell types this explanation satisfies all the observations, since kinetochore stretching leads directly into anaphase with the separation of sister chromatids. However, there are well-documented cases from other cell types in which spindle activity gradually decreases during metaphase, so that eventually the force acting on sister chromatids becomes zero. Activity then resumes at sister chromatid separation and normal anaphase movements occur. This condition has been termed 'full metaphase' and cells which enter anaphase directly are considered not to have reached full metaphase.[29]

Some observations are not explained by the above picture of metaphase, but a lengthy discussion of them will not be undertaken here. In particular, the behaviour of some univalents (chromosomes possessing only one set of kinetochore fibres) is puzzling. Theoretically they should be transported to one pole and stay there. However, they are frequently observed to oscillate between the pole and the equator and end up lying near or at the equator at metaphase.[29]

ANAPHASE Anaphase movement is directly dependent on the presence of intact kinetochore fibres, but not on their attachment to the poles. Again this can be shown by ultraviolet microbeam experiments in which kinetochore fibres are dissociated.[29, 227] Treating fibres close to the kinetochore leads to a complete cessation of chromosome movements, while dissociation close to the pole has no effect. Clearly the length of individual kinetochore fibres decreases during anaphase, but this is not due to a contraction of the fibres. For example, microbeam damage results in the formation of a spot showing decreased birefringence under polarized light. Recovery involves regrowth of the kinetochore fibres to a length sufficient for chromosome movement and thus increases the kinetochore-to-spot distance. Then the chromosome and spot move to the pole at the same speed; there is never any decrease in the distance between the two corresponding to fibre contraction. Intact fibres appear to dissociate at the pole since they do not project beyond it into the cytoplasm. One elegantly direct method for showing that anaphase movements do not require microtubule polymerization has now been devised.[110a] An isolated mitotic apparatus has been shown to be capable of spindle elongation and chromosome movements *in the absence* of added tubulin.

The poles of the spindle are therefore only responsible for the initiation and formation of the spindle structure; they play no part in subsequent chromosome movements; rather it is the spindle as a whole which is responsible for these activities.

THE FORCE-GENERATING SYSTEM Extensive investigations on thin sections with the electron microscope have revealed only one structural element in the spindle, that is, the microtubules. Arguments for the presence of another, as yet undetected, component have been advanced on the grounds that there are insufficient microtubules present to account for the mass of the spindle and the level of birefringence observed in polarized light. However, there is now some evidence that other components (e.g., cross links and possibly enzymes) are associated with the microtubules in sufficient quantity to account for these discrepancies. It is of interest that the level of birefringence falls when cells are fixed for microscopy, suggesting that some components are either destroyed or displaced from their original ordered positions.[456]

Since microtubules are the only structural element detected by microscopy, it is not surprising that they are thought to move chromosomes. The problems arise when detailed explanations are sought. Basically two types of proposal have been put forward, one in which the microtubules and associated force-generating systems form a two-component system;[499, 615] and the other in which the microtubules themselves are responsible for generating all the forces involved (a one-component system).[170, 171, 172] A variation on the two-component system visualizes the force-generating system as being attached to the microtubules in a regular manner (cross links or mechano-chemical arms).[544]

In the two component models the microtubules are assumed to have a definite polarity, for example microtubules initiated from opposite poles will have opposite polarities[499], so will those from sister kinetochores. Thus the kinetochore tubules will have an opposite polarity to the tubules coming from the pole in their half of the spindle. Either the microtubules are embedded in a 'matrix' of force-generating elements, or these elements are in the form of mechano-chemical arms. Activity of these elements results in the development of a force between oppositely polarized microtubules. Whether both sets of tubules are moved past each other, or only the kinetochore tubules past the stationary 'continuous' tubules, is still a matter of debate (see later). The attraction of the mechano-chemical arms hypothesis is that these can be envisaged as binding to specific sites on the oppositely polarized microtubule. Movement would be generated by attachment of the arm to the next site along the other microtubule. Thus the rate of movement would depend on the rate of this reaction which, if constant, would lead to constant speeds being developed, so long as the force generated was always larger than that required to move the chromosome. This proposal is in agreement with observations on anaphase movements (see page 163).

In the one-component model the microtubules are assumed to grow out into a matrix of microtubule subunits in which they are anchored. The polymerized structures are in dynamic equilibrium, with subunits capable of being removed and replaced along the length of the tubule as well as at the ends. Anaphase shortening of the kinetochore fibres is then envisaged as resulting from a shift in the equilibrium favouring removal of subunits from the tubules but not their replacement, so that the tubule undergoes continuous internal rearrangement to preserve its structure, becoming shorter in the process. Several observations may be cited in support of this type of mechanism. For example the *rate* of kinetochore microtubule depolymerization at anaphase has been shown to be linearly correlated with the anaphase velocity of the chromosomes regardless of the absolute amount of polymerized microtubule present in both anastral (plant, *Tilia*) and astral (animal, *Asterias*) spindles.[251] Similarly, increasing the hydrostatic pressure on cells, leading to an artificial depolymerization of the microtubules, induces chromosome movements at rates proportional to the rate of microtubule depolymerization.[727] Direct observation shows that the number of microtubules attached to each kinetochore in *Haemanthus* endosperm cells declines as soon as anaphase movements are initiated, and by late anaphase has fallen to less than half of the peak number at metaphase.[397a]

While it is not possible to choose between these alternatives at the present time, it is becoming apparent that insufficient attention has been paid to the 'continuous' tubules. None of the models fully accounts for the transport properties of the spindle in respect of the behaviour of cytoplasmic particles, which is thought to be a function of the continuous tubules. The observations suggest that two half spindles are present, each with oppositely directed transport behaviour. The assumption that microtubules are polarized does not account for this since it is further believed that the tubules are continuous from pole to pole, so that each half spindle contains tubules from both poles. However, the evidence from *Lithodesmium*[523–526] suggests that continuous tubules are a minority, the spindle consisting mainly of two sets of interdigitating tubules. If such a finding were substantiated for a wide range of species it would considerably simplify the interpretation of spindle behaviour and avoid such complex explanations as polarity reversal of the tubules at the equator. Comparisons of plant and animal spindles in this respect will be particularly interesting in view of the differences in their spindles and spindle formation (anastral *versus* astral[499]).

The fate of continuous tubules during anaphase is also not understood. While some theories suggest that they remain *in situ* and even lengthen (to account for the overall increase in pole-to-pole distance that occurs at anaphase), others suggest that they are transported (slide) in the opposite direction to the kinetochore tubules and end up as the interzonal fibres between the two telophase nuclei.[544] Some quantitative studies have been undertaken on animal[85, 249a, 545] and plant[397] cells with a view to distinguishing between these various possibilities. The number of microtubules present in cross sections of the spindle at different levels were counted during metaphase and anaphase. The difficulties of such work have been emphasised previously and the authors of these studies state the need for caution in interpreting their results since they are based on the analysis of relatively few cells. However, they provide strong support for the view that the number of continuous tubules is relatively small, probably in the region of 30%. There is some support for the view that anaphase spindle elongation occurs by increase in the length of the continuous tubules. However, the interpretation of the data is complicated by uncertainties about the proportion of tubules which interdigitate at the equator and the extent to which they penetrate into the other half spindle. A comparison of cell division in chick embryo epithelial and mesenchymal cells has established that spindles with few continuous tubules elongate little at anaphase, those with many undergo extensive elongation.[9]

Clearly a full appreciation of spindle behaviour will have to wait until more is known about microtubules and their properties in general. It should then be possible to decide whether spindle function can be explained in terms of this structure alone, or whether some additional factors (structures?), hitherto undiscovered, interact with the microtubules.

12.6 Cytokinesis

12.6.1 Introduction

That cytokinesis, i.e., the division of the cytoplasm into two new cells, can be a separate process was only realized after the major events of mitosis had been clarified. In 1909 van Wisselingh demonstrated that movement of dividing nuclei of *Spirogyra* filaments to one end of the cell by centrifugation did not affect the position in which the cross wall developed. Even to-day cytokinesis is all too frequently treated as if it were a closing stage of mitosis rather than an event to be considered separately. Whilst mitosis routinely leads to the formation of two identical nuclei, cytokinesis does not necessarily lead to the formation of two identical cells, especially in multicellular organisms. The differences may be gross and readily detectable, such as the formation of a large and small daughter cell, or there may be small, even initially undetectable, differences between the cytoplasms such as the exclusion of certain components (e.g., *Bulbochaete* hair cell, Plate 27a), or the separation of different types of a component (e.g., sorting out of mutant plastids, page 133). In addition, especially in plants, where the spatial relations of neighbouring cells are fixed by the enclosing walls, the plane of division followed by growth of the cell determines the shape of the tissues.[152, 496, 792] Thus division in only one plane leads to long filaments of cells as in *Spirogyra*, division in two planes leads to sheets of cells as in the alga *Enteromorpha*, and in three planes gives 3-dimensional blocks of cells.

Whilst the products of cell division contain two identical nuclei, the individual cells do not necessarily follow the same developmental paths. Thus division of cambial cells may give rise to a cell that will differentiate into vascular tissue and a cell that remains as the cambium; division of a cell in a vascular trace may give rise to a small companion cell and a large sieve element; and division of a root epidermal cell may give rise to a cell that forms a root hair and one that does not. The future development of the newly formed cell seems to be determined by its relative position in the plant body, although all cells have identical genetical constitutions. Thus cytokinesis is seen to be a most important event in the proper development and differentiation of cells to form the whole plant body.

12.6.2 Cell plate formation

Although cell plate formation has been studied in a number of tissues the root tips of plants remain the most convenient and popular material for studies of fixed and sectioned cells. Again studies of living *Haemanthus* endosperm cells have proved invaluable for studying dynamic aspects of the process.[28] However, cytokinesis in this multinucleate mass of protoplasm can be delayed and may then occur over considerable lateral distances and also between non-sister nuclei. The growth of these multiple cell plates is remarkable in that their alignment is so precise that they approach and fuse with each other edge-on without error. However, this behaviour is not typical of that occurring in the remainder of the plant body and so cytokinesis in liquid endosperm tissue may also be atypical.

In higher plants division of the protoplast into two is effected by the centrifugal growth of a disc of wall material (the *cell plate*) that originates in the centre of the cell and eventually reaches and fuses with the side wall. This is associated with a system of microtubules lying at right angles to the cell plate and parallel with the old spindle fibres; together the plate and tubules constitute the *phragmoplast*. By contrast, animal cells divide by the formation of a cleavage furrow which pinches in at the mid-line and eventually cuts the daughter cells free from each other.[108] A similar process is claimed to occur in certain algae[175] (e.g., *Chlamydomonas*[407]) although here the microtubules lie parallel to the plane of the division, forming a 'phycoplast',[647] rather than at right angles to it, as in animal cells and higher plant phragmoplasts. In *Spirogyra*[647] cytokinesis involves a combination of ingrowth of the side walls and phragmoplast formation, with a cell plate growing out from the centre.

In living cells of higher plants the phragmoplast appears as a barrel-shaped body in the cytoplasm, with the phragmoplast fibres lying parallel to the old spindle fibres and closely resembling them. The microtubules that make up the phragmoplast fibres appear to be derived directly from the mitotic spindle; however they can arise independently in the cytoplasm where the time interval between nuclear and cell division is prolonged and the spindle fibres have disappeared. Running across the middle of the phragomoplast is a thin layer, the cell plate (Plate 44j-l). Although early electron micrographs seemed to suggest that the endoplasmic reticulum is involved in formation of the cell plate,[664] it is now thought that vesicles derived from dictyosomes are the major component[341, 458, 585, 887], despite the fact that evidence for this is somewhat circumstantial.[598] It is possible that in some tissues[154, 337, 475] the endoplasmic reticulum is also closely involved, but not all observers agree with this view.

Studies of living material have shown that phragmoplast initiation commences as early as mid-anaphase, and by the time nuclear envelope re-formation is completed the phragmoplast has extended across the cell and is approaching the side walls. The phragmoplast fibres originate in the centre of the spindle and spread progressively outwards between the separating groups of chromosomes in *Haemanthus* (Plate 44j-l).[29, 458] Like spindle fibres, the phragmoplast fibres seem to generate movement of cytoplasmic particles within their immediate vicinity. This activity generally results in movement of small particles towards the forming cell plate. Also the fibres seem to maintain the separation of telophase nuclei since application of colchicine and other compounds that dissociate microtubules frequently result in fusion of the two nuclei. Such a force, acting towards the poles from the cell plate can be detected at the earliest stages by lateral movements of trailing anaphase chromosome arms and some spindle fibres, which are forced sideways and up away from the spindle region (Plate 44 i, j).[28] Dictyosomes are found in the interzonal region at late anaphase—early telophase (Plate 44j) among the phragmoplast fibres where they oscillate and move about, apparently at random. Droplets accumulate at the central region and spread out across the phragmoplast. Coalescence of these to form the new wall occurs first at the centre and then spreads outwards.

Longitudinal sections of fixed and embedded cells show that the microtubules making up the phragmoplast fibres are not continuous but overlap at the central region (Plate 48b).[341, 585] Here adjacent microtubules appear to be bound together in groups by an amorphous darkly staining material.[25, 337] In root tip cells the dictyosomes are infrequently encountered amongst these fibres but lie a short distance from them, just beneath each daughter nucleus (Plate 48a). Both smooth and coated vesicles are found associated with the dictyosomes, among the microtubules and forming a layer at the cell plate.[341] It has been suggested that they are attached to the microtubules by links and are moved from the region of the dictyosomes to the cell plate by the microtubules. However, such links are hard to demonstrate and such conjectures have been criticised.[598] At successively later stages the layer of vesicles extends further and further out, while coalescence of the central vesicles to form the new wall commences. The microtubules are predominantly associated with the edges of this expanding disc, where they appear to position or stabilize the incoming vesicles in the plane of the plate. Similarly, the dictyosomes move out away from the nuclei and eventually are found close to the side walls.[341]

The importance of microtubules to the proper development of the cell plate has been demonstrated by treating dividing cells with colchicine.[99a, 645] Loss of the microtubules inhibits movement of vesicles to the cell plate, and its growth stops. In addition the cell plate loses its position and drifts around the now binucleate cell. Even if the cell plate has fused with the side wall around part of its circumference it cannot complete cytokinesis in the absence of microtubules and remains as a flap of cell wall material sticking into the cell. These observations suggest that the cell plate is in some way different from the surrounding plasma membrane and cell wall, since dictyosome vesicles normally move to the cell surface without the assistance of microtubules.

The idea that the developing cell plate receives all of its membranes in the form of dictyosome vesicles is attractive but unproven.[598] The dictyosomes do not always become unusually abundant or hypertrophied during cytokinesis (Plate 48a). Indeed the familiar appearance of growing cell plates with vesicles at their expanding edges comes very largely from sections cut normal to the plane of the plate (as Plate 48a), while a very different picture has emerged from the few available micrographs of sections that have been cut parallel to the plane of the plate. Obviously, it is very hard to obtain such views of the cell plate.[341] They tend to show that much of the cell plate consists of irregularly anastomosing membrane-bound tubules (Plate 48c), as if vesicles had given rise to branching arms that grow out in the plane of the plate to form a meshwork that gradually coalesces. In other words the membrane of the cell plate, i.e. the future plasma membranes, may be capable of some independent growth. A further complication is that parts of the cell plate, just as parts of the plasma membrane elsewhere in the cell, can be seen bearing the fuzzy coat characteristic of coated vesicles

(Plate 48d, compare Plate 28a). Whether this signifies an input *to* the growing plate, or retrieval of membrane (etc.) *from* the plate, is unknown.

The new disc of wall material initially lacks an extensive fibrillar system (Plate 48d, f) but can be shown to contain polysaccharide. It will become the site of the middle lamella when cellulose is laid down on each side of the cell plate, usually after the completion of cytokinesis. The membrane bounding the cell plate disc appears indistinguishable from the plasma membrane of the adjacent walls (Plate 48e), but as we have seen, may be derived from the membranes of the dictyosome vesicles. It is interesting to note that in certain animal cells which resemble plant cells by virtue of their strong intercellular connections between adjacent plasma membranes, the Golgi apparatus apparently contributes directly to increasing the surface area of the plasma membrane during cytokinesis[687]. More general aspects of the origin of the plasma membrane from the Golgi apparatus are described in Chapters 7 and 13.

Although elements of the endoplasmic reticulum may or may not contribute directly to cell plate formation, they are present as expanses of flattened cisternae in the phragmoplast (Plate 48d). Some of these become trapped by the advancing vesicle network, assuming a tubular conformation where they pass through the cell plate (Plate 48d and f). This could be the origin of the plasmodesmatal connections between the daughter cells, with the tubular derivative of the endoplasmic reticulum persisting as the axial strand in the plasmodesmatal canal (Plates 48f and 14a). As remarked on page 28, it is important to discover how the frequency and position of plasmodesmata are determined.

12.6.3 *The Plane of Cell Division*

As the plane of cell division is normally coincident with the equator of the metaphase spindle it might be concluded that positioning of the cell plate is merely another property of the mitotic apparatus. That this is not always the case was ably demonstrated by van Wisselingh (1909) who centrifuged the dividing nuclei of *Spirogyra* filaments to one end of the cell and observed that this did not alter the position of the cell plate, even when carried out before spindle formation was completed. In higher plants divisions at the meristems occur in small cells with few vacuoles, where the formation of relatively large symmetrical spindles could be assumed to ensure that symmetrical divisions will occur. However, this does not adequately explain the precise patterns of planes of division that are frequently found in stem apices and root tips. Division in large vacuolate cells, where the nucleus lies in a peripheral band of cytoplasm, must involve active determination of the position of the nucleus and plane of division. In these cells the phragmoplast is seen to develop in a thin layer of cytoplasm (the phragmosome) that pushes across the vacuole at considerable lateral distances from the mitotic spindle.

The discovery of a band of microtubules several layers deep running around the periphery of certain wheat leaf cells just before the onset of prophase (the *preprophase band*) led to the suggestion that it determined the position of the dividing nucleus and hence that of the cell plate (Plate 42c).[654, 656] These cells are involved in the series of asymmetrical divisions that lead to the formation of stomatal guard cells and subsidiary cells. Here the nuclei first migrate to a position of extreme asymmetry, that is to the side wall of the cell adjacent to the guard mother cell, mitosis ensues and a strongly curved cell plate is formed, cutting off a hemispherically shaped cell next to the future guard cell. Subsequent experiments and observations on multinucleate cells (induced by caffeine treatment) showed that preprophase bands were not concerned with orientation of the nuclei, but still predicted the position of the cell plate.[649, 654] Observations on cell division in other plants suggest that the preprophase bands are formed after the general polarization of the cytoplasm, leading to the commencement of prophase,[96, 648, 654] and so may reflect cytoplasmic asymmetry rather than cause it. The formation of root hair cells is also accompanied by an asymmetric division. In some cases preprophase bands are present, but these are claimed to lie symmetrically in the cell and therefore do not predict the position of the cell plate,[654] while in others they are absent altogether.[154]

Another site of asymmetric divisions in the plant is the anther where divisions in developing pollen grains lead to the formation of a small generative cell and nucleus within the vegetative cell.[353] Prior to this division the nucleus migrates and comes to lie close to the side wall. A spindle is formed in the normal way with the equator lying parallel to and a short distance from the wall. At early anaphase the spindle fibres beside the wall (that is those of the future generative cell) curve inwards and point towards the pole which is occupied by a dense granular structure about 0.6 μm in diameter. The phragmoplast initially develops normally between the vegetative nucleus and the generative nucleus (which lies against the side wall), but then curves sharply round the latter and fuses with the wall, forming a hemispherical cell plate, composed of callose, which encloses the generative cell.

Observations on animal cells show that asymmetric divisions are normally achieved in a similar manner to plants, that is through displacement of the nucleus at an early stage. Differential growth of the astral microtubules occurs at the poles, so that the spindle is displaced to one end of the cell and the metaphase plate and subsequent cleavage furrow are off-centre. A similar situation can be induced in the normally symmetric divisions of fertilized sea urchin eggs by experimentally displacing the spindle to one side. However, after metaphase, such displacements do not alter the position of the cleavage furrow and a binucleate and an anucleate cell are formed. Thus the position of the division again appears to be fixed in the cytoplasm and independent of the mitotic apparatus.

In conclusion it can be said that despite intensive efforts the mechanism that determines the position of the nucleus remains unknown at the structural level. Undoubtedly the behaviour of individual cells is influenced by the neighbouring cells and by orientation forces that pervade the tissue as a whole. Thus cells isolated from the plant body and cultured *in vitro* do not retain an orderly division pattern but multiply at random, forming a callus. Order can be imposed by providing diffusion gradients of certain plant hormones through the callus. It has been suggested that similar diffusion gradients exist in the normal plant body which interact with oxygen and sucrose gradients to maintain normal development.[152]

13 The Cell as an Integrated Unit

13.1 Introduction

Most of this book is devoted to a chapter by chapter dissection of the cell, assessing the structure and function of individual cell components. It remains to redress the balance by concluding with a summary emphasizing that the cell is an integrated system, and not a collection of independent functional units.

The cell is such a highly unified entity that it is artificial to seek out categories of integration within it. For convenience, however, the subject will be examined here under three headings.

There is integration that is based on flows of metabolic products and intermediates from compartment to compartment. The relevant compartments are often spatially distributed in a way that can be interpreted in terms of mutual supply and demand of raw materials, so the first type of integration to be examined is headed *metabolic and spatial*.

The second heading is *genetic integration*. The eukaryote cell contains two (in animals) or three (in plants) genetic systems, housed in the nucleus, the mitochondria, and the plastids. The situation is more complex than an ecological system with two or three genetically different organisms, influencing each other merely by competing for light or nutrients. In the cell the different genetic systems interact and collaborate to control the expression of selected parts of the separately-housed sets of genetic information.

The third heading embraces elements of the first two, but goes beyond them. It is *developmental integration*, and concerns the way in which the collaborating cell components form, are maintained, and can change in the course of time.

13.2 Metabolic and Spatial Integration

No living cell, either prokaryote or eukaryote, exists without containing some measure of division of labour. Even the simplest bacterium has distinct components, e.g., a plasma membrane, ribosomes, nucleoid regions, and cannot in any sense be regarded as a soup of molecules interacting in free solution. The nature of the much more complex division of labour in eukaryote cells has been the major subject of this book so far, and in the course of describing it, several of the sorts of advantages that accrue as a result of compartmentation of the cell have become apparent.

Restricting particular enzymes and substrates to a small compartment helps to ensure that the random processes of diffusion bring them into contact sufficiently frequently to maintain rapid rates of reaction. Indeed calculations of 'collision frequencies' in compartments of various sizes, containing different amounts of enzyme and substrate, suggest that a size of about 1 μm represents a critical volume. Unless the reactants are more concentrated than is the case in most cells, cells up to 1 μm in diameter (e.g., many bacteria) need not be compartmented; above it, compartmentation is necessary to prevent the reactants from becoming too dilute, and the reactions too slow.[659]

Given a compartmented system, other possibilities are available for exploitation. Incompatible substances and reactions can be employed, because they can be kept separate from one another. Entire metabolic systems can be segregated and given optimal microenvironments by adjusting the acidity, alkalinity, solute concentration or water potential within the compartments. It is feasible for an enzyme reaction to occur in more than one compartment, with the additional possibility that slightly differing forms of the same enzyme can evolve in relation to the specific properties and functions of each compartment. This seems to have been of considerable selective advantage, judging by the abundance of forms of enzymes differing from one another in specificity to activating substances (e.g., ion-stimulated ATPases thought to be important in solute transport), or in affinity to substrates and regulatory molecules (e.g., many *isoenzymes*, or *isozymes*). Isozymic forms of some enzymes are found in different parts of the organism, and, more relevant to the present topic, in other cases a distribution amongst the different compartments within cells has also been observed. In only a few cases, however, are the functional advantages understood.

The means that has survived the tests of natural selection to be adopted in the course of evolution as the basis of intracellular compartmentation is, of course, the system of cell membranes. Cell membranes are not, however, mere barriers between compartments. They have been diversified as much as the compartments that they bound—in their three-dimensional shape, in their spatial deployment in the cell, and in their molecular make-up. They provide a skeletal framework for multi-enzyme complexes, conferring overall efficiency by allowing substrates to be passed from one enzyme to another. Some substrates are released on the same side of the membrane that they entered, others may traverse the membrane in the course of the reactions, and other membrane bound enzymes may be influenced by molecules approaching from both sides of the membrane.[136]

Specificity in the trans-membrane transport of raw materials, intermediates, and products is all important in integrating the metabolic activities of the different compartments in the cell. It is easy to write glib statements, as, for example, that chloroplasts, peroxisomes, mitochondria, and the free cytoplasmic phase in which they lie, all collaborate in the metabolism of glycolic acid emanating from the CO_2-fixing reaction of photosynthesis, and much less easy to envisage the numerous participating solutes moving in particular directions across two chloroplast envelope membranes, two mitochondrial membranes, and the peroxisome membrane. There is no need to labour the point by listing other examples of specific trans-membrane transport between compartments, but it is worth gleaning some generalizations from the preceding chapters about the types of specific transport mechanism that have evolved.

In general, specificity of trans-membrane transport implies that the solute being transported is recognised by a specific binding site on a membrane transport protein. Some processes are uni-directional, or vectorial, thus driving import or export to or from the compartment. Subsequent reactions may metabolize the solute, or it may be stored unchanged. Other processes are bi-directional exchange reactions, as for instance the supremely important mechanism that exports ATP from mitochondria, simultaneously importing ADP: this is an example of a transport process that can regulate whole areas of metabolism, not only in the mitochondrion, but in other compartments. If it does not operate, the cell is liable to run short of primary energy.

Metabolic integration is not entirely based upon transport of individual molecules across membranes. Some materials, having entered or been formed in particular compartments, are moved in membrane-bound vesicles to their destination. In such cases, specificity of transport is achieved by membrane recognition systems, and there are the advantages of transport in bulk in a manner that avoids the possible dangers of the material being dissipated in the cell and exposed to harmful enzymes.

The earlier chapters have provided many examples of spatially organized transport between compartments. It is seen in the association of chloroplasts with peroxisomes, lipid droplets with glyoxysomes, and when endoplasmic reticulum cisternae lie at sites of cell wall syntheses or glycogen metabolism, or wrapped around plastids, mitochondria or vacuoles. In all of these, and similar, cases, it seems clear that donor and receptor compartments are brought into close juxtaposition, thus facilitating the exchange of metabolites. It may even be that the membranes of the juxtaposed compartments become locally specialized in relation to transport phenomena. As already stated, the mechanism by which compartments are brought together side by side and maintained in that position is not understood. The existence of additional local specializations of the membranes, such as membrane-to-membrane cross bridges, has been suggested.[238] The very shape of some membrane systems, particularly the enormous surface to volume ratios seen in the endoplasmic reticulum, speaks of spatial organization of transport across the membrane, in that the greater the area, the greater the trans-membrane flux that can be accommodated.

When cells are broken and sub-cellular components are separated, all metabolic and spatial integration is disrupted. Some of the isolated components can nevertheless be induced to display their metabolic potential if they are supplied with some of the raw materials they normally would receive when in the cell. Reactions can be identified, and a picture of the division of metabolic labour within the cell obtained. Clues that betray collaboration of components can also be discovered, as when the addition of mitochondria to a mixture of fatty acids and glyoxysomes is seen to allow the completion of sequential reaction pathways that otherwise are cut short.

What cannot yet be achieved in preparation of isolated cell components, however, is long-term stability. Sooner or later deterioration sets in. It is a measure of the degree of collaboration that exists in the cell that artificial culture media which mimic intracellular conditions and allow subcellular components to survive (let alone grow) have not yet been devized. What the cell supplies is, so far, too complex to be matched in artificial conditions. The compartments not only feed each other when in the living cell, but they regulate each other's activities.

13.3 Genetic Integration

Brief descriptions of the DNA of mitochondria and plastids were included in chapters 8 and 9 respectively, but with no discussions of its role. Its discovery posed fundamental questions about the nature and origin of the cell, and about the operation in the cell of the sub-cellular components that possess their own genetic systems. One immediate possibility was that plastids and mitochondria are virtually cells-within-cells, i.e., autonomous bodies, genetically independent, and dependent upon the host cell only in respect of supply and demand of nutrients. It was in fact clear even before the discovery of the cytoplasmic DNA that plastids and mitochondria are *not* completely autonomous. Genetical experiments, particularly on chloroplast pigmentation, had demonstrated that many nuclear genes influence their development and biochemistry. Now that the size of the DNA is known, it is even more obvious that complete autonomy is out of the question. Mitochondria from various sources have between 15 000 and 75 000 base pairs in their DNA, and so could code for protein equivalent in total length to 5 000–25 000 amino acids, e.g., some 16–80 proteins of molecular weight 40 000, each with 300 amino acids. Since not all of the DNA is available for specifying proteins, some being taken up by RNA synthesis, and probably some in regulatory functions, there is simply not enough DNA to confer autonomy on a mitochondrion. It takes between 50 and 250 times as much DNA as there is in a mitochondrion to meet the requirements of autonomous existence in a bacterium. For chloroplasts the conclusion is not quite so clear-cut, for the greater size of each set of DNA in a plastid would in theory be sufficient to specify 200–300 proteins, each with 300 amino acids.[428]

Complete autonomy being impossible, what of the opposite extreme—that plastid and mitochondrial DNA confers no degree of autonomy at all? Like the idea of full autonomy, this extreme has also been ruled out. Genetical experiments showed that mutations can take place outside the nuclear DNA, influencing the development and functioning of both plastids and mitochondria. More recent studies of the molecular biology of plastid DNA add weight to the conclusion from genetical experiments that it is indeed functional. It can be shown to duplicate in the same manner as nuclear DNA,[123] and that the whole apparatus for mediation of protein synthesis by nucleic acids and ribosomes is present in plastids and in mitochondria.[912]

The conclusion that nuclear genes operate collaboratively with those in plastids and mitochondria, is inescapable. This is why the topic is considered here, in a chapter concerned with integration in the cell. Several approaches—genetical, biochemical and structural—have contributed to our knowledge of the interactions that are involved. Since the theme of this chapter is a general one, it is not appropriate to review in detail the many discoveries that have been made; rather, a few illustrations will be given to show the extraordinary complexity that has been achieved in the course of the evolutionary integration of the genetic systems of the cell.

There is no doubt but that the genetic system of the nucleus dominates the development of both plastids and mitochondria. In barley plants, the number of nuclear genes that are known to affect plastids approaches 100.[857] Taking higher plants together with lower plants, nuclear genes have been detected which control many of the steps of the biosynthetic pathways leading to formation of chlorophylls and carotenoids in chloroplasts, etioplasts, and chromoplasts,[430] formation of starch in amyloplasts,[726] formation of several of the components of the photosynthetic electron transport chains,[483] and other thylakoid proteins[450] and the production of some of the proteins of plastid ribosomes[71] and part of the protein of the major enzyme, ribulose diphosphate carboxylase[770], amongst other Calvin cycle enzymes.[483] There could be no more direct demonstration that the cell nucleus can affect plastids than that obtained using individual cells of the giant alga *Acetabularia*. There differences are seen when ribosomal proteins are extracted from plastids of different species and compared. If the nucleus is removed from species A and replaced by a nucleus from species B, the plastid ribosomal proteins change under the influence of the new nucleus from those characteristic of species A to those characteristic of species B, the change taking place as the plastids multiply by fission.[434]

How do nuclear genes control the placement of a particular molecule in a plastid or a mitochondrion? Two possibilities are open if the molecule is a protein. It could be made on cytoplasmic ribosomes and passed through the bounding membranes to its destination, or it could be that the messenger RNA that arises by transcription of the nuclear genes itself traverses the membranes, to be translated by plastid or mitochondrial ribosomes. The difference between the two possibilities is in the site of translation. As has been emphasized in the previous section on metabolic integration, other ramifications are possible if the molecule is a product of enzyme action. An enzyme made in the cytoplasm could stay there and make a product which is taken into the plastid or mitochondrion, or the enzyme could move and make its product closer to its final site. Products of photosynthesis are obvious examples of molecules which are made by enzymes within plastids, and move in the converse direction, from plastid to cytoplasm.

Amongst the several differences noted earlier to exist between plastid and cytoplasmic ribosomes, one is differential sensitivity to various drugs. Chloramphenicol, spectinomycin, and lincomycin inhibit translation on the small plastid ribosomes, while cycloheximide selectively attacks the large cytoplasmic ribosomes. These substances can therefore be used to discover which of the two possible sites of translation are employed during synthesis of a given material. Suffice it to say that there is good evidence that some proteins coded for by nuclear genes are made on cytoplasmic ribosomes (e.g., the coupling factor and cytochrome c of mitochondria,[842] and ribosomal proteins and chlorophyll biosynthetic enzymes of chloroplasts),[483] while for others, the messenger RNA is ignored in the cytoplasm, traverses the envelope membranes, and is translated on plastid or mitochondrial ribosomes (e.g., several membrane proteins of *Chlamydomonas* chloroplasts).[396a]

Although the evidence that nuclear genes govern the synthesis of plastid and mitochondrial components is overwhelming, it is certain that non-nuclear genes are also functional. The conclusion has been reached from breeding experiments in which non-Mendelian ratios indicative of partially or completely maternal inheritance of cytoplasmically-carried traits are obtained;[724]

from experiments on selective inhibition of translation of the different classes of ribosome,[888] and from the use of another drug, rifampicin, which does not prevent the RNA polymerase enzyme of the nucleus from transcribing nuclear genes, but does stop initiation of transcription by the corresponding enzymes of mitochondria and plastids (and bacteria).[802]

The picture that is emerging is one of extremely subtle interactions between the various sets of genes, and if the observed phenomena have selective advantages, these cannot yet be discerned. For example, the plastid genes are responsible for making plastid ribosomal RNA,[802] yet some plastid ribosomal proteins (above) are made in the cytoplasm under the direction of nuclear genes. The plastid DNA duplicates in the plastid, but the enzyme that catalyses the process, DNA polymerase, is a product of nuclear gene and cytoplasmic ribosomes.[802] The same probably applies to the special, rifampicin-sensitive, RNA polymerase that transcribes the plastid DNA.[483]

The enzyme ribulose diphosphate carboxylase demonstrates in its biosynthesis another type of collaboration between nucleus and plastid. It is a large protein, composed of two types of subunit, and the evidence from higher plant material is that one subunit, the larger of the two, and the one carrying the site that catalyses carbon dioxide fixation, is coded for and made in the plastid, while the other, smaller, subunit is coded for in the nucleus, and made on cytoplasmic ribosomes.[724a, 770] The two subunits presumably come together in the chloroplast stroma.

Moving on from collaboration at a *sub*-molecular level, as in the production of ribulose diphosphate carboxylase, the assembly of thylakoids exemplifies collaboration in the formation of a *supra*-molecular structure. Thylakoids too are derived from both plastid and nuclear genes. Most of their lipids are apparently synthesized within the chloroplast, and their proteins are synthesized both inside and outside the chloroplast.[282] The presence of polyribosomes on growing thylakoids will be recalled. One of the membrane proteins is especially interesting in that, although made outside the chloroplast, its production is normally geared to the rate at which chlorophyll is made inside.[373] It may be that the chloroplast produces a chemical signal capable of regulating the rate at which the gene coding for the membrane protein is transcribed.

The implication of the above biochemical researches is that the plastid genes and the nuclear genes operate in harmony. Only the merest hints at the nature of the signalling system between the two have emerged thus far, but indications of its specificity and importance have been available for a considerable time, especially in the results of experiments on the inheritance of defects of pigmentation in chloroplasts. The variety of interactions is bewildering.[430] There are nuclear genes which can determine mutations in plastid genes, and nuclear genes which can govern the expression of plastid genes. The best demonstration that plastid and nuclear genes need to be adapted to one another comes from studies of evening primroses (*Oenothera*), where in one sub-genus, species differences have been shown to be based on different combinations of six nuclear gene sets and five plastid gene sets. Only some of the many possible combinations are 'harmonious'. 'Disharmonious' combinations can, however, be created, whereupon the disharmony is made manifest in many ways, from defects in plastid pigmentation and in the production and architecture of thylakoids, with attendant effects on photosynthesis, to production of functionless pollen, to abnormalities of growth and development of the plants. Under natural conditions many or all of the disharmonious combinations would no doubt die out, but if they are kept alive, it can be shown that the plastid genes remain stable for many generations of plants, and are able to resume a harmonious relationship if eventually they are recombined with a compatible set of nuclear genes—a very convincing proof of the genetic continuity of plastids from generation to generation.[168, 430, 753]

The integration that has been described in this section of the chapter is quite clearly of a different nature from the mere supply and demand of metabolites that was considered in the first section. The need for metabolic integration followed from the need for compartmentation in the cell, a need that in turn was related to the advantages accruing from division of labour in the cell. A fundamental question is posed. Is the evolution of chloroplasts and mitochondria to be viewed in the same way as the evolution of other compartments of the cell, such as vacuoles or the endoplasmic reticulum? Whilst all types of compartment are biochemically specialized, chloroplasts and mitochondria, unlike the others, have been provided in the course of evolution with their own genetic systems. Two possible interpretations are being debated.[134]

The first interpretation[680] is that delegation of partial autonomy to chloroplasts and mitochondria was advantageous in view of the great complexity of the processes of photosynthesis and respiration, perhaps allowing for better regulation and responses to environmental change, or for greater diversity in developmental potential, than would have been the case if the genetic system had been entirely nucleus-based. Proponents of this view note that many of the genes in the cytoplasmic components are controlled in the nucleus, that defects in the cytoplasmic genes can often be corrected by 'restorer' genes in the nucleus,[485] and that much of the cytoplasmic DNA is similar enough to portions of nuclear DNA to hybridize with it, at least partially.[816] In short, they view the current situation as being the result of migration of selected portions of DNA from nucleus to cytoplasmic components, with retention of master copies, or controlling genes, in the nucleus.

One of the difficulties of the above 'delegation' hypothesis is that both the DNA of plastids and mitochondria, and the apparatus of protein synthesis that goes along with it, differ from the corresponding systems in the nucleus. It has already been noted that the DNA is histone-free, as in prokaryotes, and that the RNA polymerase and the ribosomes also have many features in common with those of prokaryotes. While conceding that it is reasonable to provide any DNA that migrates from nucleus to cytoplasm with its own information-processing equipment, critics of the idea of delegation find it difficult to see a rationale behind a eukaryote nucleus delegating a non-nuclear type of genetic system, and even more, find it difficult to accept the similarity of the cytoplasmic systems to those of prokaryotes as mere coincidence. The second of the two interpretations is, in fact, that plastids and mitochondria are evolutionary relics of once independent prokaryotic micro-organisms that have entered into a symbiotic association with a host cell.[532] In the course of evolution, the host and the invaders have become mutually inter-dependent and their respective genetic systems highly integrated. The host may even be in the process of taking over the genetic attributes of the invaders, the current situation being that the nucleus has asserted a substantial, but incomplete, measure of overall control.

We are not likely to see any definitive means of proving either the delegation or the symbiosis hypotheses right or wrong. Meanwhile both of them highlight the theme of genetic integration in the eukaryote cell. They also highlight the differences between mitochondria and plastids on the one hand, and the remaining components of the cytoplasm on the other. The third type of integration, now to be considered, resumes the story of the latter.

13.4 Developmental Integration

Earlier chapters concerned with the plasma membrane, vacuoles, endoplasmic reticulum, the Golgi apparatus, and microbodies, have included information on development, and all that is needed now is to try to draw the threads together. To present the conclusion before the evidence, it appears that most of the membranes of these components can be regarded as parts of a functional and developmental continuum, that has been christened the *endomembrane system* of the cell.[573a, 574]

The various constituents of the endomembrane system often appear to be separate from one another when a cell is examined by means of static micrographs. We have already seen, however, that various types of link can exist between them. The membranes may join one another physically. Nearly all of the possible categories of membrane continuity have in fact been seen or claimed at one time or another,[151] but with few exceptions they are so rare that they must either be present for only very short periods of time, and therefore hard to fix and to find, or be chance artefacts. The only connections that are likely to have much permanence and stability are seen within categories of cell component, e.g., connections between rough and smooth endoplasmic reticulum, or between endoplasmic reticulum and the outer membrane of the nuclear envelope. It is thus hard to assess what contribution the continuity of different membranes makes to integration within the cell. There is in fact considerably more evidence that a different type of link has major functions in integrating different membrane systems: a link based upon the formation and shuttling of vesicles from system to system.

Linkage of cell components *via* vesicles involves two processes. One is the obvious one of transporting the vesicle from one place to another, and it may be described in general terms as a mechanism of *membrane-flow*, or *membrane transfer*. The other is less obvious, but still crucial. It is *membrane differentiation*, or *membrane transformation*. At its simplest the latter might merely be the acquisition of a recognition system by which the vesicle can find a given destination and fuse with a receptor membrane. In more complex forms it includes the transformation of one type of membrane into another, the prime example being what happens in the dictyosomes of the Golgi apparatus, where the membranes of individual cisternae undergo maturation processes involving alterations to their dimensions and to their lipid and protein composition.

Given the ingredients of membrane continuities (some relatively permanent, some transitory), membrane flow, and membrane differentiation, what can be said about the development of the endomembrane system of the cell?

The endoplasmic reticulum, it would appear, is at the heart of the system.[239] If radioactive substrates are supplied to the cell, its membranes are the first to become labelled. This in itself is of biosythetic but not developmental significance. However, if only a short pulse of radioactive substrate is supplied, it is then seen that radioactivity leaves the endoplasmic reticulum to appear in other membranes.

We have, of course, structural observations on which to interpret the onward movement of radioactivity. Transitional vesicles are thought to move from the endoplasmic reticulum (or nuclear envelope) to give rise to juvenile cisternae at the forming face of dictyosomes. In cells where the endoplasmic reticulum is not growing very rapidly, it might be anticipated that the transitional vesicles, and in due course the whole dictyosome, would receive preferential treatment in the handing on of the available precursors for membrane synthesis. In accord with this expectation, the radioactivity that is passed on to dictyosomes proves to be more concentrated (i.e., more radioactivity per unit of membrane) than that initially taken into the endoplasmic reticulum *as a whole*. The relatively low concentration in the endoplasmic reticulum probably reflects the fact that the production of transitional vesicles is a fairly local event, restricted (to some extent) to specialized formative areas. If these could be isolated from the remaining, less active, reticulum, they would presumably be found to be at least as radioactive as the dictyosome membranes that arise from them.

So far then, biochemical and structural observations are in agreement. There is a developmental pathway from endoplasmic reticulum to Golgi apparatus, mediated by a flow of vesicles (and possibly also *via* membrane continuities, which have also been seen interconnecting the two systems). The radioactivity moves further, however, and once again ultrastructure and biochemistry are in accord. After membrane differentiation in the dictyosomes, membrane flow by vesicle formation and movement (with further differentiation), leads to two possible destinations. The experiments on the movement of radioactivity have been carried out using onion cells and liver cells. In both, one destination is the plasma membrane. In the plant the other destination is the tonoplast,[536] and in the animal it is the lysosome system. We have already seen that plant vacuoles can have lysosomal functions, so the plant and the animal are not all that dissimilar. The structural evidence that secretory vesicles carry dictyosome products and membrane to the plasma membrane, osmotically active material to the tonoplast, and 'primary lysosomes' to the tonoplast, has been described in chapter 7.

Already, then, several of the membranes of those cell components that do not contain their own genetic systems have been seen to be linked in a developmental and functional continuum, leading from endoplasmic reticulum (and nuclear envelope) *via* the Golgi apparatus to plasma membrane and tonoplast. Quantitative relationships considered previously show that at least one of the links—the formation of plasma membrane from dictyosome vesicles—is sufficiently dynamic to more than account for observed rates of growth of the end product. Whether this applies to all links is uncertain. If it does, it might then be suggested that the endoplasmic reticulum is the *only* site of membrane synthesis (but by no means the only site of membrane differentiation), all of the other membranes stemming from it by membrane flow. If it does not, then it would follow that other membranes must be capable of incorporating molecules and growing independently. Even then some of the incorporated molecules, e.g., phospholipids, might have to be made in the endoplasmic reticulum before being passed to their destination.

The developmental origin of microbodies has not yet been resolved satisfactorily. Evidence for a derivation directly from endoplasmic reticulum has, however, been presented. Some authors advocate that yet other membranes should be added to the developmental continuum. On the theory that plastids and mitochondria are relics of once independent organisms, the presence around each of a double envelope can be rationalized as follows. The inner membrane would be the relic of the plasma membrane of the invader, and the outer would be a relic of a layer of host plasma membrane that would, by analogy with present-day mechanisms of endocytosis, have become wrapped around the invader. Formation of the inner envelope membrane would thus originally have been the responsibility of the invader, while formation of the outer membrane would have been managed by the host. After the association became stable, the two membranes would have become modified in the light of their new functions, so that we should not expect them to have retained the characteristics of plasma membranes. In fact the outer envelope membranes sometimes connect to cisternae of smooth endoplasmic reticulum, and several similarities between endoplasmic reticulum and the outer mitochondrial membrane have already been mentioned. It is, therefore, conceivable that in terms of developmental pathways, the endoplasmic reticulum is the source of the outer envelopes of the semi-autonomous cell components, as well as the source of the endomembrane system in general.

Although only developments stemming *from* the endoplasmic reticulum have been considered thus far, it is clear that the continuum cannot be entirely one-directional. It has been pointed out, for instance, that the observed over-production of plasma membrane demands that there is a mechanism for *returning* membrane components to centres of membrane synthesis and differentiation within the cell. It is probable that all membranes undergo turnover, though the rate may seldom be as rapid as in the case of the plasma membrane. Several mechanisms of return flow have been postulated. One is based on the retrieval of individual molecules from a membrane, perhaps involving breakdown processes such as the conversion of proteins to amino acids, and the formation of free fatty acids from membrane lipids; however, the overall integrity of the membrane is not destroyed. A contrasting mechanism is the wholesale breakdown of membrane, as for instance in a lysosome or vacuole; the individual breakdown products would later be retrieved. A third possible mechanism is the retrieval of intact expanses of membrane in the form of vesicles, such as, for example, the enigmatic coated vesicles about which there is so little information. Since the development of membrane systems is not fully understood, it goes without saying that very little indeed is known about either qualitative or quantitative aspects of the breakdown of the various categories of cell membrane.

In chapter 7, a great deal of emphasis was placed upon dynamic aspects of one cell component, the Golgi apparatus. It is now appropriate to expand the principles and conclusions reached there to the wider context of the whole cell. It is not only the membranes of the dictyosomes that undergo turnover. *All* of the cell membranes are in a dynamic state. A mature leaf cell, or liver cell, may look very much the same no matter when the tissue is sampled for electron microscopy. Nevertheless its apparent stability arises only because it is in a state of equilibrium, with compartments remaining the same size, against a background of continual inflow and outflow of components, including components of the membranes by which the compartments are so easily recognized. By contrast, in developing systems, the compartments appear different at each sampling time because they are in non-equilibrium conditions, in which either the inflow of components predominates (in which case they grow), or the outflow predominates (in which case they diminish in size).

The rates of throughput on the developmental continuum that links the various membrane systems of the cell are significant in relation to the arguments that have cropped up, chapter by chapter, on whether compartments arise *de novo*, or are passed from cell to cell during cell division. With the background of a developmental continuum in mind we can see that what is of fundamental importance is whether the ultimate source—the endoplasmic reticulum—is passed from cell to cell. We know in fact that it is. The process of nuclear division has built in to it a mechanism ensuring that at least one cisterna, albeit a specialized one, the nuclear envelope, passes with every newly-formed nucleus into every newly-formed cell. Obviously, every new cell will also receive a quota of any other components that happen to become segregated by the developing cell plate. That, no doubt, is convenient, but perhaps not fundamental. It is not to be viewed in the same way as the partitioning of the population of mitochondria and plastids, reproduction of which is normally geared to the divison of the cell in which they lie.

13.5 The Cell as an Integrated Unit

What has been learned about the nature of the cell since the early attempt at a definition of its nature in chapter 2?

Something of the vast body of knowledge that exists on the structure and the function of the various categories of cell component has been seen, together with something of the limitations of the various approaches to the subject. Even the most ardent microscopist would be bound to agree that no description of structure, however complete, will lead to a full definition of the cell and its mode of operation. All that can be allowed is that visualization of structure provides a vital, though incomplete, part of the overall picture. It also serves as a useful starting point for more general enquiries—the philosophy underlying the whole of this book. In prosecuting those enquiries, it helps to define problems and to confine interpretations of function to the realities imposed by the facts of structural organization.

As with microscopy, so too with biochemistry. The cell is not definable in terms of a catalogue of enzymes, substrates, and reactions, any more than it is in terms of structure. But as has been shown again and again, the two approaches do complement one another. The time dimension, with which the electron microscopist finds so much difficulty in coping, is revealed to the biochemist in studies of reaction kinetics. The dimensions of space, which the biochemist so often has to ignore or to disrupt, are visible to the microscopist. Together, they can visualize the temporal and spatial organization of the cell.

Most biologists have their own mental image of what a cell is and what it does. It is, however, extremely difficult to progress from a mental picture to a definition that expresses with precision the nature of the cell. To attempt this it is necessary to look at how the cell operates, what the sum total of its operations achieves, and to place upper and lower limits on its degree of complexity.

The word 'system' is useful. It implies a collection of parts, an essential feature of which is that they are organized, either through being connected to one another, or through their functional collaboration. A growing crystal is not, in this sense, a system, but cells, tissues, and organisms are. What, then, constitutes a cellular system? One way to approach a definition starts by examining the entities that, without thinking too much about the meaning of the word, we have all along been referring to as cells.

All of the cells that we have met possess a boundary layer or surface membrane, biochemical systems for making proteins and nucleic acids (etc.), means of manipulating and utilizing chemical energy, and, all important, a store of genetical information. We have also seen that some cells are more complicated than others. It seems reasonable that any definition should recognize common ground and avoid making use of the variables that are seen when prokaryotes, eukaryotes, plants, and animals are surveyed. Viruses, which have not been considered thus far, contain genetical information and in some cases have a bounding membrane, but they rely heavily on the biochemical systems of host prokaryotes or eukaryotes to process their information and hence to drive their reproduction. The definition of the cell developed below begins by excluding viruses. It recognizes the central position of the genetical information, and requires a capacity for autonomous processing of that genetical information. This is taken as the common factor encompassing all types of prokaryote and eukaryote.

Test tube mixtures which process, i.e., transcribe and translate, genetical information can be prepared. Mitochondria and plastids can also transcribe and translate their DNA. None of these 'systems', however, conforms to the average mental image of the cellular state, and further definition is necessary. One obvious consideration is that cells are alive. This can be combined with the requirement used to exclude viruses, thereby avoiding the contentious question of whether viruses should or should not be regarded as living systems. Substituting other words in an attempt to express what is meant by 'alive' we have: a cell is a system that contains, and can process, genetical information endowing it with a capacity to reproduce using raw materials and energy obtained from its environment.

The above expression is still inadequate. It covers anything from the simplest prokaryote to the most complex of eukaryotic organisms. It is necessary to introduce the idea of the cell as a unit which can be used to build multicellular systems. This can be done by saying that a cellular system, the attributes of which are defined, contains no sub-systems that possess the same definied attributes. In other words, whereas multicellular systems can be broken down into sub-systems, i.e., cells, without loss of the features defined above, cells themselves *cannot*. To attempt to break down a cell into simpler units is to destroy it, and its life. The definition becomes: *a cell is a system that contains, and can process, genetical information endowing it with a capacity to reproduce using raw materials and energy obtained from its environment, and containing no sub-systems that possess this attribute.*

Bearing in mind the enormous diversity of living systems that has been thrown up in the course of evolution, it is perhaps unreasonable to expect an arbitrary definition to cover all possibilities. A brief look at what is and is not catered for by the above adds some insight into the nature of the cellular state and serves to emphasize that there are many manifestations of it.

First, a closer look at the cell as a unit system can be introduced by describing a system that is *not* a unit. In situations where many nuclei arise within a common plasma membrane by karyokinesis without accompanying cytokinesis, a system is created which does have the potential of being divided up into sub-systems, each capable of reproduction using raw materials and energy obtained from its environment. The marine alga *Bryopsis*, for example, can be broken to produce small masses of protoplasm which, provided that at least one nucleus is present, can regenerate a cell wall and, in time, a new plant.[812] Similar phenomena are seen in some fungi and slime moulds, and whilst they do clarify the meaning to be placed on the terms 'system' and 'sub-system', they also create a problem, for if the definition is to be followed strictly, they should not be called cells, or even multinucleate cells—hence the use of the terms *coenocyte* (page 41) or *plasmodium*.

From the formal, and rather cumbersome, definition of a 'unit', which speaks of the absence of sub-systems that match the properties of the total system, it follows that there can be no such thing as a fully autonomous sub-cellular component. Part of the present chapter has in fact been devoted to showing that plastids and mitochondria are not autonomous, though they do contain, and can process, genetic information. They require more than just raw material and energy. Not even the nucleus is autonomous in the sense of the definition. It contains most of the information of the cell, but it requires the assistance of the remaining parts of the cell to process that information. In short, subcellular components are inter-dependent rather than independent. They participate in a group existence which has evolved by division of labour within the cell, and in which the degree of collaboration is so effective that the participants *as a group* not only survive, but can grow and reproduce *as a group*.

The nature of the collaboration between the parts of the cellular system has been the main theme of the present chapter. We have seen that all parts of the system interact as donors and recipients in a cross-traffic of nutrient molecules. In addition, some of the parts interact by the more subtle processes of information transfer and regulation of the expression of information. Choice between the options of survival, growth, and reproduction is largely a matter of modulating the developmental interactions that exist between the parts of the system. All parts are continually being formed and broken down in the cyclical process summarized in the term *turnover*. Some grow by taking in precursors of molecular dimensions, while others are formed by modification of elaborate pre-formed materials that mature along branches of a developmental continuum leading from one part to another. If the opposing processes of synthesis and breakdown balance one another, the system as a whole is merely maintained. It survives. Despite the continual alteration of the participating components, the system preserves its character.

A preponderance of formative processes over breakdown processes, on the other hand, permits growth and reproduction. Qualitative changes can lead to differentiation, sometimes utterly changing the appearance and function of the cell. In a few cases, such as the loss of the nucleus in developing sieve elements and red blood corpuscles, the differentiation process is so drastic as to take the cell beyond the limits of the definition: neither mature sieve elements nor mature red blood corpuscles are capable of reproduction, though thanks to the activities of other cells in the organism, they may survive for years, and may even grow.

The extent to which other categories of cell differentiation removes 'the capacity to reproduce using raw materials and energy obtained from the environment' is rather uncertain. Irreversible differentiation is probably quite common in the highly specialized tissues of animals. With the simpler nutritional requirements of plant cells, it has proved possible to show that most types can *de-differentiate* and then reproduce. To use the word introduced on page 132, they are totipotent. No matter whether differentiation is or is not reversible, however, the important point is that the cell retains the genetical information that it had prior to its differentiation. In theory it still has the capacity described in the definition, and the definition is only violated to the extent that the capacity to reproduce may be a potential that in practice is not realized.

The word 'environment', as used in the definition, is worthy of further comment. The nature of the environment can vary enormously, and with it, so too can the nature of the 'raw materials and energy' that the cell obtains from it. Amongst the more extreme cases are those parasites and symbionts which can live *only* in the living environment provided by their host or partner. They do not violate the definition: their genetical information still expresses its capacity to organize reproduction, albeit in a highly specific environment. Nevertheless the distinction between these cases and the case of a plastid or a mitochondrion in a cell becomes very tenuous, one of quantity and not of kind. A plastid or a mitochondrion may be said to be in general much more dependent upon the genetical information of its 'environment' than is a parasite or a symbiont, and whereas no plastid has been found to have a sustained capacity to reproduce outside a cell,[269, 270] a good many (but by no means all) parasites and symbionts can be free-living at particular stages in their life-cycle, or if given appropriate culture conditions.

It is in fact only unicellular organisms that can be exposed exclusively to a non-living environment. As soon as a multicellular condition is attained, part of the environment of any given cell consists of one or more other cells. This, in some very primitive organisms, may have very little effect upon the individuality of the cells. More usually, division of labour at a supracellular level creates a collaborative group existence of the type seen within a cell, but at the level of organization of tissues and organs (Plate 49). The constituent cells become partially inter-dependent. The 'environment' from which such cells obtain their raw materials and energy is thus partly non-living, and partly the living neighbouring cells. As described above, this does not detract from the capacity of the cells to reproduce, given an environment in which they can assert their independence.

There is a fundamental distinction between the collaborative group existence shown by cells in tissues, organs and organisms, and that shown by components within cells. It is often stated that the cell is greater than the sum of its parts. It is greater by virtue of the fact that collectively, the parts establish a degree of permanence and developmental potential that can be recognized as life. The cell is the simplest group existence to achieve this vital breakthrough. By contrast, the higher levels of organization and collaboration seen in the tissues and organs of multicellular organisms go far above the basic essentials of life, outside the definition of the cellular state, and beyond the scope of this book. They, but not cells, can be broken down into sub-systems which retain the basic essentials. To that extent, the definition presented here is merely an expansion of that other often quoted statement: the cell is the unit of life.

THE ILLUSTRATIONS

The following micrographs provide the visual background for the descriptions, generalizations and principles that have been given in the text. Since many of the pictures illustrate a multiplicity of points, and in that sense do not belong in any one section or chapter, it has been thought preferable to group all of them together. Also, most of them do more than merely illustrate the text. They are specific examples and they contain much specific information. In order to exploit to the full the high information content of each picture, the legends incorporate descriptions and interpretations that do not appear elsewhere. Thus the legends supplement as well as complement the text. In order to help integrate the two, references have been given throughout the text to lead the reader to the appropriate micrographs. Additionally, a cross referencing system has been incorporated in the legends. The numbers in their margins are the text page numbers that should be consulted for background information relevant to the content of that part of the legend.

Plate 1

The Plant Cell (1). Light Microscopy

These large cells in the meristematic region of a broad bean (*Vicia faba*) root tip are viewed by phase contrast microscopy. The section was reacted with acriflavine, following oxidation in periodic acid, to stain carbohydrates yellow (e.g. in cell walls and starch grains), and subsequent immersion in another yellowish reagent—iodine in potassium iodide—gave a generally-stained preparation, best examined using blue light. The magnification is × 4200, i.e. 4.2 mm represents 1 μm, and since the thickness of the section was about 1 μm, we are in effect looking through a slice 4.2 mm in thickness, rather than at an infinitely thin 2-dimensional picture.

Each cell is outlined by its wall (CW). There are no intercellular spaces in this particular group of cells. Within the walls the major visible compartments of the cells are the numerous empty-looking vacuoles (V), the cytoplasm around the vacuoles, and the nuclei, which are numbered N1 to N4.

The nuclei are separated from the cytoplasm by the nuclear envelope (NE), best seen in nucleus (N-1), which has been caught at an early stage of division. In the other nuclei the speckles represent stained chromatin (CH); prior to division this condenses to form discrete chromosomes (CHR in N-1), leaving the nuclear envelope relatively isolated and conspicuous. After division the chromosomes uncoil again to regenerate the chromatin condition, and a stage of this process is seen in nucleus N-2. The large dense bodies in the nuclei are nucleoli (NL). The nucleolus in N-1 is lobed and irregular, but in the non-dividing nuclei its circular outline is indicative of a more-or-less spherical shape. Pale nucleolar 'vacuoles' occur in most nucleoli. No nucleolus is seen in nucleus N-3, but this does not mean that none is present. Sections are statistical samples of cells and tissues, and it is not to be expected that any one view of a cell will contain all of the components. Consider the dimensions of nucleus N-4, and assume that it and its nucleolus are spheres, both sectioned across their diameters, which at 4.2 mm representing 1 μm, are about 10 μm and 4 μm respectively. It would take 10-11 consecutive 1 μm sections to pass from one face of the nucleus to the other, and only 4-5 of these would include portions of nucleolus.

The most clearly resolved cytoplasmic components are proplastids (PP), but these can only be identified with certainty if starch grains (e.g. arrow, top right) can be detected inside them. It would appear from their varied profiles in the section (some elongated, some less so, some mere dots) that there is a population of randomly oriented, more-or-less sausage shaped proplastids in the cells. Only rarely does the full length of a 'sausage' lie within the thickness of the section. Other cytoplasmic structures can be discerned but they cannot be identified with certainty. The less densely stained particles doubtless include mitochondria (M?), and the very faint convoluted shadows (e.g. connecting the small arrows above N-1) are probably cisternae of endoplasmic reticulum, but this is a statement which can only be made with the benefit of hindsight by an observer who has studied these and similar cells with the electron microscope.

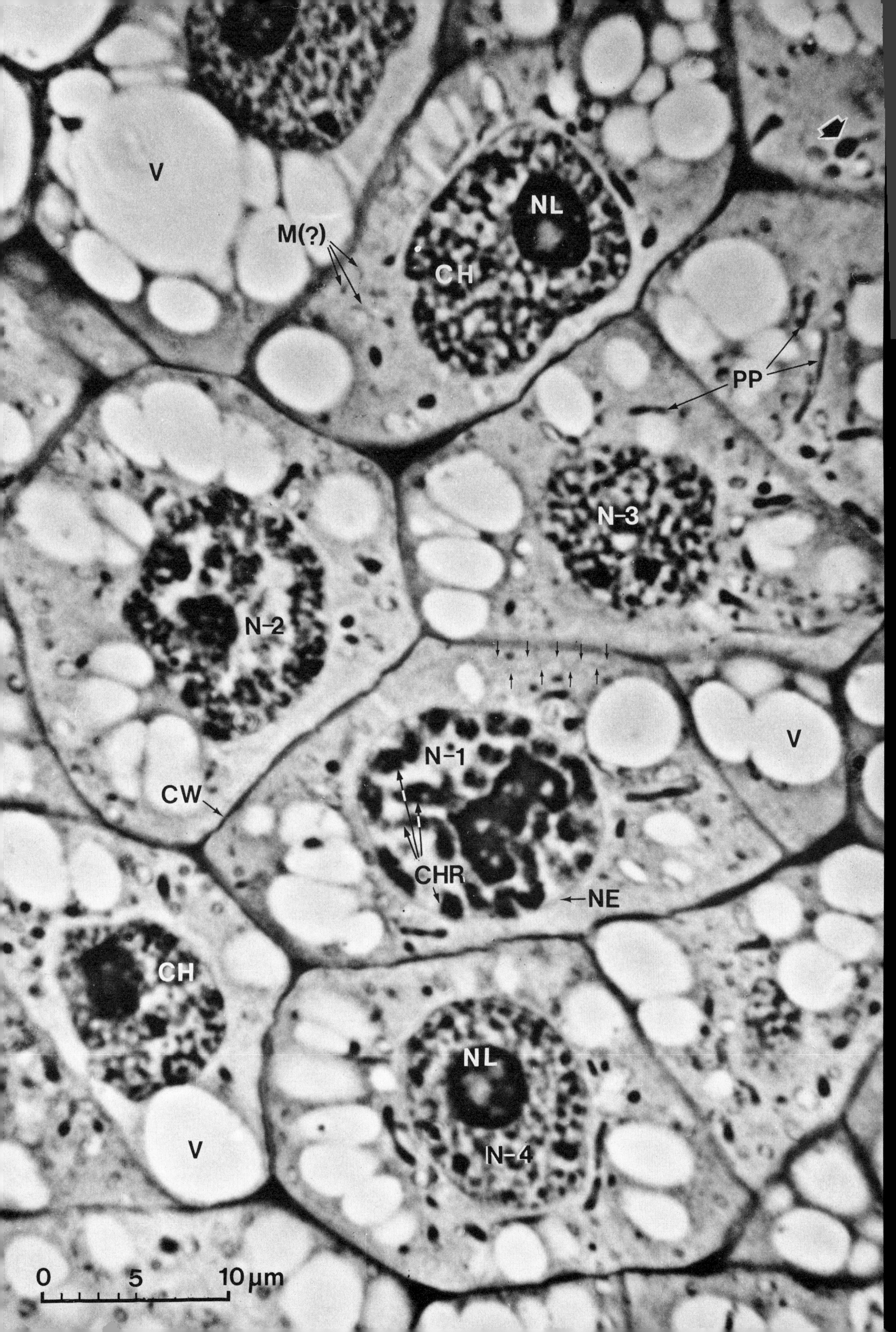
V
M(?)
NL
CH
PP
N-3
N-2
V
N-1
CW
CHR
CHR
NE
CH
NL
V
N-4
0
5
10 µm

Plate 2

The Plant Cell (2)

This section sliced through the mid-region of a cell in a root tip of cress (*Lepidium sativum*), and is viewed here by electron microscopy at × 20 000. The section was about 75 nm thick, so at this magnification the slice is about 1.5 mm thick—relatively thin compared with Plate 1.

The resolution of the electron microscope brings to light many features not seen in Plate 1. The plasma membrane (PM) is clearly distinguishable from the wall external to it (CW). By and large the plasma membrane lies at right angles to the plane of the section, so that we are looking at it edge-on. It therefore appears as a dark line in the section. The plasma membranes of adjacent cells pass through the intervening cell wall at plasmodesmata (PD). If we could see these extensions of the plasma membrane end-on, as in a section cut in the plane of the wall, they would appear as round holes. Again remembering that the section is of finite thickness, we can estimate roughly how many plasmodesmata pierce the wall. There are, for instance, 11 in the 250 mm of the left hand upright wall, and since at × 20 000, 20 mm represents 1 μm, this means 11 in an area of wall equal to 12.5 μm multiplied by the section thickness, i.e. about 75 nm. From this we can calculate that there are up to 1200 plasmodesmata per 100 μm^2 of this type of wall. 28

The vacuoles (V) are sparse and small in this view (cf. Plate 1). They lie in a densely particulate cytoplasm, each tiny particle being a ribosome. Ribosomes either lie free or else are attached to the membranes of the rough endoplasmic reticulum cisternae, giving them a characteristic beaded appearance (ER). In these cells most of the cisternae are narrow, and it should be realized that only those which lie at right angles to the plane of the section or nearly so are clearly visible. 10

Other components of the cytoplasm are present: mitochondria (M), proplastids (PP), dictyosomes (D), and a microbody (MB). Note the difference in the electron-density of mitochondria and proplastids. A small region (outlined in black at top left) is shown at magnification × 40 000 in the insert (lower left) in order to make the cross sections of microtubules present there more obvious (arrows).

The inner and outer membranes of the nuclear envelope (NE) are resolved. The outer, like the endoplasmic reticulum, bears ribosomes. The nuclear contents, perhaps most conspicuously different from the cytoplasm in the absence of membranes, consist of nucleolus (NL) and chromatin (CH) suspended in the general ground substance, or nucleoplasm. The nucleolus is an unusually large example, with fewer and smaller 'vacuoles' than those in Plate 1. It exhibits differentiation into regions that are predominantly granular (solid stars) and others in which substructure cannot be seen at this magnification, being composed of closely packed fine fibrils (open stars). The open arrows point to portions of strands of chromatin that ramify through a network of channels in the body of the nucleolus. 45

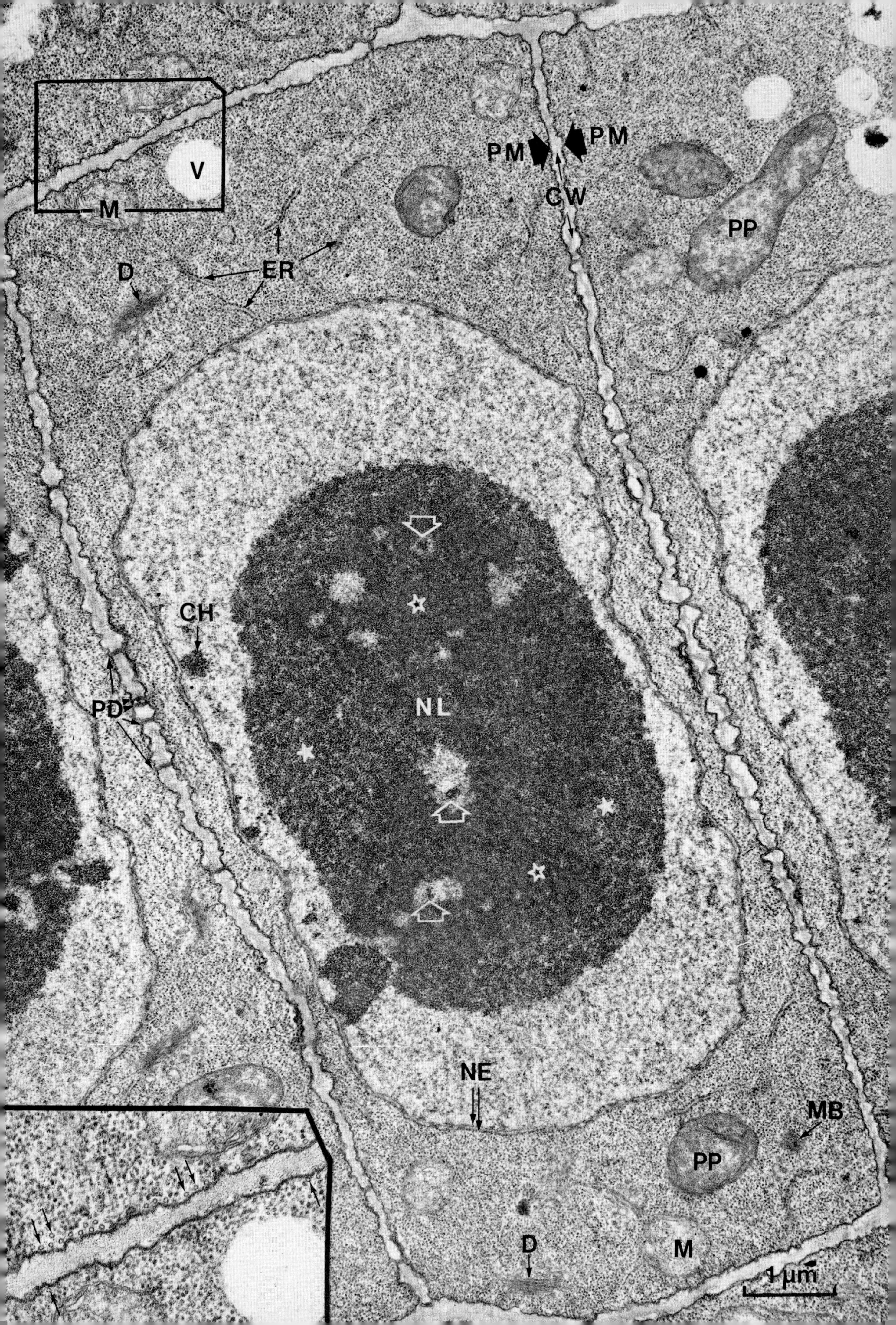

V
M
PM
PM
CW
PP
ER
D
CH
PD
NL
NE
MB
PP
D
M
1 μm

Plate 3

The Plant Cell (3)

Plate 3a The same material as in Plate 2 is seen here at a higher magnification (× 45 000). Parts of two cells are shown, separated by a wall (CW), lined by the two plasma membranes (PM). Some microtubules (arrows) lie just internal to the plasma membrane, and two (circled) are unusually deep in the cytoplasm. The cisternae of endoplasmic reticulum (ER) are all 'rough', with ribosomes studding their membranes and also lying free in vast numbers. One cisterna (ER*, lower right) is distended by the presence of accumulated contents. The left hand cell is not as free of endoplasmic reticulum as might appear at first glance. It is merely that many of its cisternae happen to lie in or close to the plane of the section. The arrowheads indicate the limits of one, barely discernible, obliquely-sectioned cisterna.

A small portion of a nucleus (N) is included at the right hand side, and the two membranes and pores (open arrows) of the nuclear envelope (NE) are visible. The successive cisternae of the dictyosome (D) lie at right angles to the plane of the section, and there are many associated vesicles nearby in the cytoplasm, or else attached to the cisternae. The double envelope of both proplastids (PP) and mitochondria (M) can be seen (squares). The tonoplasts of the two vacuoles that (only just) enter the lower edge of the picture again illustrate the difference between the crisp profile of a membrane lying 'edge on' (T) and an indistinct, obliquely sectioned membrane (T*).

Plate 3b This high magnification picture (× 100 000) illustrates one aspect of the substructure of cell membranes. The material (a young cortical cell of the root tip of *Azolla*) is especially instructive for two reasons. One is that the cells contain some substance which spreads (possibly during fixation) to the surfaces of all membranes and highlights their dark-light-dark appearance when seen edge-on in an ultra-thin section. The other is that the cells accumulate much lipid, and in older examples the vacuole (V) contains massive deposits. A small droplet of lipid in the form of a myelin figure (MF) is half way between cytoplasm and vacuole. It shows concentric layering, formed by back-to-back alignment of bimolecular layers. The cell membranes themselves show somewhat similar layering, but because they are either isolated, or else not stacked as closely as the lipid, the dark-light-dark 'tramlines' are seen individually.

The plasma membrane (PM) lines the cell wall (CW) and pierces it at plasmodesmata (PD). Other membranes included are the tonoplast (T), and cisternae of endoplasmic reticulum (ER) and a dictyosome (D).

The inserts show enlarged details from the same micrograph (× 250 000). From left to right: bimolecular layers of lipid in the myelin figure; plasma membrane; tonoplast; endoplasmic reticulum; and dictyosome cisterna. In each case the circles enclose one short portion of triple-layered membrane.

13

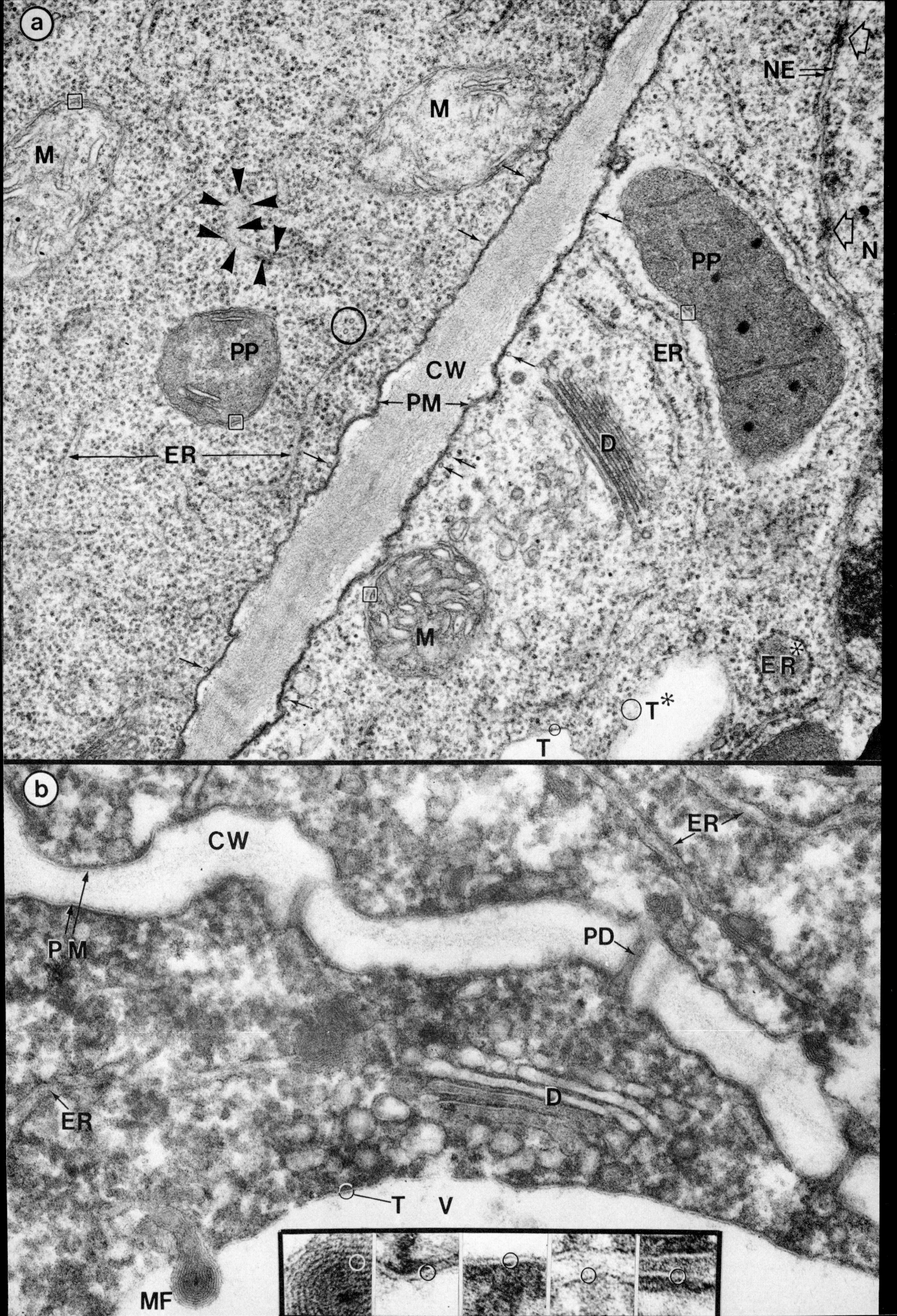
a
M
M
NE
N
PP
PP
ER
CW
PM
ER
D
M
ER*
T*
T
b
CW
ER
PM
PD
ER
D
T
V
MF

Plate 4

Plasma Membrane, Microfibrils in the Cell Wall

Plate 4a This face view of part of the plasma membrane (upper left) and cell wall (lower right) of a cell in an *Asparagus* leaf was obtained by the freeze-etching technique. Layers of microfibrils can be seen in the wall, and the scattered dark holes may be plasmodesmata (PD). Many particles about 10 nm in diameter lie in the plasma membrane, sometimes in rows (e.g. arrow). The particulate texture of some areas (asterisks) is different from the surrounding membrane. Semi-crystalline patterns can occur (not seen here). At one place the whole thickness of the plasma membrane has ripped away to expose another membrane surface within the cytoplasm (star). The particles in the expanse of membrane are probably proteinaceous, while the smooth areas between the particles represent an internal surface composed of lipid molecules, exposed when the membrane fractured along its mid line.

3 25 13

Magnification × 95 000. (We thank Drs. H. Richter and U. Sleytr and Verlag G. Fromme & Co., Wien, for permission to produce this micrograph from *Mikroskopie*, **26**, 329-346, 1970.)

Plate 4b and c The cellulose microfibrils illustrated here by means of shadow casting are in pieces of primary cell wall from which matrix materials have been extracted (cf. Plate 4a). Plate 4b is a wall viewed from the plasma membrane side. The most recently deposited microfibrils have not been pulled far out of their original orientation (which corresponds to the axis at right angles to the long axis of the cell) but the older, deeper microfibrils are more randomly oriented. The scattered sites of individual plasmodesmata are neatly circumscribed by curved microfibrils (PD) (elongating pith cell of *Ricinus*, × 22 000). Plate 4c shows (i) randomly arranged microfibrils (top centre), (ii) specific delimitation of pit fields where plasmodesmata are clustered (asterisks, see also Plate 14e), and (iii) parallel orientations (running across the lower part of the micrograph) at a strengthening rib, all in an oat coleoptile cell (× 15 000). (We thank Dr. K. Muhlethaler and Buchler und Co. AG. for permission to reproduce these plates from *Ber. Schweiz. Botan. Ges.* **60**, 614, 1950).

17 28 17

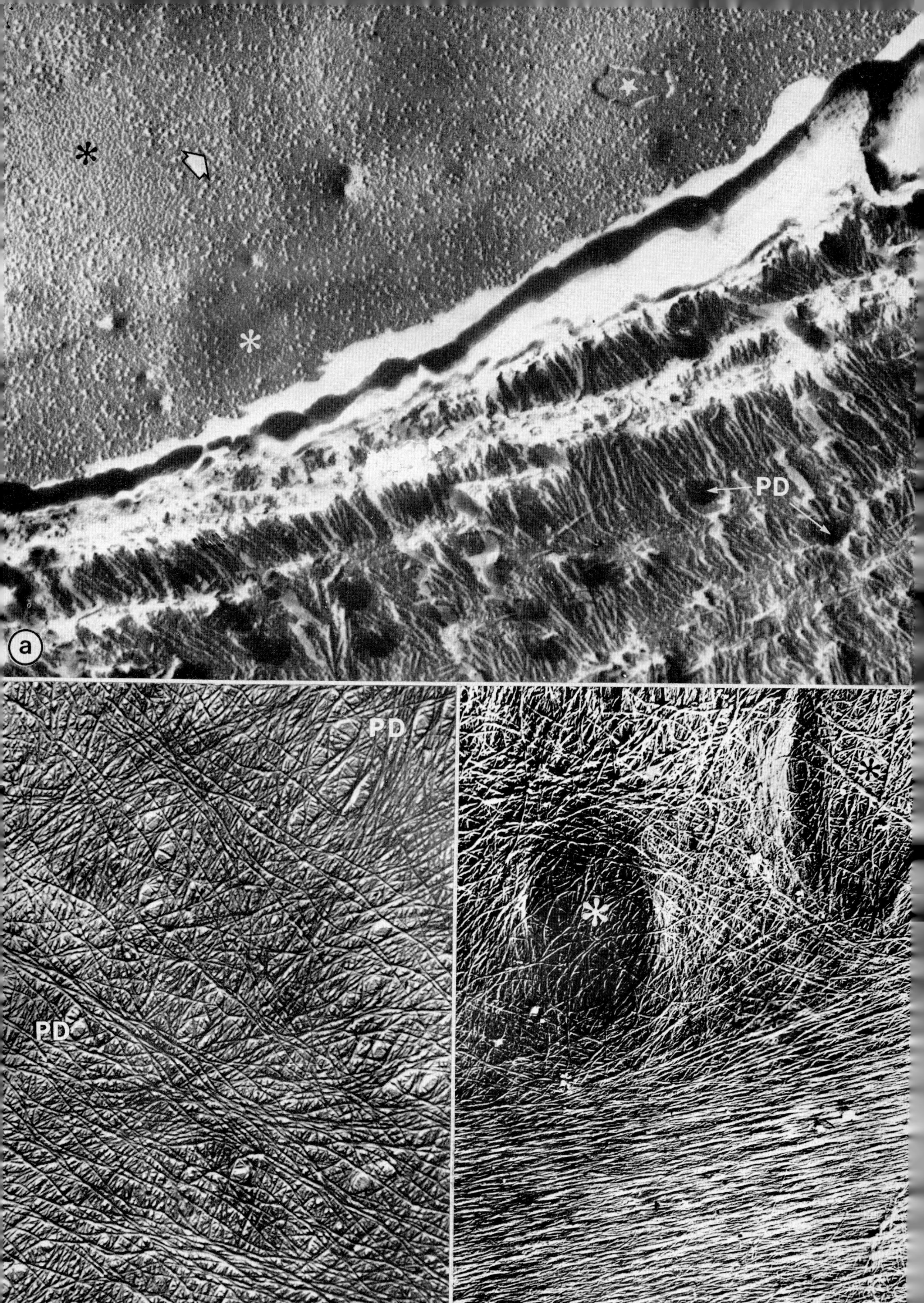
PD
a
PD
PD
b
c

Plate 5

Xylem (1): Developing Xylem Elements

Xylem is the tissue which carries the transpiration stream from root to shoot. It consists predominantly of dead xylem elements that are aligned in rows to form vessels or tracheids, together with the living xylem parenchyma cells responsible for transferring solutes into and out of the non-living conduits. Developing xylem elements (not yet dead and empty) are shown in Plate 5, and mature systems are illustrated in Plate 6.

Plate 5a This micrograph shows a longitudinal section of xylem elements in the vascular tissue of an *Azolla* root (see Plate 49 for transverse section). PX is a protoxylem element, recognizable as such because the lignified rings have been pulled apart (arrows) by elongation of the surrounding tissues after the protoxylem was itself mature. The black areas (asterisks) represent parts of cells that bulge into the protoxylem and so have been included in the plane of the section. MX is a metaxylem element, developing much later than the protoxylem. The many wall thickenings are seen (e.g. arrows), as is an end wall (W) separating one xylem element from the next in the file. Nucleus (N) and cytoplasm are still present. × 4300.

Plate 5b Part of the wall between two developing xylem elements is illustrated here in a section of leaf vein. The uppermost cell still has its cytoplasm: three dictyosomes (D) and their associated vesicles are prominent. The tonoplast (T) is intact. Scrutiny of the same field of view at higher magnification showed the position of the microtubules which overlie developing xylem thickenings. Since the microtubules are too small to be seen here, their positions are marked by arrows, 34 in all. Their association with the lignified 153 bands is very clear.

The lower cell is more advanced in its development. Its tonoplast has ruptured and only a few recognizable 37 cytoplasmic components remain. Although its cytoplasm has been largely digested, breakdown of the primary wall has not yet commenced. All but the cellulose microfibrils will become digested from zones between the thickenings, to give a mature condition as in Plate 6c. Magnification × 16 500. *Lupinus* leaf.

Plate 5c and d Developing xylem in *Phaseolus* (French bean) at × 1800, viewed by light microscopy. Plate 5c shows the onset of lignification in the thicken- 19 ings (pale areas, L, contrasting with the darker stained non-impregnated hemicellulosic parts (H) of the young thickenings). The primary wall (PW) is still present, with some cytoplasm. Plate 5d shows a later stage, with no cytoplasm, loss of the primary wall (PW) except where protected by the lignin, and uniformly lignified thickenings. Note the back to back arrangement of the 148 thickenings in neighbouring cells.

Plate 5e Developing xylem of *Galium* (goosegrass), viewed in transverse section at × 13 500, shows many of the features seen in Plate 5a and the upper half of Plate 5b, including microtubules (arrows) at the developing thickenings. In this plane of section the microtubules are seen in side view rather than transversely cut. There are several dictyosomes (D) and numerous vesicles (thought to originate from the dictyosomes) in the cytoplasm.

Plate 5f, g and h Three stages in the dissolution of the end wall (W) of vessel elements in *Phaseolus* are seen here by light microscopy at × 900. The mature vessel (derived from vessel elements) is thus an open pipe, through which the transpiration stream flows from 19 root to shoot.

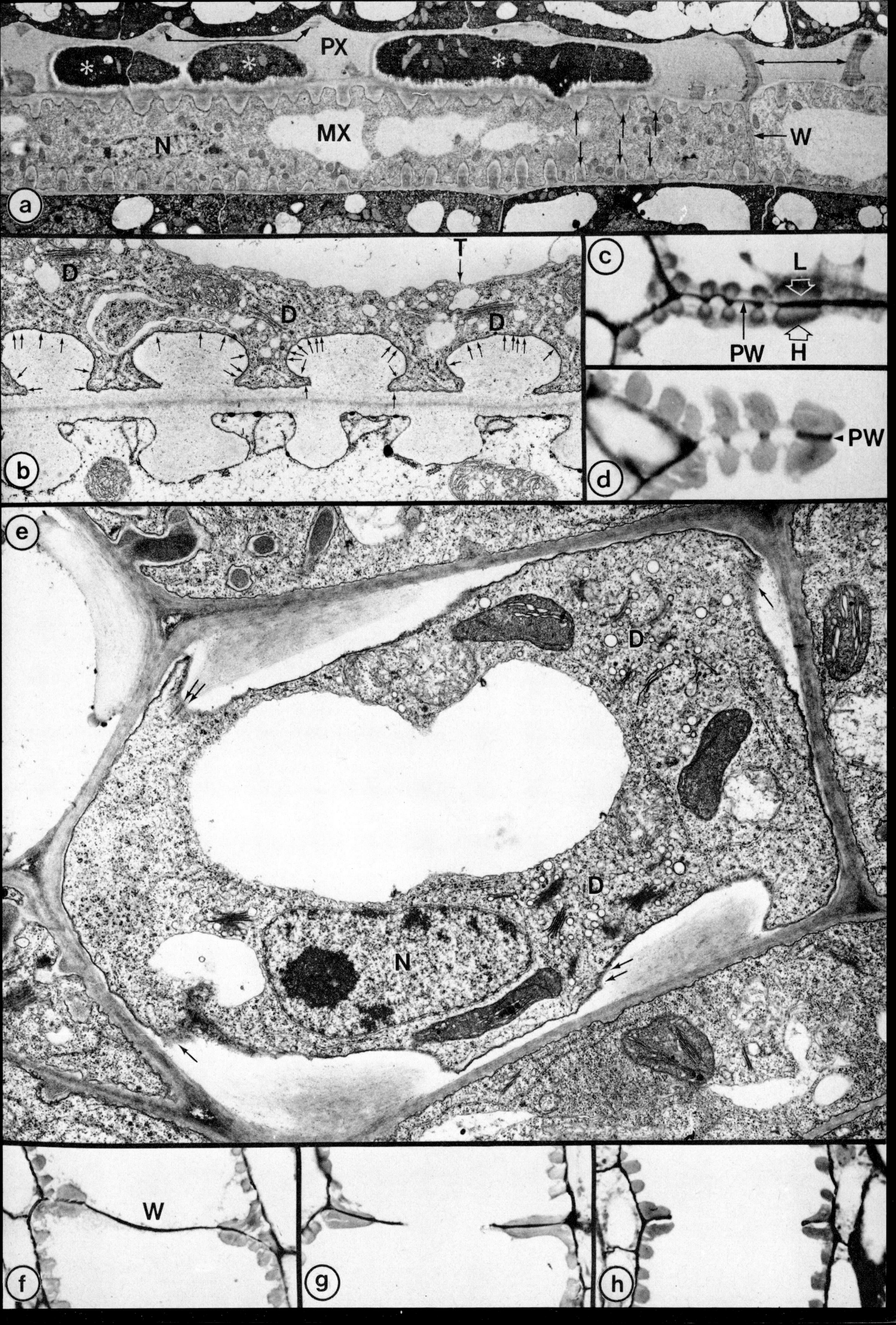

PX
MX
N
W
a
T
D
D
D
D
b
c
L
PW
H
PW
d
e
D
D
N
W
f
g
h

Plate 6

Xylem (2): Mature Xylem and Xylem Parenchyma

Plate 6a Part of a long xylem element, isolated from a lettuce leaf vein by digestion with snail gut enzymes, is viewed here by scanning electron microscopy at × 3250. Except in a few places the remains of the 19 primary wall (PW) have been digested away, and only the reticulate lignified thickenings (L) remain. Snail digestive juice contains enzymes that destroy cellulose and other wall carbohydrates.

Plate 6b In this scanning electron micrograph (× 3400) xylem elements in a pea leaf vein that was partially isolated after treatment with the enzyme pectinase are viewed (as in 6a) from the outside, though one can see into the lumen at the broken end of the topmost element (white arrow), where both outside and inside faces of the lignified thickenings are visible. The continous wall between the thickenings has suffered tearing (asterisks) during specimen preparation. It is probably composed largely of cellulosic remnants of the primary wall, and is retained here more than in 6a because of the absence of cellulose-digesting enzymes during preparation.

Plate 6c In this ultra-thin section (× 7200) of part of two xylem elements in a pea leaf vein, living xylem parenchyma cells just enter the picture at top and bottom. The lignified thickenings and the wisps of cellulose microfibrils (arrows) that interconnect them 37 are the only components to survive the auto-digestion processes of xylem maturation. These wisps presumably constitute the expanses of continuous wall seen at much lower resolution in Plate 6b. The thickenings of the upper element are much more widely spaced than those in the lower, probably indicating that it matured earlier, and became passively stretched, as in the protoxylem seen in Plate 5a.

Plate 6d This transverse section of primary xylem tissue in the stem of a fumitory (*Fumaria*) seedling includes the following features of xylem elements (X): lignified thickenings, cellulosic remnants of primary wall in exposed localities (arrowed circles) and more complete survival in areas *underlying* lignified thickenings (arrows). An important feature not previously illustrated is that living xylem parenchyma cells lie around and between the xylem elements. They are responsible for loading the xylem sap with solutes such as mineral nutrients, amino acids, and some hormones, and also for absorbing solutes from the xylem sap. Note the many cisternae of endoplasmic reticulum in the xylem parenchyma cells. Such cells can be shown to absorb amino acids from the xylem sap and to manufacture protein. Magnification × 2700.

The inserts show scanning electron micrographs: of primary xylem elements comparable to those in the section, with helical lignification (insert at lower left, × 2200); and of secondary xylem, where the lignification is much more extensive, and covers the whole wall except at lens-shaped pits (insert at upper right, 19 × 2200). (Both inserts are of castor bean (*Ricinus*) stem, and were kindly provided by Drs. J. Sprent and J. Milburn).

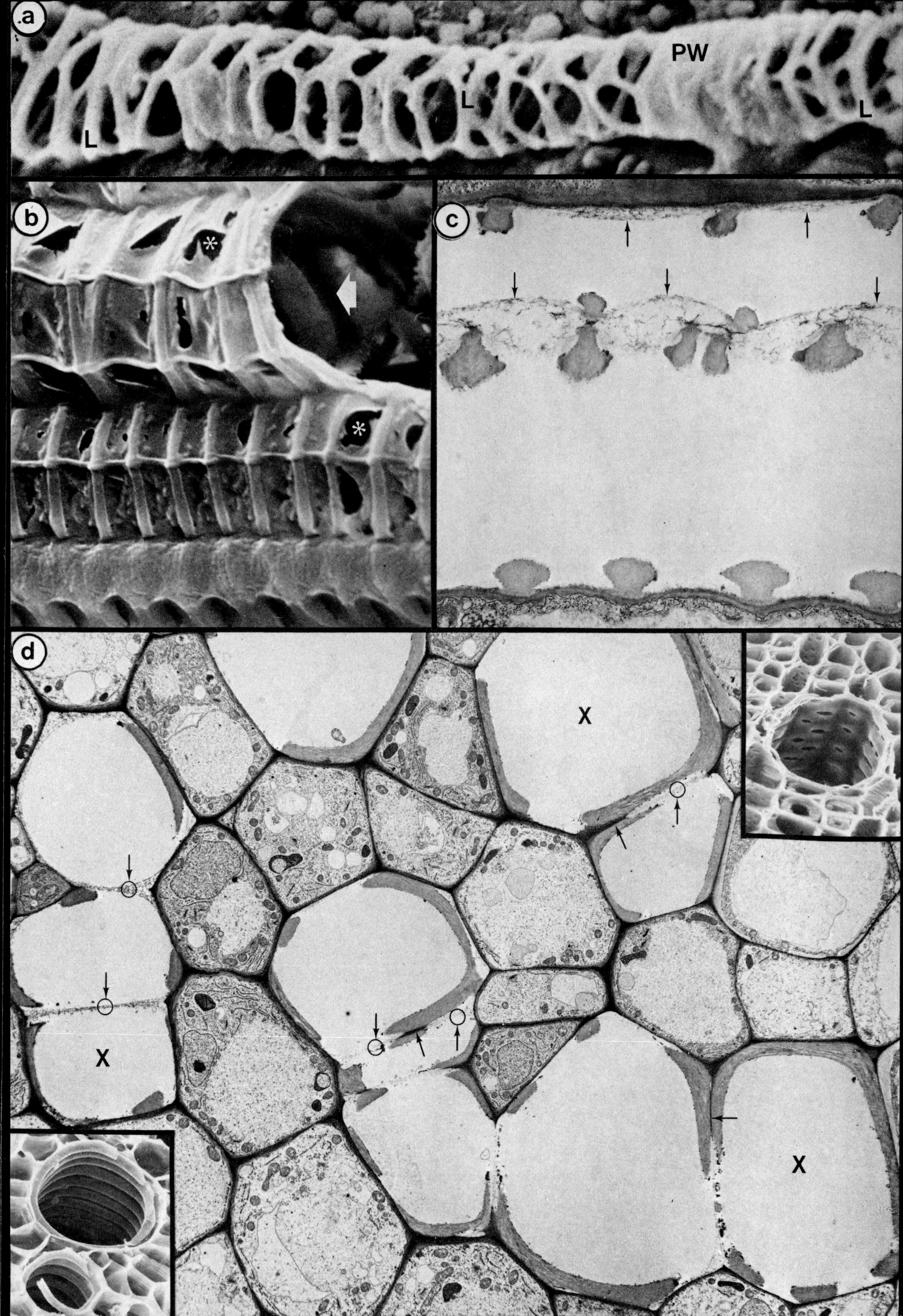
a
PW
L
L
L
b
*
*
c
d
X
X
X

Plate 7

Phloem (1): Sieve Element and Companion Cell

Phloem is the tissue responsible for long distance transport of metabolites, principally sugars, from sites of production to sites of consumption in plants. This micrograph illustrates in cross section the two principal types of cell concerned with the transport process. They are the sieve elements (lower cell) which are *apparently* empty conducting cells, and the parenchyma cells, some of which, the companion cells (upper cell) are derived from the same parental cells as the sieve elements. In flowering plants, mature sieve elements are interconnected in series to form sieve tubes (Plate 8).

20 Sieve elements mature following the breakdown of their tonoplast and much of their cytoplasm (Plate 24). The lumen of the sieve tube then contains remnant mitochondria (M) and plastids (P) and a system of protein fibres ('P-protein', arrows). In contrast to xylem cells (Plates 5 and 6) the plasma membrane of the sieve tube (PM) remains intact, in fact transport will only occur if this membrane survives. Its maintenance is one of the suggested functions of the companion cells, whose own plasma membrane is continuous with that of the sieve tubes *via* numerous plasmodesmata (PD). 27 Plasmodesmata between sieve elements and companion cells are typically compound (see also Plate 14h), and 20 incorporate linings of electron-transparent callose (C). The plasmodesmata may provide a route whereby the companion cells can function in the loading and unloading of sugars to and from the sieve elements at different points around the plant. Companion cells may also provide energy for the transport process itself (in addition to maintenance of the sieve tube plasma membrane). They are richly endowed with mitochondria.

The relative size of the two cell types differs according to the nature of the transport system. Where loading predominates over longitudinal movement, as in the minor veins of leaves, which drain photosynthates from the mesophyll tissue, the companion cells are of much greater diameter than the sieve elements. Conversely, in stems, where longitudinal movement is the major process, it is the sieve elements which are by far the larger. The example shown here is intermediate in character. It depicts part of the phloem in a vascular strand in the ovary of the grape hyacinth, *Muscari*, × 3000.

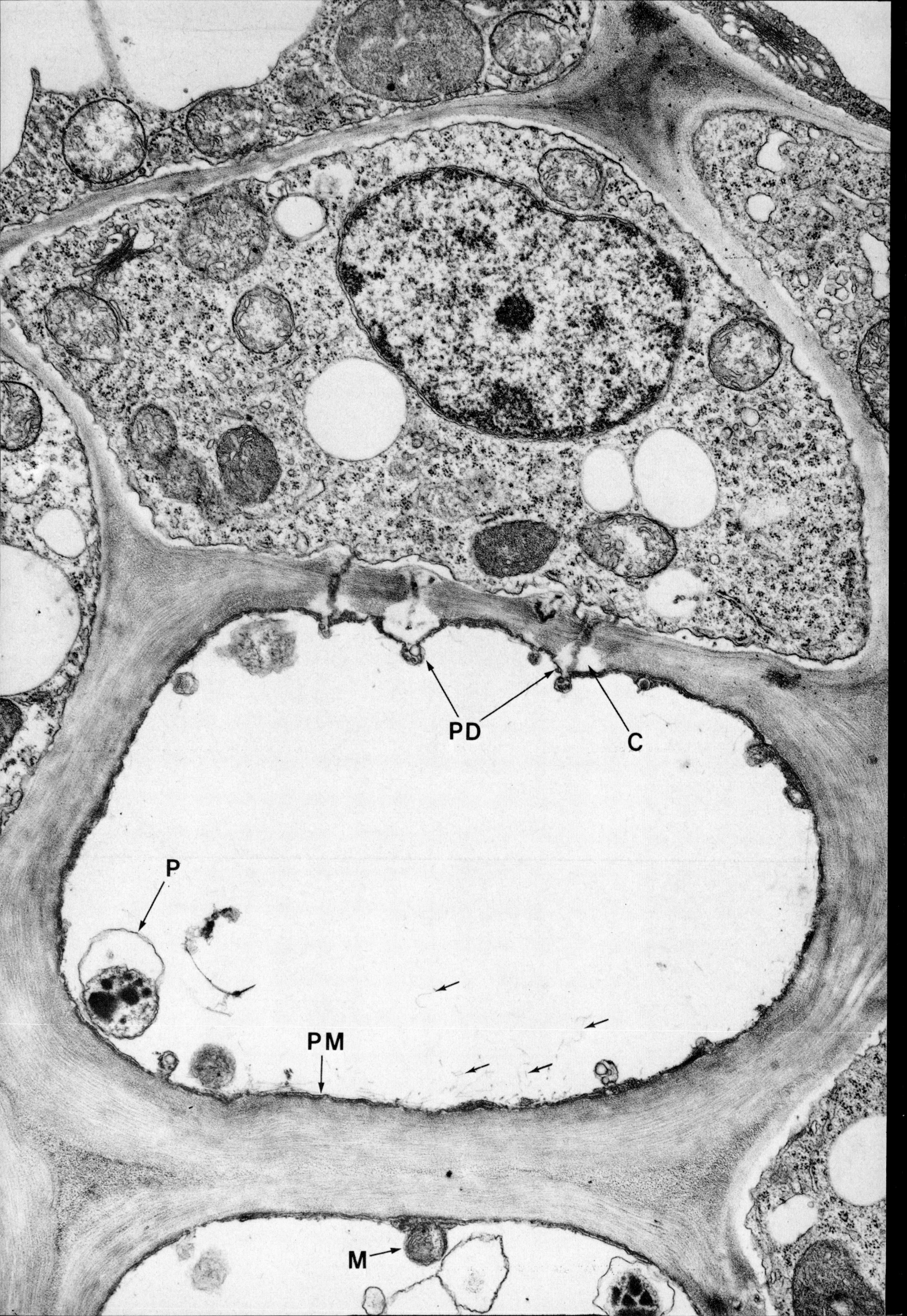
PD
C
P
PM
M

Plate 8

Phloem (2): Sieve Plates and Sieve Pores

Plate 8a In longitudinal section sieve tubes (ST) are seen to be interrupted at intervals by perforated cross walls, the sieve plates (SP), which were originally the walls between successive cells in the file of differentiating sieve elements. The pores develop from plasmodesmata 27 in the primary cross walls. In this particular tissue some of the accompanying companion cells have become specialized by the formation of wall ingrowths (see also Plate 13). Note the abundance of plasmodesmata between companion cells and sieve tubes (arrowheads). Such plasmodesmata may be the paths through which the sieve tubes are loaded or unloaded. White lupin petiole, × 3500.

Plate 8b, c and d It is possible that none of the pictures presented here is a faithful representation of the 20 sieve plate of a functional sieve tube. Sieve plates and their pores present a number of different appearances in thin sections due to changes that take place after material has been sampled from the intact plant and during the subsequent fixation period. These changes are brought about by injury-response systems, probably evolved as being advantageous in the prevention of leakage of solutes from the sieve tubes after wind or insect damage to the plant. In the condition thought to be closest to the normal state (8b) only a very small amount of callose (C, the electron-lucent ring around each pore) is present and the pores contain relatively few P-protein fibrils (P). Since sieve tubes are under considerable hydrostatic pressure, due to the osmotic effect of their sugary content, damage to the system results in a sudden release of pressure and a surge of liquid down the sieve tube. This carries masses of the P-protein fibrils (normally dispersed throughout the sieve tube) on to the sieve plates and into the pores (8c). A temporary blockage of the pores is thus effected while another reaction takes place, the formation of massive callose plugs which seal the pores (8d). Callose formation is extremely rapid and the response system is sensitive to chemical as well as to mechanical damage to the phloem. Some insects, notably aphids, do manage to circumvent these response systems, probably by creating only a small pressure drop in the parts of the sieve tubes tapped by their stylets, so they can successfully rob the plant of its food supplies. A specialized form of endoplasmic reticulum found in sieve elements is marked 66 ER in (b), (see also Plate 24c). Same tissue as in 8a. 8b, × 27 000; 8c, × 16 000; 8d, × 14 000.

Plate 8e Face view of part of a sieve plate in a similar condition to that in 8a, that is with a copious formation of electron-transparent callose pinching off the pores, which are in addition occluded by P-protein fibrils. The weft of microfibrils that delimit the pores in the sieve plate can be seen. *Coleus blumei* stem, × 19 000.

The structure of the sieve plate, and the state of its pores, are key factors in considerations of the mechanism of translocation in the phloem. At the plates, the available pathway for movement is not the lumen of the sieve tube, but the sum of the lumina of a set of very much smaller pores, each one occupied by fibrillar protein in an amount that is difficult to gauge. Longitudinal flow of sugar (etc.) along sieve tubes must be constrained more at the sieve plates than elsewhere. Are they, however, more than mere obstacles to translocation? Provision of a plugging mechanism is one likely *raison d'etre*, but some authors postulate a deeper significance, as sites at which a motive force for phloem translocation might be generated or applied to the sieve tube contents.

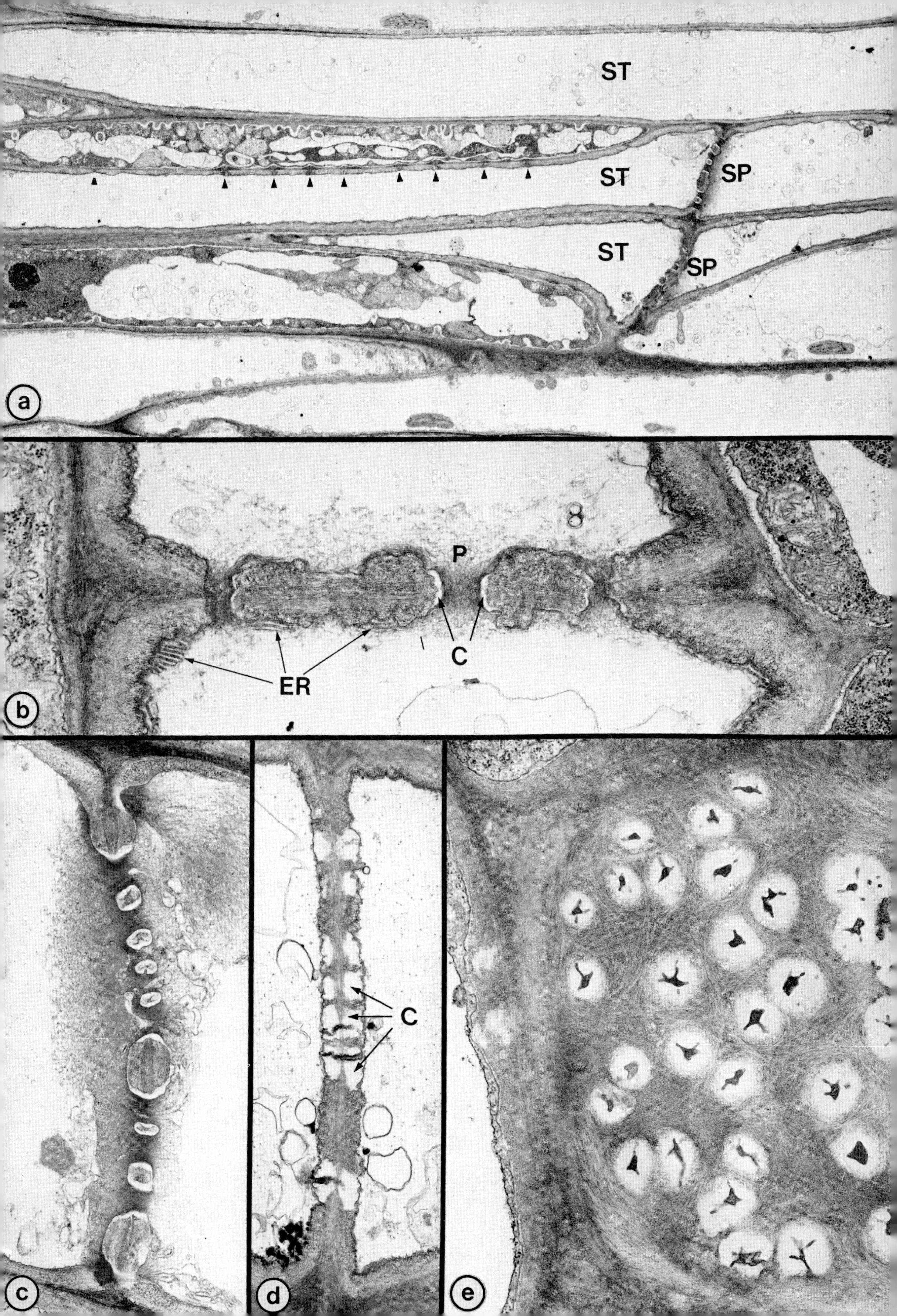
ST
ST
SP
ST
SP
a
P
C
ER
b
c
C
d
e

Plate 9

Wax and Cuticle

The outer surface of the plant is usually covered by a layer of protective material that is synthesized by the outermost cells. This plate illustrates some of the range of form exhibited by waxes and cuticles, and shows that they are not confined to the exterior of the plant, but can also develop on the surface of internal cells.

Plate 9a and b Wax formations on the lower surface of a banana leaf (scanning electron microscopy; 9b, × 320; 9a shows the central region of 9b magnified to × 1500). Curled threads of wax completely clothe the epidermal cells, except for the outer cuticularized caps of the stomatal guard cells around the slit-shaped stomatal apertures. The extruded wax threads are 25-50 μm long. (We thank Dr. P. J. Holloway and Mr. E. A. Baker of the Long Ashton Research Station, University of Bristol, for these micrographs).

21

Plate 9c and d Ridged cuticles on epidermal cells, viewed by ultra-thin sectioning (9c) and scanning electron microscopy (9d). The ultra-thin section in 9c shows the cuticle layer (C) external to the microfibrillar wall (W). The 3-dimensional shape of the ridges is better displayed in 9d, at upper right and upper left. A smooth cuticularized cover of a pair of guard cells is the major feature of 9d, with the central, slit-shaped stomatal aperture. The ability of individual cell types to regulate the form of their adcrusting layers is highlighted by comparing the specific patterns of wax distribution and cuticle morphology shown by guard cells and epidermal cells (9a-d). 9c, daffodil flower corona, × 22 000; 9d, *Auricula* sepal, × 3000.

21

Plate 9e and f This section (9e, part of a hair from the dead-nettle, *Lamium*, × 15 000) and scanning electron micrograph (9f, a hook-shaped clinging hair from the goosegrass, *Galium aparine*, × 1000) show another common type of cuticle morphology, in which the outer surface is thrown into warty projections. Note also the abundance of longitudinally oriented endoplasmic reticulum cisternae in the section.

Plate 9g The wall seen here is lining one angle of an intercellular space (IS) in the floral nectary of the broad bean (*Vicia faba*). It is heavily cuticularized. Since these cells secrete nectar, it is interesting to note the presence of very numerous branched channels leading from the microfibrillar part of the wall (W) through the cuticle layer (C) towards the intercellular space. In this nectary the secreted sugar solution emerges from intercellular spaces to the exterior *via* modified stomatal apertures. Magnification × 16 500. (Micrograph provided by R. F. Brightwell.)

22

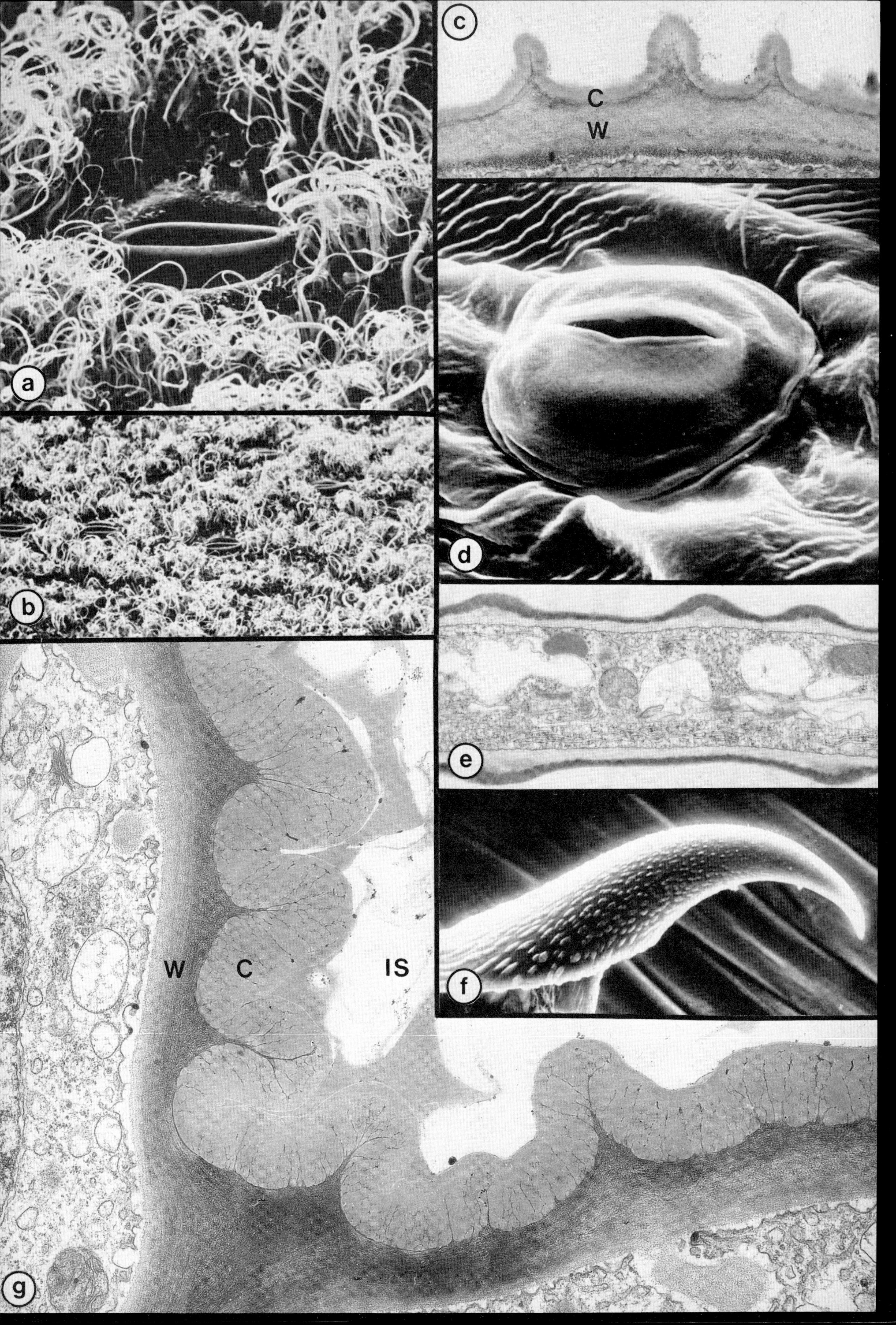
c
C
W
a
d
b
e
W
C
IS
f
g

Plate 10

A Capitate Gland

The cell wall covering gland cells and the stalks of glands often shows specialized features.

Plate 10a This gland was found on a young leaf of a dead-nettle (*Lamium*). The secretory cells surmount a stalk cell (S), which itself sits on an extension from a modified epidermal cell. Whilst variation from plant to plant in the chemical nature of substances that are secreted is reflected in corresponding variation in the degree to which *cellular* components are developed, the overall morphology seen here seems to be typical of many glands. They occur on almost all conceivable surfaces of above ground parts of plants. Another example is shown in Plate 23.

Raw materials for the secretion enter the glands *via* 29 the numerous plasmodesmata on the walls of the subtending cell and the stalk cell (arrows). The product passes through the inner, microfibrillar zone of the cell wall of the secretory cells (W), and it is common to find that the cuticle (C) is detached from this layer (asterisks). Presumably the product accumulates in the sub-cuticular space prior to its final exit, which may be 22 through specifically located pores (not seen here) or a more general but slower percolation. × 4800.

Plate 10b The cuticle over the gland is continuous with that (CE in 10a) of the epidermal cells. A specially 21 impregnated zone in the side walls of the stalk cell (stars in 10a) prevents leakage of the secretory product through the microfibrillar layer of wall back down to the subtending epidermis. This wall region is shown at higher magnification in 10b. Whether the impregnation is of cutin or suberin is open to question. Two groups of microtubules (arrows) lie at the extremities of the impregnated zone—an observation which raises the question of whether they might, earlier in the development of the gland, have had anything to do with specifying the distribution of whatever cytoplasmic components were responsible for synthesizing and/or depositing the impregnation. At any event, it seems clear that in this region there is no *non*-impregnated microfibrillar wall outside the plasma membrane (PM), through which secreted products might leak, or through which water might escape and evaporate from the leaf to the exterior. × 30 000.

Plate 10c This micrograph shows details of part of the surface of one of the secretory cells, including the cuticle and the underlying microfibrillar layer, with the space between the two. The plasma membrane (PM) undulates, and, as is common in secretory and absorptive cells, cisternae of endoplasmic reticulum (ER) approach very close to it (arrows). This close 65 juxtaposition may well be related to the transport of solutes between internal compartments of the cell and the extra-cytoplasmic spaces. In other gland cells, marked changes in the disposition of endoplasmic reticulum cisternae occur when secretion is initiated or ceases. Dictyosomes (D) are also present. × 70 000.

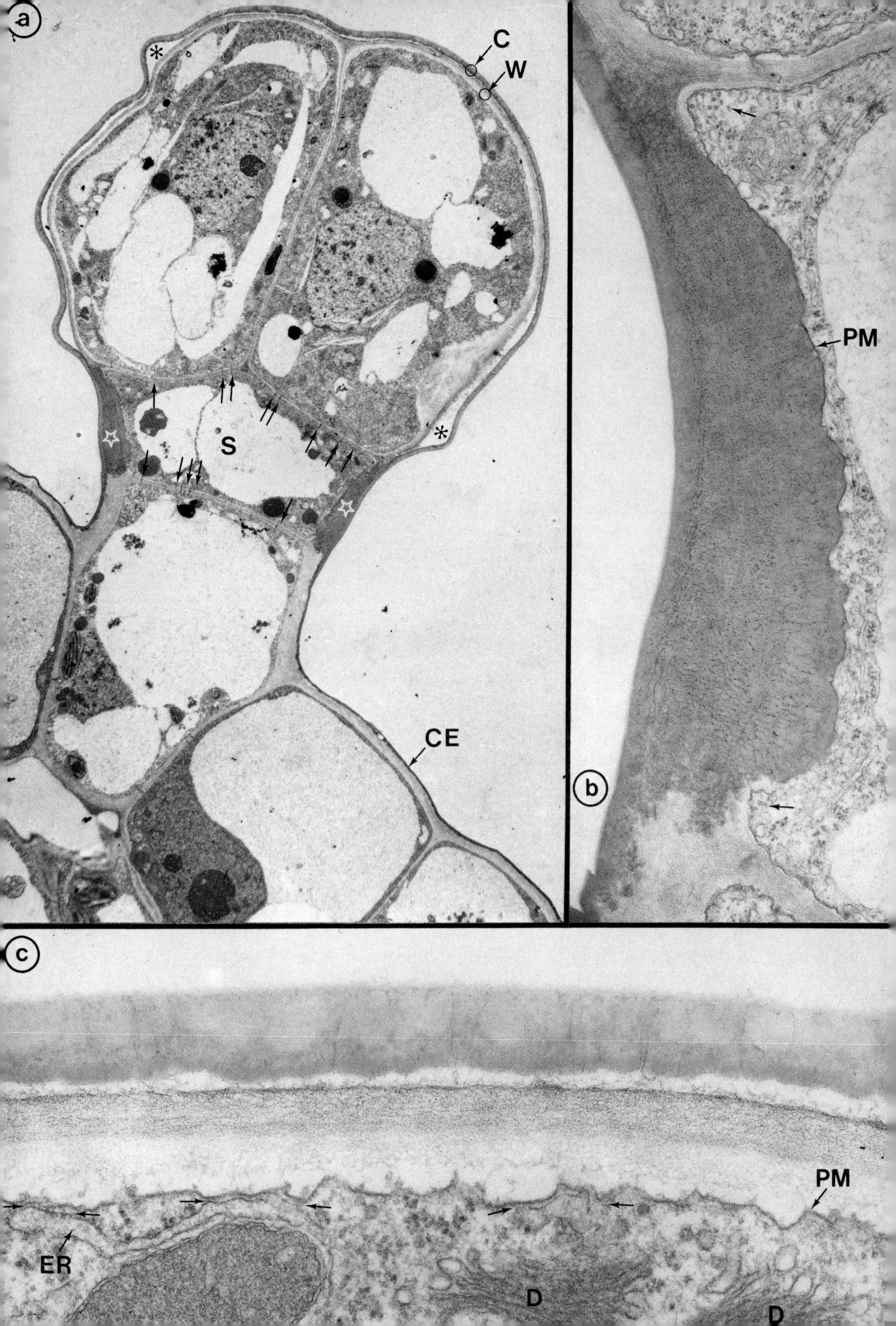
a
C
W
S
CE
b
PM
c
PM
ER
D
D

Plate 11

Pollen Grains (1): Developmental Stages

Plate 11a In a very early stage of pollen development, the pollen mother cells of the sporogenous tissue of the anther become isolated by deposits of callose (C) so massive that the primary wall is obliterated, apart from the middle lamella (circles). As in sieve plates and plasmodesmata (Plates 7, 8, 14h) the callose appears virtually structureless and electron-transparent. Initially the pollen mother cells are interconnected by cytoplasmic bridges (PD*), facilitating rapid distribution of nutrients and synchronizing the development of the cells. Remnants of true plasmodesmata (PD) are also present. Both types of intercellular connection are eventually sealed by callose deposition. × 22 000.

Plate 11b and c As in 11a, these micrographs show developing pollen of the cultivated oat (*Avena sativa*), but at a more advanced stage of development in which the callose walls have been digested away and the mature walls laid down. 11b is an enlarged (× 16 500) detail from 11c (× 2800). In 11c, maturing pollen grains are seen in the anther loculus (L), the wall of which consists of three layers of cells (stars) (the innermost layer partially crushed). The loculus is lined by a layer of nutritive cells, the tapetum (T) (considered in detail in Plate 22).

The tapetal cells have almost lost their primary walls, leaving their plasma membrane exposed (PM in 11b). Their nuclei are typical of many cells of very specialized function in that the bulk of the chromatin is extremely condensed, appearing electron dense. It is thought that the remaining dispersed chromatin controls the specialized functions of the cells, which in this case primarily concern the production of nutrients for pollen grain development, and the production of a carpet of orbicules. These irregularly spherical bodies (arrows, 11b) are seen lining the anther loculus on the tapetal cells. They become coated with the very resistant compound sporopollenin. Superficially they resemble units of the pollen grain wall, but there is evidence that the latter is manufactured *in situ*, and *not* by transfer of orbicules from the tapetum as 'building blocks'.

The pollen grain wall is made up of a thin layer called the intine (I), of microfibrils and matrix, together with the exine (E), composed largely of sporopollenin. The exine is subdivided into an inner platform, the nexine (N) from which rods, or bacula (B), extend like columns. A roof, or tectum (T) surmounts the bacula. Oat pollen grains, like various others, have a single, circular pore (see 11b), where the exine bulges in a rim around a separate lid, or operculum (O), of sporopollenin sitting on a thickened pad of intine. This pad in particular, and the intine in general, are major sites of the hayfever antigens that diffuse rapidly out of moistened, mature pollen grains. The pore is also the site of pollen tube emergence during germination of the grain, and it may also function in the water economy of the grain—opening in moist conditions and remaining tightly shut in dry conditions.

At the stage illustrated the pollen grains are vacuolated and have not yet accumulated food reserves. They have not long been released from the tetrads formed at meiosis. Their irregular shape may be an artefact. The pore arises on that face of each grain that lies outermost in the tetrad—a morphogenetic feature that recurs in a somewhat different form in Plate 12a.

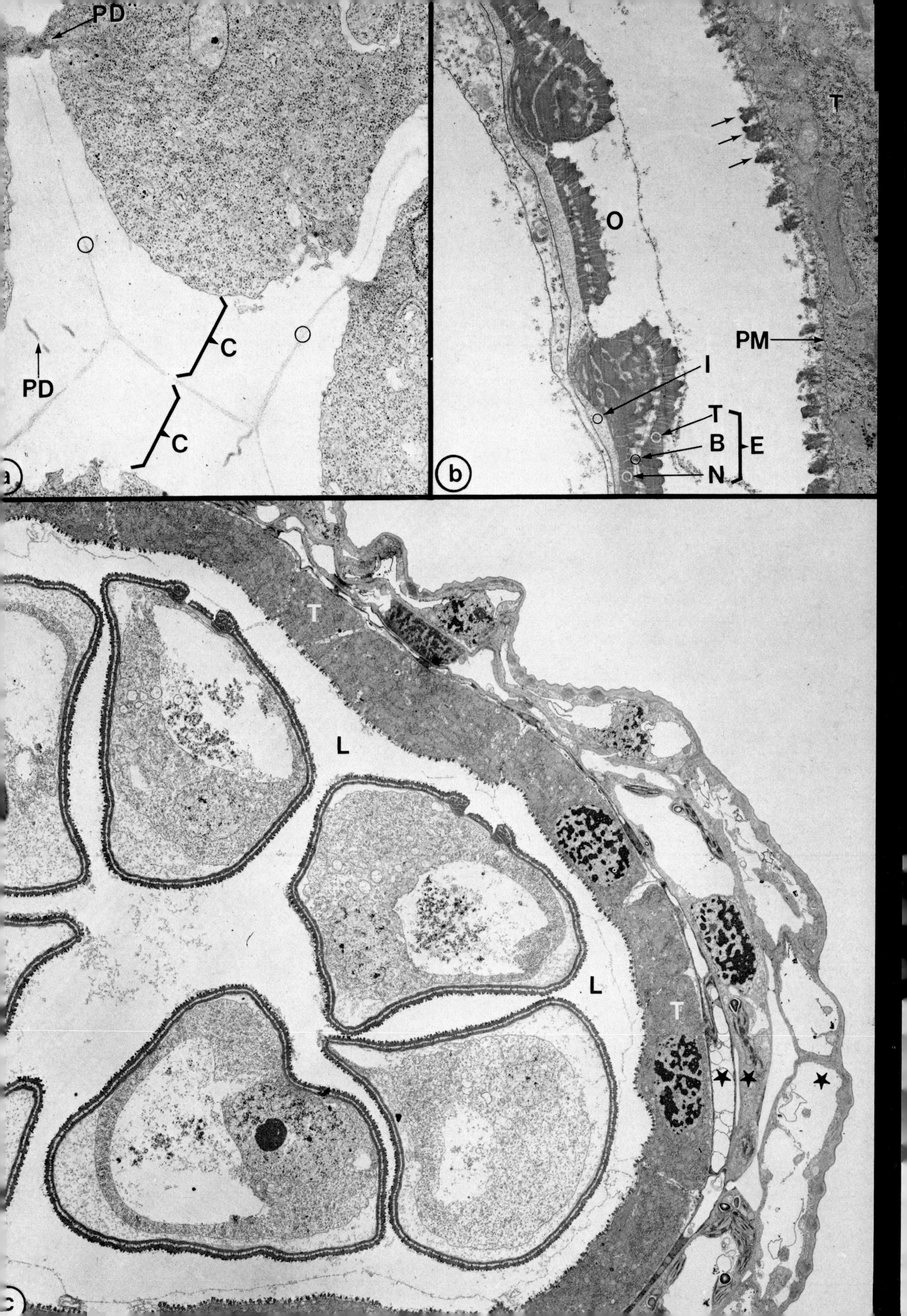

PD
PD
C
C
a
O
T
PM
I
T
B
N
E
b
T
T
L
L
c

Plate 12

Pollen Grains (2): The Mature Wall

The scanning electron microscope is an ideal instrument for examining the suface topography of pollen grains. The examples shown here are from lily (*Lilium longiflorum*).

Plate 12a Pollen grains are seen lying in an anther loculus at the time of dehiscence. The remains of the tapetum and the loculus wall lie along the bottom of the picture. The organization of the pollen grain wall is determined at a very early stage of development. In lily pollen there are two distinct regions of wall. The smooth, unpatterned area, the colpus, forms on the outward-facing wall of each grain in a tetrad. The remainder of the wall is sculptured in a polygonal pattern, described further below. The colpus seems to be the spatial and functional equivalent of the pore seen in oat pollen in Plate 11. Its exine is relatively thin, and several examples showing its inward collapse are included. There is also a very small and shrivelled sterile grain. × 640.

66

Plate 12b This view shows part of a pollen grain (upper half of micrograph) lying on a carpet of orbicules (more or less spherical bodies, coated with sporopollenin, seen in the lower half of the micrograph). The layer of orbicules may aid the dispersal of the pollen when the anther loculus matures and splits open. Their appearance in ultra-thin section is shown in Plates 11b, c and 22. × 4500.

22

Plate 12c The architecture of the patterned part of the pollen grain wall is seen here at × 13 500. The major difference between lily and oat (Plate 11) exines is that, in the former, the columnar bacula are not crowded all over the surface, but are restricted to the sides of irregular polygons. The tectum, or 'roofing layer', on top of the bacula, of necessity lies in the same polygonal pattern. In the central regions of the polygonal areas one can see down to the relatively smooth outer surface of the nexine. It is remarkable that whilst the position of the colpus is determined in relation to the cell axes in the tetrad, the genes that govern the detailed pattern of bacula and tectum exert their influence earlier, in the pollen mother cell, that is at a stage similar to that in Plate 11a, prior to the appearance of either intine or exine. Once the genes have been transcribed, the information they contain is present in the cytoplasm, ready to be utilized at the appropriate time during pollen maturation. Thus experimentally produced fragments of cytoplasm from mother cells or young pollen will develop the patterned exine even if nuclei are absent.

66

The architecture shown in these micrographs is characteristic of lily pollen, and similarly, pollen grains of other species develop their own specific shapes and patterns. Some of the wide range of possibilities should be evident from the differences observable between oat (Plate 11) and lily (Plate 12).

(We thank Drs. J. Heslop-Harrison and H. Dickinson for these micrographs, reproduced by permission from: 12a, *27th Symposium Soc. Dev. Biol. Suppl.* 2, 1968, Academic Press; 12b, *Planta*, **84**, 199–214, 1969, Springer-Verlag; 12c, *Society for Experimental Biology Symposium*, **25**, 1971).

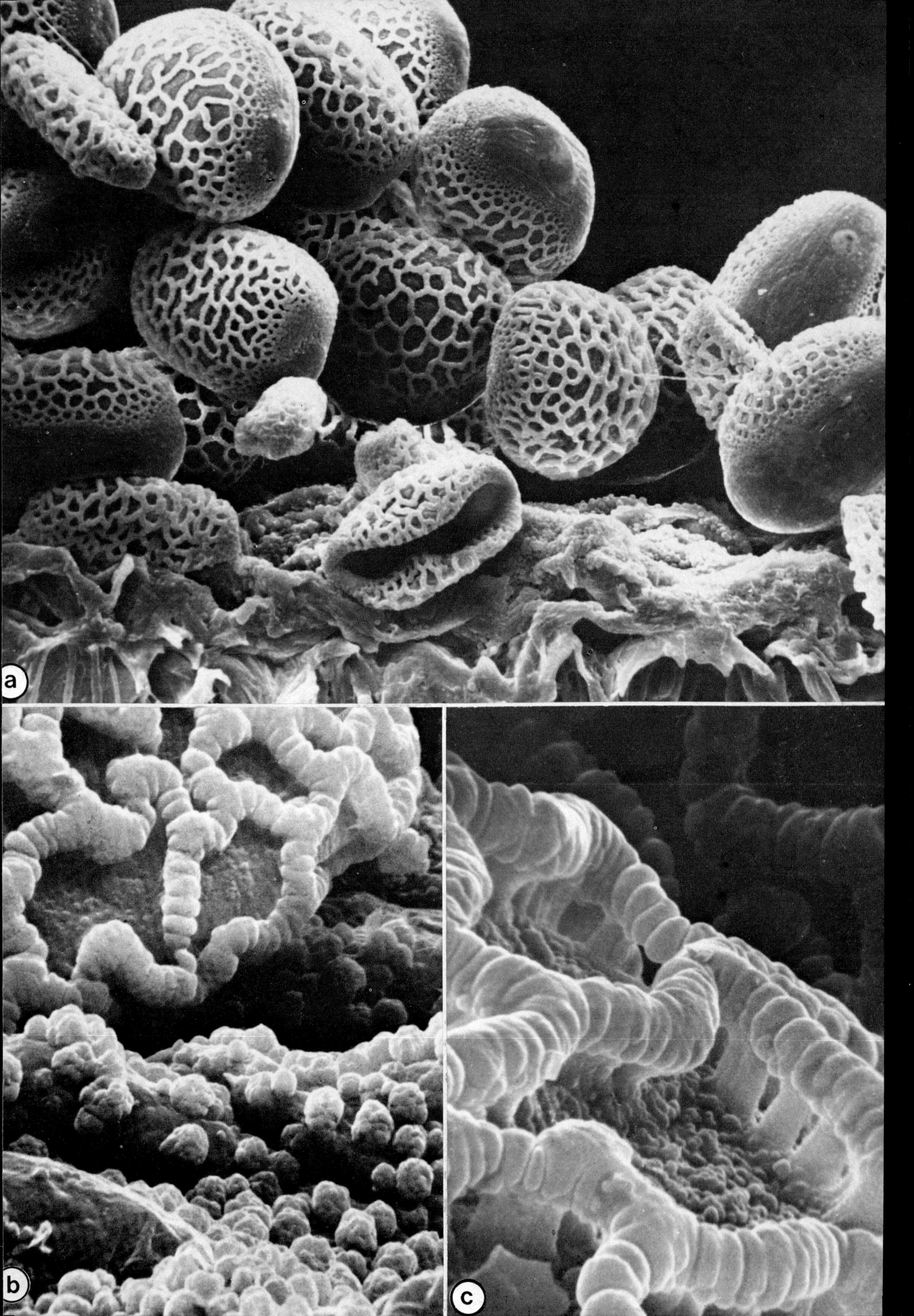
a
b
c

Plate 13

Transfer Cells

In transfer cells, the surface area of the plasma membrane is augmented by the development of ingrowths of wall material which protrude into the living contents of the cell. 13a illustrates their appearance in the light microscope and 13b in the electron microscope. The functional significance is very different in the two cases selected for inclusion in this Plate.

Plate 13a Transverse sections of vascular bundles in the root nodules of nitrogen-fixing leguminous plants show that each bundle is surrounded by a layer of endodermal cells, identified in the picture by the [31] circlet of arrowheads pointing to the Casparian strips (see also Plate 16). Within the endodermis the pericycle cells nearly all have a fuzz of darkly-stained cell wall ingrowths (e.g. circles). Ingrowths are not prominent near the phloem (P) but do back on to the xylem (X). There is a complex two-way traffic of solutes through these cells. Sucrose enters the nodule *via* the phloem and passes through plasmodesmata to the cells (not shown) where nitrogen-fixation occurs. It is utilized in respiration and to provide carbon skeletons for the manufacture of nitrogenous amino acids and amides. The latter diffuse through plasmodesmata back to the vascular bundles. Within the endodermis the pericycle transfer cells are thought to secrete the nitrogenous solutes into the extra-cytoplasmic spaces, mainly cell walls, from where they diffuse into the xylem elements, [31] and are flushed out of the nodule towards the root to which it is attached, by osmotic uptake of water. Sweet pea (*Lathyrus*) nodule, × 600.

The legume nodule is, in effect, a gland with an internal duct (the xylem). The secretory activities are [30] carried out by the pericycle transfer cells, presumably aided by the large surface area of plasma membrane that they possess.

Plate 13b In this example, the transfer cells are xylem parenchyma cells modified by the development of wall ingrowths. They surround two xylem elements in a vascular bundle, sectioned just at the level at which it leaves the stem and passes out into a leaf. Such 'leaf traces' often display spectacular arrays of transfer cells. By using radioactive tracers it has been found that they can absorb solutes from the xylem sap, but it may be the case that they can also load it with solutes. It appears that departing leaf traces are for some reason especially active zones for these exchanges between the xylem and the surrounding living cells. Parts of the xylem other than at departing leaf traces (e.g. Plate 6d) have xylem parenchyma cells unmodified as transfer cells.

The way in which the wall ingrowths enhance the surface area of the plasma membrane (PM) is clearly visible. The ingrowths can branch and anastomose. Because they are finger-like projections from the wall and bend in all directions, they often appear in the section as apparently isolated profiles. Transfer cells of all types characteristically have many mitochondria (e.g. M in cell at upper right). [30]

The two xylem elements themselves display several of the features also seen in Plate 6—microfibrillar remnants of the primary wall (arrowed circle), survival [19] of microfibrils plus matrix in zones protected by lignin (arrows), and lignified thickenings (seen both in transverse and longitudinal section here because the xylem elements have been sectioned obliquely). *Galium aparine* seedling, × 12 000.

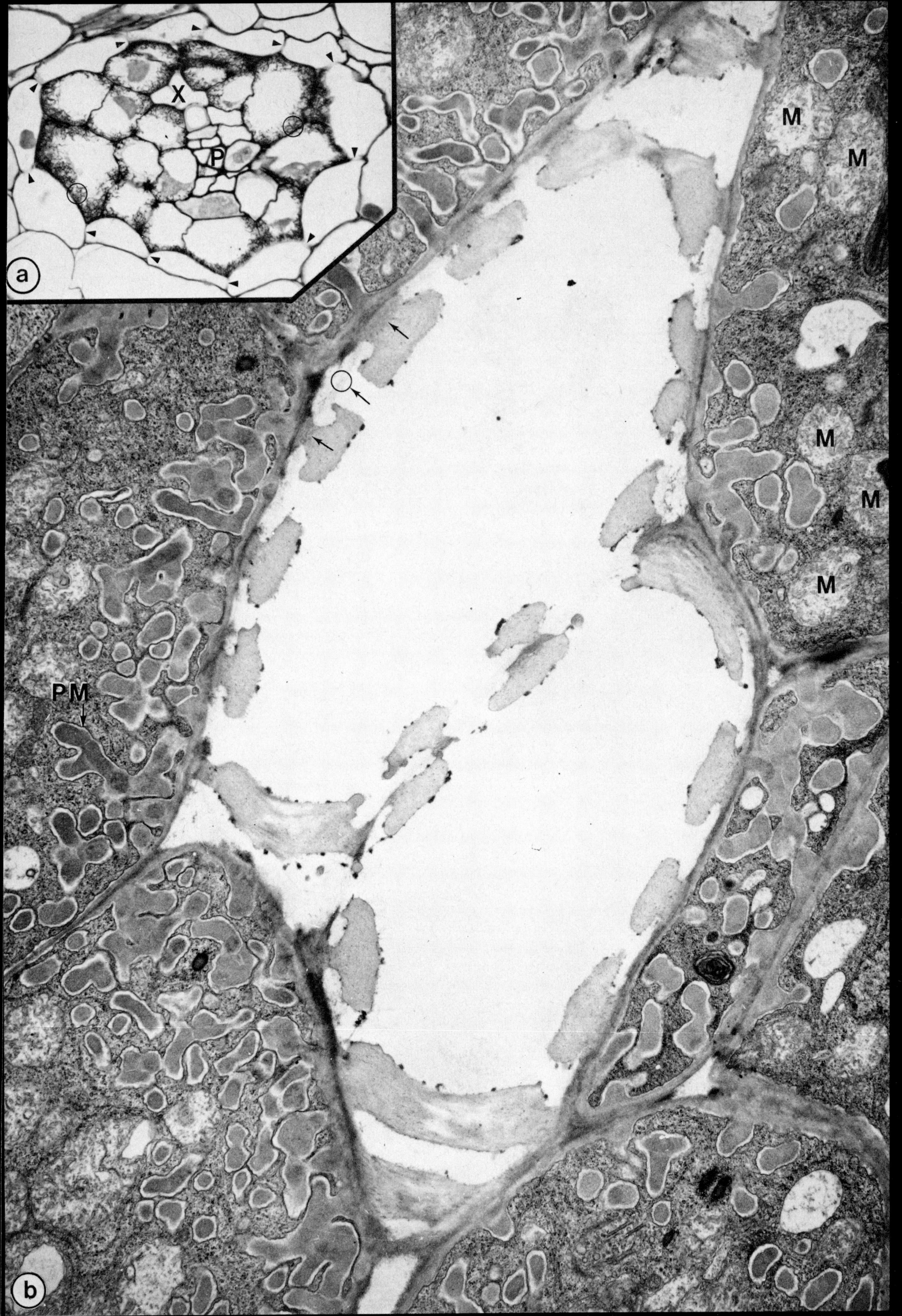

a
X
P
M
M
M
M
M
PM
b

Plate 14

Plasmodesmata

This collection of micrographs illustrates several aspects of plasmodesmatal ultrastructure. Plasmodesmatal canals are seen 'end on' (14b, d) as well as in side views (14a, c, f-h). When interpreting the latter it is well to remember that the diameter of the canals is not very different from the thickness of the sections. Also the canals can pass out of the thickness of the section and so apparently disappear. Their appearance depends greatly on whether they lie symmetrically *within* a section or whether the knife edge slices through them so that parts of an individual are present in two adjacent sections.

Plate 14a Three plasmodesmata are shown here piercing the wall between two meristematic cells in a cabbage root. Small arrowheads follow the path of the plasma membrane from cell to cell in the lowermost example. In each plasmodesma there is a narrow axial strand (large arrowheads) almost certainly continuous with cytoplasmic cisternae of endoplasmic reticulum (followed by the small arrowheads in the topmost example). Such strands have been called desmotubules. × 45 000.

Plate 14b End-on views of very numerous plasmodesmata in a section that grazes a curved cell wall in the stele of an *Azolla* root meristem. The magnification is × 12 000, so a convenient way of counting the number of plasmodesmata per unit area is to cut a 12 × 6 mm window in a piece of paper, lay it over different areas, and obtain an average count. The true area of the window is 1.0 × 0.5 μm. The same method can be applied to the nuclear envelope pores (e.g. white arrows). The plasmodesmata tend to occur in rows (e.g. the row of five between the arrowheads). N, nucleus.

Plate 14c, d and e Plasmodesmata are often grouped in primary pit fields.

In 14c (oat leaf, × 31 000) the group is at a point of contact between mesophyll cells; 14d (pea leaf, × 66 000) is a similar situation, seen in oblique/surface view; 14e (root tip of maize, × 16 000) is a shadow-cast preparation showing the cellulose microfibrils delimiting the pit area and the individual plasmodesmatal pores. Microfibril patterns can also be discerned in 14d, an image which can be compared with that in Plate 8e.

14c and d both show the presence of axial desmotubules (arrows) in the plasmodesmatal lumina. In 14c the plasma membrane is seen to be closely constricted around the desmotubules at the cytoplasmic extremities of the plasmodesmata (e.g. circles). Note also the strange convolutions (arrowheads) in the central part of these plasmodesmata.

Plate 14f, g and h These three micrographs illustrate compound plasmodesmata, in which *several* canals meet in the interior of the wall. In 14f (between two transfer cells in the phloem parenchyma of a lupin leaf vein, × 35 000) many canals radiate in both directions from the centre of the wall. Because they radiate at a variety of angles, serial sections would be needed to demonstrate continuity of cytoplasm between the two cells. In 14g (vascular parenchyma of *Polemonium* stem, × 95 000) a side passage interconnects two plasmodesmata. The dark-light-dark configuration of the plasma membrane is shown to be retained throughout. In 14h the upper, empty-looking cell is a sieve element and the lower its companion cell. Many plasmodesmata funnel from the latter towards fewer canals leading into the former. On the sieve element side the canals are lined with electron-lucent callose (*Lupinus*, × 35 000). In all three examples the wall is swollen at the site of the compound plasmodesmata. These swellings can easily be seen with the light microscope, especially those on the walls of sieve elements.

(We thank Dr. K. Muhlethaler and Elsevier Publishing Co. for permission to reproduce Plate 14e from *Ultrastructural Plant Cytology*, 1965).

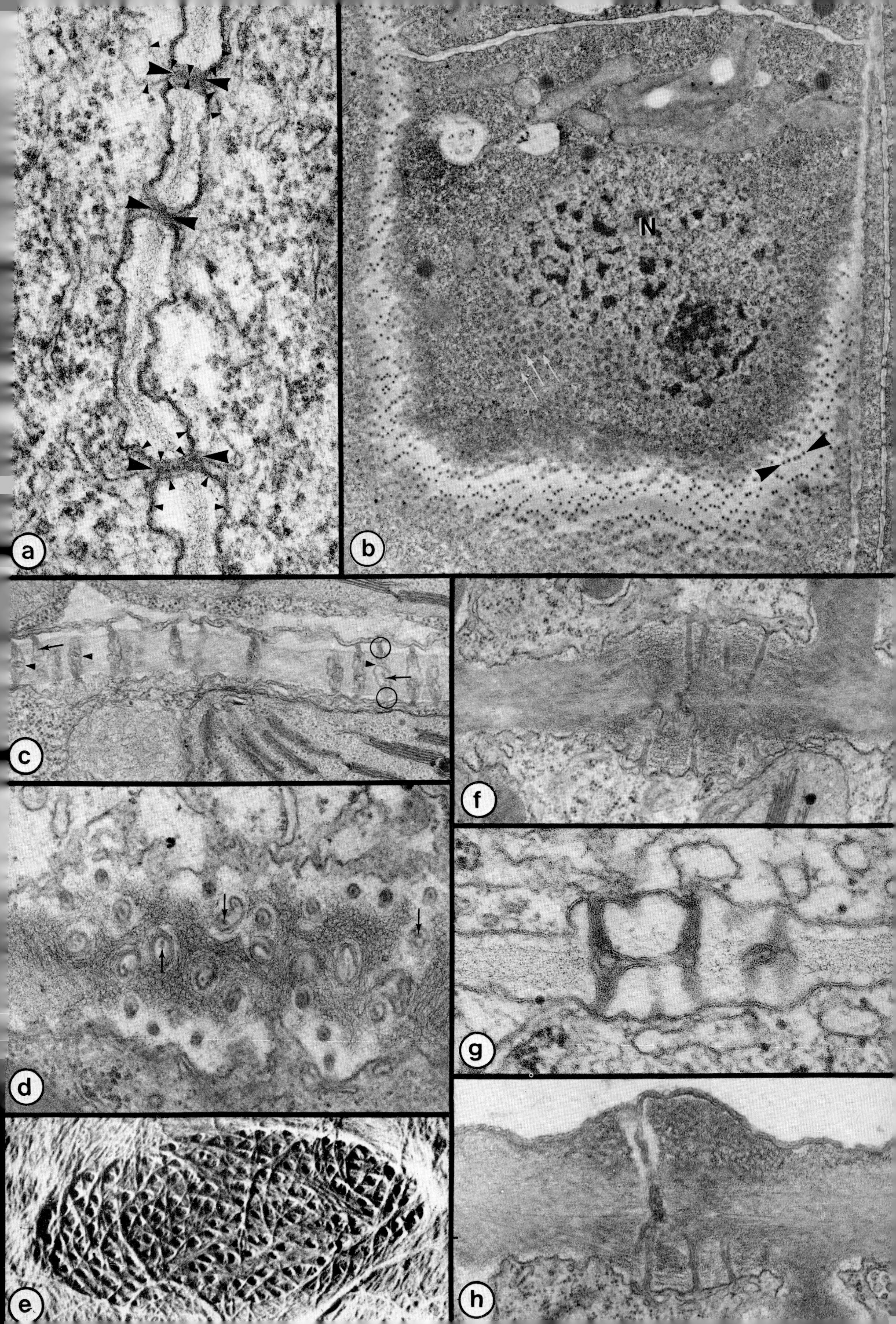
a
b
N
c
d
e
f
g
h

Plate 15

Pits

This collection of micrographs depicts two very different types of pit connection between cells that have massive secondary walls. One type (15a, b) is 'simple', and in fibres that remain alive for a considerable period; the other (15c–e) is 'bordered', and in tracheids that die at maturity. The former functions in movement of substances between living cells, the latter in movement of xylem sap in wood.

Plate 15a and b The two fibres (top and bottom halves of the picture) of 15a are still relatively young. The close approach of the plasma membrane to the middle lamella (ML) in the pit itself shows how thin the primary walls of the two neighbours originally were. 18 Subsequently the cells deposited very thick secondary walls (S) everywhere *except* in the immediate vicinity of the plasmodesmatal connections (PD) in the pit regions. The secondary walls are microfibrillar at this stage, and not yet impregnated with lignin. The more they become lignified, the more important the symplastic connections are if the cells are to remain alive. A full complement of cytoplasmic components is present—chloroplast (C), mitochondrion (M), dictyosomes and associated vesicles (D), and endoplasmic reticulum (ER).

The layer of secondary wall does not overarch the pit, as shown both in the section and in the scanning electron microscope view (15b). The pit is therefore described as *simple*. Plate 15a, pea leaf, × 48 000; Plate 15b, oat leaf, × 7000.

Plate 15c, d and e These scanning electron micrographs show structures in the side walls of xylem tracheids in a piece of pine wood. The wood was split open longitudinally to expose the insides of the cells to view. The pits occur in rows, and occupy a large fraction of the side walls. The diameter of the tracheid in Plate 15c is approximately the distance between the two black sloping lines. Three pits are included in this image. That on the left has retained its cowl, or border, of secondary wall; by chance the other two have had the border split off. Structural components lying within the border are seen at higher magnification in 15e. When they in turn are removed the inner face of the border on the opposite side of the pit is revealed (15d). Thus these bordered pits consist of two opposed cowls, with a central *pit membrane* between them.

Whereas the pit 'membrane' (it is a wall structure and not a true cell membrane) is at the level of the middle lamella and primary wall, the overarching borders are composed of heavily lignified secondary wall. The pit membrane consists of a lens shaped disc of lignified wall material (the *torus* (T)) suspended in the cavity between the pair of borders on bundles of microfibrils largely arranged like the spokes of a wheel (MF). The diamter of the torus is somewhat greater than that of the apertures of the borders. It can move on its slings of microfibrils so that it is either free in the centre, or it can be pushed sideways to lie snugly against one or other of the apertures. In other words it acts as a valve. When the interconnected tracheids are both full of columns of sap ascending the tree the valve will be in the central, open position. Lateral transfer through the large gaps in the microfibrillar sling is possible. If the negative tensions that 'pull' the sap up the tree lead to cavitation and air bubble formation in one tracheid, the change in pressure will cause the valve to swing across and seal off the damaged element.

Formation of the pit membrane (torus plus sling) during the maturation of tracheids involves phenomena similar to those described for maturation of primary xylem elements. The primary wall is hydrolysed *except* where it is protected by lignin, to leave only matrix- 19 free cellulose—in this case the microfibrils of the sling. The warty excrescences seen through the sling in 15e and in full view in 15d also arise in the final stages of maturation. They may derive from remnants of cytoplasm that come to lie scattered over the inner face of the wall.

Plate 15c, × 1500; 15d, × 3400; 15e, × 7600.

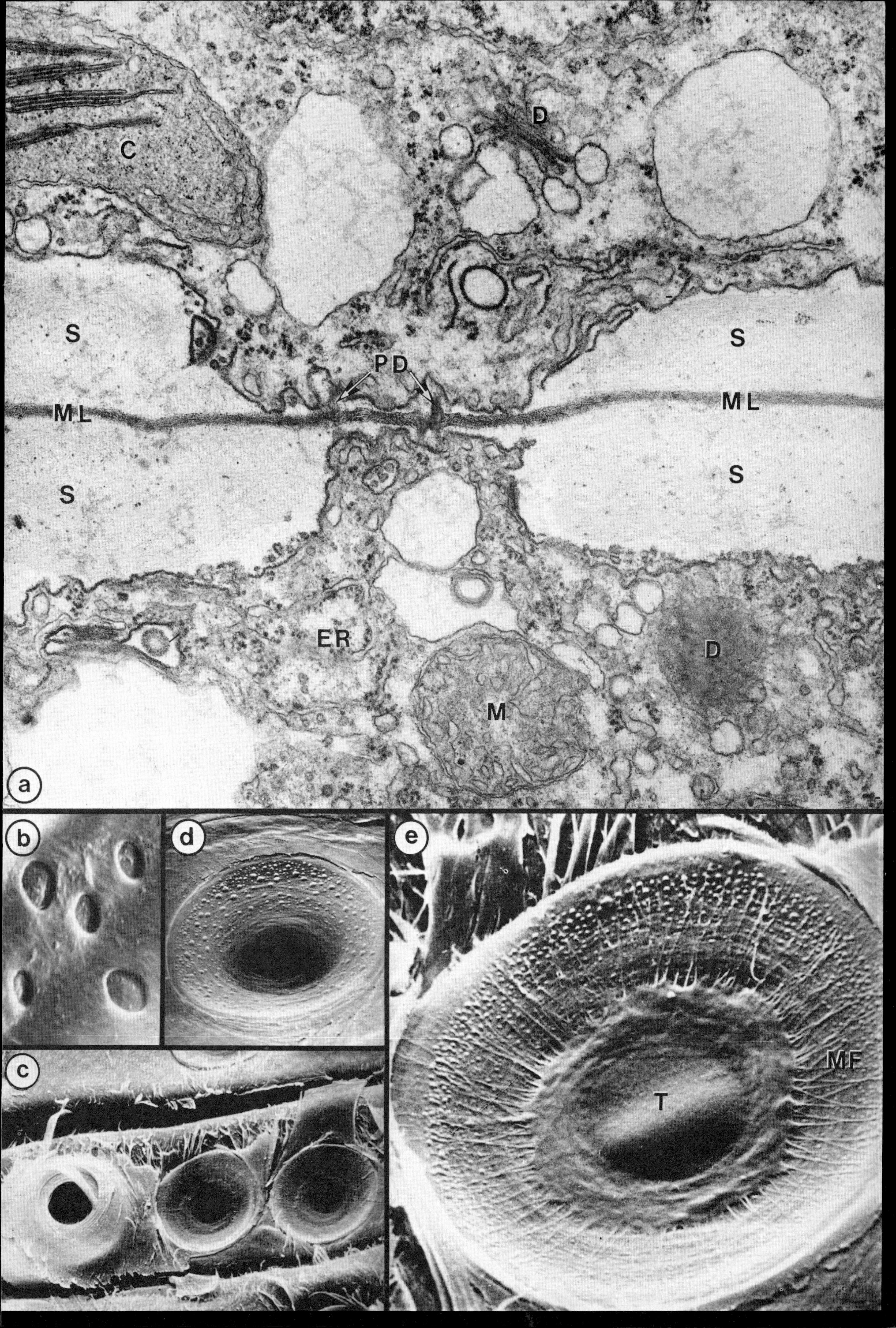

C
D
S
S
PD
ML
ML
S
S
ER
D
M
a
b
d
e
c
MF
T

Plate 16

Endodermis and Casparian Strip

The location of the endodermal cell layer in a root is shown in a light micrograph, and the two electron micrographs then focus on the structure of the Casparian strip in the radial walls of the endodermal cells.

Plate 16a Transverse section of the central stele (vacuolar tissue) of an *Azolla* root. The cell types are labelled so that reference can be made to Plate 49, which shows the same types in the same spatial arrangement in an immature part of the root. E, endodermal cell; P, pericycle cell; S, sieve element; PP, phloem parenchyma; PX, protoxylem; MX, metaxylem. Arrowheads point to the Casparian strips, which were differentially stained by the technique used. The lowermost is the most clearcut example. × 1700. 31

Plate 16b The radial wall separating two endodermal cells (E) runs across the bottom of this micrograph. Part of a cortex cell is seen at the left, and part of a pericycle transfer cell at the right. The material is the same as that viewed in Plate 13a (a vascular bundle in a legume root nodule), indeed the orientation of the cells is much as at the arrowhead on the extreme left of that picture.

The Casparian strip lies between the open arrows. Even at this low magnification it can be seen to have a more homogeneous texture than the neighbouring wall. For instance the middle lamella (ML) almost disappears in the strip. Notice too that the undulations of the plasma membrane become smoothed out at the Casparian strip. × 15 000.

Plate 16c A portion of Casparian strip (top) is here contrasted with a portion of normal cell wall (bottom) that happened to lie in the same field of view. The smooth texture of the strip is presumably due to impregnation of all of the spaces between the microfibrils with cutin or suberin. The plasma membrane (PM) shows the dark-light-dark appearance on both walls, but very conspicuously so at the Casparian strip because it is unusually flat, and, in this case, lies at right angles to the plane of the section. It has been found that if endodermal cells are plasmolysed, the plasma membrane sticks tenaciously to the impregnated Casparian strip. The membrane can even rupture down its centre line, the cytoplasmic face pulling away with the cytoplasm and the wall face remaining attached to the strip—much as when frozen tissue is fractured as part of the freeze-etching process, where again membranes tend to rupture along their mid plane. 23

Other membranes are present. The narrow compartment across the middle of the picture is a flattened vacuole (V), bounded by its tonoplast (T). The thickness of the tonoplast more or less matches that of the plasma membrane, and by contrast the membrane of the endoplasmic reticulum (ER) is seen to be much thinner—its 'tramline' configuration is only just detectable (circle). Material as in 16b. × 120 000. 33

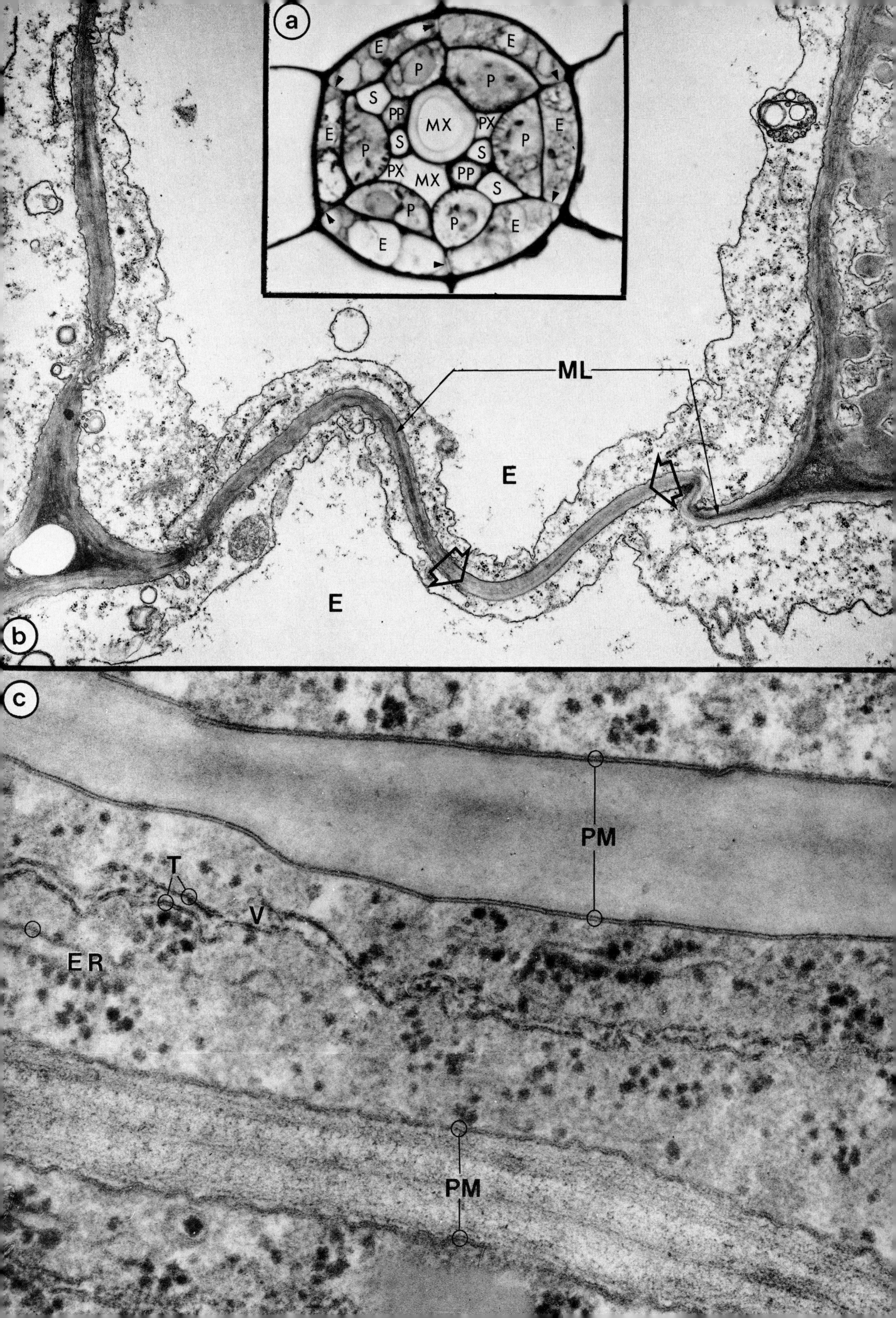
a
E
E
P
P
S
PP
MX
PX
E
E
P
P
S
S
PX
MX
PP
S
P
P
E
E
b
ML
E
E
E
c
PM
T
V
ER
PM

Plate 17

Vacuoles

Hardly any low magnification view of plant cells is without vacuoles. They are present in most of the micrographs in the book. The three pictures collected here illustrate specific features.

Plate 17a These views of the tonoplast were obtained by the freeze-etching technique. The fracture plane has passed *over* one vacuole (left) and *under* its neighbour (right), probably along the mid line of the tonoplast membrane. The resulting convex and concave surfaces 33 are therefore the vacuolar (V) and cytoplasmic (C) halves of the tonoplast, viewed from its cleaved interior. The tonoplast is asymmetric. Particles lie scattered or in short rows in the convex vacuolar half, while in the concave cytoplasmic half they are so crowded that no smooth surface can be seen between them. It is not known whether the particles in the two faces are chemically similar or not, or whether proteinaceous solute pump molecules are included amongst them.

Parts of other cell components lie in the cytoplasm surrounding the vacuoles—notably cisternae of endoplasmic reticulum (ER).

Magnification × 60 000. (We thank Dr. B. A. Fineran and Academic Press for permission to reproduce this micrograph from *J. Ultrast. Res.*, **33**, 574–586, 1970).

Plate 17b The tonoplast (T) of this vacuole (V) in a soybean root tip cell passes around a complex invagination, connected to the cytoplasm at a narrow isthmus. 33 The tripartite dark-light-dark substructure of the tonoplast is revealed, along with its asymmetry, the vacuolar face being darker and carrying a surface deposit.

Whorls of membranous material within the invagination also show dark-light-dark triple layering where their orientation with respect to the plane of the section is appropriate. In places it can be seen that a thin zone of non-membranous material is sandwiched between adjacent whorls, alternating with clear intermembrane spaces. This inclusion could therefore have originated from a set of concentrically arranged cisternae by a process of autophagic digestion. The 36 figure shows but one example of many irregular forms of membranous vacuolar inclusion. While most such inclusions can be interpreted in terms of digestion, it is unfortunate that they could also be artifacts produced by bad fixation. In the present case the remainder of the cell and tissue showed no signs of damage, so the inclusion is at least likely to be 'real', whether or not it depicts autophagic digestive activity. × 125 000.

Plate 17c Many of the compounds stored in vacuoles are of low molecular weight and leak out of the tissue when it is being prepared for electron microscopy. Some, however, are retained and appear in electron micrographs as non-membranous vacuolar inclusions. 35

The flocculent particles seen here in the vacuole of a potato leaf cell probably consist of a protein that inhibits protein-digesting enzymes (e.g. trypsin, chymotrypsin) of animals. Many organisms, from bacteria to flowering plants, produce this class of protein, and species in the family Solanaceae (like potato and tomato) accumulate one type, known as 'Inhibitor-1'. They do so particularly when the leaves are bruised or, as in the example shown, cut off the plant. It may be that Inhibitor-1 is one of the extremely numerous 'defence chemicals' of the plant kingdom, in this case conferring the biological advantage of reducing the palatability of the plant to herbivorous animals. Potato leaf × 18 000.

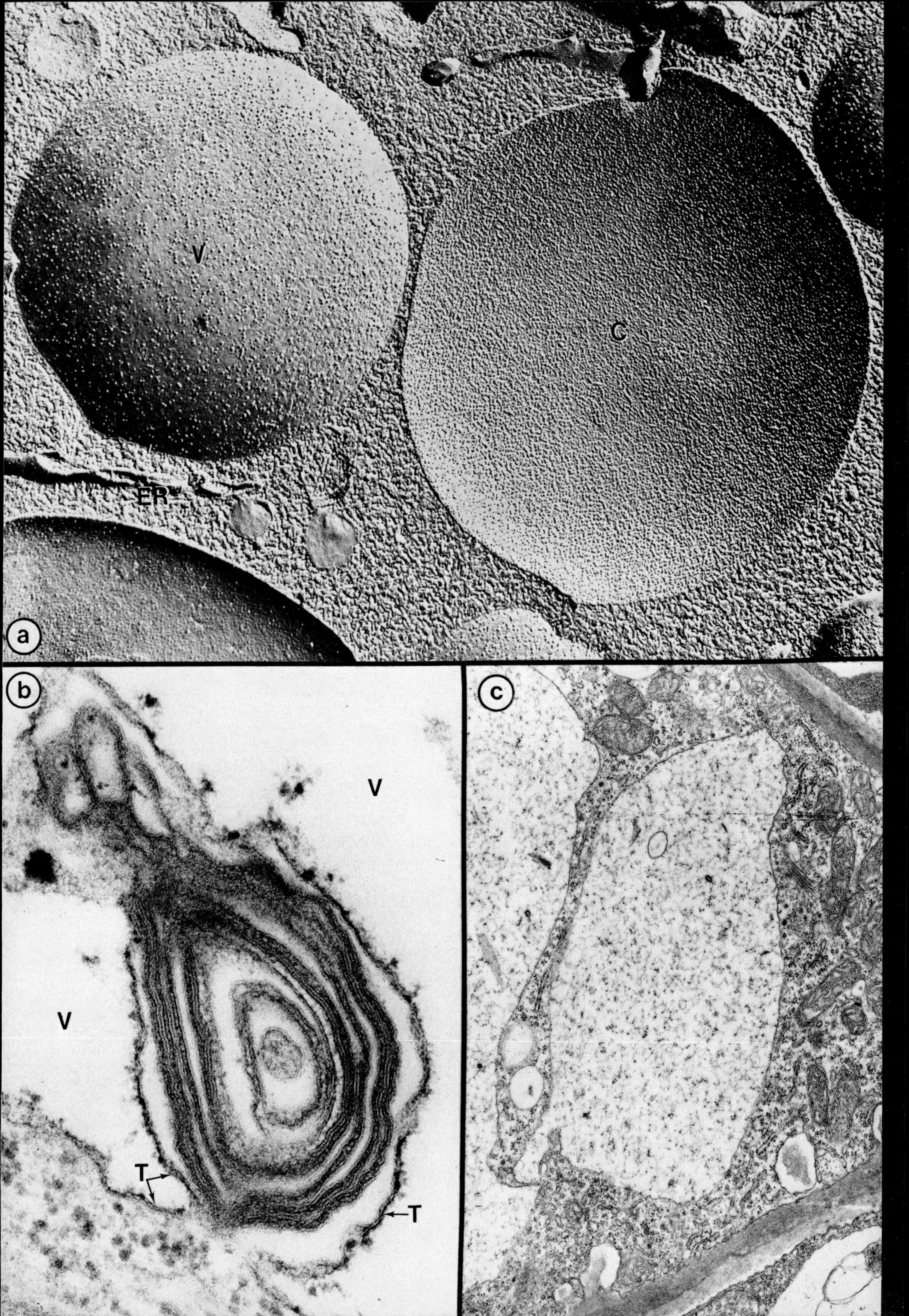

V
C
ER
a
b
V
V
T
T
c

Plate 18

The Nuclear Envelope and its Pores

Plate 18a The fracture in this freeze etched preparation of the nuclear region of a *Selaginella kraussiana* cell has passed along the surface of the nuclear envelope, revealing the pores in surface view (lower half of picture), and has broken through the inner (I) and 51 outer (O) membranes of the envelope. The upper half of the picture shows relatively featureless nucleoplasm (N). The pores are also seen in side view (arrows) at intervals along the cross-fractured nuclear envelope membranes. A continuity between endoplasmic reti- 51 culum (ER) and nuclear envelope may occur at the point marked by the star. The helical symmetry with 54 which these nuclear pores are arranged is unusual. Magnification × 24 000. (We thank Drs. B. W. Thair and A. B. Wardrop for this micrograph, reproduced by permission from *Planta*, **100**, 1-17, 1971).

Plate 18b This three-dimensional reconstruction of part of the surface of a nucleus in a cress root tip cell was made by projecting the electron microscope images of nine adjacent sections on to sheets of polystyrene, to give a final magnification of × 150 000. The outline of the nuclear envelope in each section was cut out, the pores were marked, and the nine layers superimposed in the correct alignment. The scale represents 1 μm. The nuclear surface undulates, as indicated by the irregular 'contours' traced out by each layer of the model, and the pores lie in no obvious pattern, at a 'pore density' 42 of 15-20 per μm^2 of nuclear envelope.

Plate 18c Where the nuclear surface is irregular, a single section can, as here in another cress root cell, include both side (S) and face views of pores in the nuclear envelope (NE). In some of the face views 54 (arrows) a granule is seen in the centre of the pore. Although regions of heterochromatin (C) touch the inner membrane of the nuclear envelope, a halo of 53 clear nucleoplasm lies around the pores (lower left). × 32 000.

Plate 18d Several details of pore structure can be seen in this glancing section of a nuclear envelope. The appearance of a pore depends on its position and 53 orientation within the thickness of the section, which at this magnification (× 100 000) is a slice about 5 mm thick. The pore perimeter is octagonal, and the flat edges of the octagons are visible in several cases (pores 1, 3, 4, 8, 11). Other features shown include: particulate components of the inner annulus (identifiable by proximity to chromatin) (upper parts of pores 2 and 4) and outer annulus (identifiable by proximity to polyribosomes) (pores 5 and 9); fine fibrils traversing the pore lumen (pores 1, 7, 8, 10) and apparently nearly occluding some pores (pores 4, 9, 11); pores arranged equidistant from a clump of chomatin (C) (e.g. pores 6, 7, an un-numbered pore, 8, and 10); fibrils passing between chromatin and the pore margins (arrows); polyribosomes (P) on the outer surface of the nuclear envelope; particles in the centre of some (1, 2, 4, 6, 7, 8-11) but not all pores. *Vicia faba* root tip.

Plate 18e Side views of pore complexes are seen here, in the same material and at the same magnification as in 18d. The nucleoplasm is at the top of each picture. All examples show continuity of inner (I) and outer (O) envelope membranes at the pore margins, but other features vary. Part of the variation is due to the pores lying at slightly different levels with respect to the section thickness. Units of the annulus are seen at the inner margin (e.g. single arrows) and outer margin (e.g. double arrows). Particles thought to be preribosomal in nature (large open arrows) are: absent 56 (top left); in the nucleoplasm near the pore (top centre); at the inner part of the pore lumen (top right); both inside and outside the pore (lower left); in the pore and just outside in the cytoplasm (lower centre); and in the pore as well as both inside and outside (lower right). These particles are smaller than cytoplasmic ribosomes, visible in the lower part of each micrograph. As in 18d, strands (chromatin?) are sometimes seen connected to the inner annulus.

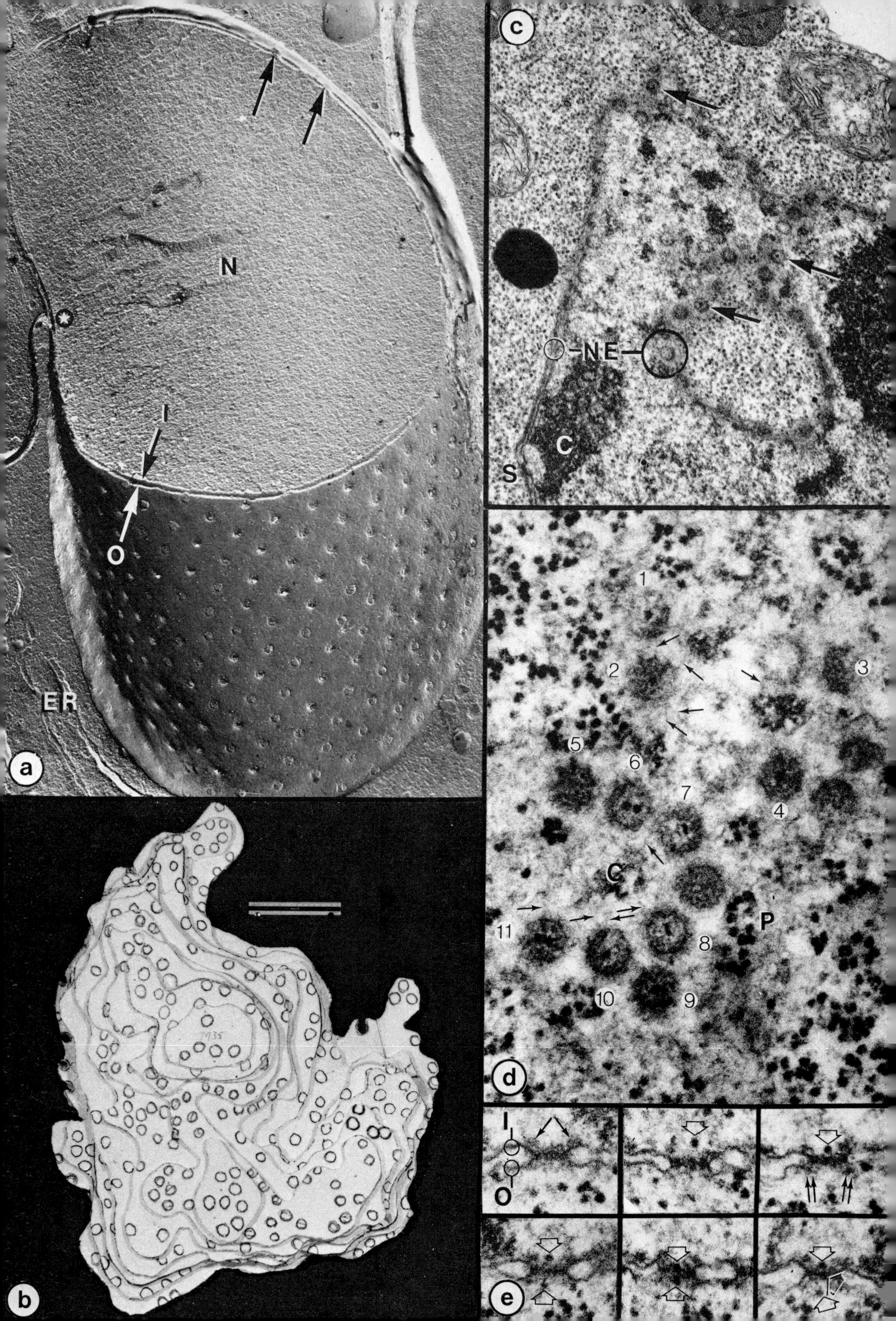

a
N
I
O
ER
b
c
NE
C
S
d
1
2
3
4
5
6
7
8
9
10
11
C
P
e
I
O

Plate 19

The Nucleolus

The nucleolus is a characteristic feature of the nuclei of eukaryotic cells. Major constituents are a repeated sequence of genes (DNA), and a mass of products (RNA) of the activity of those genes. Most of the RNA is a precursor of the RNA of ribosomes, and the dynamic processes observable in nucleoli centre on the synthesis of this material, first in fibrillar and later in granular form, and on the transport of the pre-ribosomal RNA from the nucleolus to the cytoplasm.

Plate 19a-f These six photomicrographs illustrate dynamic processes in a nucleus of a cell growing in artificial culture, the cultured tissue having been derived originally from a tobacco plant. The nucleolus (centre of each picture) is surrounded by a bright halo, due to the phase-contrast optical system used. At the start of the sequence (19a) the nucleus (margin outlined by arrows) is ellipsoidal. The nucleolus, about 7 μm in diameter, contains a large nucleolar 'vacuole' about 70 μm^3 in volume. One minute later (19b) the 'vacuole' is connected to (arrow), and apparently emptying into, the nucleoplasm. After another 15 seconds (19c) the 'vacuole' is much smaller. It is just detectable after a further 15 seconds (19d), but cannot be seen (19e) 2 minutes after the first picture was taken. By this time the whole nucleolus has shrunk by about the volume of the 'vacuole'. One hour later (19f) another 'vacuole' has formed in the nucleolus, which has regained its original total volume, and the nucleus itself has become more rounded. All × 1400. (We thank Dr. J. M. Johnson for these micrographs, reproduced by permission from *Amer. J. Bot.*, **54**, 189-198, 1967).

44 45

Plate 19g Part of a nucleolus in a *Vicia faba* root tip cell nucleus is shown here magnified × 58 000. A central 'vacuole' (V) is enclosed within the nucleolus (as in 19a, f). Granular (G) and fibrillar (F) zones constitute the dense material, along with small areas of chromatin (arrows) ramifying through electron-transparent channels. Condensed (CC) and dispersed (DC) chromatin is seen in the nucleoplasm; the former at one point touching and penetrating into the nucleolus (dashed lines). The 'vacuole' contains scattered fibrils and particles, and it is presumably they that are discharged into the nucleoplasm when 'vacuoles' pulsate.

45 49

Plate 19h This high magnification (× 150 000) view shows 6-8 nm nucleolar fibrils (F) and 12-14 nm granules (G) near the periphery of a nucleolus (*Vicia faba* root tip). The granules (see also Plate 18e) are smaller than cytoplasmic ribosomes. Some granules are attached to, or associated with, fibrils in a manner suggesting that they might be formed by folding and condensation of fibrils. Convoluted fibrils of DNA-histone are seen in the chromatin (C), near which lies a perichromatin granule (arrow), with angular profile, 50-60 nm in diameter.

45 52

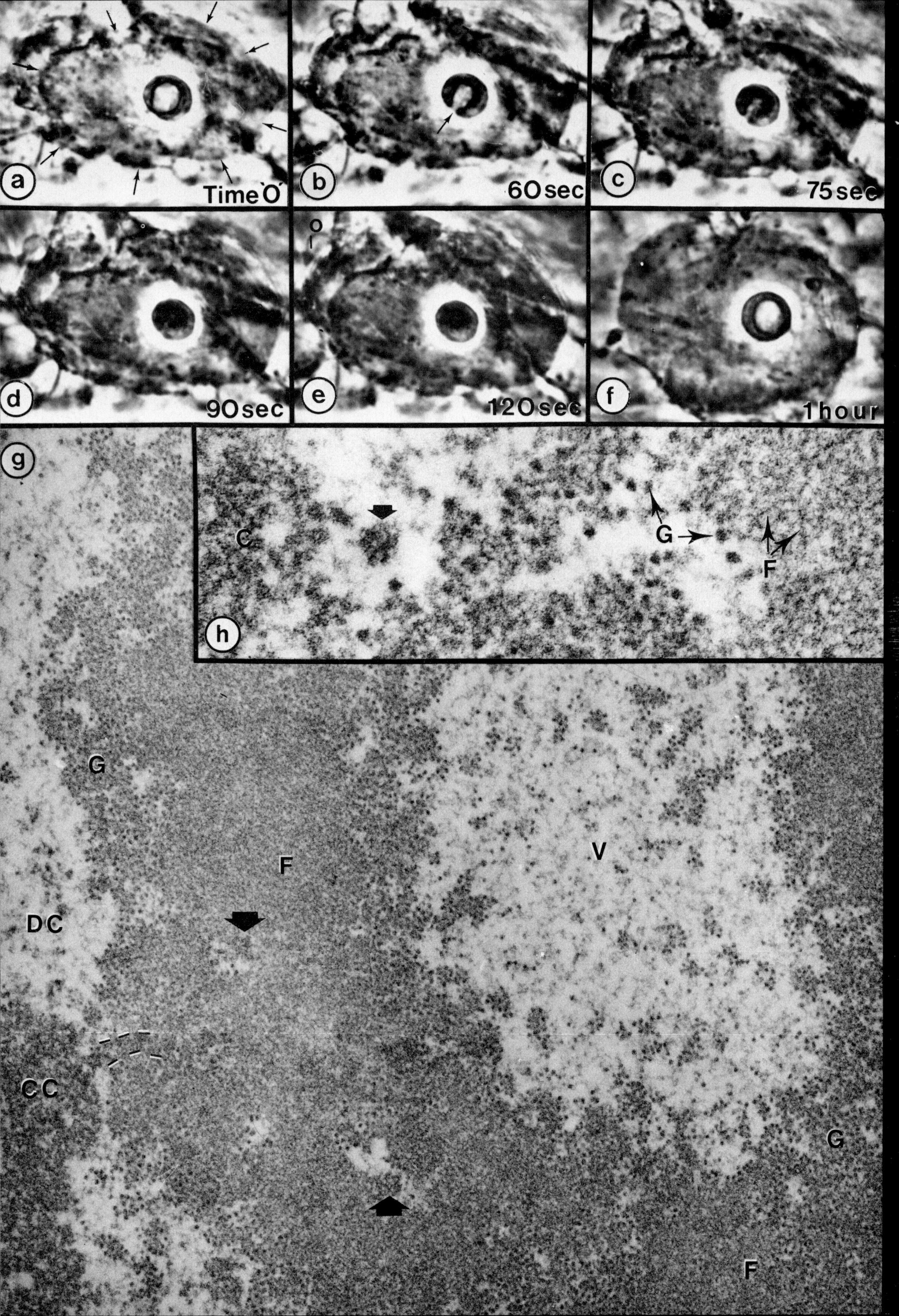
a
Time 0
b
60 sec
c
75 sec
d
90 sec
e
120 sec
f
1 hour
g
h
C
G
F
G
F
V
DC
CC
G
F

Plate 20

Plates 20–24 depict a range of specialized plant cells, and are grouped together in order to illustrate the diversity of structure and function exhibited by the endoplasmic reticulum.

The Endoplasmic Reticulum, Polyribosomes, and Protein Synthesis in Cotyledon Cells

The cotyledons of legume seeds manufacture and store large amounts of protein. Developing and nearly mature cells are shown here.

Plate 20a This light micrograph shows a typical 61 cotyledon cell at an advanced stage of development, stained by the periodic acid—Schiff's reagent procedure (which reacts with carbohydrates, as in the cell wall and in starch grains) and with bromophenol blue (which stains basic materials such as proteins). The central nucleus (N), containing chromatin and a nucleolus with 'vacuoles', is surrounded by large starch grains (S) (stained pink in the original preparation, appearing dark grey here with a lighter central region) and protein bodies (PB) of various sizes (stained dark blue in the original, black here). These are interspersed with a vacuolated cytoplasm which contains an extensive endoplasmic reticulum system (open arrows). Note also the intercellular spaces and the darkly stained material (probably proteinaceous) deposited in the corners of the walls lining these spaces. *Lathyrus* (sweet pea) seed, × 660.

Plate 20b Cotyledon cells at an early stage of development contain an extensive endoplasmic reticulum, the cisternal membranes of which are covered with spiral 60 polyribosomes. These ribosomes assemble the storage proteins of the cotyledon, principally *vicilin* and *legumin*. As the length of a polyribosome spiral corresponds approximately to the length of the messenger RNA molecule that is being translated, it is possible to estimate the molecular weight of the equivalent protein. The longest polyribosomes in this micrograph contain 20 ribosomes in a total spiral length of about 500 nm. Three nucleotides are required to code for one amino acid, and they occupy a length of approximately 1.0 nm of the RNA strand, implying a possible total of about 500 amino acids, and if each has an average molecular weight of 120, the total molecular weight of the protein is estimated to be about 60 000. In fact, biochemical analyses have shown that legumin and vicilin are complex proteins, each made up of several amino acid chains, the largest of which have molecular weights within 10% of 60 000. Magnification × 26 000 *Vicia faba,* (the broad bean), young cotyledon.

Plate 20c As the protein accumulates it is transferred by an undetermined route to large protein bodies (PB) from the endoplasmic reticulum (ER). Individual 61 protein bodies have been shown to contain both vicilin and legumin molecules. It is thought that protein bodies arise by the accumulation of protein within vacuoles. In addition, lipid stores begin to accumulate in the form of droplets (small arrows), frequently associated with the endoplasmic reticulum, but external to the cisternae (e.g. beside the asterisk). Note also the dictyosome (D) which has become so reduced as to be scarcely recognizable. In this micrograph and the previous one, the paucity of free ribosomes is evident. (*Vicia faba*, nearly mature cotyledon, × 14 000).

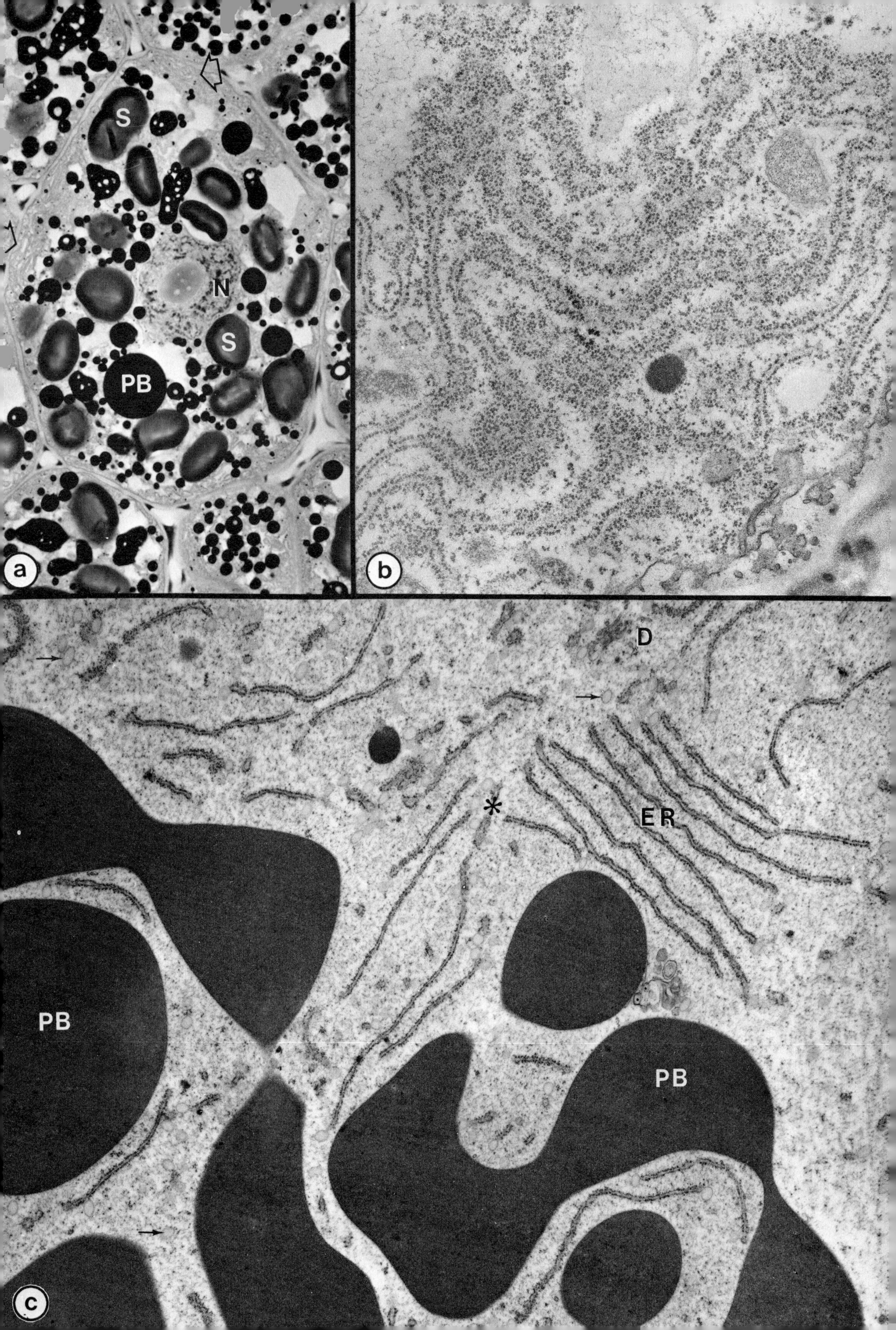
S
N
S
PB
a
b
D
ER
*
PB
PB
c

Plate 21

The Endoplasmic Reticulum and Polyribosomes

Plate 21a This unusually regular arrangement of the endoplasmic reticulum is found in surface glands (trichomes) on leaves of *Coleus blumei*. The 12 to 14 parallel cisternae are stacked so closely that the intervening cytoplasm is restricted to very thin layers, which, however, expand in some places (stars). The cisternae are interconnected at branch points (arrows) and in a few places by swollen intracisternal spaces (S), which contain diffusely flocculent material, perhaps the product synthesized by the system. The product (Pr) secreted by the gland passes through the inner layer of the cell wall (CW) and accumulates beneath the cuticle (C). The endoplasmic reticulum is continuous with the outer membrane of the nuclear envelope (open arrow, lower right). The stroma of the plastids (P) is filled with a material which (after specimen preparation) is very dense to electrons. The gland probably secretes mono- or di-terpenes. × 22 000.

Plate 21b One of the spaces into which the cisternae open is shown here enlarged from 21a. The membranes of the endoplasmic reticulum have a just discernible tripartite substructure (black and white arrow) and are thinner than the adjacent tripartite plasma membrane (black arrow).

It is clear that the intra-cisternal cavity (Ci) is not in open communication with the space outside the plasma membrane (Co), but several clues in the micrograph suggest that this arrangement of membranes may alter with time, and that cisternal contents can be released to the exterior. The clues are: similarity in size and flocculent contents between Ci and Co, suggesting that the former could be a reservoir which is filled up and discharged to the exterior; the presence of fragments of membrane in Co, as if they had been cast off during the previous discharge; one fragment (open arrow) not only has a tripartite construction of the same dimensions as the plasma membrane, but also encloses several ribosomes, perhaps derived from the narrow bridge of cytoplasm that separates Ci and Co. It should be noted that although the endoplasmic reticulum and plasma membrane are in close proximity over much of the periphery of the cell (21a), there is no evidence that the two can fuse. Rather, presence of fragments of membrane in Co suggest that discharge of Ci may be by explosive rupture of the membranes following increase in internal pressures, rather than by fusion of endoplasmic reticulum and plasma membrane. On this hypothesis, the two classes of membrane would heal their ruptures after the pressure had been released, again without fusion of unlike membranes. These speculations illustrate the way in which a static image can be used to hypothesize about dynamic processes; it is, however, important to recognize that to confirm or reject the ideas requires further experiments. × 75 000.

Plate 21c Two polyribosome conformations which differ from the spirals seen in Plate 20b are shown here. In the main picture, which shows oblique surface views of cisternae of rough endoplasmic reticulum in a hair cell of the alga *Bulbochaete*, the polyribosomes often lie in parallel chains. Polyribosomes are also seen on the outer surface of the nuclear envelope (N-nucleus). × 67 000. (We thank T. Fraser for this micrograph).

The insert shows a more common configuration, which occurs free in the cytoplasm, unattached to membranes. It consists of a tight helix of ribosomes. The example shown is from a bean root tip cell, and because the helix is much longer than the section is thick, there is no guarantee that the 0.4 μm length included here represents the total length. Comparable polyribosomes in liver cells have been shown to be based on unusually long strands of helically wound messenger RNA. × 90 000.

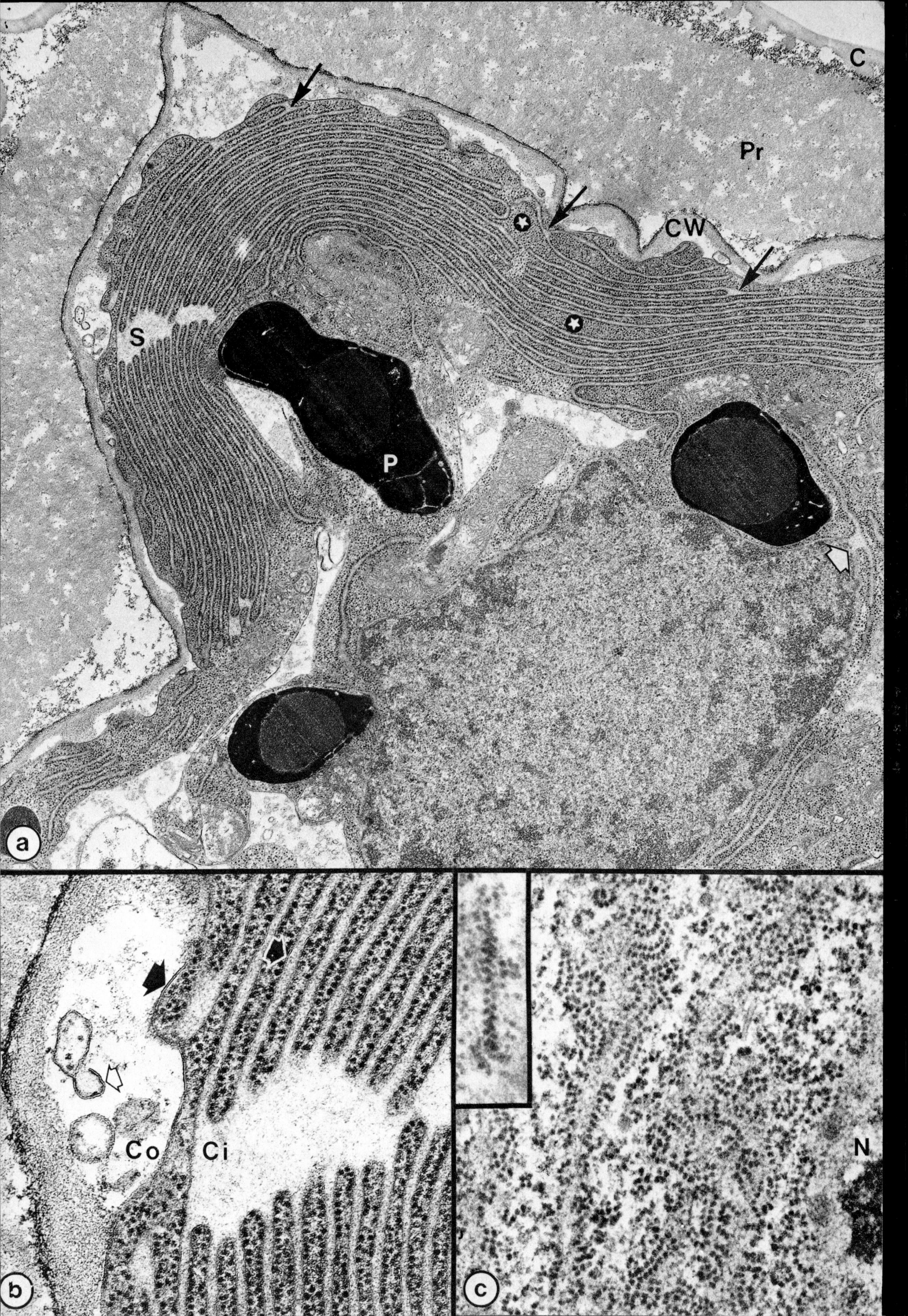
C
Pr
CW
S
P
a
Co
Ci
b
N
c

Plate 22

The Cytoplasm of Tapetal Cells

Cells of the tapetum have already been introduced in Plate 11, and further details of their cytoplasm are shown here using as an example an anther of *Avena strigosa*, containing developing pollen grains.

Plate 22a-c These cells are most unusual in that they 33 lack vacuoles and (at maturity) cell walls. In several respects they resemble an animal epithelium, with one face (on the right in (a)) directed towards the anther locule (LO), and the other (on the left) based on the cells comprising the anther wall (see also Plate 11). A small portion of a pollen grain intrudes at upper right.

Orbicules (O) and remnants of cell wall material (arrows) (see Plates 11 and 12) are present on the locular side of the tapetal plasma membrane (PM). The cytoplasm is filled with plastids (P), mitochondria (M), dictyosomes (D), and both rough and smooth endoplasmic reticulum (ER).

The rough endoplasmic reticulum tends to lie in 63 aggregates of fairly parallel cisternae (as in liver cells). Its polyribosomes are spiral. The rough cisternae are continuous with smooth regions, as shown at the dashed lines in (a) and (b). Both types of cisternae contain a material of moderate electron density, but the smooth parts tend to be dilated into sausage-shapes or spheres (22c, arrowhead) except in the case of smooth 65 cisternae that ensheath plastids and mitochondria (best seen in 22b—arrowheads).

The cytoplasm contains numerous large vesicles (V) with flocculent contents. Other micrographs suggest that these are derived from the dictyosomes, and profiles such as the one marked by an asterisk (lower left) that they liberate their contents at the plasma membrane by reverse pinocytosis (not necessarily at 75 any one face of the cell).

The cell wall structure seen on the left hand side of 22a is complex. The plasma membrane (PM) is not in close contact with the wall, which probably consists of an electron-transparent lipoidal deposit (L) on an otherwise conventional type of wall (open arrow). Some plasmodesmata (PD) pierce both layers. The composite wall may be a seal that reduces outward losses of nutrients leaking from the locule between the tapetal cells.

Provision of nutrients for developing pollen is known to be a major function of the tapetum. It also synthesizes certain proteins which become incorporated into the exine of the mature pollen, and it manufactures the cores of orbicules which lie on the locular surface, where, like the pollen grains themselves, they become coated with sporopollenin. The sub-cellular source of the precursors (probably carotenoids) of the sporopollenin has not been identified with certainty: a, × 28 000; b, × 48 000; c, × 48 000.

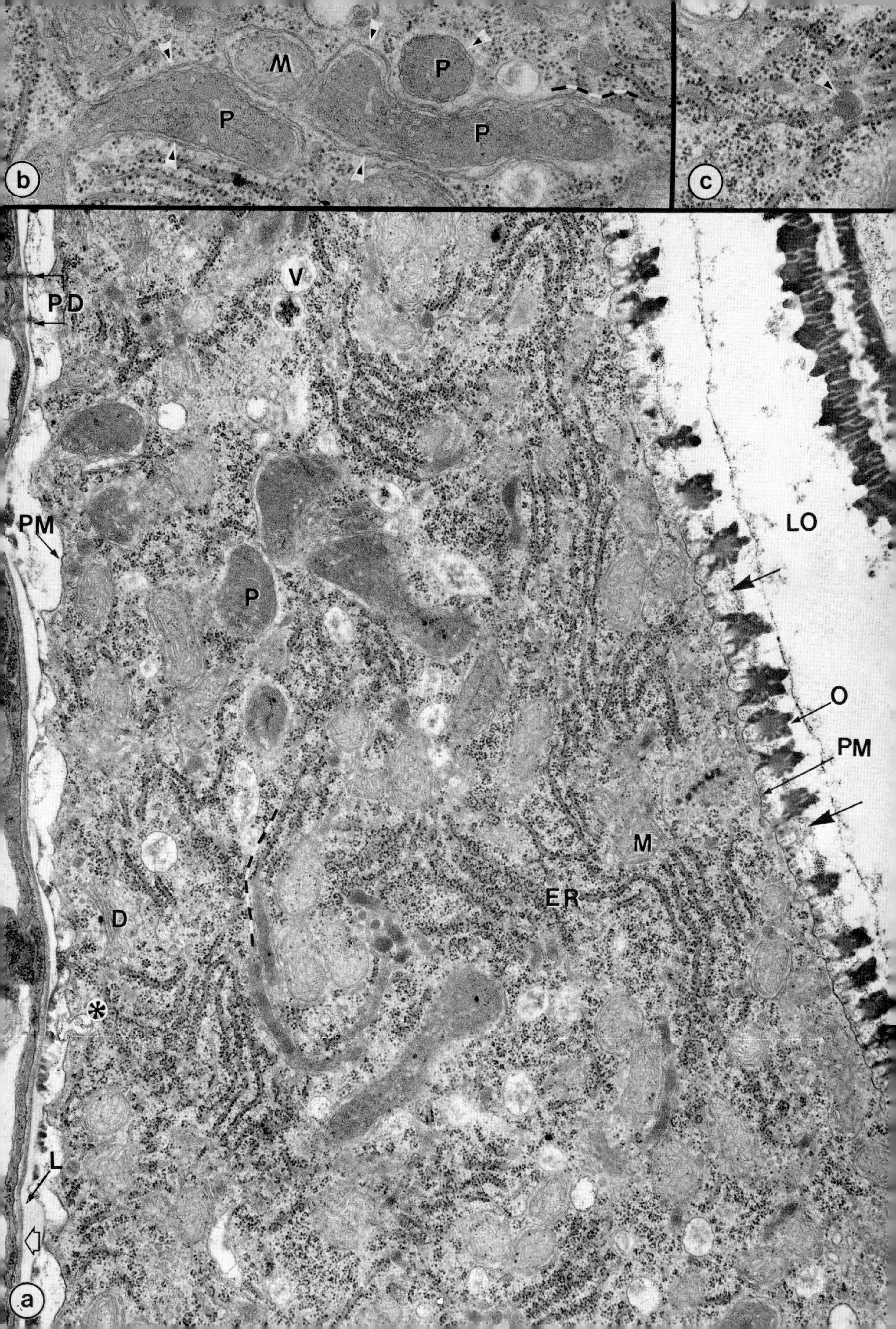
M
P
P
P
P
b
c
PD
V
PM
P
LO
O
PM
M
ER
D
*
L
a

Plate 23

Smooth Endoplasmic Reticulum in 'Farina' Glands

Plants manufacture and secrete an enormous diversity of chemical substances. The example illustrated here, a white floury material, or 'farina', is produced by single celled glands developed particularly by members of the flowering plant family *Primulaceae*. Farina glands are one of several categories of plant gland in which a system of smooth endoplasmic reticulum membranes is exceptionally well developed.

Plate 23a and b These scanning electron micrographs show the floury appearance of the surface of an *Auricula* sepal, visible to the eye as a yellow powder, and at low magnifications ((a), × 1100) as mounds of fine crystals. Each mound ((b), × 5800) is a gland cell, covered by its secretion product in the form of contorted ribbon-shaped or linear crystals radiating from the cell surface. A mixture of substances is present, the major components being flavonoids. Some people are allergic to the 'flour'.

Plate 23c In the course of preparation for ultra-thin sectioning, this material (a farina gland on a young petal of *Primula kewensis*, × 40 000) had its secretion product dissolved away. Consequently none is seen outside the cell wall (CW). The main feature of the cytoplasm is the system of smooth endoplasmic reticulum tubules, each one 60-100 nm in diameter, ramifying through the cytoplasm. Not many branched tubules are visible (arrows), and it is obvious from the varied profiles displayed in the section that the tubules do not lie in a regular pattern. The ground substance of the cytoplasm is granular, and contains a rather sparse population of ribosomes, usually in clusters, presumably polyribosomes (e.g. circle). The mitochondria (M), being nearly circular in outline, are probably nearly spherical, but the electron-transparent areas within them suggest that they might have become swollen during preparation. Microbodies (MB), with granular contents and single bounding membrane, are conspicuous. The plastids (P) are small and simple. An electron-dense deposit lines parts of the tonoplast of several of the vacuoles (V). In a number of places (asterisks) the configuration of the tonoplast suggests that material such as the small droplets of electron-dense material that are present throughout the cytoplasm (especially at lower left) associated with the endoplasmic reticulum, may be being incorporated into the vacuoles.

59 88 135

The significance of this cytoplasmic organization in relation to the synthetic and secretory activity of the farina gland is obscure. The plasma membrane (PM) is fairly smooth, and there is no evidence that products of secretion leave the cytoplasm in the form of vesicles by reverse pinocytosis (*granulocrine* secretion). It has been suggested that the outward transport is *eccrine*, that is, by a flux of free molecules (not in vesicles) across the plasma membrane, and thence to the exterior by diffusion through the cell wall. Crystallization occurs upon exposure to the air. The labyrinths of smooth endoplasmic reticulum are thought to be associated with the synthesis of terpenoid substances, for they are present in a variety of plant glands secreting fats, oils, and fragrant essential oils. Flavonoid molecules are in part derived from the same precursors as terpenoids.

64

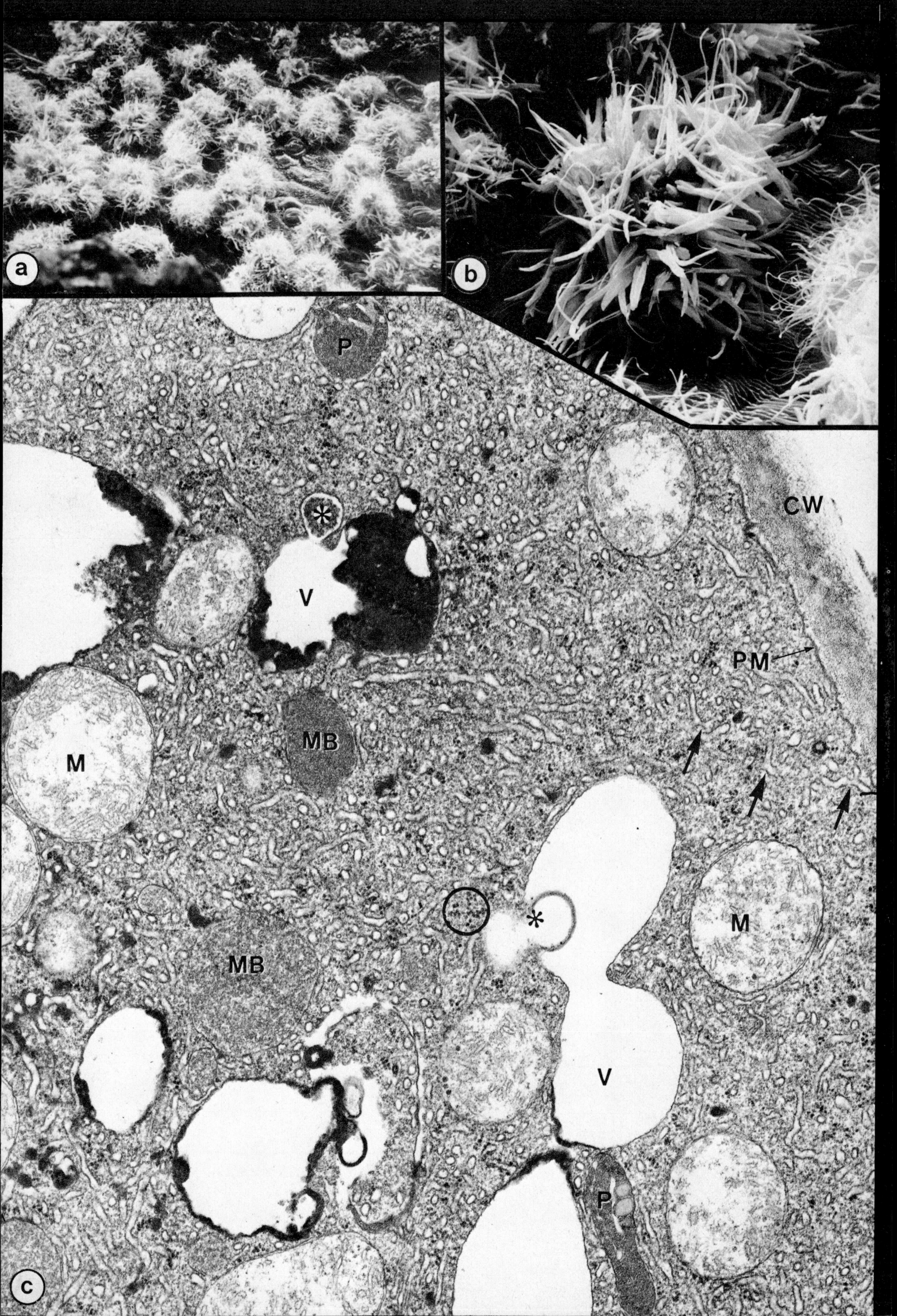
a
b
P
CW
*
V
PM
MB
M
*
M
MB
V
P
c

Plate 24

Developmental Changes in the Endoplasmic Reticulum of Sieve Elements

The endoplasmic reticulum can alter greatly during maturation of a cell. This plate illustrates one such developmental sequence, taking as an example the specialized endoplasmic reticulum found in sieve elements (the 'sieve element reticulum'). Plate 8b shows the reticulum (ER) in a mature sieve element, and should be examined before the more detailed pictures presented here.

Plate 24a Young sieve elements possess a full complement of cell components, though few are represented in this transverse section (*Coleus blumei* stem apex). The tonoplast (T) delimits the vacuole (V) at this early stage of development, but is destined to break down, so that cytoplasmic and vacuolar material will intermingle as what has been called *mictoplasm*. Irregular 66 cisternae of rough endoplasmic reticulum (ER) are present in quantity throughout the cytoplasm, and remain while the cell synthesizes masses of a proteinaceous material (P), formerly described as 'slime bodies' and now as P-protein bodies (P being an abbreviation for phloem). An unusual type of vesicle is found in these cells, but is not restricted to them. It is spherical (see circled example in insert), about 50 nm in diameter, and spines, from which the name *spiny vesicle* is derived, radiate from its surface. *Coated vesicles* (e.g. Plates 27 and 28) are similar, but with a less distinctive coat. Spiny vesicles are usually present in clusters, and in this micrograph they lie (arrows) between massed endoplasmic reticulum cisternae and a growing P-protein body. It is conjectural whether they have a role in transferring protein (perhaps as 'spines'?) from the one component to the other. P-protein persists in mature sieve elements (a few dispersed strands are visible in Plate 7, and larger quantities around sieve plates in Plate 8). × 37 000, insert × 80 000.

Plate 24b Later in the development of sieve elements the cytoplasm and vacuole mix, the nucleus (frequently) disappears, and the endoplasmic reticulum metamorphoses to generate the *sieve element reticulum*. 63 Cisternae of rough endoplasmic reticulum (RER) aggregate in stacks, and in the process the ribosomes are lost from all but the outermost faces of the stack (arrows). *Vicia faba*, root tip, × 72 000.

Plate 24c The stacks of sieve element reticulum are usually associated with the plasma membrane. As the contents of the mictoplasm are dissipated, the reticulum becomes more complex in shape. With the eventual loss of *all* ribosomes, it can be difficult to identify and distinguish the intracisternal spaces (IC) from the mictoplasm. In this example five cisternae are seen adjacent to the plasma membrane (PM), compared with which their tripartite construction is thinner (see circled portions of membrane). Cell wall—CW, lumen of sieve element—L. *Vicia faba*, phloem supply to floral nectary, × 200 000.

The functions, if any, of the sieve element reticulum are not understood. One suggestion that has been put forward is that it might prevent mitochondria or plastids from being swept into the sieve plate pores.

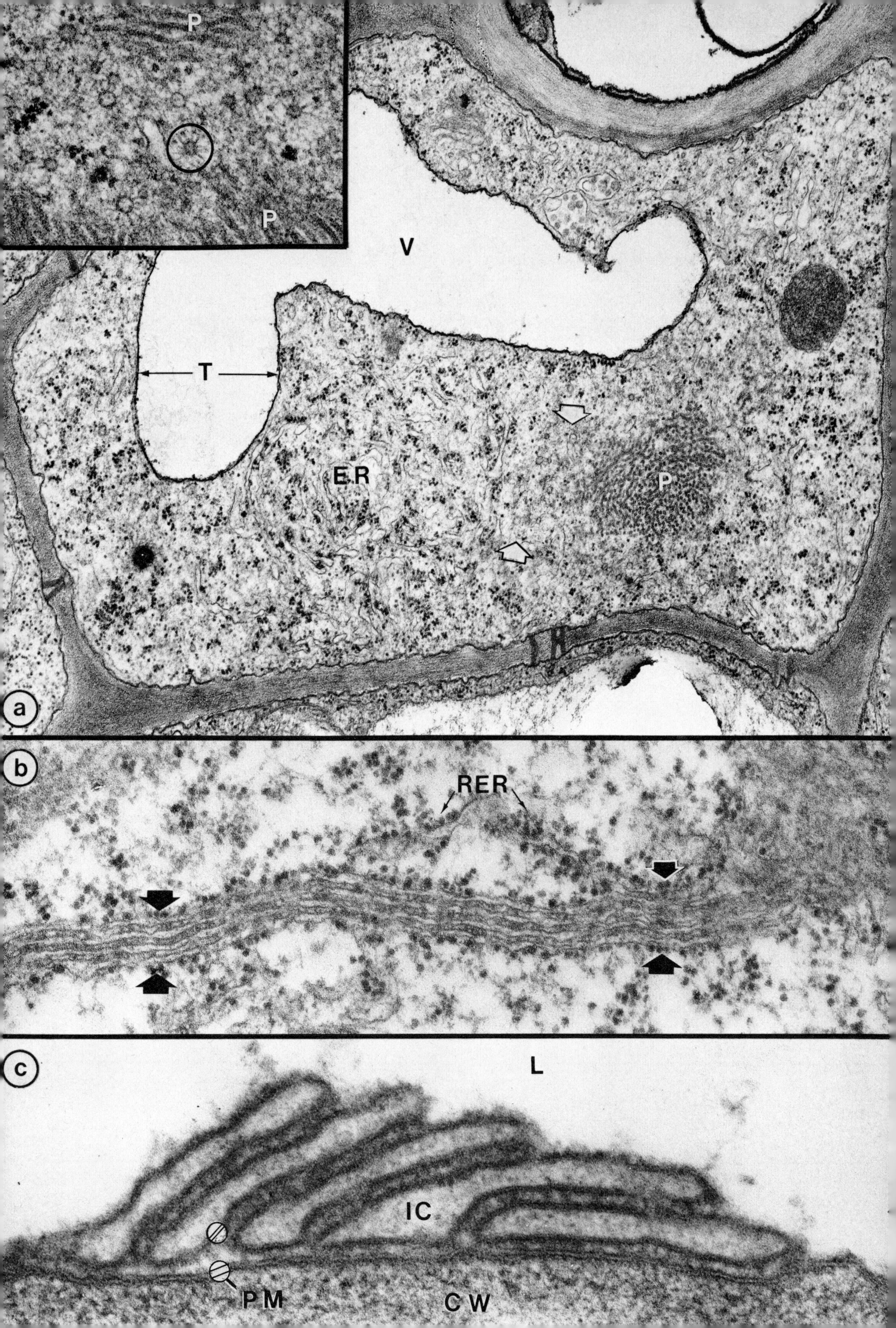
P
P
V
T
ER
P
a
b
RER
c
L
IC
PM
CW

Plate 25

Plates 25-28 present a collection of micrographs illustrating the structure and function of dictyosomes, the units of the Golgi apparatus of the plant cell.

The Membranes of Dictyosomes

Plate 25a The membranes of this freeze-etched dictyosome were revealed by a fracture which passed along an expanse of membrane (upper right), through a collection of vesicles, then steeply down in a succession of narrow steps through 6-8 cisternae, and finally along a more level plane where it exposed face views of a cisterna (centre) and numerous vesicles lying nearby in the cytoplasm.

Comparisons with ultra-thin sections from the same material (cells of the green alga *Micrasterias*) allow identification of the small vesicles at the upper right as 75 transitional vesicles (TV) at the *forming face* of the dictyosome. The first dictyosome cisterna in the stack is pierced by a cleft (arrow), and had the cell not been killed, it is probable that later-formed cisternae would also have been split. In other words, this cleft may be the first sign of a division process which would have 84 resulted in the production of two dictyosomes. The larger vesicles (V) at the bottom and lower left of the micrograph represent the membrane-bound products formed at the *maturing face* of the dictyosome. Other small vesicles lie at the margins of the successive cisternae.

It is difficult, by inspection of the fracture face, to distinguish surfaces that might be external faces of membrane, from surfaces that might represent the internal architecture of a membrane that has been split along its hydrophobic interior.

The extensive surface labelled M probably represents the interior of the membrane of a mature cisterna. Smaller, but similar, views can be seen amongst the ledges above. Other micrographs have shown that, moving from the forming to the maturing face of the dictyosome, the number of particles per unit area of membrane increases. There are about 7000 particles per 73 square μm in the mature cisternae (as at M). The majority of the particles is restricted to the central area of the cisterna, and a marked decline in their frequency is seen towards the periphery, where the membrane surface begins to vesiculate. The vesicles at the maturing face (V) have relatively few particles, a condition that could arise either by removal of the particulate component(s) of the membrane, or by 'dilution' of the particles by extension of the non-particulate areas. × 82 000.

(We thank Drs. L. A. Staehelin and O. Kiermayer and Cambridge University Press for permission to reproduce this picture from *Journal of Cell Science*, **7**, 787-792, 1970).

Plate 25b and c The forming face of this dictyosome (in another green alga, *Bulbochaete*) is uppermost, where the bifacial endoplasmic reticulum (ER), transitional vesicles (TV), and coalescence of transitional vesicles to form the first cisterna (arrow) can be seen. Vesicles with visible contents are present at the margins of the dictyosome and near the maturing face (V). The upper six cisternae have granular contents, both in the central regions and at their periphery, but in the next three cisternae the central regions are different. There the membranes are clearly tripartite (see (c), a higher 72 magnification view of the same dictyosome), and the intra-cisternal compartment is very thin, thus restricting the contents to the margins, which is where vesicles are formed. In the most mature cisternae the membranes have moved apart again, and they appear to be empty. Presumably the former contents have been packaged into vesicles. The central membranes may break down 79 into fragments (F).

Magnifications: b, × 70 000; c, × 155 000, (micrographs provided by T. W. Fraser).

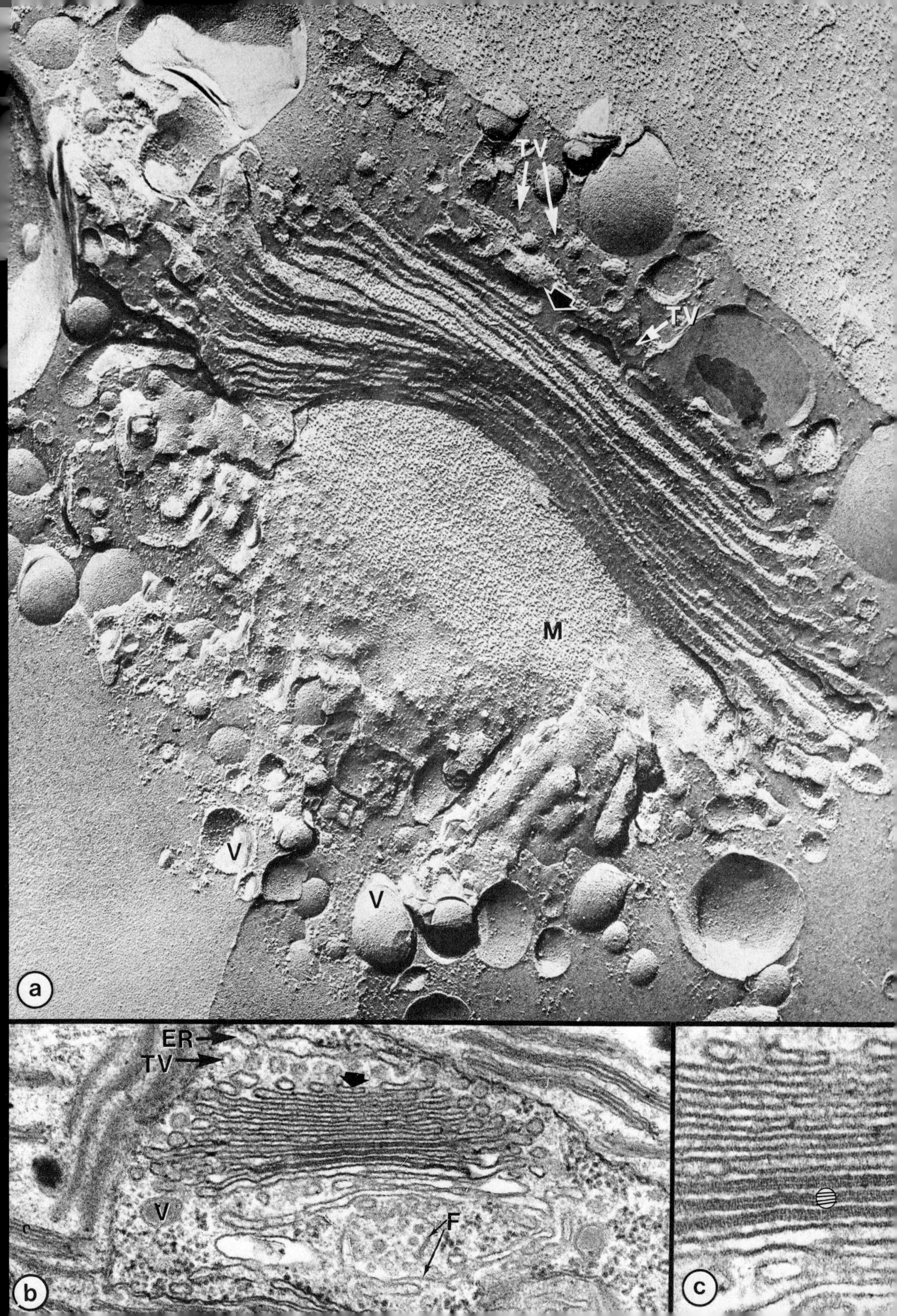
TV
TV
M
V
V
a
ER
TV
V
F
b
c

Plate 26

Production of Scales in the Golgi Apparatus

These micrographs, all of members of a genus of planktonic flagellates in the Haptophyceae, called *Chrysochromulina*, illustrate a number of central principles relevant to the functioning of the Golgi apparatus. (a) – (f) depict the end product and (g) – (h) the specificity and mode of its formation.

Plate 26a *Chrysochromulina* produces walls composed of scales. An empty discarded scale case of *C. pringsheimii* is seen here in a shadow-cast whole mount. Two types of scale are visible—the long 'pins' at the ends of the case, and the smaller type, of which a few have broken away from the case (arrow).

Plate 26b - d Plate 26b is an ultra-thin section of *C. chiton*. The scale case in incomplete, nevertheless two types of scale are easily distinguished. Those lying on the outside (O) have a flat base and a curved rim. A shadow-cast example is shown in (d), with the inner face (that closest to the plasma membrane) exposed to view, revealing an intricate pattern of fibrils. The inner scales (I), lying between the outer scales and the cell itself, are slightly curved, with a small recurved rim. In (c) they are viewed from the inside (upper specimen) and also from the outside (lower specimen). The pattern is visible on the inner face of the inner scales, but is overlain with amorphous material on the outer face. The shadow-cast outer face shows the recurved rim, formed from circumferential microfibrils. (b) also shows cell contents: nucleus, N; chloroplast, C, with stalked pyrenoid, P; mitochondria, M; and the single dictyosome, D, with the maturing face at the top.

Plate 26e, f These two sections (along with (a) and (b)) illustrate genetic specificity in the shape of the scales. (e) is a genetically deviant form of *C. chiton,* and differs from that in (b) only in the shape of the outer scales (O), which have the same type of base, but lack the curved rim. The inner scales (I) are like those of 'normal' *C. chiton.* (f) shows outer (O) and inner (I) scales of *C. camella,* in which the outer scales are shaped like cups with four rings of perforations on the sides. Numerous other distinctive species-specific forms could be illustrated.

Plate 26g, h Dictyosomes from the 'normal' form of *C. chiton* (scales as in (b) (c) and (d)), and from the deviant form (scales as in (e)) are shown in (g) and (h) respectively. Both micrographs show: (i) *the dorsiventrality of the dictyosomes:* endoplasmic reticulum (ER), transitional vesicles (between open arrows), and the forming faces are at the *lower* part of the picture. (ii) *formation of scales in dictyosome cisternae:* the first visible signs of scales are indicated by the arrows. Recognizable outer (O) and inner (I) scales are present, each type within its own cisterna. The example of a 'normal' outer scale in (g) shows that the shape of the cisternal membrane matches the shape of the scale (arrowheads). Cisternae at the maturing face, close to the plasma membrane (PM), contain scales of mature appearance. Liberated scales are shown in (h). (iii) *dorsiventrality of cisternae:* in every instance where the plane of section is suitable, it can be seen that the scales in the successive cisternae are oriented in the same way, that is, with their outer faces towards the maturing face of the dictyosome. (iv) *expression of genetic specificity in the Golgi apparatus:* the character which distinguishes the two forms of *C. chiton,* i.e. the shape of the scales, is seen to be developed within the dictyosome cisternae.

The micrographs thus illustrate formation of highly distinctive species-specific structures by the Golgi apparatus. The cisternae are shown to possess individual synthetic capabilities, and, like the dictyosome as a whole, to be dorsiventral.

(a) × 1750; (b) × 10 000; (c) – (h) all × 30 000.

(We thank Professor I. Manton for providing all of these micrographs. (d) and (g) are previously unpublished: (f) is reproduced by permission from *Arch. Mikrobiologie*, **68**, 116, 1969; (a) from *Jour. Marine Biol. Assoc. U.K.*, **42**, 391, 1962; (b), (c) and (e) from *Jour. Cell Sci.*, **2**, 411, 1967; (h) from *Jour. Cell Sci.*, **2**, 265, 1967).

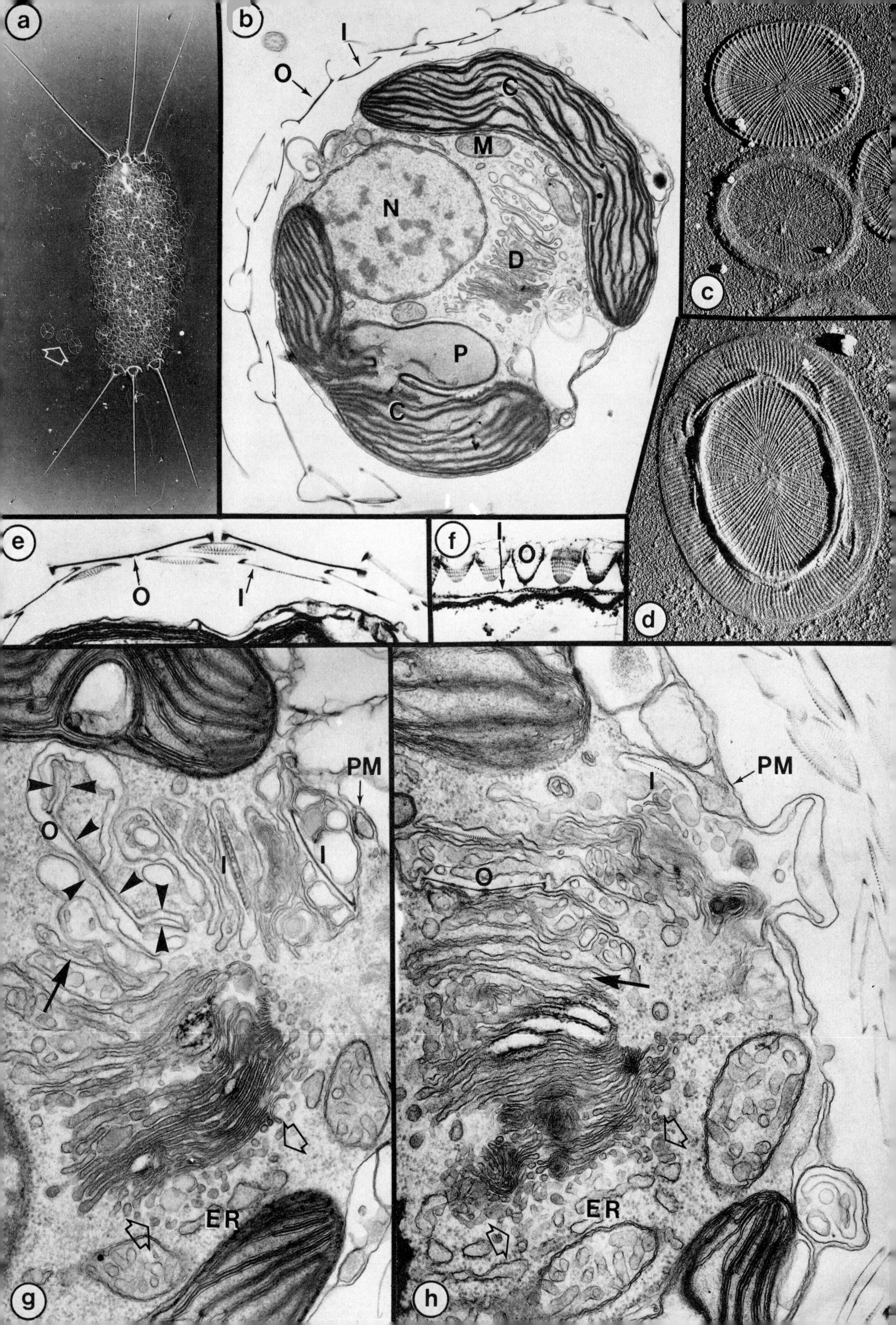

a
b
O
I
C
M
N
D
P
C
c
d
e
O
I
f
I
O
g
PM
O
I
I
ER
h
PM
I
O
ER

Plate 27

Relationships between Dictyosomes, Endoplasmic Reticulum, and Nuclear Envelope

Spatial relationships between the above components of the cell are especially obvious in certain algal cells.

Plate 27a The filamentous green alga *Bulbochaete* possesses 'hair cells' (chaetae) containing a conspicuous Golgi apparatus consisting of numerous discrete dictyosomes (D 1-3), between which ramify cisternae of rough endoplasmic reticulum (ER). The dictyosomes are oriented with respect only to the subjacent endoplasmic reticulum, thus in D-1 the forming face is lowermost, in D-2 uppermost, and in D-3 at the left hand side. In all three the reticulum is bifacial, lacking ribosomes in the zones (between arrows at D-2 for example) nearest to the forming faces. Transitional vesicles (TV) are present, but so are very numerous other vesicles. The larger type (V), more or less spherical, with granulo-fibrillar contents, are formed at the margins of cisternae at the maturing face. As shown in the insert at top right, their bounding membrane is tripartite (arrow), and of the same dimensions as the plasma membrane. The insert at lower right is part of a section adjacent to that used for the main picture, and it shows dictyosome D-2. Whereas D-2 (main picture) merely shows transitional vesicles (bearing a fuzzy coat on their membrane), D-2 (insert) shows what can best be interpreted as a stage of formation of a coated transitional vesicle (upper arrow) and a stage of coalescence of another such vesicle (lower arrow) with the first dictyosome cisterna.

The endoplasmic reticulum is again bifacial where it lies alongside (between arrows) the vacuole (Va). Hair cells in *Bulbochaete* have no chloroplasts (in fact no plastids of any kind), yet they produce abundant Golgi vesicles, and clearly require raw materials. It may be that the latter enter the hair cell *via* plasmodesmata (PD) from photosynthetic cells on the other side of the cell wall (CW). The vacuole could be a storage reservoir for raw materials, and the closely juxtaposed tonoplast (T) and bifacial endoplasmic reticulum a device whereby entry of nutrients into the reticulum is facilitated. Once in the reticulum, nutrients could pass to the bifacial regions subjacent to dictyosomes. (Micrographs courtesy of T. W. Fraser. Main picture and lower insert × 52 000; upper insert × 120 000.)

Plate 27b In many algae, as here in *Tribonema* at × 60 000, the dictyosomes lie alongside the nucleus, and transitional vesicles (TV) are seen between the nuclear envelope (NE) and the forming face. The examples in the micrograph are used to illustrate presumed stages in formation of a transitional vesicle from the outer membrane of the nuclear envelope, as in the sequence 1-7. The region of the nuclear envelope that produces vesicles has a thick (up to 100 nm) layer of grey-staining fibrillar materials on its outer surface (e.g. star). This is of unknown significance but since it is not seen elsewhere on the nuclear envelope it may be part of an apparatus for vesicle production. For example a system of contractile microfilaments may serve to distort the membrane surface and pinch off the vesicles. (Micrograph provided by Dr. G. F. Leedale, and reproduced by permission, from *Brit. Phyc. Jour.*, **4**, 159-180, 1969).

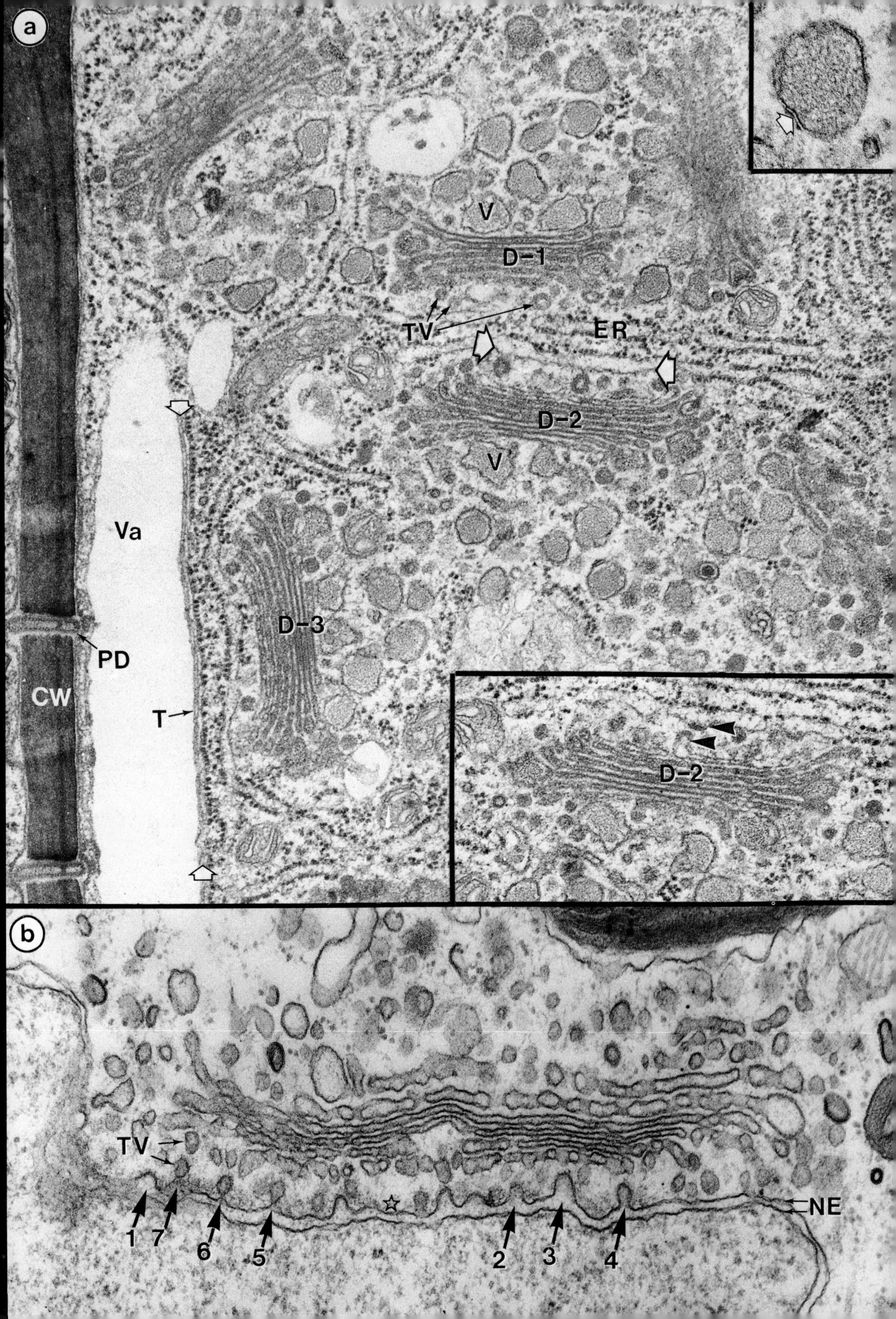
a
V
D-1
TV
ER
D-2
V
Va
D-3
PD
CW
T
D-2
b
TV
NE
1
7
6
5
2
3
4

Plate 28

The Golgi Apparatus and Mucilage Secretion by Root Cap Cells

The Golgi apparatus of cells in the root cap has been extensively studied. It mediates the production of a polysaccharide mucilage that is exported by reverse pinocytosis through the plasma membrane and extruded to the surface of the root tip. Two grasses—timothy (*Phleum*) and corn (*Zea*)—are used in this plate because they illustrate different types of vesicle formation.

Plate 28a The periphery of this root cap cell of *Phleum pratense* shows part of the nucleus (N) and the thick, mucilage-laden cell wall (CW). The cytoplasm contains numerous dictyosomes (D), in which the cisternae contain fibrillar material, visible even close to the forming face (F). The cisternae remain flat, with no vesicles at their periphery, and increase in thickness (see numbered sequence in the dictyosome at the right), eventually rounding off at the maturing face to form single, large vesicles, apparently without fragmentation into small vesicles (compare *Zea*, below). Similar large vesicles (V) appear close to the plasma membrane (PM), and the plasma membrane bulges over packages of fibrillar material (e.g. black arrows) that closely resemble the contents of the vesicles. It is reasonable to conclude that within this single micrograph we are seeing many stages in the manufacture, packaging, movement, and delivery of the fibrallar material to its final destination in a more dispersed form (probably as a result of hydration) in the wall (asterisk).

Other noteworthy features include: (1) No obvious relationship between endoplasmic reticulum and the forming face of the dictyosomes, and no identifiable transitional vesicles (compare Plate 27a and b). (2) The forming dictyosome cisternae are composed of branched tubules (T), visible in a dictyosome that lies with its forming face in the plane of the section. The insert shows the appearance of the same forming face in the adjacent section, with extensions of the tubular system. (3) Other micrographs have shown coated vesicles near dictyosomes, and two are seen here at the plasma membrane (circled), either being formed at it or else having fused with it. × 30 000.

Plate 28b The ultra-thin section shown here (the periphery of a root cap cell of Zea *mays*) was treated by a procedure involving oxidation with periodic acid to produce aldehyde groups in polysaccharides; coupling of the reagent thiocarbohydrazide to the aldehydes, and of silver proteinate to the thiocarbohydrazide. The end result is that polysaccharides are rendered electron-dense and so can be located. They occur, as expected, in the cell wall (CW), and conspicuously in dictyosome cisternae, thus confirming the suggestion that the latter are sites of polysaccharide accumulation. The insert shows that the polysaccharide does not have the same appearance in the cisternae and nearby vesicles as in the wall (bottom right hand corner of insert), thus indicating that a maturation, or perhaps more simply a hydration, process takes place between the two locations.

Vesicle formation at the maturing faces of *Zea* dictyosomes is not the same as in *Phleum* (above). The dictyosome under the insert shows stained polysaccharide in the successive cisternae and then, at the maturing face, its migration to form *peripheral* vesicles (solid arrows), leaving a fragment of membrane (open arrow) derived from the central region of the cisterna. As in *Phleum*, there are no clear associations between endoplasmic reticulum and dictyosomes. × 15 000, insert × 36 000.

(Micrograph (b) and insert provided by Dr. M. Rougier; (b) reproduced by permission from *Journal de Microscopie*, **10,** 67-82, 1971).

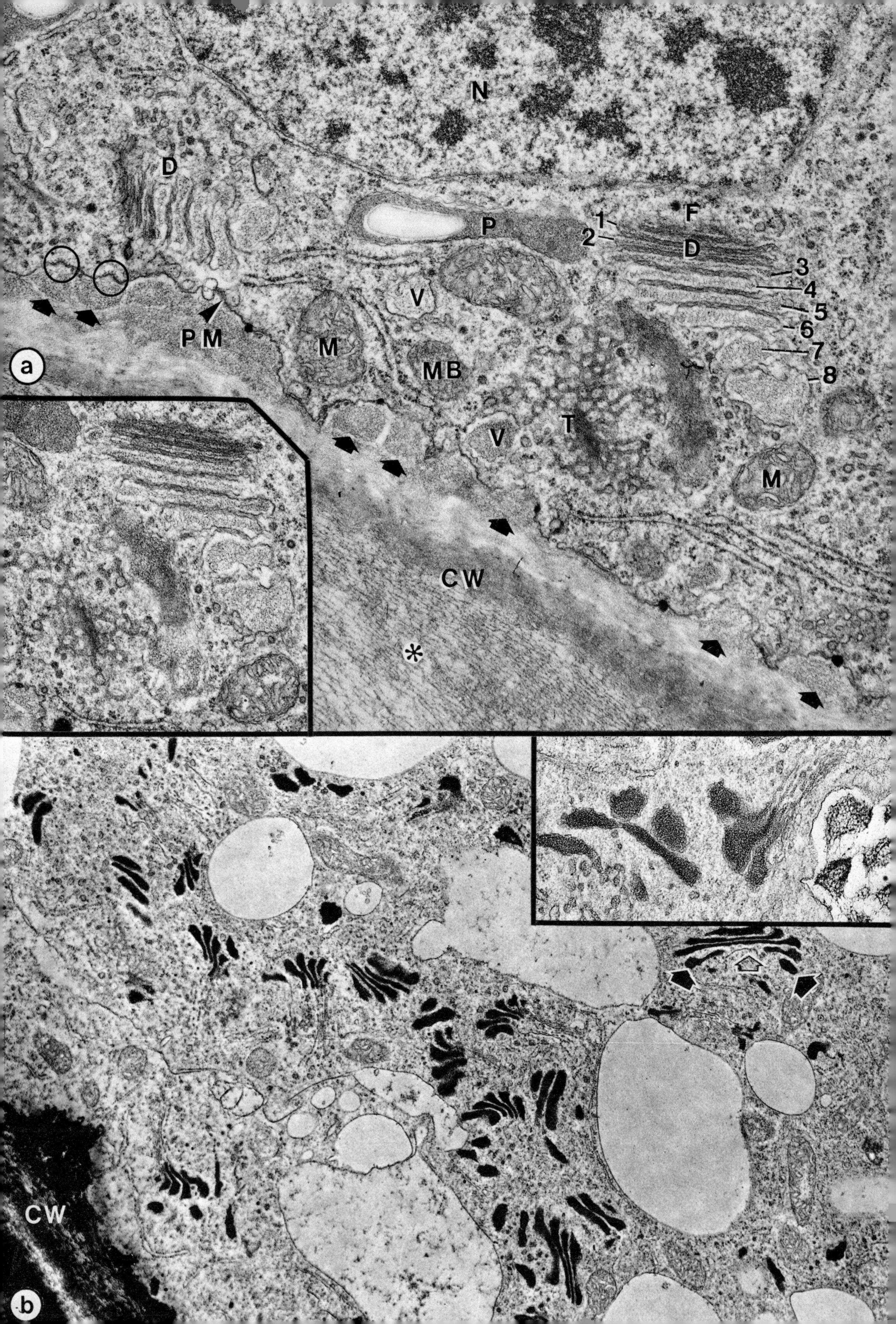

N
D
F
P
1
2
D
3
4
5
6
7
8
V
PM
M
MB
a
V
T
M
CW
*
CW
b

Plate 29

Mitochondria (1)

Mitochondria are constituents of all plant cells, and their number and conformation in a cell are related to the respiratory activity of that particular cell. The micrographs on this page illustrate some of the more commonly encountered forms, as seen in cells having a range of respiratory activities.

Plate 29a This group of mitochondria is lying in the cytoplasm of a parenchyma cell within the floral nectary of broad bean (*Vicia faba*). Although it is difficult to ascribe a precise function to this particular cell, the tissue as a whole secretes nectar, a solution containing mainly sucrose, in an energy-requiring process involving active transport across the plasma membrane. Most of the mitochondrial profiles seen here are circular, suggesting that the mitochondria approximate to spheres 0.75 – 1.5 μm in diameter. Individual mitochondria are seen to be surrounded by [89] an outer (O) and an inner (I) membrane, the latter being infolded (arrowheads) to form the cristae (C). In the matrix of many of the mitochondria there are electron-transparent areas, the nucleoids (N), which contain fine [90] DNA-fibrils (F). Mitochondrial ribosomes (R) lie in the more densely stained regions of the matrix, but are inconspicuous. Mitochondrial granules (G) are more obvious, and probably consist of calcium phosphate. One of the mitochondria (asterisk) is linked to another (star) by a continuous outer membrane (O). This could [94] represent a stage of mitochondrial division, but the same appearance could also be obtained in certain sections of a Y-shaped mitochondrion.

Other features are the plastid (P), and the dictyosome (D) (seen in face view with its associated vesicles). There are numerous dictyosome vesicles, some quite large (DV), and containing a fibrillar material. Many of the free cytoplasmic ribosomes are organized in polyribosomal helics (PR) (see also Plate 21c). × 28 000.

Plate 29b This transfer cell (see also Plate 13) is located alongside two xylem elements (X) in the cotyledonary node of a lettuce seedling (*Latuca sativa*). The cytoplasm adjacent to the wall-membrane apparatus contains many mitochondria with densely packed cristae (C), which nearly completely obscure the nucleoids (N); otherwise they are similar to the [89] mitochondria in (a) above. The high rate of respiration indicated by the number and conformation of these mitochondria may be connected with consumption of energy in the pumping of solutes across the plasma membrane of the transfer cell. × 24 000.

Plate 29c Many cells are less active (in terms of respiration) than either nectary or transfer cells, and their mitochondria are correspondingly simpler, as illustrated here using part of a young endodermal cell of an *Azolla* root. Although the mitochondria are similar in size and shape to those in (a) they have a much reduced system of cristae (C). The outer (O) and inner (I) membranes are quite distinct, but neither shows the tripartite construction as, for example, found in the plasma membrane. The outer membrane (and the cytoplasm in general) is densely stained, probably as a result of liberation of phenolic compounds during fixation. The mitochondrial ribosomes (R) are smaller than their cytoplasmic counterparts. As in the other [90] micrographs on this plate, the number of profiles of cristae that are seen far exceeds the number of visible connections between the cristae and the inner mitochondrial envelope (e.g. open arrows). The inference is that the connections are small, and that the cristae expand from small 'necks' into the central compartment of the mitochondrion. × 106 000.

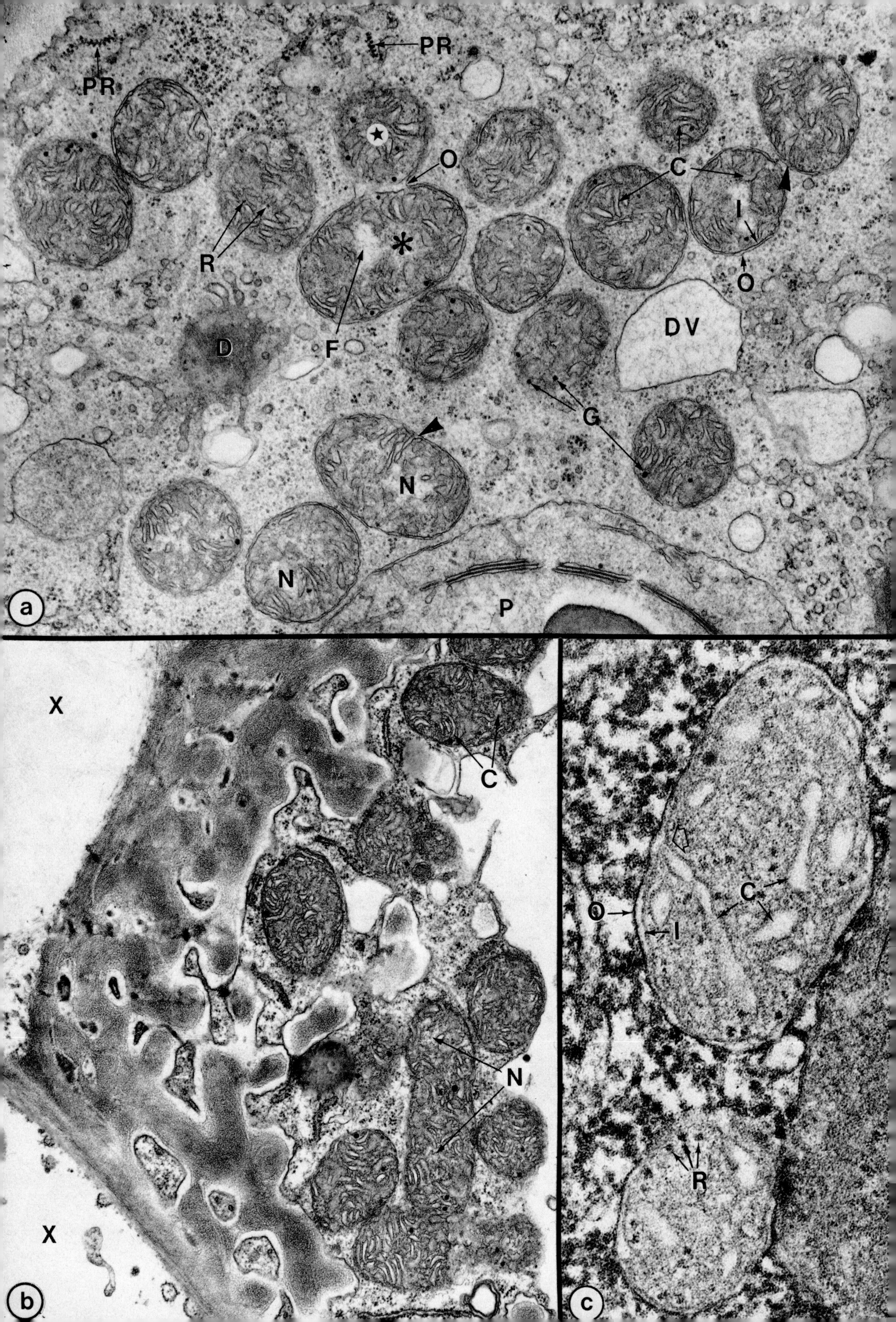

PR
PR
O
C
R
F
D
DV
I
O
G
N
N
P
a
X
C
X
N
b
O
I
C
R
c

Plate 30

Mitochondria (2)

Although the examples presented in Plate 29 are typical of mitochondria in many plant cells, other configurations can also be found. Some are displayed here.

Plate 30a These unusual mitochondria were found in some root tip cells of white lupin (*Lupinus albus*). 10 Interpretation of the three-dimensional configuration is clearly impossible from a single micrograph. One mitochondrial profile (M-1) forms an irregular but continuous ring, with another (M-2) lying quite close to it. M-1 may indeed be ring-shaped, but it is more likely that both it and M-2 are parts of a single plate that undulates up and down through the plane of the section. Support for the latter proposal comes from other micrographs such as the insert, which demonstrates that arms of the mitochondrion are connected (asterisk). If all parts of M-1 and M-2 are connected in this manner, the plate so formed must be at least 15 μm × 4 μm. The insert also shows the numerous small cristae (C) and mitochondrial ribosomes (MR) (smaller than cytoplasmic ribosomes (CR)) lying in the mitochondrial matrix. Magnifications: × 16 000, insert × 36 000.

Plate 30b and c The micrograph shown in (b) (× 20 000) is one of a sequence of 43 that together encompassed the whole of a young cell of the unicellular green alga *Chlorella fusca* var. *vacuolata*. The complete sequence was used in building the three-dimensional reconstruction of the mitochondrion and chloroplast shown in (c). The model is viewed as from the bottom of (b) looking upwards; (b) corresponds to the level in the model marked by the arrows X and Y. Components are labelled as follows: nucleus (N – dotted line in (c), nucleolus seen in (b)); vacuoles (V – containing electron-dense polyphosphate bodies); chloroplast (C – 130 containing a pyrenoid (P) which is shown exposed in (c) by cutting away the layer of chloroplast that covers it); microbody (MB – lying in the cytoplasm, but in 137 close proximity to the pyrenoid, a regular feature of these cells); centriole pair (CP – not included in (b)).

The main feature displayed in the reconstruction is that the mitochondrion (M), seen as quite separate profiles (1-8) in (b), is a 3-dimensional continuum of loops and branches – a mitochondrial reticulum. Some 88 branches lie close to the cell surface (2 and 3 in (b) = A and B in (c)). Another branch (7 in (b), C in (c)) penetrates a cytoplasmic channel that runs through the chloroplast, emerging at D. Very few cells or organisms have been reconstructed in three dimensions, and it is therefore not known how commonly the mitochondrion is in the form of a reticulum.

The three layered (dark-light-dark) structure (W) in (b) is the remains of the wall of the parent cell. In this (but not all) species of *Chlorella*, the layer of the parent cell wall that survives contains a substance closely resembling sporopollenin, the resistant material of 22 pollen grain walls (Plates 11, 12).

Plate 30d One bizarre mitochondrial conformation that is commonly found is a ring-shaped profile that probably results when a cup-shaped mitochondrion is sectioned across the 'cup'. The two layers of the mito- 88 chondrial envelope are visible at both inner *and* outer faces of the ring (circles). In this example, from a developing cotyledon of *Phaseolus multiflorus*, the crista is unusual in that it is very extensive, perhaps even forming a continuous cisterna within the mitochondrion. × 60 000.

Plate 30e In contrast to the mitochondria of normal vegetative cells these four examples, from a developing pollen grain of *Avena fatua*, are very small (none is more than 0.5 μm in diameter) and relatively undifferentiated. Each contains a few cristae (C) and mitochondrial ribosomes (R) within the inner (I) and outer (O) membranes. Both mitochondria and plastids can become very rudimentary during development of pollen. × 55 000.

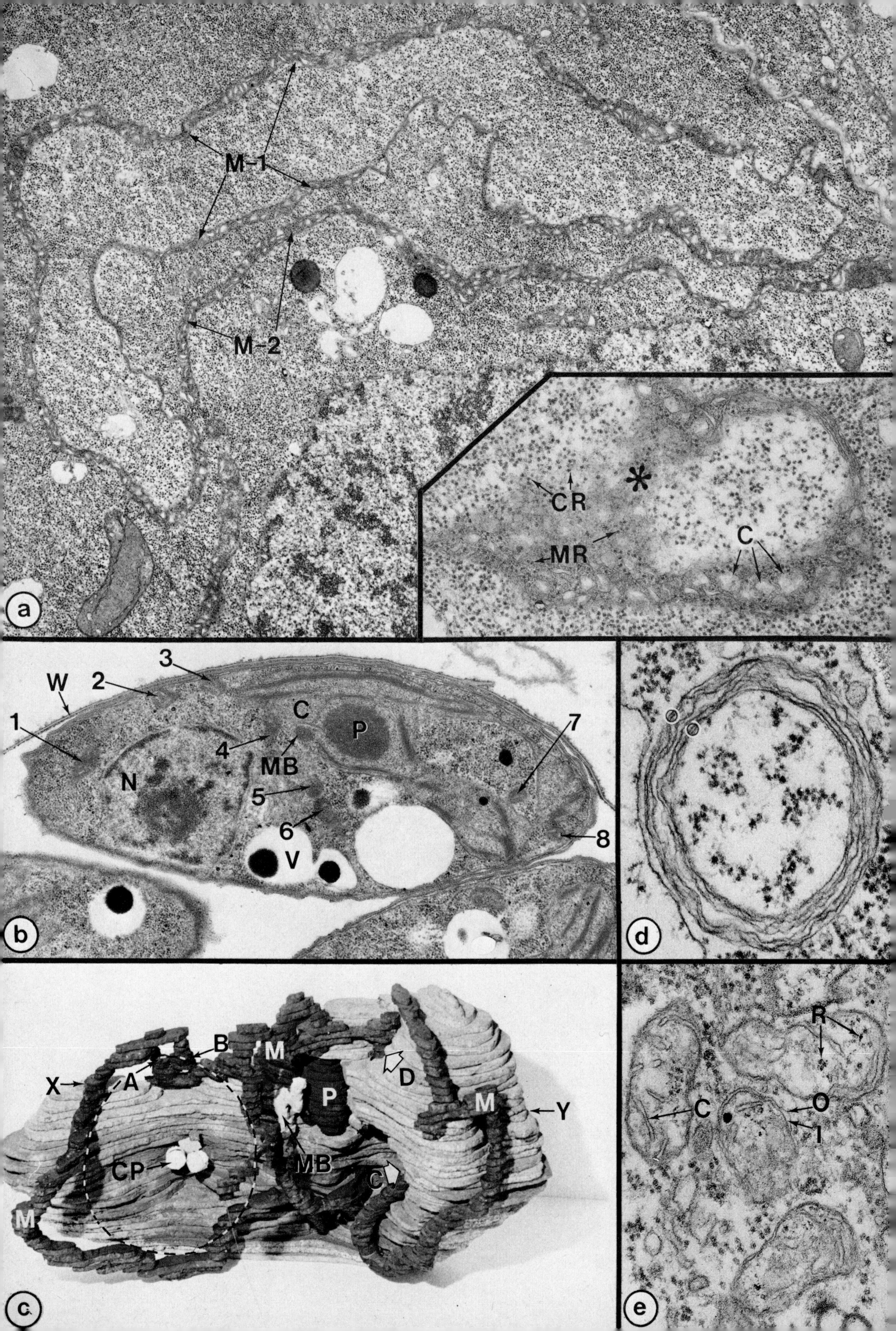
M-1
M-2
a
CR
MR
C
W
2
3
1
C
P
4
MB
N
5
6
V
7
8
b
d
X
B
A
M
P
D
M
Y
CP
MB
C
M
c
R
C
O
I
e

Plate 31

Plates 31-40 are concerned with the structure, development, and inter-relationships of plastids.

Plastids I: Proplastids and their Development to Etioplasts and Chloroplasts

Plate 31a Plastids with rudimentary internal structures, but with the capacity to develop into more [98] complex types, are described as proplastids. They are illustrated here (P) in a cell at the outer surface of a stem apex, and in part of an underlying cell (bottom left of micrograph). They are similar in both cells, despite the fact that maturation and differentiation of the two will produce very different plastids—rudimentary types in the mature epidermis, and well developed chloroplasts in the underlying tissue.

A nucleus (N), with nucleolus, a large vacuole (V) with flocculent contents, several smaller vacuoles, and the thin cuticle (C) are also seen. Oat (*Avena sativa*) stem apex, × 5300.

Plate 31b The proplastids of meristematic cells at the stem apex (31a) are similar to those in root tip cells. Root tip proplastids, however, do not normally develop into chloroplasts. The one shown here displays the two concentricmembranes of the envelope (visible, e.g., within the circles), with occasional invaginations of the inner membrane (arrows, left hand side). A small starch grain (S) is present in the section, partly ensheathed by internal membranes (thylakoids). The internal membrane system is sparse and not organized. A few particles which may be plastid ribosomes (smaller than cytoplasmic ribosomes) are seen (e.g. within square) and nucleoid areas (large arrows), with fine fibrils, presumably containing DNA, are also present in the stroma. *Vicia faba* root tip, × 77 000.

Plate 31c This micrograph is representative of the appearance of cells which are differentiating in the light [110] from the meristematic condition shown in 31a to become mesophyll cells. The vacuoles (V) are large, and there are intercellular spaces (IS) in the tissue. Much of the chromatin in the nucleus (N) is heterochromatic (dense areas).

The plastids (P) have developed complex internal membrane systems, and by now are young chloroplasts, though grana are not yet obvious. Some starch grains are present in them (S).

Avena sativa, young leaf of illuminated seedling, × 8000.

Plate 31d The same tissue as for 31c was used to obtain this micrograph, but the seedling was grown in darkness so that it became etiolated. Vacuoles (V), intercellular spaces (IS), and nucleus (N) are much as in the light-grown seedling, but the plastids have developed into young etioplasts (E). The main feature of [115] etioplasts is the semi-crystalline lattice of membranous tubes known as the prolamellar body (open arrows). The example numbered 1 is of the type illustrated further in Plate 37h, and number 2 is of the type shown in Plate 36e and g. Number 3 is probably of the same type as Plate 37g. If the plant were to be illuminated, etioplasts would pass through developmental stages of the type shown in Plate 38, and become chloroplasts.

Avena sativa, young leaf of etiolated seedling, × 8000. Reproduced by permission from *Biochemistry of Chloroplasts,* II, 655–676 (Ed. T. W. Goodwin, Academic Press, 1967).

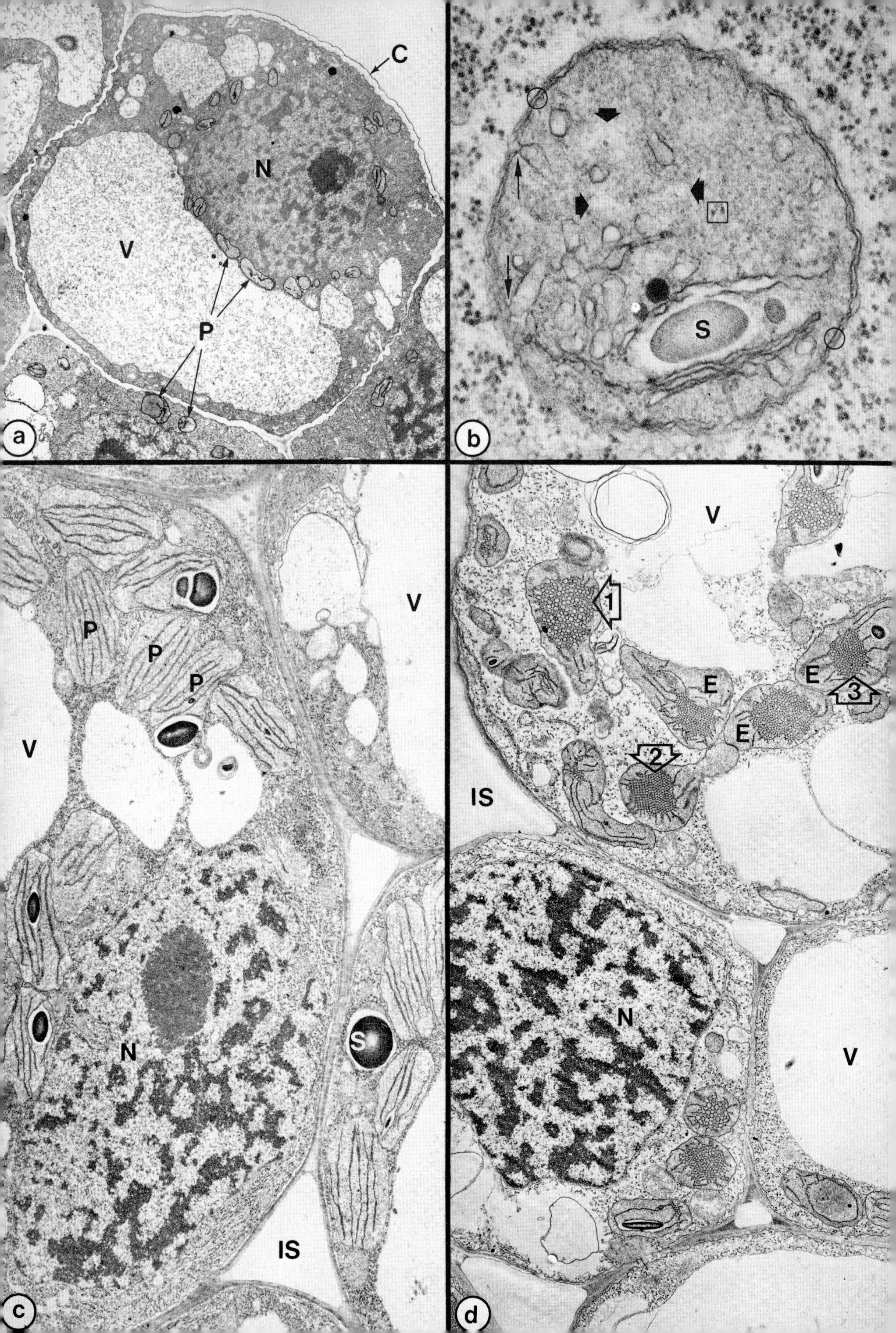

C
N
V
P
a
S
b
P
P
P
V
V
N
S
IS
c
V
1
E
E
3
E
2
IS
N
V
d

Plate 32

Plates 32–35 illustrate chloroplasts.

Plastids II: Chloroplasts (1)

Plate 32a A transverse section of a typical leaf is viewed here by light microscopy. The cells of the upper and lower epidermis (lowermost and topmost cell layers, respectively) do not contain well developed chloroplasts. These are visible, however, in cells of both the spongy mesophyll (S) and the palisade layer of columnar cells (P). Two small veins are included, and in the one on the right, it can be seen that chloroplasts are not obvious in the bundle sheath cells (stars) (compare Plate 34a). *Hypochaeris radicata*, × 165.

108

Plate 32b This is a high magnification light micrograph of the same material as in (a). One palisade cell is shown. The section (1-2 μm thick) has passed through the peripheral layer of chloroplasts in the cell. Numerous grana (the dark grains) are resolved within each chloroplast. The hemispherical shape of the chloroplasts can be seen from the two types of profile present, the 'side views' corresponding to the plane shown in (c), the 'top views' to that in (d). × 1100.

99

Plate 32c and d The electron microscope reveals finer details of chloroplast structure: the two membranes of the chloroplast envelope (E); 'side' (S) and 'top' (T) views of grana; densely staining ribosomes in the stroma (circled) (note in (c) that they are smaller than cytoplasmic ribosomes); starch grains (G), which for some unknown reason have become stained very differently; the system of fret membranes (F) that interconnect the grana (better seen in (c) than in (d), where most of them are present in oblique or face view); plastoglobuli (P). (c) *Avena ventricosa*, × 33 000; reproduced by permission from *Canadian J. Genetics and Cytology*, **12,** 21-27. (d) *Zea mays* mesophyll chloroplast, × 18 500.

Plate 32e and f Further details of *Zea mays* grana are illustrated here in 'top view' (e) and 'side view' (f). The grana contain very many more layers (discs) than those in (c), and the stroma (S) is very much subdivided by the frets (F) passing between the grana. A few ribosomes (circled) can be seen in the stroma. The fret channels are somewhat swollen in (f), making their presence and their course easier to distinguish. In (e) the frets are in oblique view, and they appear merely as grey shadows, some of which are marked. The granum at top right appears to have been sectioned in the plane of the discs. The striations across the other grana indicates how oblique the section is in each case—each striation representing an oblique slice through one disc. Numerous frets (white arrowheads) can be detected entering the granum at top right, and since the section is only thick enough to accommodate about 3 granum discs, this means that *each* disc must develop *many* connections to frets. (e) × 55 000, (f) × 75 000.

100

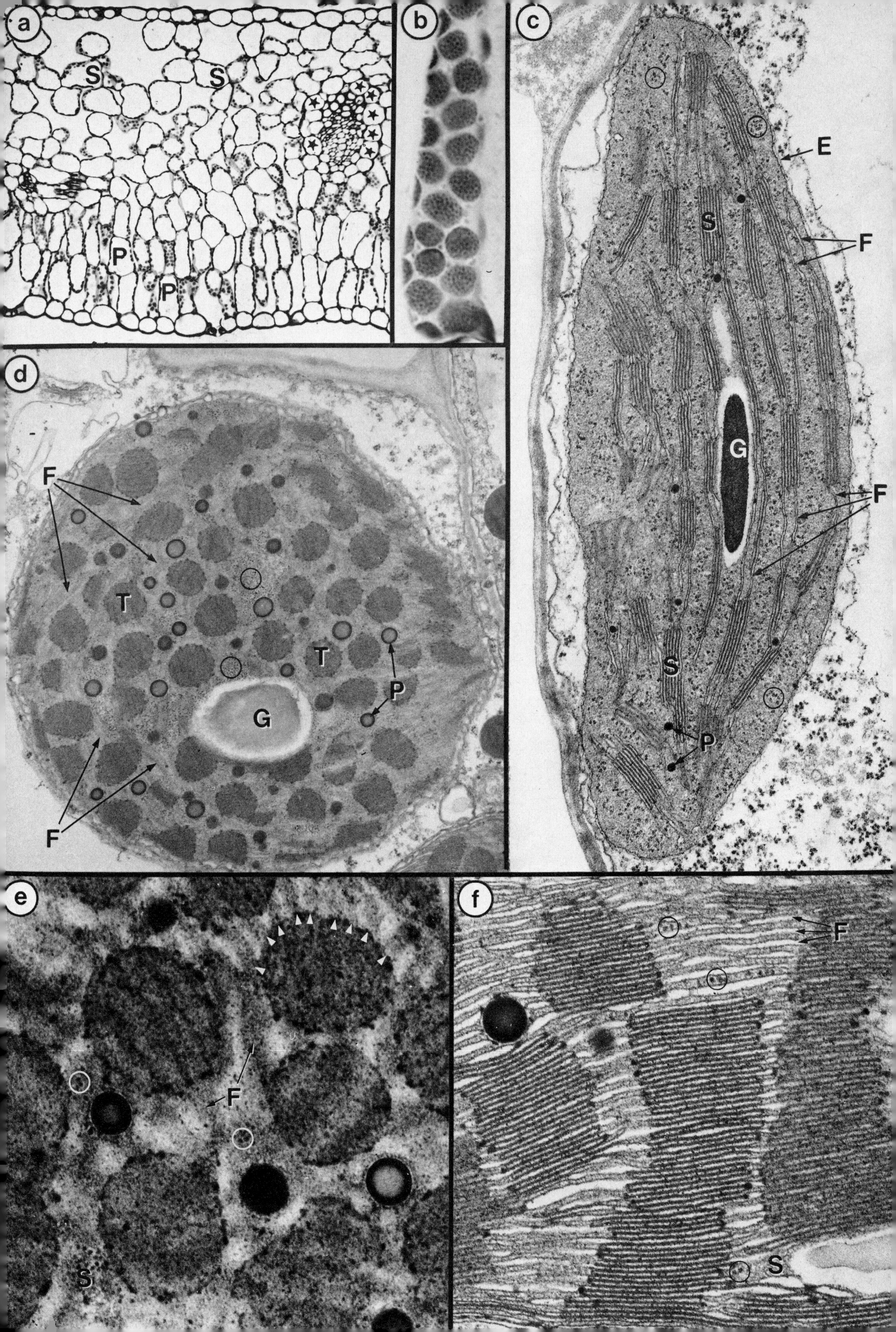
a
S
S
P
P
b
c
E
S
F
G
F
S
P
d
F
T
T
P
G
F
e
F
S
f
F
S

Plate 33

Plastids III: Chloroplasts (2): Details of Chloroplast Membranes

Plate 33a This ultra-thin section of part of a lupin (*Lupinus* sp.) leaf chloroplast shows several details of 121 both the chloroplast envelope (E) and the internal membrane system. The inner of the two membranes of the envelope is invaginated in places. No subunits can 107 be seen in the membranes, but the 'A-space' between granum discs is visible where the plane of the section is precisely at right angles to the plane of the discs (white arrowheads). Where the section has included the points at which fret connections enter granum discs, continuity between the fret channels and the disc loculi is shown (open arrows). × 140 000.

Plate 33b Grana with very numerous discs are examined here by freeze etching. The fracture plane has passed down two grana (G, lower left and upper 105 right) approximately at right angles to the plane of the discs. A third granum is in the centre of the micrograph, and it lay obliquely to the fracture, so that surface views of membranes were exposed. The clear white bands are 'steps' where the fracture has descended from one membrane to another. The intervening particle-studded areas represent the interior of the disc membranes. One face carries large particles (the B-face), the other larger numbers of smaller particles (the C-face). It can be seen that on passing from the discs to the system of frets that interconnects the 3 grana, the large B-face particles become less frequent (arrows from labelled B-face). The small C-face particles do not diminish in numbers from disc to fret (arrows from labelled C-face). *Alocasia*, × 67 000.

Plate 33c In this high magnification view of a freeze etched preparation, about two thirds of the area of a granum disc membrane is seen in the bottom part of the picture, adorned with the scattered large particles typical of the B-face. Numerous frets radiate away from the circumference of the disc, and the fracture has 102 exposed examples of both their B- and their C-faces (B and C respectively). Several fret B-faces are continuous with the B-face of the disc (arrows), and as in (b), the frequency of the larger particles is seen to be much lower in the frets. By contrast, the fret C-faces have very numerous small particles. Judging by the number of large particles present, the area at the upper left of the picture may be the edge of another granum, in which small areas of many successive B- and C-faces are exposed. *Lomandra longifolia*, × 105 000.

(We thank Dr. D. Goodchild for providing (b) and (c).)

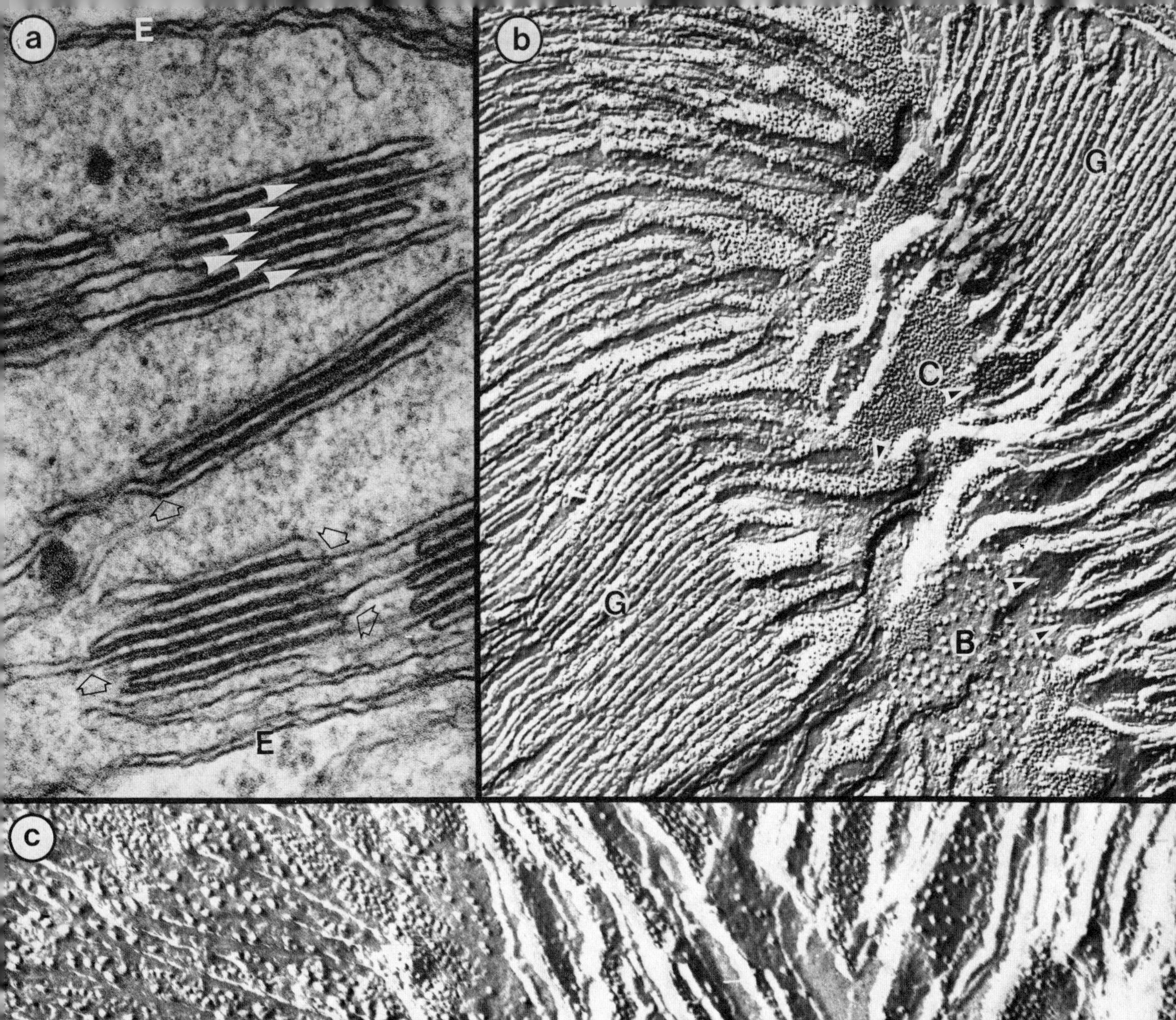
a
E
E
b
G
C
G
B

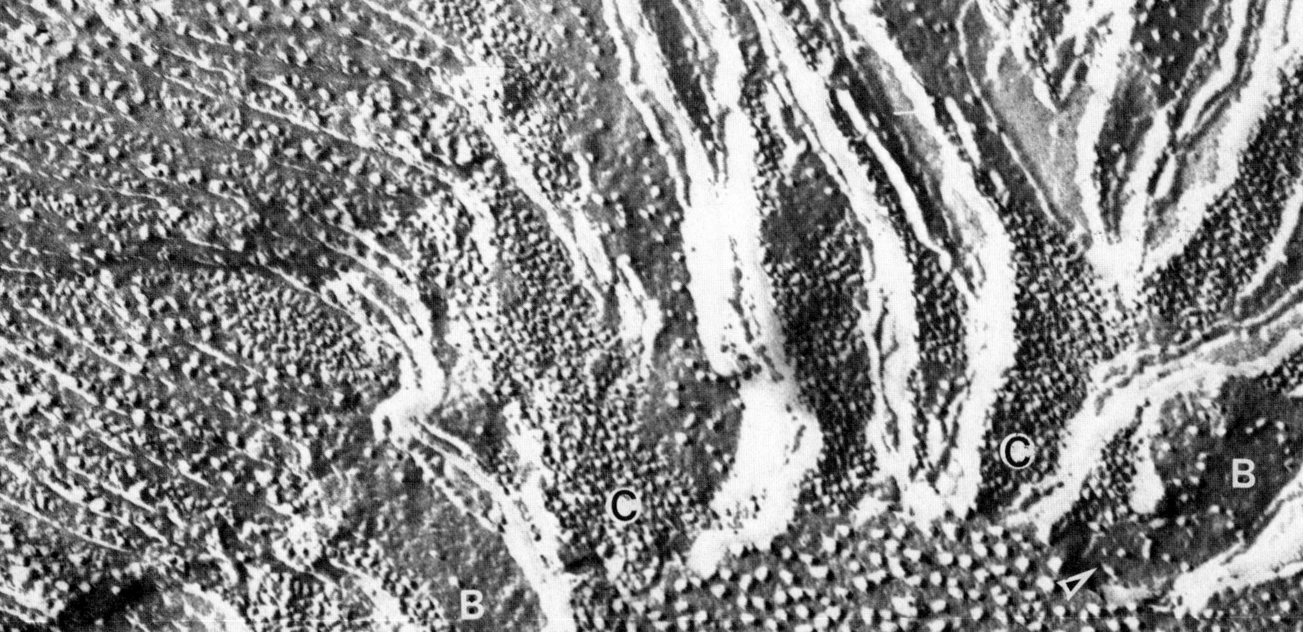
c
C
C
C
B
B
C
B

Plate 34

Plastids IV: Chloroplasts (3): Dimorphic Chloroplasts in the C-4 Plant, *Zea mays*

Plate 34a The two types of chloroplast found in C-4 plants are distinguishable even in the light microscope, as here in a section of a maize leaf. The xylem (X) and phloem (P) of each vein are surrounded by a ring of bundle sheath cells—5 in this example. The bundle sheath is in turn surrounded by mesophyll cells. Grana show up darkly stained in the mesophyll chloroplasts (e.g. at large arrowheads). Starch grains are unstained (e.g. small arrows), and are abundant in the agranal bundle sheath chloroplasts. × 1750.

108

Plate 34b Bundle sheath (BS, lower right) and mesophyll (M, upper left) chloroplasts are compared in this electron micrograph. The former contain starch grains (G) in the stroma between the simple, agranal, internal membranes. The latter possess grana and frets (seen in more detail in Plate 32). Both types have chloroplast ribosomes (smaller than cytoplasmic ribosomes), plastoglobuli, and the usual double membrane envelope. The leaf was still growing when it was fixed, and the chloroplasts had not completed their development, as indicated by the presence of a small region of prolamellar body lattice (solid arrow), perhaps a relic of the membrane growth that had taken place in darkness during the night before the early morning harvesting of the material.

115

The cell wall passing diagonally across the micrograph contains the layer of suberized material that surrounds each bundle sheath cell (open arrows). Plasmodesmata (P, parts of two groups included) interconnect the bundle sheath and mesophyll cell protoplasts. The tonoplast (T) of each cell can be seen, separated from the vacuolar face of the chloroplasts by only a very thin layer of cytoplasm. × 27 000.

31

Plate 34c Parts of bundle sheath (BS, on left) and mesophyll (M, on right) chloroplasts are shown here at higher magnification, along with the intervening wall, with its suberized lamella (open arrows, here obliquely sectioned) and a group of plasmodesmata (P). Both of the chloroplasts possess a feature not shown in the previous figure—invagination of the inner of the two envelope membranes to form a 'peripheral reticulum' in the chloroplast. The invaginations (large solid arrowheads) lead into somewhat dilated sacs, pierced by many perforations through which the stroma penetrates (small arrows). This structure is not restricted to C-4 chloroplasts.

The bundle sheath chloroplast does not have a completely 'unstacked' membrane system. Two and three disc rudimentary grana are seen (stars), survivals from early development when grana were larger and more abundant. × 64 000.

107

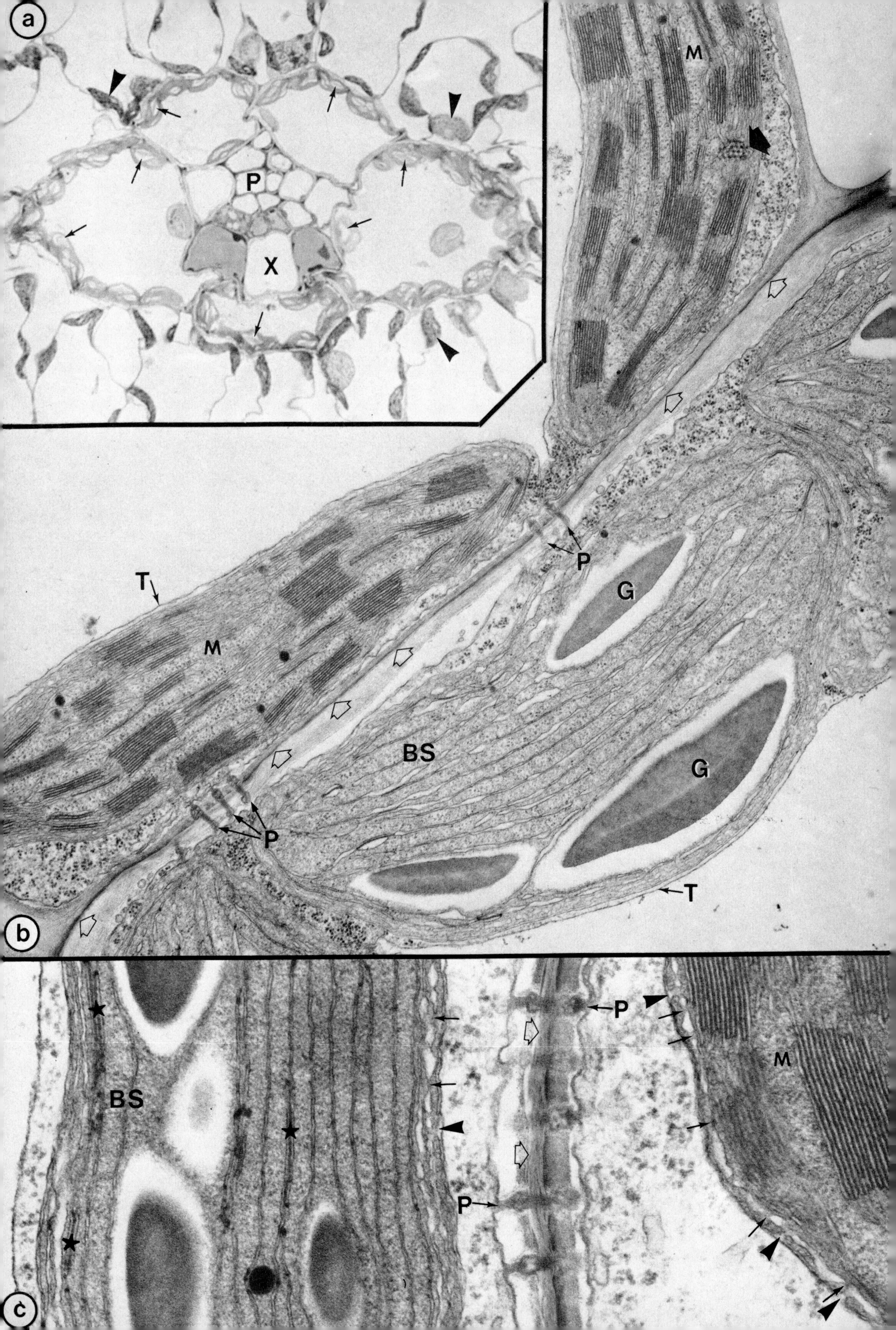

a
P
X
M
P
G
T
M
BS
G
P
T
b
BS
P
M
P
c

Plate 35

Plastids V: Chloroplasts (4): Components of the Stroma

Plate 35a Several components of the stroma are displayed in this micrograph of a greening oat leaf (see also Plate 38c, d).

125 *Ribosomes:* More densely stained than the surrounding material, the plastid ribosomes may be scattered or in chains (rectangle) or clusters (between stars). Some lie in contact with, and probably are bound to, the thylakoids (arrowheads). They are smaller than cytoplasmic ribosomes, seen at lower left outside the plastid envelope (E).

123 *Nucleoid:* The nucleoid (N) sectioned here contains fine fibrils, 2-3 nm in thickness (small arrows). These are probably the histone-free DNA of the plastid. Other fibrillar material is present, as are granules (double arrows) which could be developing plastid ribosomes.

Proteinaceous ground substance: The general background in the plastid is finely particulate (except in the nucleoid, which presumably contains some material that excludes the rest of the stroma). The particles cannot be identified, but many must consist of the 99 CO_2-fixing enzyme, ribulose 1,5-diphosphate carboxylase, some molecules of which are shown negatively stained (pale against a dark background) at high magnification in the insert (one example ringed). The enzyme is a very large protein, with a molecular weight of half a million, each molecule being a particle of side about 10 nm.

Crystals and spherulites (ordered aggregates) of proteinaceous material are quite commonly found in the stroma, particularly when the plastids have been subjected to physiological stress. The example shown here (S) occurs in *Avena* plastids under natural conditions. It consists of bundles of fibrils, which in turn are built up of small units, perhaps molecules of ribulose diphosphate carboxylase. × 94 000, insert × 430 000.

Plate 35b The iron-containing protein, ferritin, occurs in most types of plastid, occasionally in con- 128 spicuous masses, as here in a lupin leaf chloroplast. Each electron dense point represents the several hundred iron atoms present in the centre of each protein molecule. The protein shell around the iron core is not visible. Some parts of the mass of molecules show a measure of symmetry. When ferritin crystallizes in the plastid the arrangement of molecules is as shown in the insert.

Several dark, spherical plastoglobuli are present at the periphery of the ferritin aggregate. × 120 000.

Plate 35c-g Chloroplasts of most algae contain one or more pyrenoids, which are easily seen, even with 130 the light microscope, and in the green algae are rendered especially conspicuous by the development of ensheathing grains of starch. In the electron microscope the pyrenoid is seen as a dense granular mass, occasionally with a crystalline lattice, and in many species traversed by modified thylakoids. The example shown here is simple, with no thylakoids. The sequence of pictures illustrates changes that occur during the life of a *Chlorella* cell. In very young cells (C) starch (solid stars) appears as a thin plate around the pyrenoid (P) and accumulates while the cell photosynthesizes and grows (d, e). Starch grains also appear in other parts of the chloroplast stroma (open stars in d, e and f) but the pyrenoid starch is especially labile. It is eroded *before* the non-pyrenoid starch as the cell begins to divide (f), and the products of the division generally have starch-free pyrenoids (g).

(c) and (d) illustrate (as does Plate 30b, c) the proximity of the microbody (M) to the pyrenoid in this alga. 123 All × 33 000, except (f), at × 18 000. (Micrographs provided by Dr. A. W. Atkinson, Jr.)

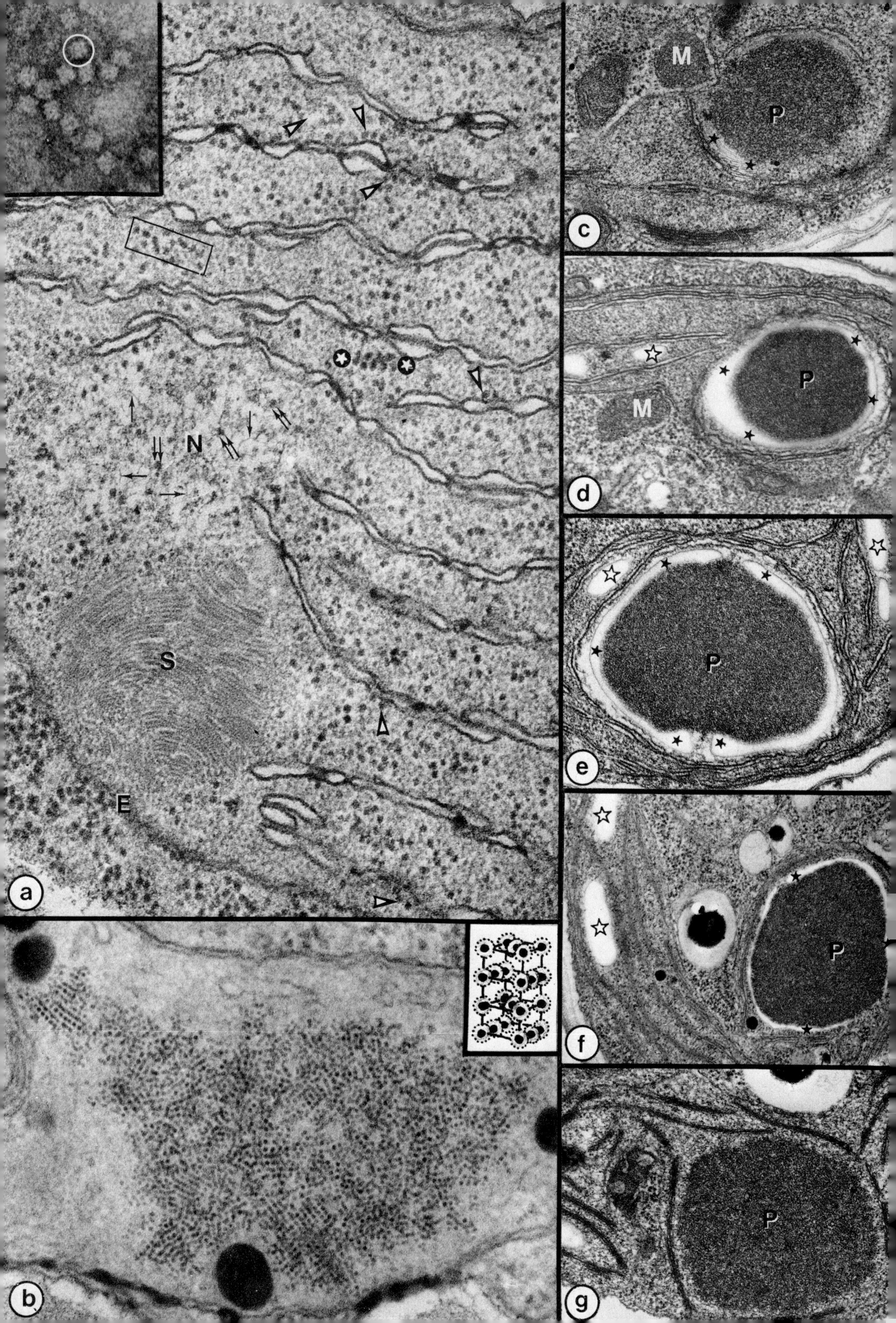

a
b
c
d
e
f
g
N
S
E
M
P

Plate 36

Plastids VI: Etioplasts and Prolamellar Bodies (1)

Etioplasts develop in leaves on plants that are growing in darkness. Their main distinguishing feature is the system of membranes known as the prolamellar body. The micrographs and models on this and the next plate illustrate some of the diversity of form of prolamellar bodies. All of the examples are from leaves of oat seedlings (*Avena sativa*).

Plate 36a These two etioplasts display the two 121 membranes of the plastid envelope (black arrow), and numerous plastid ribosomes, which are somewhat smaller than their cytoplasmic counterparts. The prolamellar bodies (PLB) can be seen to be semi-crystalline lattices, but in these examples the plane of section is not such as to display regular lattice planes to advantage. Subsequent micrographs have been selected to do this. The lattice at the bottom centre region of the left hand prolamellar body (above the white arrow) is obviously different from the neighbouring lattice (see also Plate 37h). × 47 000.

Plate 36b, c, d One, uncommon, form of pro- 111 lamellar body lattice is illustrated here by a micrograph, a 3-dimensional drawing, and a model. The lattice is composed of tubes, branched and interconnected in three axes at right angles to one another. The unit of construction is shown enclosed within the square on the micrograph. The etioplast stroma penetrates the prolamellar body *between* the tubes, but in (b) little structure is seen in the stroma component of the prolamellar body because uranyl acetate stain (which adds contrast to ribosomes) was not used. Models of the type shown in (d) will be used to illustrate the other types of prolamellar body (below): comparison with (c) emphasizes that such models represent only the orientation of the tubes, and *not* their diameter and smoothly confluent surface contours. (b) × 78 000.

Plate 36e, f Most prolamellar body membrane 112 lattices consist of tetrahedrally branched tubular units. The arrangement of the tubes in the lattice shown here is analogous to the crystal structure of wurtzite (a mineral form of zinc sulphide). The view of the model in (f) (left hand side) matches an area such as that marked on the micrograph. The other views of the model are side views—looking at an edge (centre) or at a face (right hand side) of the hexagonal 'crystal'. (e) × 70 000.

Plate 36g, h The micrograph and photographs of models depict an alternative lattice form, analogous to diamond crystals or the zincblende form of zinc sulphide. The overall crystal shape is bi-pyramidal. The views of the model were taken looking at an edge (left hand side), at a face (centre) and at a vertex (right hand side). The major difference between this lattice and the previous is that successive planes of hexagonal 'rings' are out of register. The centre picture of the model illustrates the overlapping hexagons of the lattice (compare (f) (left hand)). The section (g) is somewhat tilted relative to (h) (centre), but individual hexagons are seen in horizontal bands, each band being part of one plane of the lattice. The insert (of the area above the arrow at lower centre of the main picture) shows how the successive bands are out of register by half a hexagon. This does *not* apply to the lattice shown in (e) and (f).

Both (e) and (g) show the many etioplast ribosomes in the stroma component of the lattice. They also show continuity of lattice tubes and flattened thylakoids projecting outwards from the prolamellar body. (g) × 70 000, insert × 120 000.

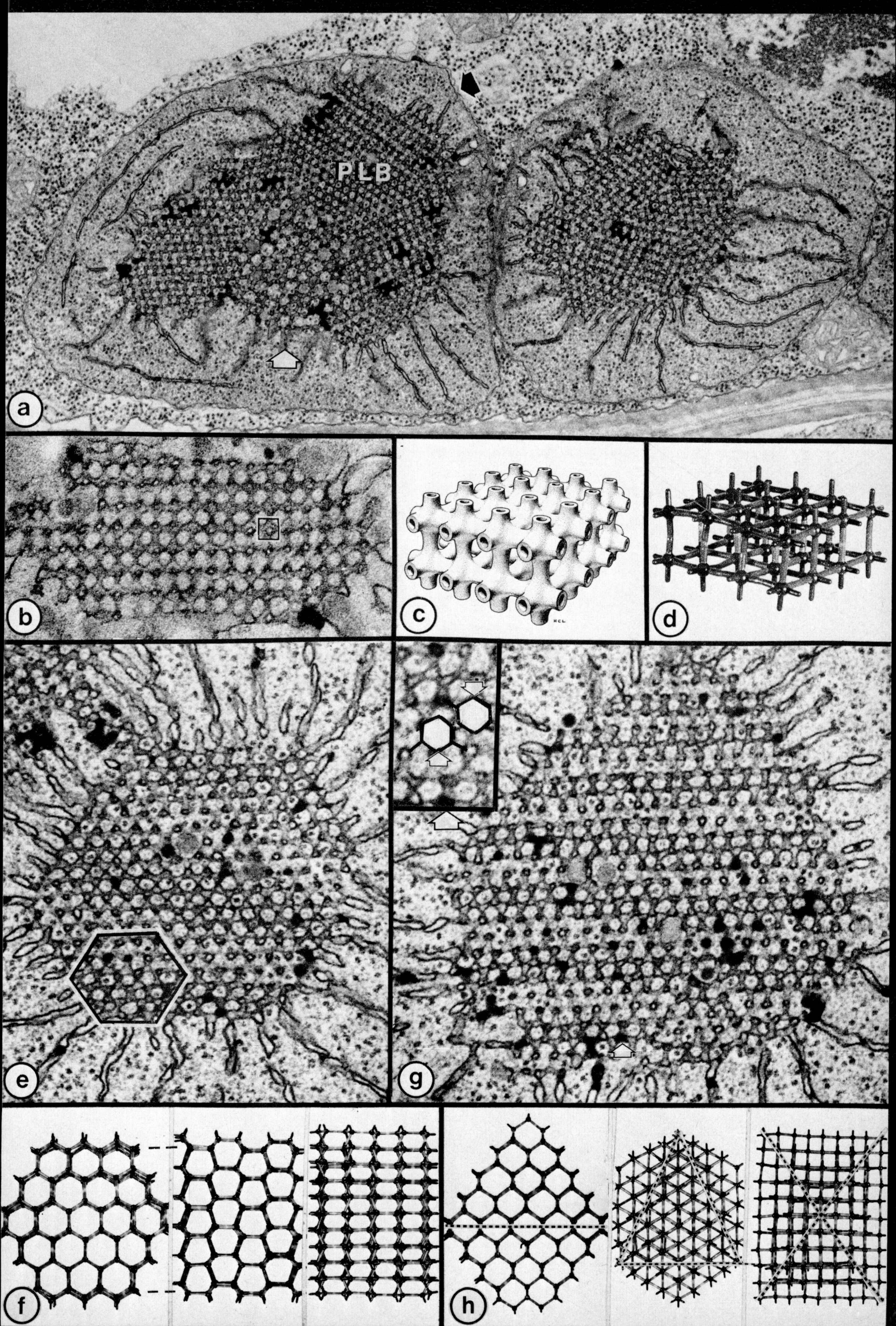
PLB
a
b
c
d
e
g
f
h

Plate 37

Plastids VII: Prolamellar Bodies (2)

Plate 37a Tetrahedrally branched models of carbon 112 atoms were used (as in Plate 36f, h) to construct this large tetrahedron-shaped constructional unit, which is the basis of the prolamellar bodies in (c), (d) and (g). It consists of 'zincblende' type lattice (as Plate 36g, h) but with a pentagonal dodecahedron at each vertex, and special 5- and 6-membered rings running along each edge. The latter arise when the large tetrahedral units join together in prolamellar bodies.

Plate 37b, c, d, e 'Centric' prolamellar bodies have 20 large tetrahedral units radiating outwards, one from each of the 20 vertices of a central pentagonal dodecahedron (seen in section at the centre of (c) and (d). In three dimensions, an icosahedral shape is generated, as seen in (b), where a model is viewed looking straight at one vertex (the edges of the tetrahedral units on the near side of the model have been drawn in). In median section (to which (c) and (d) *approximate*) parts of 10 of the large tetrahedral units are seen, each one radiating out from the centre. The 10 sectors are marked in the micrographs by the straight white lines. For comparison, a median slice of the model is shown in (e). Ribosomes are seen in the lattice. (c) × 38 000; (d) × 73 000.

Plate 37f, g Here the large tetrahedral units (see (a)) are combined in a more complex fashion—not just radiating out from a central pentagonal dodecahedron, but filling space between many evenly spaced pentagonal dodecahedra. (f) represents a slice (rather thick in relation to the ultra-thin section seen in (g)) through a model of this form of packed tetrahedra. One pentagonal dodecahedron is at the centre, 4 others are marked (arrows) at the periphery, and yet others were above and below the plane of this slice. The triangular sectors at upper left and lower right in the model represent triangular faces of units of the type seen in (a).

In the micrograph (g), lines indicate the boundaries of triangular faces of the type seen in (f). Just as the model could have been extended, this prolamellar body extends to upper left. Also certain areas are missing (e.g. pentagonal dodecahedra at or near positions indicated by asterisks. (g) × 84 000.

Plate 37h, i, j, k There is symmetry in this very complex form of prolamellar body, but the sections are so 113 much thinner than the basic units of construction that the details of the lattice are difficult to discern. An area approximating to (j) is marked on the micrograph. Much of the lattice consists of rows of pentagonal dodecahedra, interconnected at 60° to one another to form a network containing gaps which create the hexagonal pattern seen in (j). Rows inclined at 60° to one another can be seen if the micrograph is held up and viewed at a shallow angle along the three axes parallel to the sides of the white hexagon. Side views of the lattice (viewed edge-on in (i), and face-on in (k)) show how the successive strata of pentagonal dodecahedra (between brackets) are joined by other types of ring.

This type of prolamellar body can occur in isolation, or it may be joined to one of the other types (as in Plate 36a). (h) × 70 000.

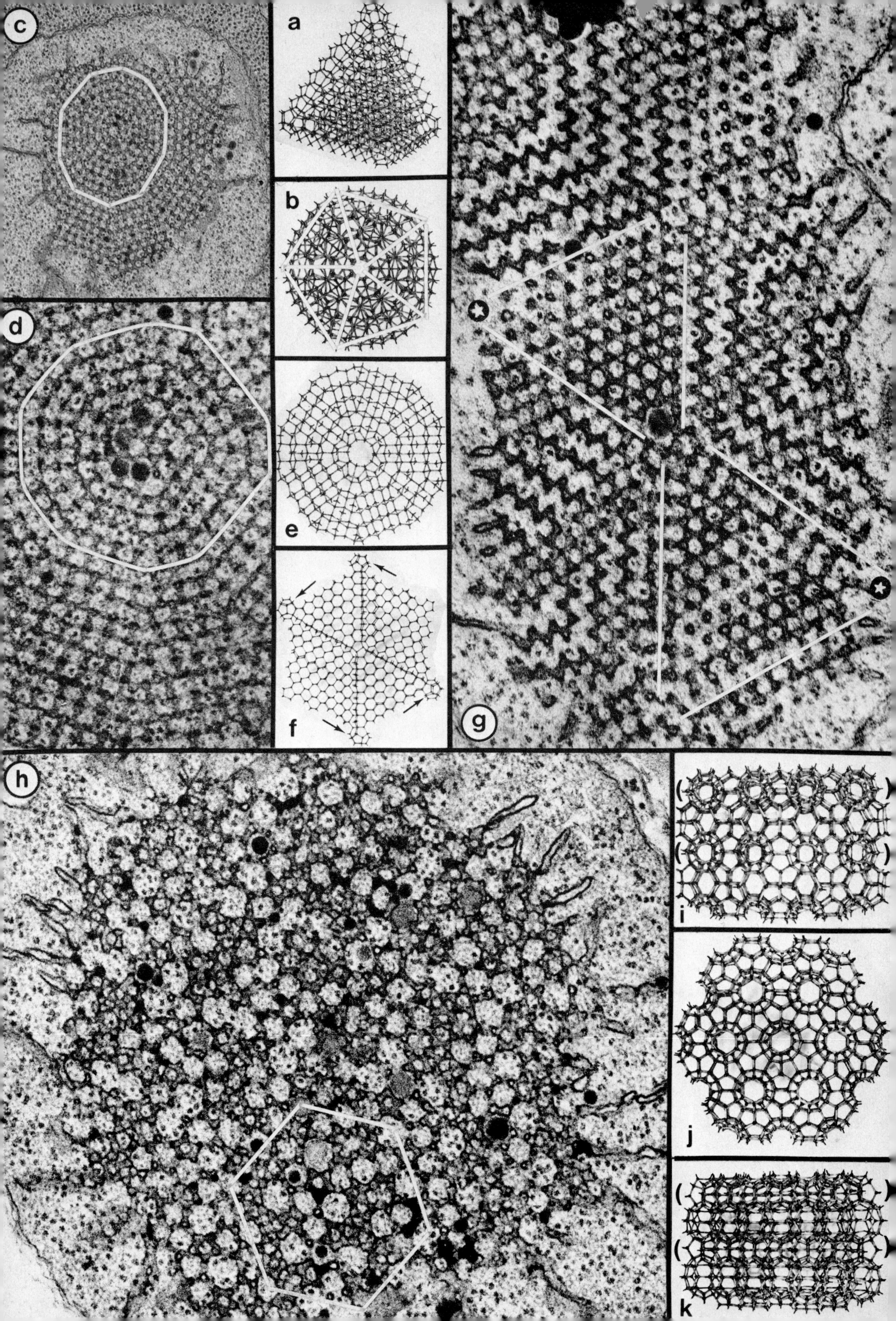
c
a
b
d
e
f
g
h
i
j
k

Plate 38

Plastids VIII: The Greening Process: From Etioplast to Chloroplast

Plate 31 illustrates the development of proplastids to chloroplasts in plants grown in the light, and their very different development to etioplasts in plants that are kept in darkness. Subsequent plates deal with chloroplasts (32-35) and etioplasts (36, 37) in more detail. The present plate shows three stages in the conversion of etioplasts to chloroplasts, as seen when a dark-grown plant is illuminated. As in Plate 31, the micrographs are of portions of oat (*Avena sativa*) leaf. Each stage is illustrated by means of sections which show thylakoids in profile view (a, c, e) and in face view (b, d, f).

131

Plate 38a, b Illumination very rapidly converts the protochlorophyll found in the prolamellar body of etioplasts to chlorophyll. Much slower structural changes are also set in train. The prolamellar body loses its crystallinity and the membrane in it metamorphoses into spaced out flattened sacs called primary thylakoids. This pair of micrographs represents a stage reached 2 hours after the dark-grown plants were brought into the light. The prolamellar body remnant (PR) has not completely dispersed. Many of the perforations that were delimited by tubular membranes in the prolamellar body still survive as small pores through the primary thylakoids (small arrows in the profile (a) and face (b) views of the thylakoids). The plastid ribosomes are mostly in clusters and chains suggestive of polyribosomes (e.g. open arrows). Invaginations of the inner plastid envelope membrane (stars) and a nucleoid area (N) are present in (a).

Plate 38c, d The amount of protochlorophyll that is converted by light to chlorophyll is very small compared with the amount of chlorophyll found in a mature chloroplast. Net synthesis of chlorophyll begins in the greening plastids after a lag period, which in the material shown here lasts 2-3 hours, and during which no new chlorophyll, and probably no new membrane, is produced. The beginning of the period of rapid synthesis is marked by the appearance of portions of membrane overlapping (large arrowheads) the primary thylakoids. This, the first stage of granum formation, is shown in both profile (c) and face (d) view (and at higher magnification in Plate 35a). One of the overlapping discs is sectioned through its fret connection (opposed arrowheads). Most of the perforations seen in (a) and (b) have by now disappeared from the primary thylakoids, though a large prolamellar body remnant still survives (right hand side of (c)). The membranes of the primary thylakoids are still continuous with those of the prolamellar body (e.g. at small arrows).

110

Plate 38e, f After 10 hours in the light the leaves are obviously green, but still not as green as they would be if they contained mature chloroplasts. Chlorophyll and membrane synthesis has progressed. The small overlaps seen in (c) and (d) have extended to become full sized granum discs (see (f)), and new disc-shaped layers have been added, producing small but clearly recognizable grana, interconnected by frets derived from the primary thylakoids (see (e)). A mass of plastoglobuli marks the remnant of the prolamellar body in each picture.

127

(c), (d), (e) and (f) all show that chains of ribosomes lie on the surface of the growing membrane (beside asterisks in all four micrographs), very like the polyribosomes found on the rough endoplasmic reticulum. These 'rough thylakoids' may be synthesizing protein molecules which, when complete, pass directly into the growing membrane surface.

126

Magnification × 36 000 in all except (c), at × 64 000.

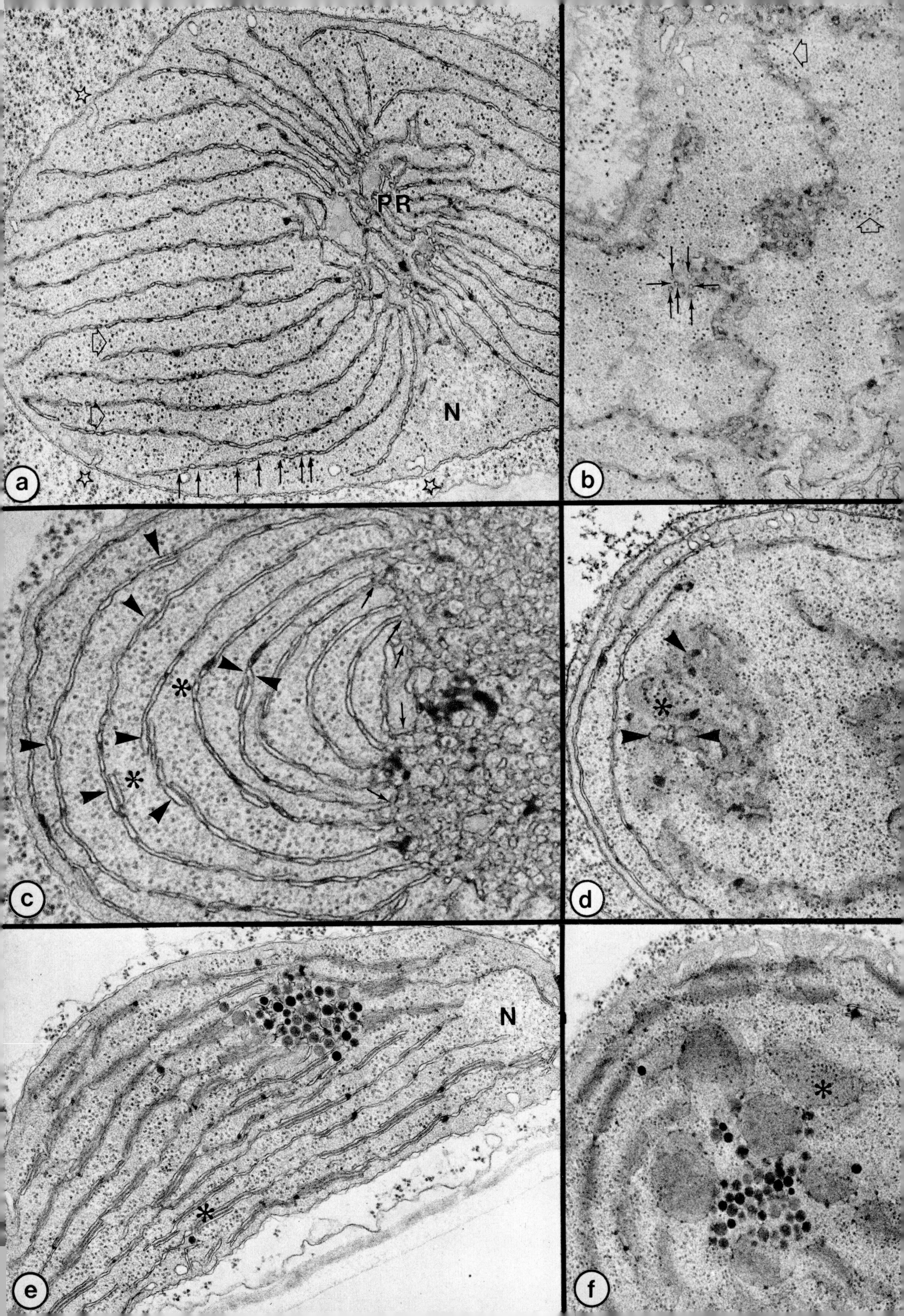

PR
N
a
b
c
d
N
e
f

Plate 39

Plastids IX: Amyloplasts

Plate 39a, b These two scanning electron micrographs show cells in a piece of potato tuber that was prepared by conventional fixation and then dried by the 'critical point' method, in which distortion is minimized. A freshly-cut surface was exposed before taking the pictures. The cells are highly vacuolate in life, indeed some parts of those in (a) appear to be nearly empty. The most conspicuous of the cytoplasmic components is the population of starch grains. The grains are ovoid when large, and nearly spherical when small.

The cell at the bottom of (a) is seen at higher magni- 116 fication in (b). Amyloplast envelope membranes are not resolved; they lie closely appressed to the starch grains. The largest grains are about 30 μm × 50 μm, and the smaller spheres about 3 μm in diameter. Numerous strands of cytoplasm, some with remains of more particulate cell components, form a network stretching out to a peripheral thin layer of cytoplasm at the cell wall. The gaps between the strands presumably represent vacuoles which in life were traversed by the strands of cytoplasm. (a) × 280; (b) × 680.

Plate 39c These three amyloplasts from a peripheral cell of a soybean (*Glycine*) root cap are very much smaller than the reserve amyloplasts in (a) and (b) and Plate 20a. Many starch grains (S) are present in each. General features of plastids that can be seen include the double membrane envelope (black arrow) and nucleoid areas (white arrow). The internal membrane system is not well developed, but stacked thylakoids occur occasionally (star). Thylakoids sometimes lie closely appressed to starch grains (open arrows), but it is not known whether this reflects membrane-activity in starch metabolism, or whether the starch grains merely became pressed against the membranes as they grew. × 23 000.

Plate 39d The starch in sieve element plastids is of an unusual type, containing a high proportion of branched 116 chains of glucose. Whereas ordinary amylase would digest potato starch grains, pretreatment with a special 'de-branching' enzyme is necessary to break down sieve element starch. The grains are unusually electron-dense, and display a granular composition approaching that of glycogen (which is also a branched polymer of glucose). The double envelope can be seen around the upper grain. *Coleus* petiole sieve element, × 15 000.

Plate 39e These amyloplasts grouped round the nucleus (N) in a young root cap cell of *Cosmea* each contain numerous starch grains (usually round in shape), and in addition accumulations of material that is extremely dense to electrons after processing for electron microscopy. The accumulations lie in distended intra-thylakoid compartments: in other words they are not in the stroma, where electron-dense plastoglobuli are found. In other material it has been found that this type of accumulation can be digested away from the section by treatment with the lipid- and protein-digesting enzymes lipase and pronase. It may therefore contain a lipoprotein. Phenolic material could also be present. Equally dense deposits are seen in the vacuoles (asterisks) and there is evidence (again from other material), that plastids can extrude phenol- 127 containing droplets to the cytoplasm and vacuoles. Other features of the micrograph include mitochondria (M) with cristae and small dense granules, and lipid droplets (L). × 26 000.

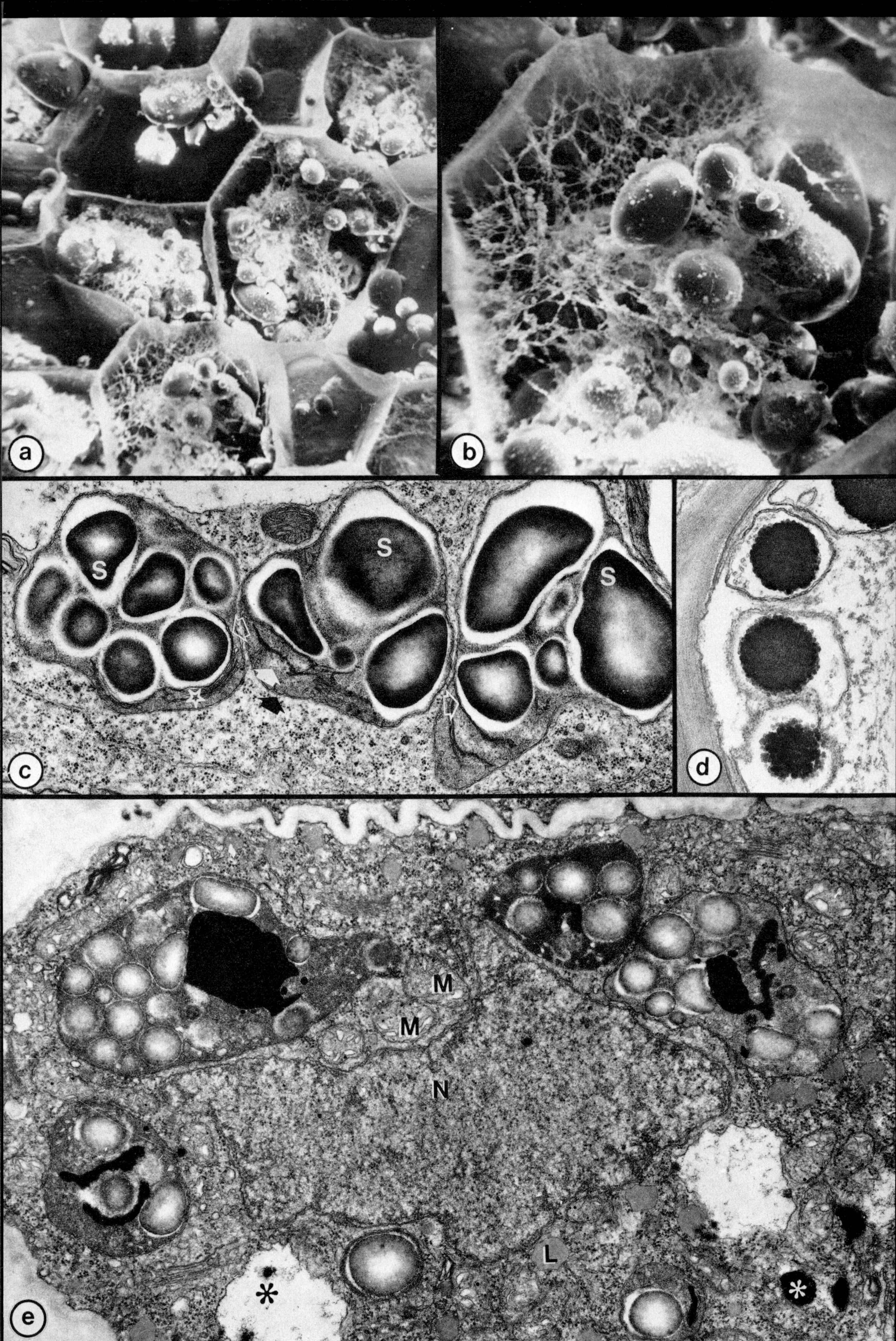
a
b
S
S
S
c
d
M
M
N
L
*
*
e

Plate 40

Plastids X: Chromoplasts

Plate 40a Chromoplasts are illustrated here in part of a cell from the orange rim of the corolla tube of a *Narcissus poeticus* flower. The upper left portion of the micrograph is occupied by part of the nucleus, which is seen to contain much dense heterochromatin, and to 42 have a high ratio of surface area to volume, being penetrated by long cytoplasmic channels (asterisks) which are lined by pore bearing (arrows) nuclear envelope. The cytoplasm is vacuolate (V), and contains mitochondria (M), cisternae of rough endoplasmic reticulum, numerous free ribosomes, and lipid droplets 135 (L).

The chromoplasts dominate the cytoplasm. Their outlines, and especially their internal membranes, are convoluted. The clear zones within them (stars) 120 represent what in life were crystals of beta-carotene. They are now electron transparent, some at least of the carotene having been extracted during dehydration of the specimen after the shape of the crystals was preserved by fixation. Numerous membranes undulate through the electron transparent areas. Many of the 120 chromoplasts contain electron dense globules—plastoglobuli, which may, like the crystals, contain chromoplast pigment, but probably also contain (as in chloroplasts) plastid quinones. × 15 000.

Plate 40b The chromoplast shown here (from a tomato fruit) is in a relatively early developmental stage. Its juvenility is shown by the presence of many small grana: later in development these disappear, leaving electron dense plastoglobuli and crystals of lycopene (lycopene is a precursor of beta carotene, which also occurs in tomato chromoplasts, but (except in certain varieties) in concentrations that are apparently too low 121 to give extensive crystallization).

The lycopene crystals (stars) have angular outlines, and are surrounded by membrane. They also contain undulating membranes, sometimes aggregated in electron dense stacks (crystal in chromoplast at left hand side). In these respects lycopene crystals resemble beta-carotene crystals (see (a)).

Other features shown in the chromoplasts include a round membrane-bound inclusion (arrow), aggregated electron dense material (open arrow, possibly remnants of a starch grain, or components of plastoglobuli undergoing crystallization). The stroma contains plastid ribosomes, and nucleoid areas (large circle). The double envelope of the chromoplast is also visible (small circles).

Tomato fruit parenchyma cells are very large and vacuolate (V, vacuole; T, tonoplast). A microbody is included in the micrograph (asterisk). × 32 000.

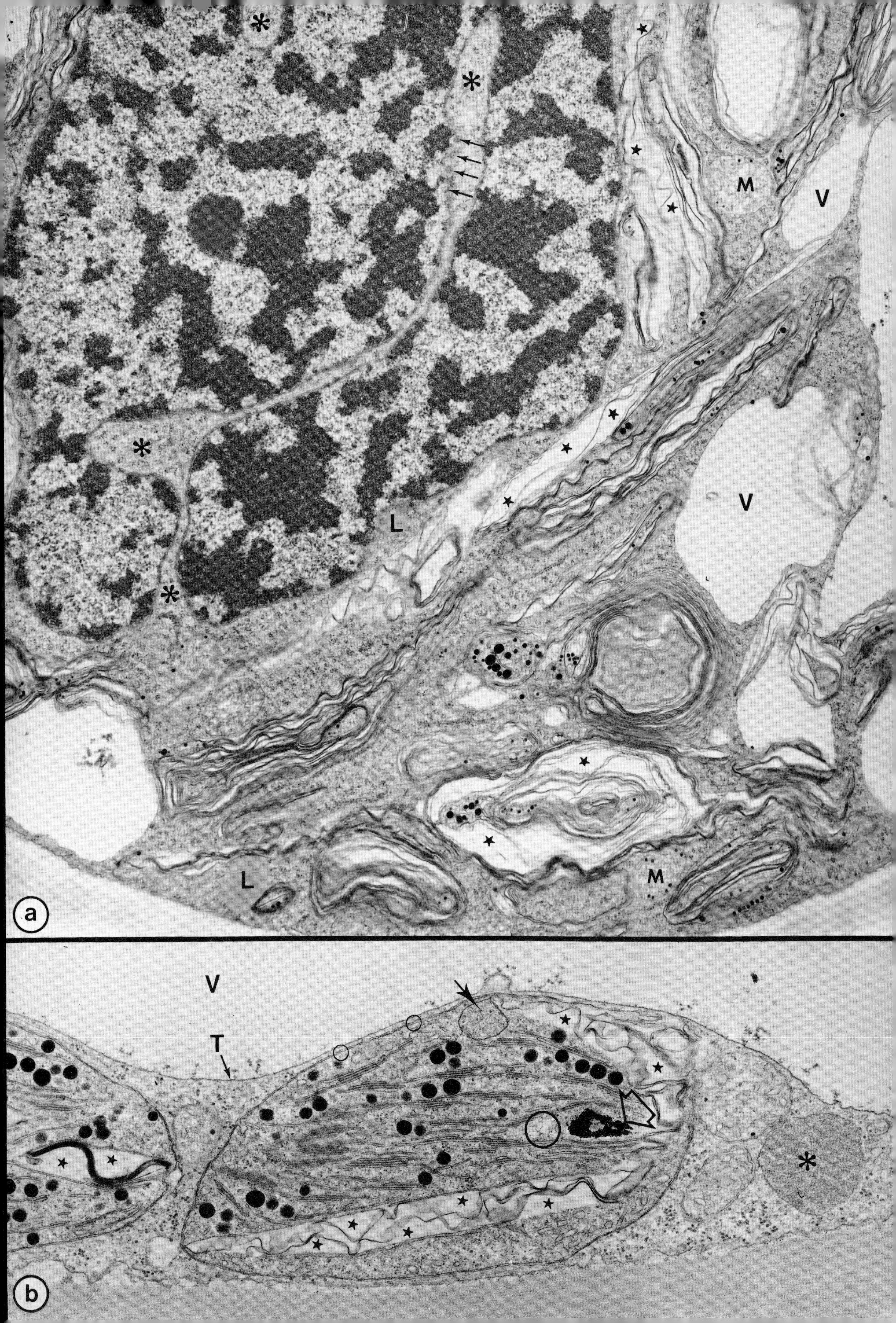
M
V
V
L
L
M
a
V
T
b

Plate 41

Microbodies

Microbodies are found in various biochemically distinguishable categories, two of which, peroxisomes and glyoxysomes, are illustrated here.

Plate 41a The microbody (MB) in this section of a [135] tobacco leaf is of the type known as a peroxisome, which functions in the enzymatic processing of glycolic acid, produced by chloroplasts, and broken down in peroxisomes yielding carbon dioxide together with other substances which can be retrieved and utilized by the plant. Microbodies (in general) are bounded by a single membrane (arrows). The peroxisome shown here contains a large crystal (CY) surrounded by a diffusely granular matrix. The single membrane is closely appressed to the outer membranes of the two adjacent chloroplasts. The latter contain grana (G), frets (F), and ribosomes (CR) which are smaller than their cytoplasmic counterparts. Note the invagination (I) of the inner membrane of the chloroplast envelope. Also present in the cytoplasm is a mitochondrion (M). The electron-transparent area to the upper right is part of the large vacuole.

[136] The presence of the enzyme catalase in the peroxisomes has been demonstrated. After normal fixation with glutaraldehyde pieces of tissue were incubated with 3.3′—diaminobenzidine and hydrogen peroxide. Visualization and stabilization of the precipitate produced by catalase activity in the presence of these substances was achieved by routine post-fixation with osmium tetroxide, whereupon the sites of enzyme activity became heavily stained. The insert shows a typical peroxisome from tissue treated in this way. Enzyme activity occurs throughout the peroxisome, but is concentrated in the crystal (CY). It is of interest that where the peroxisome membrane abuts on to the neighbouring chloroplasts (arrows), there is a stronger staining reaction than elsewhere. It may be that in these regions of close contact, there is an especially high rate [137] of transport of molecules between chloroplast and peroxisome. × 41 000; insert × 44 000.

(Micrographs courtesy of Drs. S. E. Frederick and E. H. Newcomb, reproduced by permission of the Rockefeller University Press from the *Journal of Cell Biology*, **43**, 343 (1969)).

Plate 41b and c Microbodies in the cells of cotyledons of lipid-storing seeds (e.g. the sunflower illustrated here) provide a striking example of the versatility of microbody activity. During early stages of germination and seedling growth the lipid reserves are broken down with the aid of enzymes present in the microbodies (MB). Some of the enzymes comprise the 'glyoxylate cycle', so this type of microbody has been called the glyoxysome. They lie alongside the lipid [136] droplets (L) in the cotyledon cells. 41(b) represents a stage four days after germination. After all or most of the stored lipid has been consumed the cotyledons enlarge, become green and photosynthetic, and function as leaves. 41(c) represents such a stage, seven days after germination. The microbodies (MB) in (c) are in the same cells as those in (b), but have become associated with the newly developed chloroplasts (C). The [138] glyoxysome type of microbody has been replaced by (or has metamorphosed into) a peroxisome type of microbody. The microbodies in these cells have a somewhat irregular shape, and, as in 41a, the prominent single membrane bounds a dense granular matrix, which in the present examples contains a crystal (CY).

The lipid droplets in (b) possess a surface skin of half membrane (arrows) which is probably formed as a [135] result of the orientation of those lipid molecules that come into contact with the aqueous environment of the cytoplasm. (b) × 31 000; (c) × 34 000.

(Micrographs courtesy of Drs. P. J. Gruber and E. H. Newcomb, reproduced by permission of Springer-Verlag from *Planta*, **93**, 269 (1970)).

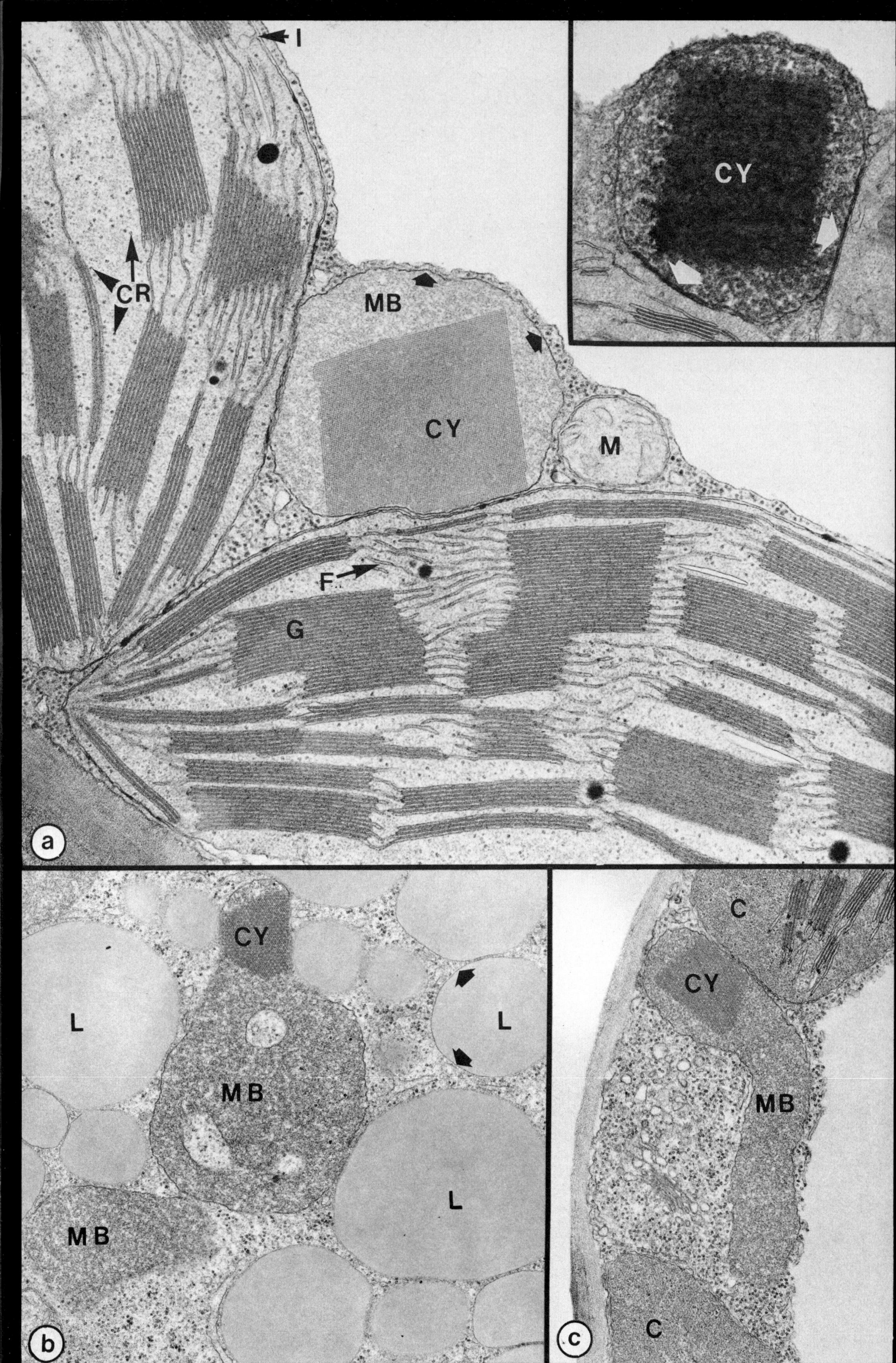
I
CR
MB
CY
M
F
G
a
CY
CY
L
L
MB
L
MB
b
C
CY
MB
C
c

Plate 42

Plates 42 and 43 show aspects of the microtubules of non-dividing cells. Those of cells that are undergoing cell division appear in Plates 44–48.

Cortical Microtubules

Plate 42a One location where higher plant microtubules are found in non-dividing cells is in the cell 147 cortex, just beneath the plasma membrane. Here they may be seen in transverse (T) or longitudinal (L) section. The tubules have a densely staining wall about 7 nm thick, surrounding a clear lumen. Often there is a 142 clear zone around the outer perimeter of the microtubule, from which ribosomes (etc.) are excluded. Cortical microtubules may be linked to one another, and to the plasma membrane, by short 'bridges' (not shown here). *Azolla* root tip, × 54 000.

Plate 42b Sections tangential to the cell surface include the cell wall (CW), plasma membrane (PM) and the immediately underlying cytoplasm (CP). A major feature that is often visible in such planes of 147 section is the correspondence between the orientation of microtubules and that of cellulose microfibrils in the cell wall on the other side of the plasma membrane. In this section from a spinach root tip (*Beta vulgaris*) the wall microfibrils (MF) are clearly visible running horizontally across the lower part of the micrograph (lower arrow). Nearby there is a similarly oriented microtubule (1). Above this there is a band of microtubules (2) running at a slight angle to (1), and again the adjacent wall microfibrils parallel this orientation (arrow at right centre). A further band of microtubules (3) runs at an angle across the main group.

The irregular nature of the cell surface (also shown in (c)) generates very complex images in tangential sections. Numerous dark-light-dark profiles of sectioned membranes are visible (see circled areas, for example), corresponding to sections through the sides of bumps and hollows in the plasma membrane. The irregular electron-transparent areas correspond to similar areas seen in transverse sections of the wall (as in (c)). Many vesicles (V), possibly of dictyosomal origin, lie in the peripheral layer of cytoplasm, amongst the microtubules. × 50 000.

Plate 42c Just before prophase of mitosis it is common for microtubules to aggregate in a bundle lying close to the wall in an equatorial position in the cell. The aggregate is known as the preprophase band, and an 170 example may be seen here between the large arrowheads. The cell wall was very much longer than the small portion illustrated in this micrograph, and the cortical cytoplasm along its length contained no microtubules except for those in the preprophase band. The bumps and hollows of the wall and plasma membrane seen in (b) are also shown here, but in transverse rather than tangential section. Cabbage root tip, × 50 000.

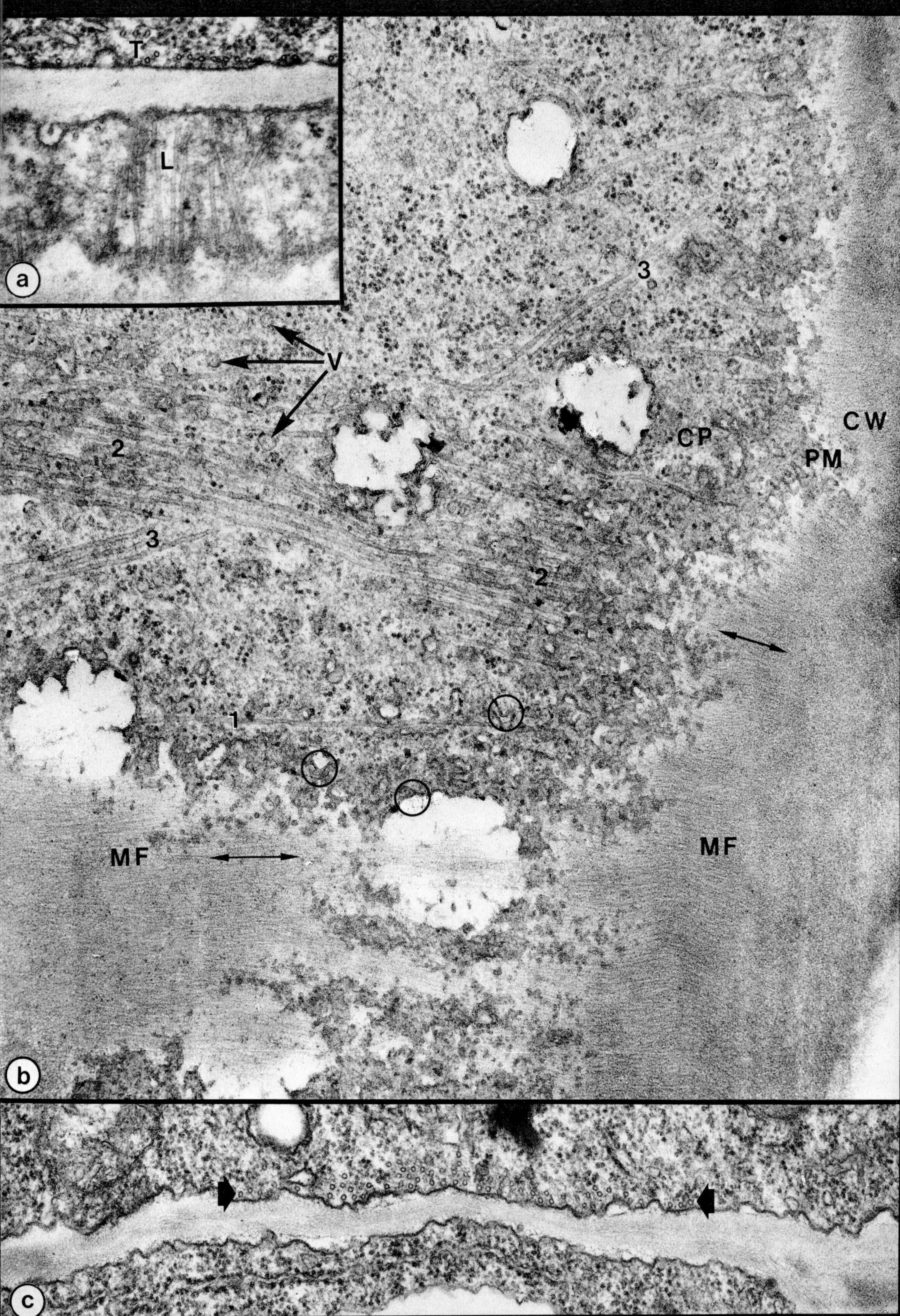

a
T
L
b
V
2
3
3
2
1
CP
PM
CW
MF
MF
c

Plate 43

Microtubules and Microfilaments

Plate 43a Certain cells accumulate compounds which can bring about progressive modification of the normal staining patterns after processing for electron microscopy (see also Plate 3b). The appearance of microtubules is especially subject to this phenomenon. A relatively minor effect is seen in the microtubules (MT) to the left of the cell wall in (a). Normally their walls are stained uniformly (as in Plate 42), but here they are somewhat uneven, probably because densely-staining material has become bound to them during fixation. The cell and the microtubules to the right of the cell wall are more severely affected. Dense material has been deposited around *and within* the microtubules, and in interstices of the microtubule wall itself, which appears pale against the dark interior and exterior. *Azolla* root tip, × 132 000.

Plate 43b - e Deposition of electron-dense material around and within the microtubule leads to a form of 'natural' negative staining, revealing details of microtubule substructure that are not normally visible. One microtubule from (a), (circled) is shown enlarged in (b) to × 860 000, and its wall clearly consists of a ring of subunits. It is not possible (in this picture) to count the precise number of subunits in the circumference. The information is, however, present in the image, and can be extracted. The image of each subunit is photographically superimposed upon that of each of its neighbours. This accentuates the regular periodicity around the circumference, while randomly distributed points in the image tend to cancel one another. The subunits can be counted when the picture is rotated just the right amount between photographic exposures to give perfect superimposition. A trial-and-error procedure is used, rotating the picture through a range of angles (e.g. for 4 subunits, 90°; for 12 subunits, 30°). Superimposed images formed by this method are shown in (c), (d) and (e). The three pictures represent tests aimed at discovering whether the tubule in (b) has 12, 13 or 14 subunits around its circumference. Only one, (d), gives an image that is free from distortion and blurring. In the others the angle of rotation did not correspond to the angle subtended by the subunits, so that no true image reinforcement occurred. The conclusion is that the wall of the microtubule contains 13 circumferentially placed subunits.

142

Plate 43f In non-dividing cells microtubules may be found elsewhere than in the cell cortex, associated in a skeletal function with cell components other than the plasma membrane. In this example it may be that they anchor the nucleus (N) in position in a specific region of the cell. The section was tangential to the nucleus, thus enabling the pores (NP) and the polyribosomes (P) of the nuclear envelope to be seen. *Bulbochaete* hair cell, × 29 000. (Micrograph courtesy of T. W. Fraser).

154

Microfilaments

Fine strands, which have become known as microfilaments, are being discovered in more and more cells, plant and animal. It is likely that they participate in the movement of cytoplasm and its components (see also Plate 27b); thus while microtubules may be likened to intracellular bones, microfilaments may be regarded as some form of intracellular muscle.

150

Plate 43g - i These phase contrast photomicrographs were taken at approximately 30 second intervals, and illustrate the same part of a living hair on a petiole of *Heracleum mantegazzianum*. One fibrous strand (microfilament bundle) is in focus in all three photomicrographs, but others are also visible. Cytoplasmic streaming is particularly vigorous near the fibres. Comparison of the three micrographs shows that a plastid (P) moved diagonally across the field, while there are much more dramatic changes in the population of mitochondria (M). × 3300; (micrographs courtesy of Drs. T. P. O'Brien and M. E. McCully, reproduced by permission of Springer-Verlag from *Planta*, **94**, 91 (1970)).

Plate 43j It is difficult to detect microfilaments by electron microscopy unless, as here, appreciable lengths lie in the plane of the ultra-thin section. In this example the microfilaments, each 2-5 nm in diameter, mostly lie parallel to one another, but the bundles frequently change direction. Spinach root tip cell, × 47 000.

150

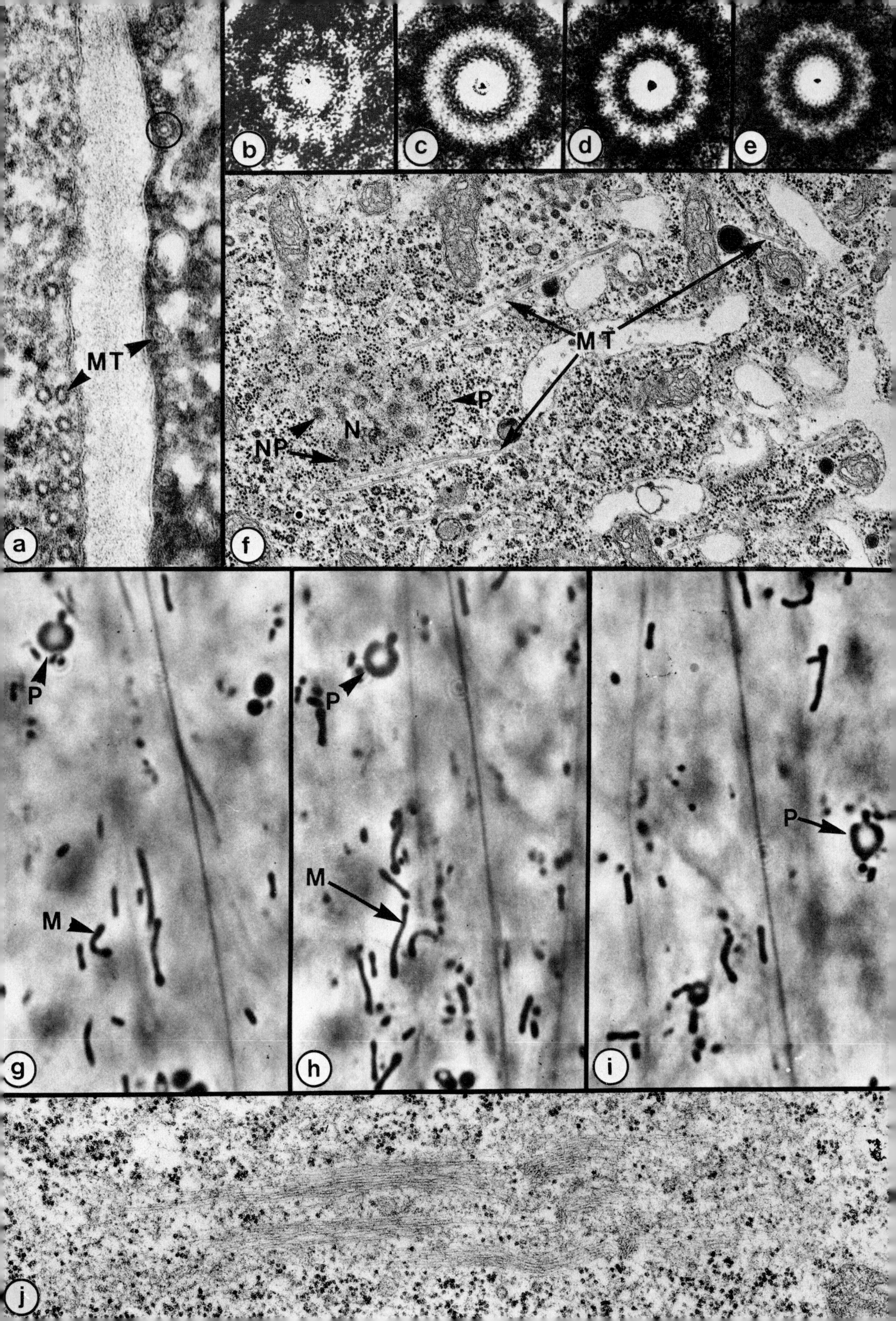

a
MT
b
c
d
e
f
MT
P
N
NP
g
P
M
h
P
M
i
P
j

Plate 44

Cell Division (1): Mitosis in *Haemanthus*

This set of photomicrographs was taken using the Nomarski interference-contrast technique. The sequence shows the course of mitosis in a single living endosperm cell from the blood lily, *Haemanthus katherinae* Bak. These cells can be removed from young fruits, and, because they have no cell wall, they can be spread on a microscope slide and sequential observations made as they undergo division. The cumulative total time from the first micrograph in the sequence (a) is recorded at the end of each caption below. All the micrographs are at the same magnification (× 700). The identification of some of the smaller objects is based on electron microscope studies of the same material.

161

Plate 44a - c *Prophase*. The nuclear envelope (NE) forms a distinct boundary between the nuclear contents (condensing chromosomes, CH; disintegrating nucleolus, NL) and the developing clear zone surrounding the nucleus (see also nucleus N-1 in Plate 1). In this case the clear zone is 3-polar (asterisks in (b)). The small protrusions of the nuclear envelope at the poles are where microtubules from different directions intermingle. As prophase progresses, the 3-polar condition gives way to the normal bi-polar division figure (c). The cytoplasmic components are arranged around the periphery of the cell: the numerous small vacuoles (V) are quite conspicuous, but others are not readily identifiable, except for a large mitochondrion (M) which can be seen in (a). Times: (a) O; (b) 15 min; (c) 22 min.

162

Plate 44d - f *Prometaphase-metaphase*. Following rupture of the nuclear envelope (between (c) and (d)), a normal bi-polar spindle (poles at asterisks in (d)) has now been formed. The spindle fibres are as yet barely discernible. Movements of chromosomes during prometaphase (d) and (e) result in the orientation of their kinetochores (points of attachment to spindle fibres) at the equator (E-E) by the time of metaphase (f). Kinetochore fibres (arrowheads) are particularly prominent in the upper half spindle. That the chromosomes are double, consisting of chromatids twisted around one another, has been visible since early prophase (a): the sister chromatids now gradually untwist ((d) and (e)), prior to their separation in the next stage of mitosis. Times: (d) 1 h. 2 min; (e) 1 h 18 min; (f) 1 h 40 min.

163

Plate 44g - i *Anaphase*. Movement of the sister kinetochores to opposite poles gives the typical trailing arm chromosome configurations of early (g) and mid (h) anaphase. At these stages the kinetochore fibres (arrowheads) are still visible. By late anaphase (i) the chromosomes have begun to coalesce at the poles. Phragmoplast fibres now develop at the centre of the spindle, and their activity results in lateral movements of the trailing chromosome arms: comparison of (i) and (h) shows how they fan out from their previous alignment along the axis of the spindle. Dictyosomes (D) begin to invade the phragmoplast region. Times: (g) 1 h 50 min; (h) 1 h 56 min; (i) 2 h 7 min. (Notice the rapidity of the movements of chromosomes during anaphase, as compared with during prometaphase and metaphase).

162

Plate 44j - 1 *Telophase – early cytokinesis*. Progressive condensation of the two chromosome masses results in the formation of two daughter nuclei. Each becomes bounded by its own nuclear envelope (NE in (l)). A comparable stage is shown in nucleus N-2 of Plate 1, where a newly-regenerating nucleolus lies amongst the uncoiling chromosomes. Between the two nuclei the phragmoplast fibres develop further (F in (j)) and material accumulates along the former equator, giving rise to the cell plate (CP in (k) and (l)). Dictyosomes (D) in side view and face view ((k) and (l) respectively) are prominent among the fibres. A very long and attenuated mitochondrion (M) lies across the cell plate in (l). Had the cells not been removed from the fruit, the cell plates formed at the division illustrated in this plate would have developed further to become the first cell walls of the previously naked endosperm tissue. Times: (j) 2 h 13 min; (k) 2 h 22 min; (1) 2 h 40 min.

(Micrographs provided by Dr. A. S. Bajer).

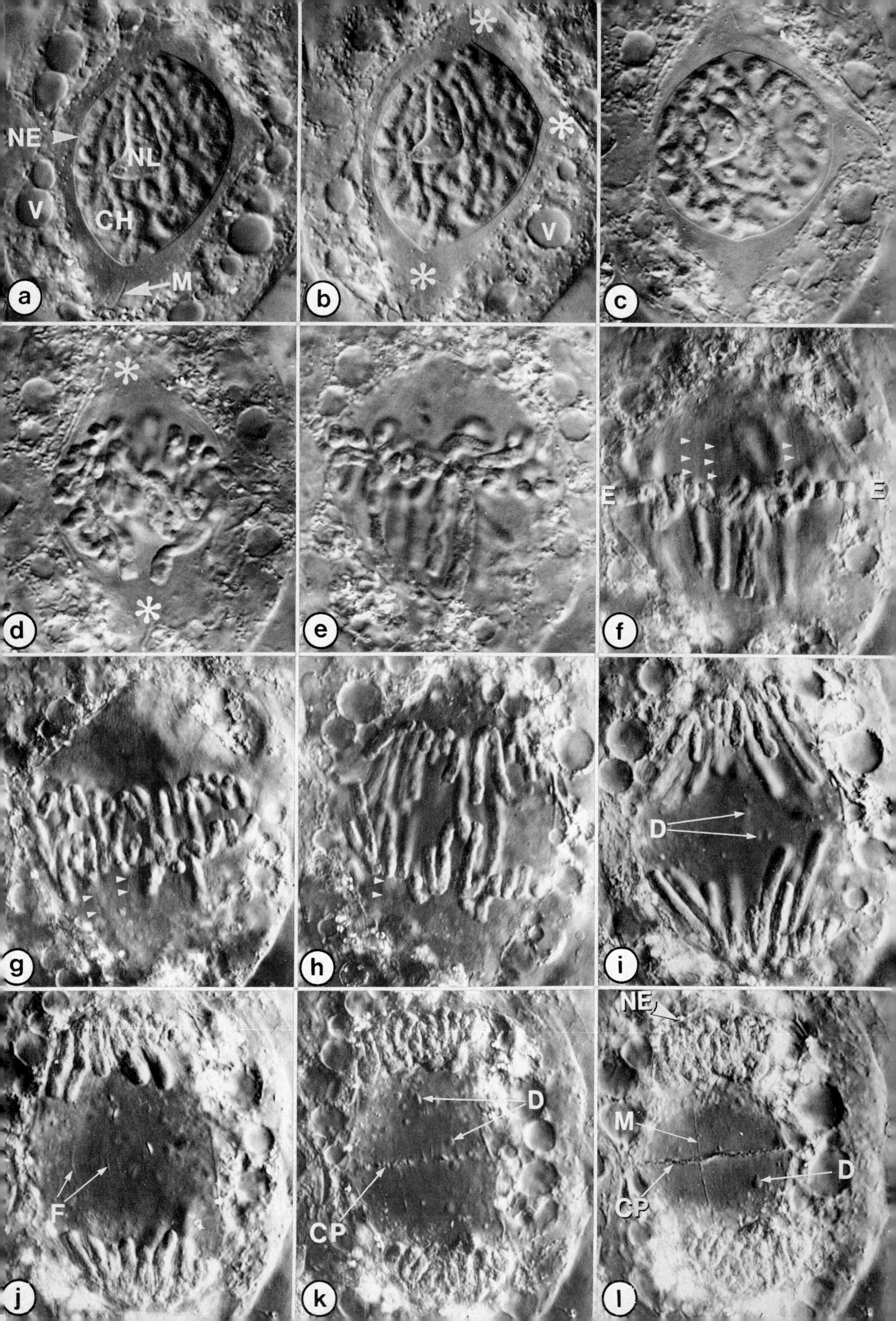

NE
NL
V
CH
M
a
b
V
c
d
e
E
E
f
g
h
D
i
F
j
D
CP
k
NE
M
CP
D
l

Plate 45

Plates 45-48 display electron micrographs of ultra-thin sections, complementary to the light micrographs of Plate 44. Unless stated otherwise, all of the sections are longitudinal with respect to the cell axis. Where possible, the micrographs are oriented so that the spindle poles lie at the top and bottom of each figure.

Cell Division (2): Prophase

Plate 45a Prophase nuclei (corresponding to Plate 44 a-c) show condensed chromatin (CH) and the initial stages of dispersal of nucleolar material (NL). Segregation of granular and fibrillar zones is evident in the nucleolus. 45 The nucleoplasm, still enclosed by the intact nuclear envelope (NE), shows little structure at this magnification. There is no obvious clear zone around the nucleus, indeed proplastids (P), mitochondria (M) and vacuoles (V) are all lying close to the nuclear envelope. × 6500.

Plate 45b The area enclosed by the rectangle in (a) is shown at higher magnification in (b). NL marks a mass of nucleolar fibrils, and nucleolar granules are at upper left. Part of a chromosome (CH) is also included. The background nucleoplasm is relatively empty at this stage (cf. below). × 28 000.

Plate 45c - e The areas enclosed by the rectangles in (c) are shown enlarged in (d) and (e). At this stage (approximately between (d) and (e) of Plate 44) the nuclear envelope (NE), with its pores (NP in (e)), is disintegrating and a number of breaks have occurred (see asterisks in (c) and (e)). Although the nuclear envelope has only been partially removed, the nucleoplasm between the chromosomes (CH) has been invaded by a large number of ribosomes, mostly in the form of polyribosomes (compare (c) and (d) with (a) and (b) above). Vesicles (VE) also appear in the former nucleoplasm (note the presence of dictyosomes (D in 162 (c)) outside the disintegrating envelope). Components of the future spindle have made their first appearance in the nucleoplasm: microtubules (MT in (d) and (e)) are now evident, near breaks in the nuclear envelope (e), as well as in the outer region of the nucleoplasm (e) and in central regions (d). The nucleolus (NL in (c) and (d)) appears more dispersed than in (a).

One of the most striking changes is in the nature of the nuclear envelope. When intact, it bears pores, and ribosomes are bound only to the cytoplasmic face of the outer membrane (Plates 18d, 43f). It has been caught here in a state in which it more closely resembles cisternae of rough endoplasmic reticulum. Thus it has been ruptured, and ribosomes are bound to both the 162 inner (solid arrows in (e)) and the outer membrane. Pores are still present. A similar state recurs at telophase of mitosis (Plate 47d).

(c) × 16 000; (d) and (e) × 50 000.

(a) - (e) all from root tip cells of *Vicia faba*.

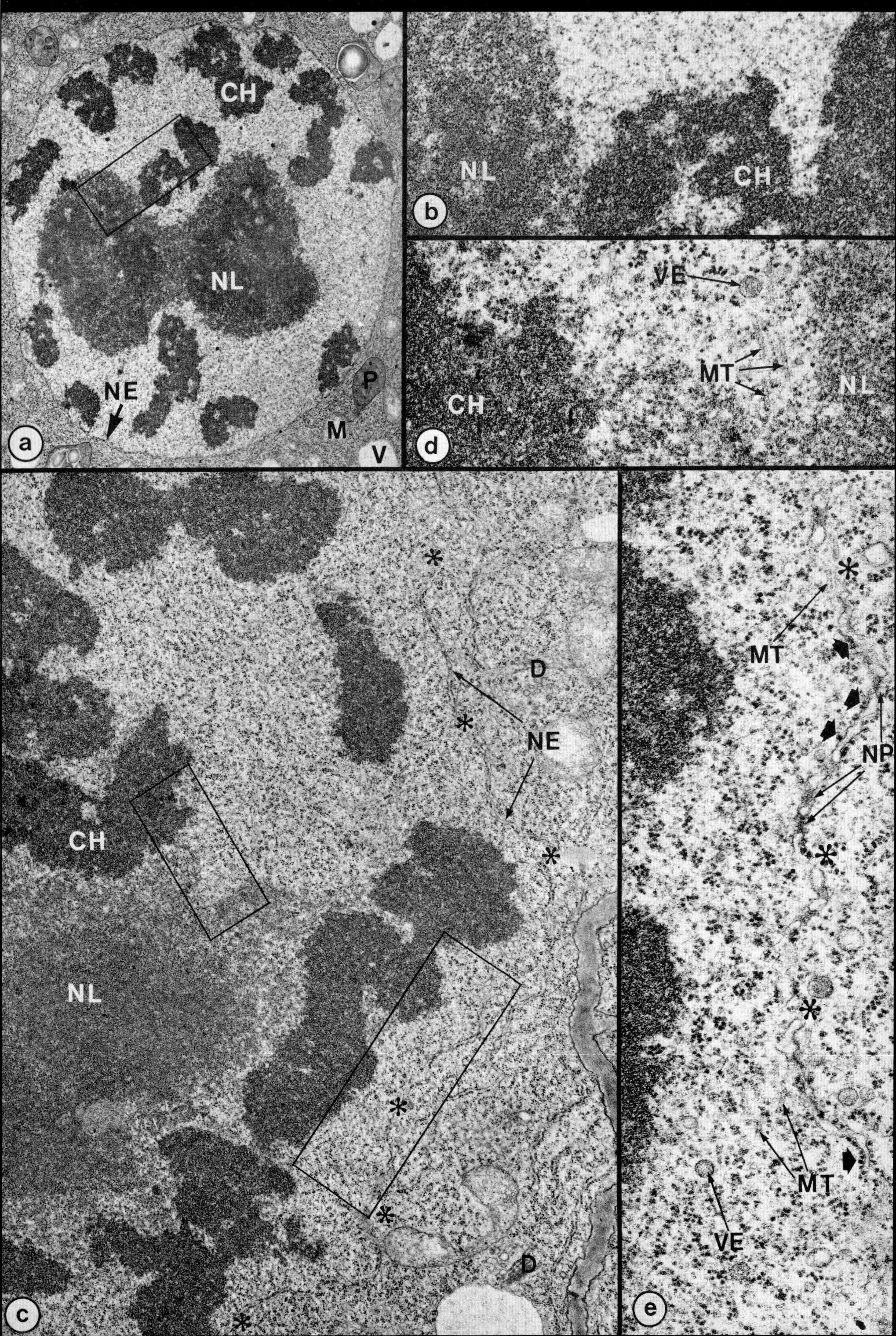

CH
NL
NE
P
M
V
a
NL
CH
b
VE
MT
CH
NL
d
*
D
NE
CH
NL
c
MT
NP
VE
e

Plate 46

Cell Division (3): Prometaphase and Metaphase

Plate 46a This micrograph continues the series started in Plate 45a and c. It is equivalent in stage to Plate 44d, that is, prometaphase, and shows a small sector of the clump of chromosomes liberated from a nucleus. Microtubules, which make their appearance in the nucleoplasm at the time of the rupture of the nuclear envelope (Plate 45d, e), are now present in extensive 163 arrays (MT). The chromosomes (CH) have not yet become aligned on the equator of the division figure. The nucleolus (NL) is so dispersed as to be scarcely detectable. *Vicia faba* root tip, × 11 000.

Plate 46b, c and d illustrate stages of mitosis in cells in root tips of white lupin (*Lupinus albus*). The most obvious difference between lupin and the broad bean (*V. faba*) used in Plates 45 and 46a is that the chromosomes (and the cells) are very much smaller in the former.

Plate 46b As was the case in (a), the chromosomes shown here were fixed during prometaphase movements, and hence lie at different, non-equatorial, levels of the division figure. The microtubule (MT) system has developed sufficiently to be called the spindle. The microtubules are in bundles, and are oriented longitudinally, in the pole-to-pole axis. Those seen in this micrograph do not, however, seem to pass from pole to pole. Rather they extend longitudinally from the chromosomes (CH). They are, in other words, the microtubules that constitute the kinetochore fibres of Plate 44f. × 36 000.

Plate 46c This off-centre longitudinal section of a metaphase cell (equivalent to Plate 44f) shows the paired chromatids (CH) lying near the periphery of the equatorial region of the spindle. Kinetochore fibres (KF, hardly visible at this magnification) run from the kinetochores, oppositely oriented on the chromatid pairs, and mix with other fibres in the spindle, all passing towards the poles. The cell components (proplastids (P), mitochondria (M), dictyosomes (D), lipid droplets (L) and vacuoles (V)) are in general excluded from the spindle region and lie between it and the cell wall (just outside this micrograph). × 8000.

Plate 46d Kinetochores vary in structure throughout the eukaryotes. Here the electron microscope reveals 159 little but a matrix (K) in which the kinetochore microtubules (KM) terminate. Other microtubules (MT) penetrate between the chromosomes (CH), and probably are examples of pole-to-pole microtubules (as distinct from pole-to-kinetochore).

The inserts present the alternative view of a kinetochore, that is, as seen in sections cut in the plane at right angles to that of the main micrograph, and at a position equivalent to the level of the arrows from the letters K. The material was a dividing *Chlorella* cell. The upper insert shows transversely sectioned microtubules (ringed), one single, and one pair, in each case surrounded by fuzzy material. In the *adjacent* section (lower insert) of precisely the same area, the microtubules are no longer visible, and more of the fuzzy kinetochore or chromosomal material is included. The microtubules clearly are of the kinetochore type, and the two sections must have spanned the microtubule termini in or on the chromatids.
× 36 000; inserts × 45 000.
(Inserts provided by Dr. A. W. Atkinson, Jr.)

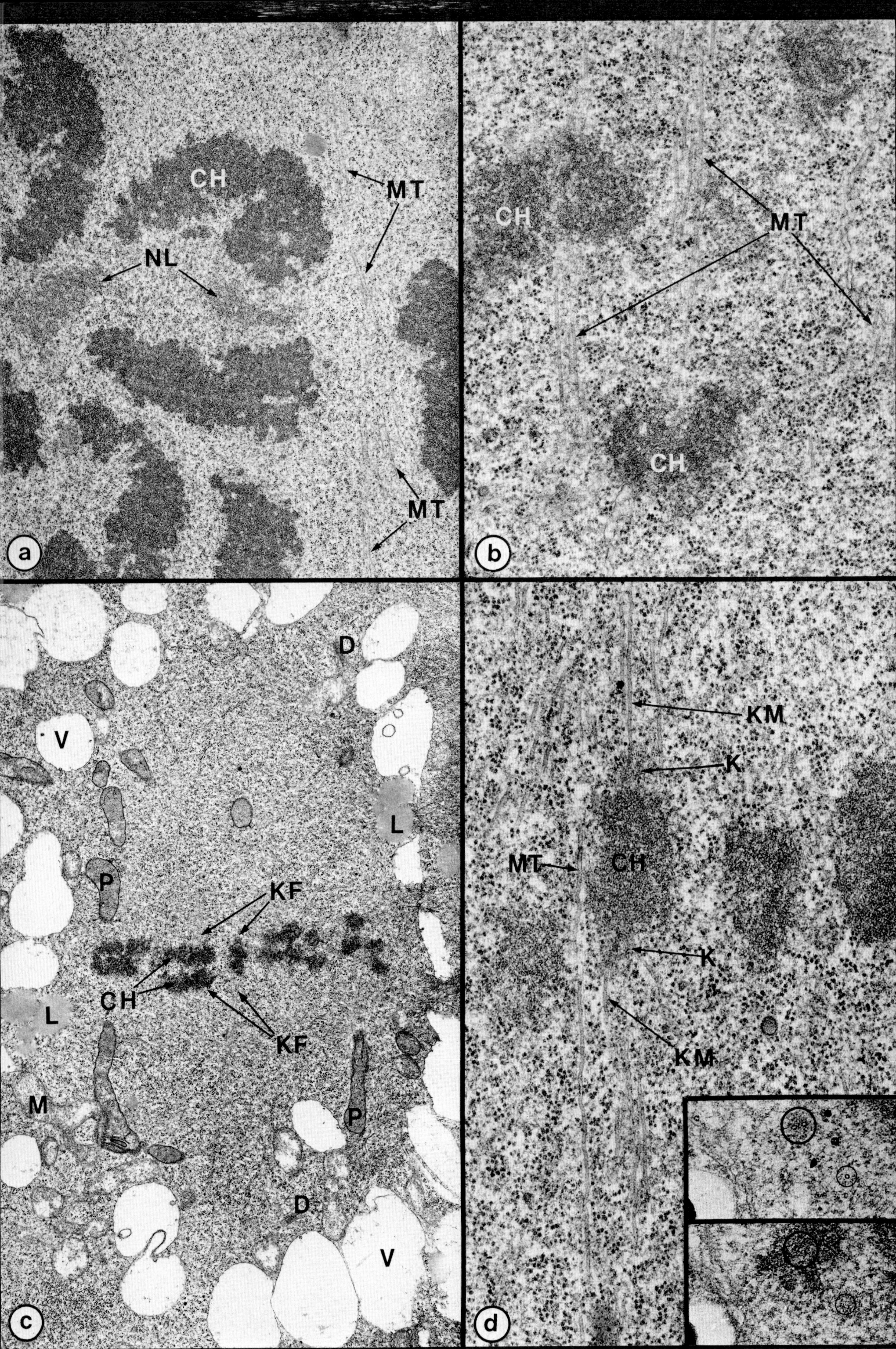

CH
MT
NL
MT
a
CH
MT
CH
b
D
V
L
P
KF
CH
L
KF
M
P
D
V
c
KM
K
MT
CH
K
KM
d

Plate 47

Cell Division (4): Anaphase—Early Telophase

Plate 47a This mid-anaphase cell (equivalent to Plate 44 g-h) from a white lupin root tip shows parts of several chromatid arms, caught within the section at various positions between the equator (which runs approximately left to right) and the spindle poles (at top and bottom). Two sister chromatids, each carrying 45 a nucleolar organizer region (NO), lie close to each other at the left hand side of the spindle. The nucleolar organizer is clearly different from the chromatid material on either side of it; light microscope staining reactions indicate that the concentration of DNA is low in the organizers. The whole spindle region, filled with numerous ribosomes and portions of sectioned spindle microtubules (MT), is surrounded by many cisternae of rough endoplasmic reticulum (ER), outside which lie other components of the cytoplasm: dictyosomes (D), mitochondria (M), plastids (P), and vacuoles (V). × 8000.

Plate 47b and c Later in anaphase, when the chromosomes begin to coalesce at the polar regions (upper parts of both (b) and (c); stage equivalent to Plate 44i), each chromosome becomes surrounded by a zone of granular material (large arrows in (b)) which is separated from the chromatin by a narrow electron-transparent space. Micrograph (b) shows a trailing chromosome arm possessing a nucleolar organizer region (NO), and a further example of the same structure is included in (c). Microtubules (MT), seemingly attached to the chromosomes, are seen in (b). Note also the vesicles (VE) amongst the trailing arms. Elements of endoplasmic reticulum modified by the development of nuclear pores (NP) are also present close to the chromo- 53 somes. *Vicia faba* root tip, (b) × 16 000; (c) × 10 000.

Plate 47d By early telophase (equivalent to Plate 44 j-k) a nuclear envelope with pores (NP) invests most of the coalescing chromosomes. The insert shows the new nuclear envelope in more detail, including a pore (large arrowhead), and, where it is not closely appressed to chromosome material, ribosomes on *both* surfaces. Some of the ribosomes on the future inner membrane are arrowed. At this stage the nuclear envelope therefore resembles the fragments present soon after the end of prophase (Plate 45e), but it is not clear whether such fragments persist throughout metaphase and anaphase 162 to contribute to the re-forming envelope at telophase, or whether the prophase fragments lose their pores to become typical rough endoplasmic reticulum, with some cisternae later on re-synthesizing pores and metamorphosing back to the nuclear envelope condition.

The nucleolar organizer region (NO) has started to 160 expand at its peripheral regions, differentiating an outer granular layer (G). Further expansion leads to regeneration of the nucleolus, as in nucleus N-2 in Plate 1, which is from the same material as the present micrograph (*Vicia faba* root tip), but is at a slightly later stage of telophase, with a larger nucleolus and with the chromosomal material beginning to de-condense towards the interphase condition.
× 32 000, insert × 52 000.

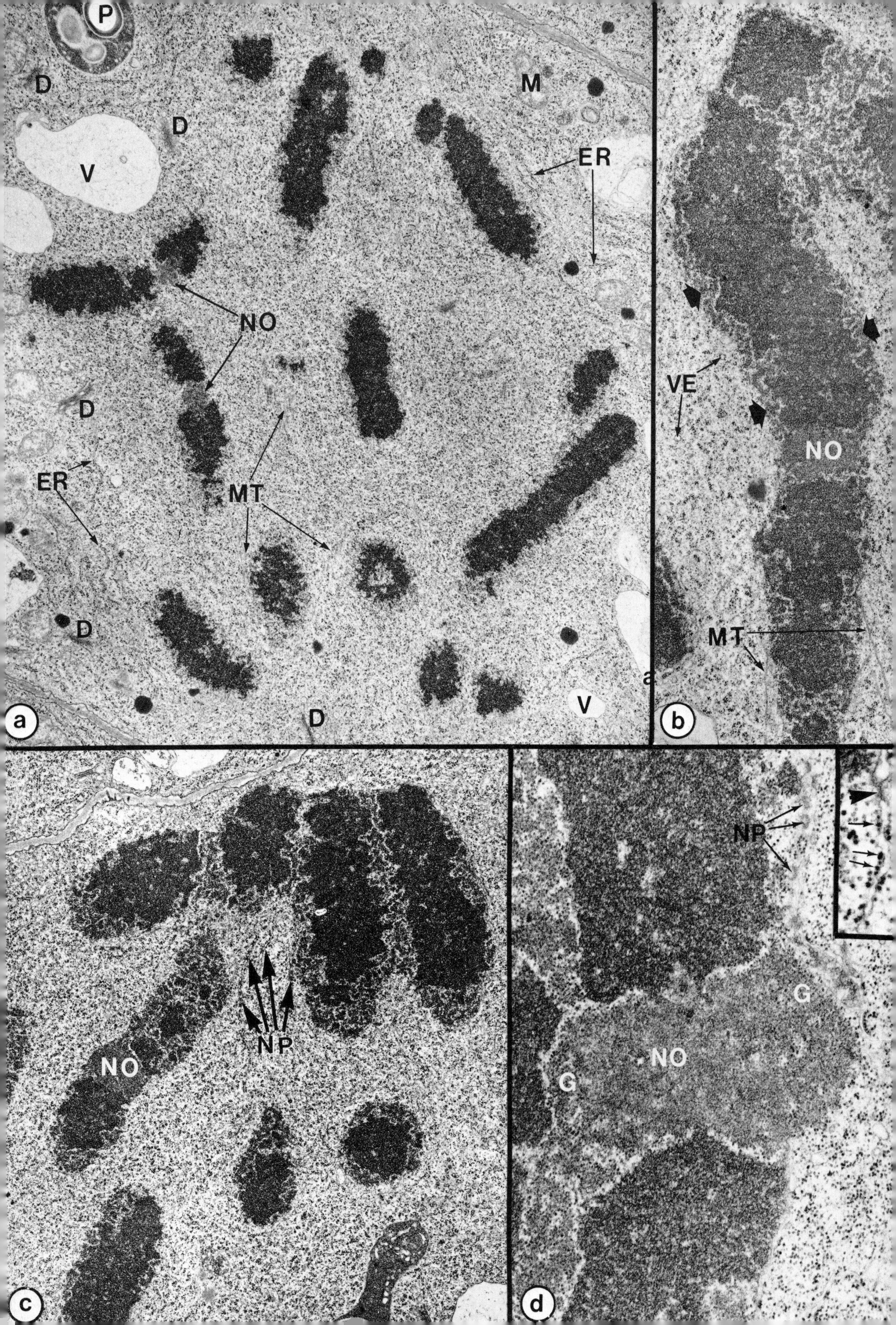

P
D
D
V
M
ER
NO
D
ER
MT
D
D
V
a
VE
NO
MT
b
NO
NP
c
NP
G
NO
G
d

Plate 48

Cell Division (5): Telophase and Cytokinesis

Plate 48a This off-centre longitudinal section of a *Beta vulgaris* root tip cell is equivalent in stage to Plate 44 k-l. It shows two daughter telophase nuclei (N), each bounded by a nuclear envelope (NE) that is by now complete. Between the two lie phragmoplast micro- 169 tubules (MT) and the developing cell plate (CP). The discrete vesicles (VE) of the young cell plate have in places begun to coalesce to form larger units (large arrow). Continuation of coalescence gives rise to the plasma membranes of the cross wall separating the daughter cells. A dictyosome (D) is present beside one of the nuclei, but is inconspicuous: none is seen (in this section) in the vicinity of the developing cell plate. × 13 500.

Plate 48b Details of phragmoplast microtubules and a young cell plate are seen here at a similar stage to that in (a). Long microtubules (MT) interdigitate (large arrows) at the layer of vesicles (VE) which form the cell plate. Vesicles are also present throughout the cytoplasm on both sides of the cell plate (small arrows). *Vicia faba* root tip cell, × 23 000.

Plate 48c It is very difficult to obtain face views of the cell plate—that is, in sections cut at right angles to that of (a). Such a section would be seen edge-on in the view shown in (a), and would be only a fraction of a millimetre in thickness at the magnification of (a). A face view of part of a cell plate at a late stage of its development is presented in (c). Due to its undulating contour, the plate passes into and out of the section, so that part of the micrograph includes coalescing vesicles of the plate, and part neighbouring cytoplasm. Some vesicles (large arrows) remain as discrete spheres between the advancing cell plate and the side wall (CW). About 150 profiles of cross-sectioned phragmoplast microtubules (arrowheads) are seen throughout the micrograph. They are most concentrated amongst the coalescing vesicles and tubules of the cell plate. *Avena sativa* anther, × 47 000.

Plate 48d, e, f Small segments of cell plates are shown here in side profile (as in (a)) in order to demonstrate certain points of detail. Regions of the coalescing membrane surface (black arrows in (d), and just to the left of ER in (f)) are coated with an array of fuzzy spikes, forming a layer some 15 nm in thickness on the cytoplasmic face of the membrane. Such surfaces resemble the fuzzy coating of coated vesicles (Plate 27a). It is not possible to say from a static image such as this one whether coated vesicles gave rise to the coated areas of the cell plate by fusing with it, or whether coated vesicles are being formed at the cell plate, and have been caught just prior to their release into the cytoplasm. A similar dilemma arose in connection with Plate 28a. Alternatively the surface coat might have properties and functions independent of vesicle arrival or formation.

A second point, illustrated in (d) and (f), is that cisternae of endoplasmic reticulum (ER) can become trapped in a position such that they interconnect the daughter cells that are being separated by the growing 28 cell plate. This could be the source of the axial structures of plasmodesmata (PD, see also Plate 14a, c, d), thought to be derivatives of endoplasmic reticulum.

A point of contact between the side wall (CW) of the dividing parental cell and the extending cell plate is shown in (e). The dark objects in the parent cell wall are remains of plasmodesmata, which look as if they may have become occluded. As in (d) and (f), the coalesced vesicles of the cell plate contain fibrillar material which is the first sign of the primary wall that will separate the two daughter cells. Similar material is seen in the protuberance which is emerging from the parent wall towards the cell plate. Nearby in the cytoplasm on both sides of the plate are numerous vesicles (VE), amongst the phragmoplast microtubules (MT). Some of the latter may be terminating at the cell plate, but as the section is at an angle relative to the plane in which most of the microtubules lie, this cannot be detected with certainty. Here and there hemispherical profiles on the membrane of the cell 169 plate (open arrows in (d) and (e)) suggest that vesicles such as those lying free (VE) have fused with and delivered their contents to the plate (but see above comments on coated vesicles).

Vicia faba root tip cell, (d) × 67 000, (e) × 55 000, (f) × 55 000.

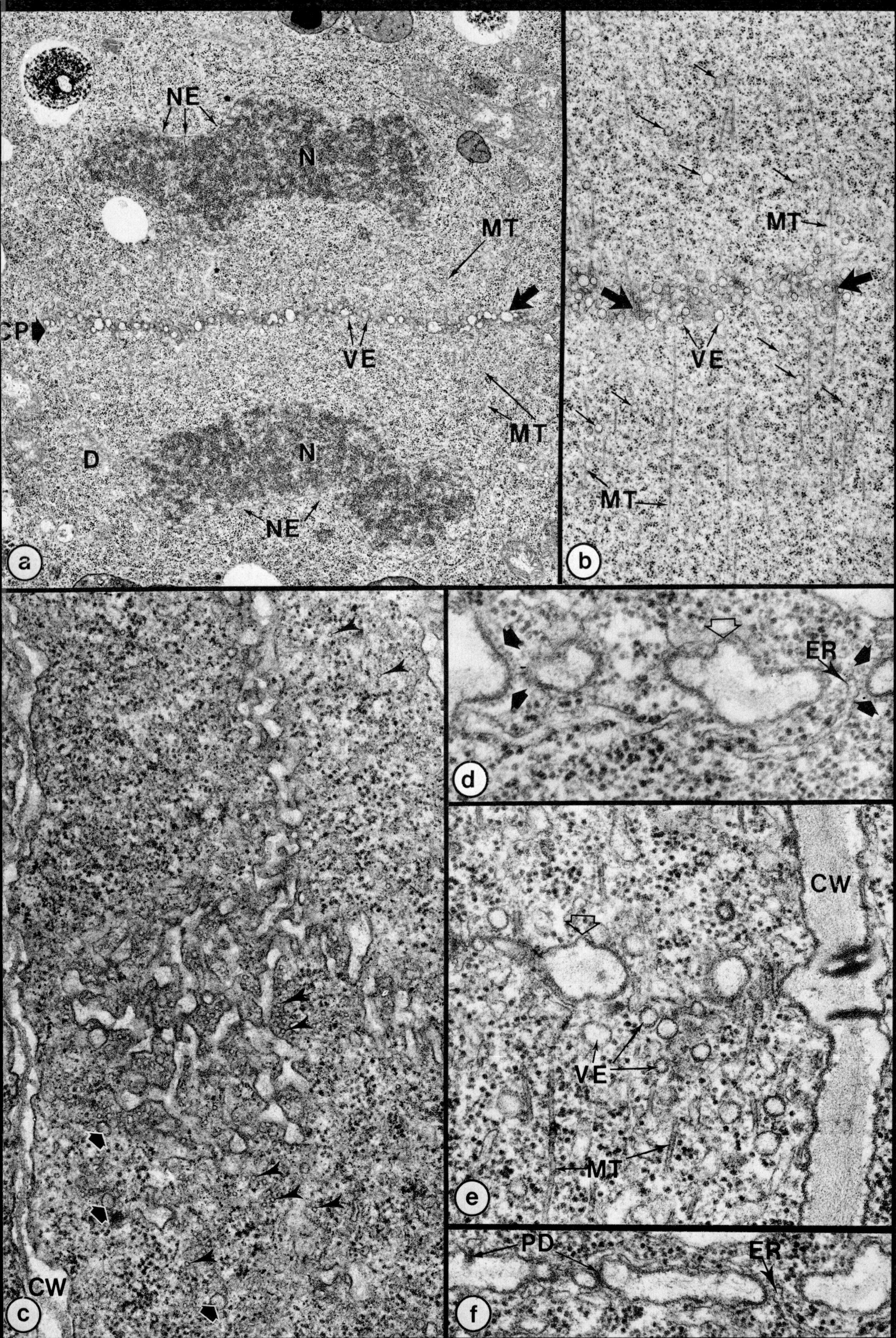
NE
N
MT
CP
VE
MT
D
N
NE
a
MT
VE
MT
b
ER
d
CW
VE
MT
e
CW
c
PD
ER
f

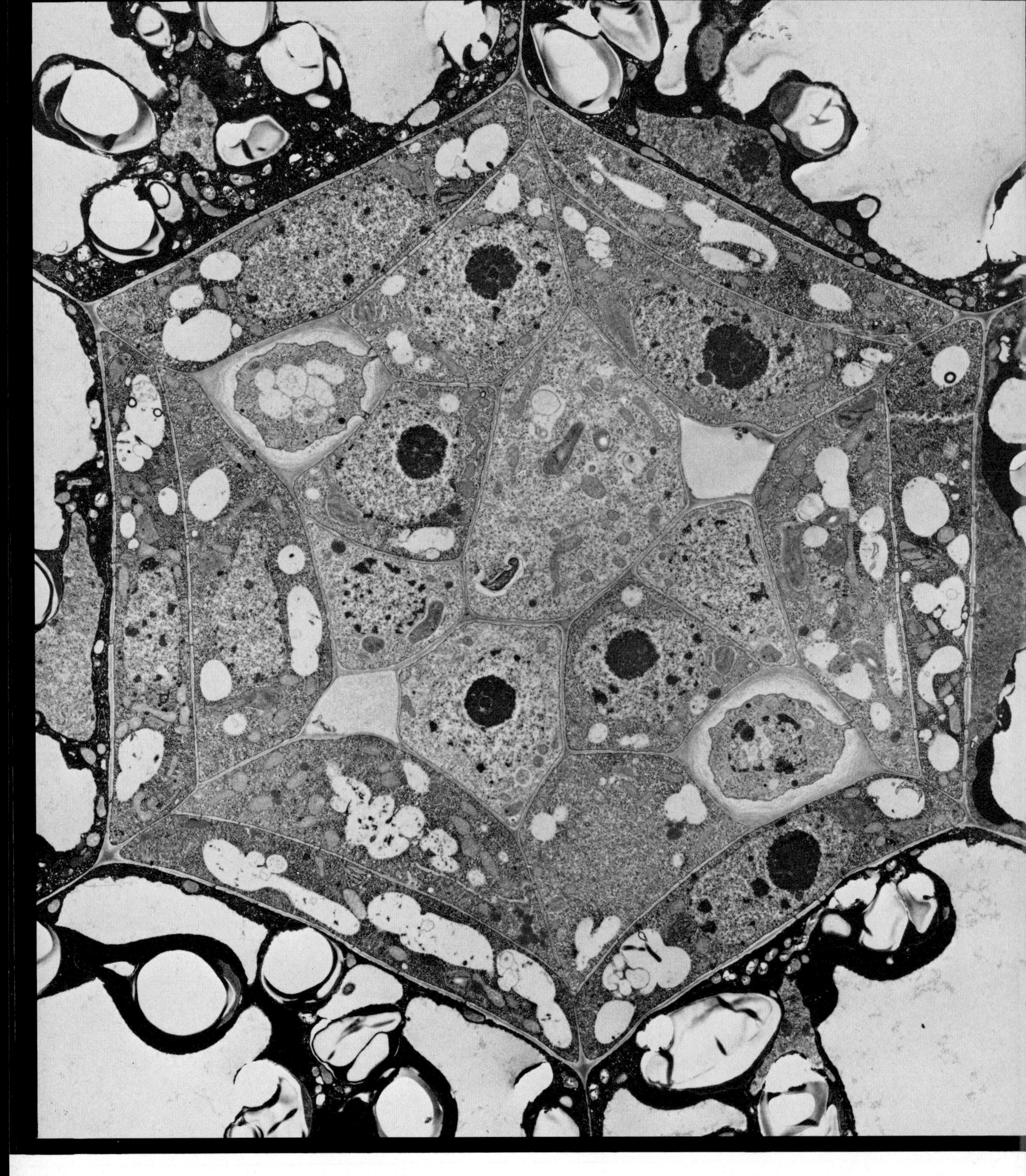

Plate 49 Plates 1–48 have illustrated some details of individual cells and components of cells. The purpose of the final plate is different, and largely symbolic: to emphasize by its precision and symmetry that there is another aspect of plant cell biology—the structural and functional integration of cells in the tissues and organs of the plant. The cells that are seen here are juvenile, in a cross section cut near the tip of the highly miniaturized root of the water fern *Azolla* (× 5800). Despite the compact nature of the overall structure, and their close proximity, the 22 cells specialize along 6 different pathways of maturation. The mature state is shown, with labels, in Plate 16a. When mature, the cells are disposed in unchanged numbers and in an unchanged geometrical relationship, in which the 6 cell types collaborate to perform the multifarious functions of the vascular system of the root. Every cell type plays its specialized part: no one could operate on its own. Just as specialized sub-cellular components participate in a collaborative group existence upon which depends the survival of the cell as an entity, so it is with the cells of the plant. Their collaboration creates functional entities at levels of organization higher than that of the cell—and beyond the scope of the present book.

Bibliography—Author Index

(Authors' text page references are indicated by italic figures in parentheses at the end of each entry)

1 ADAMS, M. and WARR, J. R. (1972). Colchicine-resistant mutants of *Chlamydomonas reinhardi*. *Exp Cell Res.*, **71**, 473–475. *(143)*

1a AFZELIUS, B. A. (1969). Ultrastructure of cilia and flagella. In *Handbook of Molecular Cytology*, Ed. Lima-de-Faria, A. North-Holland, Amsterdam and London, 1219–1242. *(149)*

2 ALBERSHEIM, P. (1965). The substructure and function of the cell wall, and biogenesis of the wall. In *Plant Biochemistry*. Ed. Bonner, J. and Varner, J. E. Academic Press, New York. *(17)*

3 ALBERSHEIM, P. and ANDERSON, A. J. (1971). Host-pathogen interactions. III. Proteins from plant cell walls inhibit polygalacturonases secreted by plant pathogens. *Proc. natn. Acad. Sci. U.S.A.*, **68**, 1815–1819. *(20)*

3a ALBERSHEIM P. and KILLIAS, U. (1963). The use of bismuth as an electron stain for nucleic acid. *J. Cell Biol.*, **17**, 93–103. *(42)*

4 ALBERTE, R. S., THORNBER, J. P. and NAYLOR, A. W. (1972). Time of appearance of photosystems I & II in chloroplasts of greening jack bean leaves. *J. exp. Bot.*, **23**, 1060–9. *(131)*

5 ALBERTE, R. S. THORNBER, J. P. and NAYLOR, A. W. (1973). Biosynthesis of the photosystem I chlorophyll-protein complex in greening leaves of higher plants. *Proc. natn. Acad. Sci. U.S.A.*, **70**, 134–137. *(131)*

6 ALFERT, M. and DAS, N. K. (1969). Evidence for control of the rate of nuclear DNA synthesis by the nuclear membrane in eukaryotic cells. *Proc. natn. Acad. Sci. U.S.A.*, **63**, 123–128. *(52)*

7 ALLEN, R. D. (1969). Mechanism of the seismonastic reaction in *Mimosa pudica*. *Pl. Physiol., Lancaster*, **44**, 1101–1107. *(35)*

7a ALLEN, R. D. (1974). Some new insights concerning cytoplasmic transport. *Symp. Soc. exp. Biol.*, **28**, 15–26. *(152)*

8 ALLEN, T. D., HAIGH, M. V. and HOWARD, A. (1973). Ultrastructure of giant plastids in a radiation induced mutant of *Osmunda regalis*. *J. Ultrastruct. Res.*, **42**, 491–501. *(133)*

9 ALLENSPACH, A. L. and ROTH, L. E. (1967). Structural variations during mitosis in the chick embryo. *J. Cell Biol.* **33**, 179–196. *(168)*

10 AMALDI, F., GIACOMONI, D. and ZITO-BIGNAMI, R. (1969). On the duplication of ribosomal RNA cistrons in chinese hamster cells. *European J. Biochem.*, **11**, 419–423. *(161)*

11 AMELUNXEN, F., THALER, I. and HARSDORFF, M. (1970). Die Struktur der Phytoferritinkristalle von *Phajus grandifolius*. *Z. Pflanzenphysiol.* **63**, 199–210. *(129)*

12 ANTON-LAMPRECHT, I. (1967). Anzahl und vermehrung der Zellorganellen im Scheitelmeristem von *Epilobium*. *Ber. dt. Bot. Ges.* **80**, 747–754. *(72, 84, 93, 99, 132, 133)*

13 ARGYROUDI-AKOYUNOGLOU, J. H. and AKOYUNOGLOU, G. (1973). On the formation of photosynthetic membranes in bean plants. *Photochem. Photobiol.*, **18**, 219–228. *(115)*

14 ARNOLD, C. G., SCHIMMER, O., SCHÖTZ, F. and BATHELT, H. (1971). The mitochondria of *Chlamydomonas reinhardii*. *Arch. Mikrobiol.* **81**, 50–67. *(88)*

15 ARNOTT, H. J. and HARRIS, J. B. (1973). Development of chloroplast substructures with apparent secretory roles in the young tobacco leaf. *Tissue and Cell.* **5**, 337–347. *(127)*

16 ARNTZEN, C. J., DILLEY, R. A. and CRANE, F. L. (1969). A comparison of chloroplast membrane surfaces visualized by freeze-etch and negative staining techniques; and ultrastructural characterization of membrane fractions obtained from digitonin-treated spinach chloroplasts. *J. Cell Biol.*, **43**, 16–31. *(106, 107)*

16a ASHTON, F. T. and SCHULTZ, J. (1971). The three-dimensional fine structure of chromosomes in a prophase *Drosphila* nucleus. *Chromosoma*, **35**, 383–392. *(159)*

17 ATKINSON, A. W. Jr (1972). Ultrastructural studies on *Chlorella*. Ph.D. Thesis, The Queen's University, Belfast. *(143, 147)*

18 ATKINSON, A. W. Jr., GUNNING, B. E. S. and JOHN, P. C. L. (1972). Sporopollenin in the cell wall of *Chlorella* and other algae: ultrastructure, chemistry and incorporation of ^{14}C-Acetate, studied in synchronous cultures. *Planta*, **107**, 1–32. *(22)*

19 ATKINSON, A. W. Jr., JOHN, P. C. L., and GUNNING, B. E. S. (1974). The growth and division of the single mitochondrion and other organelles during the cell cycle of *Chlorella*, studied by quantitative stereology and three dimensional reconstruction. *Protoplasma*, **81**, 77–109 *(34, 88, 94)*

20 AVANZI, S., MAGGINI, F. and INNOCENTI, A. M. (1973). Amplification of ribosomal cistrons during the maturation of metaxylem in the root of *Allium cepa*. *Protoplasma*, **76**, 197–210. *(50)*

21 BAILEY, C. J. and BOULTER, D. (1970). The structure of legumin, a storage protein of broad bean (*Vicia faba*) seed. *Eur. J. Biochem.*, **17**, 460–466. *(61)*

22 BAILEY, C. J. and BOULTER, D. (1972). The structure of vicilin of *Vicia faba*. *Phytochemistry*, **11**, 59–64. *(61)*

23 BAILEY, C. J., COBB, A. and BOULTER, D. (1970) A cotyledon slice system for the electron autoradiographic study of the synthesis and intracellular transport of the seed storage protein of *Vicia faba*. *Planta*, **95**, 103–118. *(61)*

24 BAILEY, R. W. and MACRAE, J. C. (1973). Hydrolysis of intact leaf starch grains by glucamylase and α-amylase. *F.E.B.S. Letters*. **31**, 203–204. *(118)*

25 BAJER, A. (1968). Fine structure-studies on phragmoplast and cell plate formation. *Chromosoma*, **24**, 383–417. *(169)*

26 BAJER, A. and JENSEN, C. (1969). Detectability of mitotic spindle microtubules with the light and electron microscopes. *J. Microscopie*, **8**, 343–354. *(163)*

27 BAJER, A. and MOLE-BAJER, J. (1969). Formation of spindle fibres, kinetochore orientation, and behaviour of the nuclear envelope during mitosis in endosperm. *Chromosoma*, **27**, 448–484. *(159, 162)*

28 BAJER, A. and MOLE-BAJER, J. (1971). Architecture and function of the mitotic spindle. In '*Advances in Cell and Molecular Biology*', Ed. DuPraw, E. J. Vol. 1 pp. 213–266. Academic Press, New York. *(162, 163, 168, 169)*

29 BAJER, A. S. and MOLE-BAJER, J. (1972). Spindle dynamics and chromosome movements. *Int. Rev. Cytol.* Suppl. **3,** *(158, 160, 162, 163, 164, 165, 166, 169)*

30 BAKER, D. A. and HALL, J. L. (1973). Pinocytosis, ATP-ase and ion uptake by plant cells. *New Phytol.*, **72,** 1281–1291. *(25)*

31 BAKER, E. A. and PARSONS, E. (1971). Scanning electron microscopy of plant cuticles. *J. Microsc.*, **94,** 39–49. *(21)*

32 BAL, A. K. and GROSS, P. R. (1964). Asynchronous synthesis of RNA in nucleoli of root meristem. *Science, N.Y.*, **143,** 808–810. *(50)*

33 BAL, A. K. and PAYNE, J. F. (1972). Endoplasmic reticulum activity and cell wall breakdown in quiescent root meristems of *Allium cepa* L. *Z. Pflanzenphysiol*, **66,** 265–272. *(67)*

34 BANKS, W. and GREENWOOD, C. T. (1973). Molecular properties of the starch-components and their relation to the structure of the granule. *Ann. N.Y. Acad. Sci.*, **210,** 17–33. *(116, 117, 119)*

35 BARDELE, C. F. (1973). Strukter, Biochemie und Funktion der Mikrotubuli. *Cytobiologie*, **7,** 442–488. *(145)*

36 BARLOW, P. (1970). Vacuoles in the nucleoli of *Zea mays* root apices and their possible significance in nucleolar physiology. *Caryologia*, **23,** 61–70. *(49)*

37 BARLOW, P. W. (1972). The ordered replication of chromosomal DNA: a review and a proposal for its control. *Cytobios*, **6,** 55–80. *(43)*

38 BAUER, H., DIETZ, R. and RÖBBELEN, C. (1961). Die Spermatocytenteilungen der Tipuliden. III. Das Bewegung sverhalten der Chromosomen in Translokationsheterozygoten von *Tipula oleracae. Chromosoma*, **12,** 116–189. *(166)*

39 BAUER, W. D., TALMADGE, K. W., KEEGSTRA, K. and ALBERSHEIM, P. (1973). The structure of plant cell walls. II. The hemicellulose of the walls of suspension-cultured sycamore cells. *Pl. Physiol., Lancaster*, **51,** 174–187. *(16)*

40 BAZZAZ, M. B. and GOVINDJEE (1973). Photochemical properties of mesophyll and bundle sheath chloroplasts of maize. *Pl. Physiol., Lancaster*, **52,** 257–262. *(109)*

41 BEAMS, H. W. and KESSEL, R. G. (1968). The Golgi apparatus: structure and function. *Int. Rev. Cytol.*, **23,** 209–276. *(71)*

42 BEAMS, H. W. and MUELLER, S. (1970). Effects of ultracentrifugation on the interphase nucleus of somatic cells with special reference to the nuclear envelope—chromatin relationship. *Z. Zellforsch.*, **108,** 297–308. *(52)*

43 BEEVERS, H. (1969). Gloxysomes of castor bean endosperm and their relation to gluconeogenesis. *Ann. N.Y. Acad. Sci.*, **168,** 313–324. *(136)*

43a BEHNKE, O. (1974). The microtubule clear space. *8th International Congress Electron Microscopy, Canberra,* **II,** 330–331. *(142)*

44 BEHNKE, H.-D. (1968). Zum Aufbau gitterartiger Membranstrukturen im Siebelementplasma von *Dioscorea. Protoplasma*, **66,** 287–310. *(66)*

44a BELCHER, J. H. (1968). The fine structure of *Furcilla stigmatophora* (Skuja) Korshikov. *Arch. Microbiol.*, **60,** 84–94. *(89)*

45 BELL, J. K. and MCCULLY, M. E. (1970). A histological study of lateral root initiation and development in *Zea mays. Protoplasma*, **70,** 179–205. *(20)*

46 BEN-SHAUL, Y. and NAFTALI, Y. (1969). The development and ultrastructure of lycopene bodies in chromoplasts of *Lycopersicum esculentum. Protoplasma*, **67,** 333–344. *(120)*

47 BEN-SHAUL, Y., TREFFRY, T. and KLEIN, S. (1968). Fine structure studies of carotene body development. *J. Microscopie*, **7,** 265–274. *(120)*

48 BERJAK, P. (1972). Lysosomal compartmentation: ultrastructural aspects of the origin, development, and function of vacuoles in root cells of *Lepidium sativum. Ann. Bot.*, **36,** 73–81. *(37)*

49 BERNHARD, W. and GRANBOULAN, N. (1968). Electron microscopy of the nucleolus in vertebrate cells. In *Ultrastructure in Biological Systems,* Vol. 3, The nucleus. Eds. Dalton, A. J. and Haguenau, Academic Press, New York. 81–149. *(47)*

50 BERNS, M. W. and CHENG, W. K. (1971) Are chromosome secondary constrictions nucleolar organisers? *Exp Cell. Res.*, **69,** 185–192. *(46)*

51 BICKLE, T. A. HOWARD, G. A. and TRAUT, R. R. (1973). Ribosome heterogeneity. The nonuniform distribution of specific ribosomal proteins among different functional classes of ribosomes. *J. biol. Chem.*, **248,** 4862–4864. *(49)*

52 BIRNSTIEL, M. (1967). The nucleolus in cell metabolism. *A. Rev. Pl. Physiol.*, **18,** 25–58. *(45)*

52a BISALPUTRA, T. and BAILEY, A. (1973). The fine structure of the chloroplast envelope of a red alga, *Bangia fusco-purpurea. Protoplasma*, **76,** 443–454. *(45, 121, 122)*

53 BISALPUTRA, T. and BISALPUTRA, A. A. (1969). The ultrastructure of chloroplast of a brown alga *Sphacelaria* sp. I. Plastid DNA configuration—the chloroplast genophore. *J. Ultrastruct. Res.*, **29,** 151–170. *(124)*

54 BISALPUTRA, T. and BISALPUTRA, A. A. (1970). The ultrastructure of chloroplast of a brown alga *Sphacelaria* sp. III. The replication and segregation of chloroplast genophore. *J. Ultrastruct. Res.*, **32,** 417–429. *(124)*

55 BISALPUTRA, T. and BURTON, H. (1969). The ultrastructure of chloroplast of a brown alga *Sphacelaria* sp. II. Association between the chloroplast DNA and the photosynthetic lamellae. *J. Ultrastruct. Res.*, **29,** 224–235. *(124)*

56 BISALPUTRA, T. and BURTON, H. (1970). On the chloroplast DNA-membrane complex in *Sphacelaria* sp. *J. Microscopie*, **9,** 661–666. *(125)*

57 BLACK, C. C. (1973). Photosynthetic carbon fixation in relation to net CO_2 uptake. *A. Rev. Pl. Physiol.*, **24,** 253–286. *(108, 109)*

58 BLACK, C. C. Jr. and MOLLENHAUER, H. H. (1971). Structure and distribution of chloroplasts and other organelles in leaves with various rates of photosynthesis. *Pl. Physiol., Lancaster*, **47,** 15–23. *(108)*

59 BLACKWELL, S. J., LAETSCH, W. M. and HYDE, B. B. (1969). Development of chloroplast fine structure in aspen tissue culture. *Am. J. Bot.*, **56,** 457–463. *(115, 132)*

60 BOARDMAN, N. K., ANDERSON, J. M., KAHN, A., THORNE, S. W. and TREFFRY, T. E. (1970). Formation of photosynthetic membranes during chloroplast development. In *Autonomy and Biogenesis of Mitochondria and Chloroplasts.* Eds. Boardman, N. K., Linnane, A. W. and Smillie, R. S., North Holland, Amsterdam. 70–84. *(131)*

61 BOASSON, R., BONNER, J. J. and LAETSCH, W. M. (1972). Induction and regulation of chloroplast replication in mature tobacco leaf tissue. *Pl. Physiol., Lancaster*, **49,** 97–101 *(132)*

62 BOGORAD, L. (1967). The role of cytoplasmic units. Control mechanisms in plastid development. *Dev. Biol. Supple.*, **1**, 1–31. (*130*)

62a BONNETT, H. T. Jr. (1968). The root endodermis: fine structure and function. *J. Cell Biol.*, **37**, 199–205. (*31*)

63 BORISY, G. G. and TAYLOR, E. W. (1967). The mechanism of action of colchicine. Binding of colchicine-^{3}H to cellular protein. *J. Cell Biol.*, **34**, 525–533. (*142*)

64 BORISY, G. G. and TAYLOR, E. W. (1967). The mechanism of action of colchicine. Colchicine binding to sea urchin eggs and the mitotic apparatus. *J. Cell Biol.*, **34**, 535–548. (*142*)

65 BORST, P. (1970). Mitochondrial DNA; structure, information content, replication and transcription. In *Symp. Soc. exp. Biol.*, **24**, 201–225. (*93*)

66 BORST, P. and GRIVELL, L. A. (1971). Mitochondrial ribosomes. *F.E.B.S. Letters*, **13**, 73–88. (*90*)

67 BOUCK, G. B. (1963). Stratification and subsequent behaviour of plant cell organelles. *J. Cell Biol.*, **18**, 441–457. (*68*)

68 BOUCK, G. B. (1972). Architecture and assembly of mastigonemes. *Adv. Cell Molec. Biol.*, **2**, 237–271. (*74*)

69 BOUCK, G. B. and BROWN, D. L. (1973). Microtubule biogenesis and cell shape in *Ochromonas*. I. The distribution of cytoplasmic and mitotic microtubules. *J. Cell Biol.*, **56**, 340–359. (*146*)

70 BOURQUE, D. P., BOYNTON, J. E. and GILLHAM, N. W. (1971). Studies on the structure and cellular location of various ribosome and ribosomal RNA species in the green alga *Chlamydomonas reinhardi*. *J. Cell Sci.*, **8**, 153–183. (*126*)

71 BOURQUE, D. P. and WILDMAN, S. G. (1973). Evidence that nuclear genes code for several chloroplast ribosomal proteins. *Biochem. biophys. Res. Commun.*, **50**, 532–537. (*175*)

72 BRACKER, C. E. and GROVE, S. N. (1970). Surface structure on outer mitochondrial membranes of *Phythium ultimum*. *Cytobiologie*, **3**, 229–239. (*89*)

73 BRACKER, C. E. and GROVE, S. N. (1971). Continuity between cytoplasmic endomembranes and outer mitochondrial membranes in fungi. *Protoplasma*, **73**, 15–34. (*89*)

74 BRADBEER, J. W. (1973). The synthesis of chloroplast enzymes. In *Biosynthesis and its Control in Plants*, Ed. Milborrow, B. V. Phytochemical Society Symposium, Academic Press, New York, **9**, 279–302. (*115, 131*)

75 BRADLEY, M. O. (1973). Microfilaments and cytoplasmic streaming: inhibition of streaming with cytochalasin. *J. Cell Sci.*, **12**, 327–343. (*152, 153*)

76 BRAM, S. and RIS, H. (1971). On the structure of nucleohistone. *J. molec. Biol.*, **55**, 325–336. (*44*)

77 BRANGEON, J. (1973). Compared ontogeny of the two types of chloroplasts of *Zea mays*. *J. Microscopic*, **16**, 233–242. (*108*)

78 BRANTON, D. (1969). Membrane Structure. *A. Rev. Pl. Physiol.*, **20**, 209–238. (*12*)

79 BRANTON, D. and DEAMER, D. W. (1972). Membrane structure. *Protoplasmatologia* II/E/1. Springer-Verlag, Berlin. (*12, 104, 105*)

80 BRANTON, D. and PARK, R. B. (1967). Subunits in chloroplast lamellae. *J. Ultrastruct. Res.*, **19**, 283–303. (*104*)

81 BRASELTON, J. P. (1971). The ultrastructure of the non-localised kinetochores of *Luzula* and *Cyperus*. *Chromosoma*, **36**, 89–99. (*159*)

82 BREIDENBACH, R. W. and BEEVERS, H. (1967). Association of the glyoxylate cycle enzymes in a novel subcellular particle from castor bean endosperm. *Biochem. biophys. Res. Commun.*, **27**, 462–469. (*136*)

82a BRETSCHER, M. S. (1973). Membrane structure: some general principles. *Science, N.Y.*, **181**, 622–629. (*12*)

83 BRIARTY, L. G., COULT, D. A. and BOULTER, D. (1969). Protein bodies of developing seeds of *Vicia faba*. *J. exp. Bot.*, **20**, 358–372. (*36, 61*)

84 BRIGGS, D. E. (1973). Hormones and carbohydrate metabolism in germinating cereal grains. In *Biosynthesis and its Control in Plants*. Ed. Milborrow, B. V. *Phytochem. Soc. Symp.*, **9**, 219–277. (*62*)

85 BRINKLEY, B. R. and CARTWRIGHT, J. (1971). Ultrastructural analysis of mitotic spindle elongation in mammalian cells *in vitro*. *J. Cell Biol.*, **50**, 416–431. (*168*)

86 BRITTEN, R. J. and KOHNE, D. E. (1968). Repeated sequences in DNA. *Science, N.Y.*, **161**, 529–540. (*44*)

87 BROOKS, J., GRANT, P. R., MUIR, M., VAN GIJZEL, P. and SHAW, G. (1971). *Sporopollenin*. Academic Press, New York. (*22*)

88 BROOKS, J. and SHAW, G. (1971). Recent developments in the chemistry, biochemistry, geochemistry and post-tetrad ontogeny of sporopollenins derived from pollen and spore exines. In *Pollen Development and Physiology*. Ed. Heslop-Harrison, J. Butterworth, London. 16–31. (*22*)

89 BROWN, D. L. and BOUCK, G. B. (1973). Microtubule biogenesis and cell shape in *Ochromonas*. II. The role of nucleating sites in shape development. *J. Cell Biol.*, **56**, 360–378. (*146*)

90 BROWN, R. M., Jr. (1972). A cellulosic glycoprotein secretory product of the Golgi apparatus. *J. Cell Biol.*, **55**, 30a. (*80*)

91 BROWN, R. M., Jr. (1973). The role of the Golgi apparatus in scale and coccolith biogenesis. *J. Cell Biol.*, **59**, 35a. (*74*)

92 BROWN, R. M., Jr. and FRANKE, W. W. (1971). A microtubular crystal associated with the Golgi field of *Pleurochrysis scherffelii*. *Planta*, **96**, 354–363. (*80*)

93 BROWN, R. M., Jr., FRANKE, W. W., KLEINIG, H., FALK, H. and SITTE, P. (1970). Scale formation in Chrysophycean algae. I. Cellulosic and noncellulosic wall components made by the Golgi apparatus. *J. Cell Biol.*, **45**, 246–271. (*76, 81*)

94 BRUSKOV, V. I. and ODINTSOVA, M. S. (1968). Comparative electron microscopic studies of chloroplast and cytoplasmic ribosomes. *J. molec. Biol.*, **32**, 471–473. (*126*)

95 BUCHHOLZ, J. T. (1947). Methods in the preparation of chromosomes and other parts of cells for examination with an electron microscope. *Am. J. Bot.*, **34**, 445–454. (*87*)

96 BURGESS, J. (1970). Microtubules and cell divison in the microspore of *Dactylorchis fuschii*. *Protoplasma*, **69**, 253–264 (*170*)

97 BURGESS, J. (1970). Interactions between microtubules and the nuclear envelope during mitosis in a fern. *Protoplasma*, **71**, 77–89. (*162*)

98 BURGESS, J. (1971). Observations on structure and differentiation in plasmodesmata. *Protoplasma*, **73**, 83–95. (*27*)

99 BURGESS, J. (1971). The occurrence of plasmodesmata-like structures in a non-divison wall. *Protoplasma*, **74**, 449–458. (*28*)

99a BURGESS, J. and NORTHCOTE, D. H. (1969). Action of colchicine and heavy water on the polymerization of microtubules in wheat root meristem. *J. Cell Sci.*, **5**, 433–451. (*166, 169*)

100 BURNS, E. R. and SOLOFF, B. L. (1972). Nucleolar vacuoles in cells cultured from lung and peritoneum of *Diemictylus viridescens. Tissue and Cell*, **4**, 63–71. (*49*)

101 BURNS, R. G. (1973). Kinetics of the regeneration of sea-urchin cilia. *J. Cell Sci.*, **13**, 55–67. (*144*)

102 BURR, F. A. and EVERT, R. F. (1972). A cytochemical study of the wound-healing protein in *Bryopsis hypnoides. Cytobios*, **6**, 199–215. (*61*)

103 BURR, F. A. and WEST, J. A. (1972). Protein bodies in *Bryopsis hypnoides:* Their relationship to wound-healing and branch septum development. *J. Ultrastruct. Res.*, **35**, 476–498. (*21, 61*)

104 BUSCH, H. and SMETNA, K. (1970). *The Nucleolus*. Academic Press, New York. (*46*)

105 BUTTERFASS, T. (1973). Control of plastid division by means of nuclear DNA amount. *Protoplasma*, **76**, 167–195. (*132*)

106 BUTTROSE, M. S. (1962). The influence of environment on the shell structure of starch granules. *J. Cell Biol.*, **14**, 159–167. (*118*)

107 BUTTROSE, M. S. (1969). The dissolution and reaccumulation of starch granules in grape vine cane. *Aust. J. biol. Sci.*, **22**, 1297–1303. (*118*)

107a BUVAT, R. (1969). *Plant Cells*. World University Library, London. (*132*)

108 BYERS, B. and ABRAMSON, D. H. (1968). Cytokinesis in HeLa: Post-telophase delay and microtubule-associated motility. *Protoplasma*, **66**, 413–435. (*169*)

108a CALLOW, M. E. and EVANS, L. V. (1974). Studies on the ship-fouling alga *Enteromorpha*. III. Cytochemistry and autoradiography of adhesive production. *Protoplasma*, **80**, 15–27. (*78, 80*)

109 CALVAYRAC, R., VAN LENTE, F. and BUTOW, R. A. (1971). *Euglena gracilis:* formation of giant mitochondria. *Science, N.Y.*, **173**, 252–254. (*88*)

110 CAMPBELL, P. N. (1970). Functions of polyribosomes attached to membranes of animal cells. *F.E.B.S. Letters*, **7**, 1–7. (*62*)

110a CANDE, W. Z., SNYDER, J., SMITH, D., SUMMERS, K. and MCINTOSH, J. R. (1974). A functional mitotic spindle prepared from mammalian cells in culture. *Proc. natn. Acad. Sci., U.S.A.*, **71**, 1559–1563. (*167*)

111 CAROTHERS, Z. B. and KREITNER, G. L. (1967). Studies of spermatogenesis in the Hepaticae. I. Ultrastructure of the *Vierergruppe* in *Marchantia. J. Cell Biol.*, **33**, 43–51. (*146*)

112 CAROTHERS, Z. B. and KREITNER, G. L. (1968). Studies of spermatogenesis in the Hepaticae. II. Blepharoplast structure in the spermatid of *Marchantia. J. Cell Biol.*, **36**, 603–616. (*146*)

113 CASS, D. D. (1973). An ultrastructural and Nomarski-interference study of the sperms of barley. *Can. J. Bot.*, **51**, 601–605. (*132, 147*)

114 CATESSON, A.-M. (1962). Modifications saisonnieres des vacuoles et variations de la pression osmotique dans le cambium d'*Acer pseudoplatanus. C.r. hebd. Seanc. Acad. Sci. Paris*, **254**, 3887–3889. (*34*)

114A CATESSON, A. M. (1974). Cambial cells. In *Dynamic Aspects of Plant Ultrastructure*, Ed. Robards, A.W. McGraw-Hill, London, 358–390 (*34, 68*)

115 CAVALIER-SMITH, T. (1970). Electron microscope evidence for chloroplast fusion in zygotes of *Chlamydomonas reinhardii. Nature, Lond.*, **228**, 333–335. (*130, 133*)

116 CECCHINI, J. P., MIASSOD, R. and RICARD, J. (1972). Processing of precursor ribosomal RNA in suspensions of higher plant cells. *F.E.B.S. Letters*, **28**, 183–187. (*49*)

117 CHAFE, S. C. and WARDROP, A. B. (1970). Microfibril orientation in plant cell walls. *Planta*, **92**, 13–24. (*147*)

118 CHEN, C. H. and LEHNINGER, A. L. (1973). Ca^{2+} transport activity in mitochondria from some plant tissues. *Archs Biochem. Biophys.*, **157**, 183–196. (*90*)

119 CHEN, J. C. W. (1973). The kinetics of tip growth in the *Nitella* rhizoid. *Plant and Cell Physiol.*, **14**, 631–640. (*153*)

120 CHEN, J. C. W. (1973). Observations of protoplasmic behaviour and motile protoplasmic fibrils in cytochalasin B treated *Nitella* rhizoid. *Protoplasma*, **77**, 427–435. (*152*)

121 CHEN, J. L. and WILDMAN, S. G. (1970). 'Free' and membrane-bound ribosomes, and nature of products formed by isolated tobacco chloroplasts incubated for protein synthesis. *Biochem. biophys. Acta.*, **209**, 207–219. (*127*)

122 CHEN, T. M., CAMPBELL, W. H., DITTRICH, P. and BLACK, C. C. (1973). Distribution of carboxylation and decarboxylation enzymes in isolated mesophyll cells and bundle sheath strands of C_4 plants. *Biochem. biophys. Res. Commun.*, **51**, 461–467. (*108*)

123 CHIANG, K.-S. and SUEOKA, N. (1967). Replication of chromosomal and cytoplasmic DNA during mitosis and meiosis in the eucaryote *Chlamydomonas reinhardi. J. Cell Physiol.*, **70**, sup. 1, 89–112. (*125, 175*)

124 CHOUINARD, L. A. (1970). Localization of intranucleolar DNA in root meristematic cells of *Allium cepa. J. Cell Sci.*, **6**, 73–85. (45)

125 CHRISTIE, A. O., EVANS, L. V. and SHAW, M. (1970). Studies on the ship-fouling alga *Enteromorpha*. II. The effect of certain enzymes on the adhesion of zoospores. *Ann. Bot.*, **34**, 467–482. (78)

126 CHUA, N.-H., BLOBEL, G., SIEKEVITZ, P. and PALADE, G. E. (1973). Attachment of chloroplast polysomes to thylakoid membranes in *Chlamydomonas reinhardtii. Proc. natn. Acad. Sci. U.S.A.*, **70**, 1554–1558. (*127*)

127 CLARKSON, D. T., ROBARDS, A. W. and SANDERSON, J. (1971). The tertiary endodermis in barley roots: Fine structure in relation to radial transport of ions and water. *Planta*, **96**, 292–305. (*29, 31*)

128 CLOWES, F. A. L., and DE LA TORRE, C. (1972). Nucleoli in X-rayed meristems. *Cytobiologie*, **6**, 318–326. (*45*)

129 CLOWES, F. A. L. and JUNIPER, B. E. (1968). *Plant Cells*, Blackwell, Glasgow. (*10, 68, 72, 77, 88*)

130 COATS, L. W. (1973). Sulfolipid control of grana membrane stacking. *J. Cell Biol.*, **59**, 59a. (*107*)

131 COCKING, E. C. (1970) Virus uptake, cell wall regeneration, and virus multiplication in isolated plant protoplasts. *Int. Rev. Cytol.*, **28**, 89–124. (*17, 25*)

132 COCKING, E. C. (1972). Plant cell protoplasts—isolation and development. *A. Rev. Pl. Physiol.*, **23**, 29–50. (*17*)

133 COHEN, I. (1937). Structure of the interkinetic nucleus in the scale epidermis of *Allium cepa. Protoplasma*, **27**, 484–495. (*51*)

134 COHEN, S. S. (1970). Are/Were mitochondria and chloroplasts microorganisms? *Am. Scient.*, **58**, 281–289. (*176*)

135 COHEN, W. D. and REBHUN, L. I. (1970). An estimate of the amount of microtubule protein in the isolated mitotic apparatus. *J. Cell Sci.*, **6**, 159–176. (144)

136 COLEMAN, R. (1973). Membrane-bound enzymes and membrane ultrastructure. *Biochim. biophys. Acta.*, **300**, 1–30. (*174*)

137 COLVIN, J. R. (1971). Structure and formation of the cellulose microfibril. *High Polymers* Vol. V. *Cellulose and Cellulose Derivatives* Part IV. Bikales, N. M. and Segal, L. (*16*)

138 COMINGS, D. E. and OKADA, T.A. (1970). Association of chromatin fibres with the annuli of the nuclear membrane. *Exptl Cell Res.*, **62**, 293–302. (*52, 53*)

139 COMINGS, D. E. and OKADA, T. A. (1970). Association of nuclear membrane fragments with metaphase and anaphase chromosomes as observed by whole mount electron microscopy. *Exptl Cell Res.*, **63**, 62–68. (*53*)

140 COSS, R. A. and PICKETT-HEAPS, J. D. (1973). The effect of isopropyl-n-phenyl carbamate (IPC) on the spindle and phycoplast microtubule organizing centres (MTOCS) in the green alga *Oedogonium cardiacum*. *J. Cell Biol.*, **59**, 65a. (*164*)

141 COSTERTON, J. W. F. and MACROBBIE, E. A. C. (1970). Ultrastructure of *Nitella* translucens in relation to ion transport. *J. exp. Bot.*, **21**, 535–542. (*25, 65*)

142 COTÉ, W. A., Jr. (1967). *Wood Ultrastructure, an Atlas of Electron Micrographs*. University of Washington Press (*19*)

143 COULOMB, C. (1973). Diversité des corps multivésiculaires et notion d'hétérophagie dans le méristème radiculaire de scorsonère. *(Scorzonera hispanica), J. Microscopie*, **16**, 345–360. (*23*)

144 COULOMB, P. and COULOMB, C. (1972). Processus d'autophagie cellulaire dans les cellules de méristèmes radiculaires de la Courge (*Cucurbita pepo* L. Cucurbitacée), mis en état d'anoxie. *C. r. hebd. Séanc. Acad. Sci. Paris*, **274**, 214–217. (*37, 67*)

145 COULOMB, P., COULOMB, C. and COULON, J. (1972). Origine et fonctions des phytolysomes dans le méristème radiculaire de la courge *Cucurbita pepo* L. I. Origine des phytolysomes. Relations reticulum endoplasmique—dictyosomes—phytolysomes. *J. Microscopie*, **13**, 263–280. (*37, 79, 83*)

146 COULOMB, P. and COULON, J. (1971). Fonctions de l'appareil de Golgi dans les méristèmes radiculaires de la courge (*Cucurbita pepo* L. Cucurbitacée). *J. Microscopie*, **10**, 203–214. (*37, 78, 79, 83*)

147 COX, G. and JUNIPER, B. (1972). Electron microscopy of cellulose in entire tissue. *J. Microsc.* **97**, 343–355. (*16*)

148 COX, G. C. and JUNIPER, B. E. (1973). Autoradiographic evidence for paramural-body function. *Nature, New Biol.*, **243**, 116–117. (*23*)

149 CRAN, D. G. and POSSINGHAM, J. V. (1972). Two forms of division profile in spinach chloroplasts. *Nature, New Biol.*, **235**, 142. (*133*)

150 CRESTI, M., PACINI, E. and SARFATTI, G. (1972). Ultrastructural studies on the autophagic vacuoles in *Eranthis hiemalis* endosperm. *J. submicr. Cytol.*, **4**, 33–44. (*67*)

150a CRONSHAW, J. (1964). Crystal containing bodies of plant cells. *Protoplasma*, **59**, 318–325. (*135*)

150b CRONSHAW, J. (1974). Phloem differentiation and development. In *Dynamic Aspects of Plant Ultrastructure*, Ed. Robards, A. W. McGraw-Hill, London, 391–413. (*20, 21, 66*)

151 CROTTY, W. J. and LEDBETTER, M. C. (1973). Membrane continuities involving chloroplasts and other organelles in plant cells. *Science, N.Y.*, **182**, 839–841. (*177*)

152 CUTTER, E. G. (1969). *Plant Anatomy: Experiment and Interpretation. Part 1. Cells and Tissues*. Edward Arnold, London. (*7, 168, 171*)

153 CUTTER, E. G. (1971). *Plant Anatomy: Experiment and Interpretation. Part 2. Organs*. Edward Arnold, London. (*7*)

154 CUTTER, E. G. and HUNG, C.-Y. (1972). Symmetric and asymmetric mitosis and cytokinesis in the root tip of *Hydrocharis morus- ranae* L. *J. CellSci.*, **11**, 723–737. (*169, 170*)

155 DAINTY, J. (1968). The structure and possible function of the vacuole. In *Plant Cell Organelles*. Ed. Pridham, J. pp. 40–46, Academic Press, New York. (*34*)

156 DAINTY, J. (1969). The water relations of plants. In *The Physiology of Plant Growth and Development*. Ed. Wilkins, M. B., published by McGraw-Hill, London. 421–452. (*24, 34*)

157 DARVEY, N. L. and DRISCOLL, C. J. (1971). Nucleolar behaviour in *Triticum*. *Chromosoma*, **36**, 131–139. (*46*)

158 DASHEK, W. V. and ROSEN, W. G. (1966). Electron microscopical localization of chemical components in the growth zone of lily pollen tubes. *Protoplasma*, **61**, 192–204. (*79*)

159 DAUWALDER, M., WHALEY, W. G. and KEPHARD, J. E. (1969). Phosphatases and differentiation of the Golgi apparatus. *J. Cell Sci.*, **4**, 455–498. (*73*)

160 DAWES, C. J. (1971). *Biological Techniques in Electron Microscopy*. Barnes and Nobel, inc., New York. (*2*)

161 DE LA TORRE, C. and CLOWES, F. A. L. (1972). Timing of nucleolar activity in meristems. *J. Cell Sci.*, **11**, 713–721. (*161*)

161a DENTLER, W. L., GRANETT, S., WITMAN, G. B. and ROSENBAUM, J. L. (1974). Directionality of brain microtubule assembly *in vitro*. *Proc. natn. Acad. Sci. U.S.A.*, **71**, 1710–1714. (*144*)

162 DEXHEIMER, J. (1966). Sur les modifications de reticulum endoplasmiques des grains de pollen de *Lobelia erinus* (L.) traites par le chloramphenicol. *C. r. hebd. Séanc. Acad. Sci., Paris*, **262**, 853–855. (*68*)

163 DEXHEIMER, J. (1966). Sur les modifications, sous l'action de diverses substances, de reticulum endoplasmique des tubes polliniques de *Lobelia erinus* L. *C. r. hebd. Séanc. Acad. Sci., Paris*, **263**, 1703–1705. (*68*)

164 DICKINSON, H. G. (1970). Ultrastructural aspects of primexine formation in the microscope tetrad of *Lilium longiflorum*. *Cytobiologie*, **1**, 437–449. (*26*)

165 DICKINSON, H. G. and HESLOP-HARRISON, J. (1971). The mode of growth of the inner layer of the pollen-grain exine in *Lilium*. *Cytobios*, **4**, 233–243. (*22, 66*)

166 DICKINSON, H. G. and LEWIS, D. (1973). Cytochemical and ultrastructural differences between intraspecific compatible and incompatible pollinations in *Raphanus*. *Proc. R. Soc. Lond.* B, **183**, 21–38. (*21*)

167 DIEHN, B. (1973). Phototaxis and sensory transduction in *Euglena*. *Science, N.Y.*, **181**, 1009–1015. (*128*)

168 DIERS, L. (1970). Origin of plastids: cytological results and interpretations including some genetical aspects. *Sym., Soc. exp. Biol.*, **24**, 129–145. (*176*)

169 DIERS, L. and SCHÖTZ, F. (1965). Über den Feinbau pflanzlicher Mitochondrien. *Z. Pflanzenphysiologie*, **53**, 334–343. (*89*)

170 DIETZ, R. (1969). Bau und Funktion des Spindelapparats. *Naturwissenschaften*, **56**, 237–248. (*167*)

171 DIETZ, R. (1972). Die Assembly—Hypothese der Chromosomenbewegung und die Verenderungen der Spindellange während der Anaphase I in Spermatocyten von *Pales ferruginea* (Tipulidae, Diptera). *Chromosoma*, **38**, 11–76. (*164. 167*)

172 DIETZ, R. (1973). Anaphase behaviour of inversions in living crane-fly spermatocytes. *Chromosomes Today, Proc. Oxford Chromosome Conf.*, Oxford 1970. (*164, 167*)

173 DOBBERSTEIN, B. and KIERMAYER, O. (1972). The occurrence of a special type of Golgi vesicles during secondary wall formation in *Micrasterias denticulata* Breb. *Protoplasma*, **75**, 185–194. (*76*)

174 DODGE, J. D. (1969). A review of the fine structure of algal eyespots. *Br. phycol. J.*, **4**, 199–210. (*128*)

175 DODGE, J. D. (1973). *The Fine Structure of Algal Cells*. Academic Press, London. (*128, 169*)

176 DONNAY, G. and PAWSON, D. L. (1969). X-ray diffraction studies of echinoderm plates. *Science, N.Y.*, **166**, 1147–1150. (*114*)

177 DÖRR, (1969). Feinstruktur intrazellular wachsender *Cuscuta*-Hyphen. *Protoplasma*, **67**, 123–137. (*28*)

177a DOUCE, R. (1974). Site of biosynthesis of galactolipids in spinach chloroplasts. *Science, N.Y.*, **183**, 852–853. (*110, 121*)

178 DOWNTON, W. J. S. (1970). Preferential C4-dicarboxylic acid synthesis, the post illumination CO_2 burst, carboxyl transfer stop, and grana configurations in plant with C_4-photosynthesis. *Can. J. Bot.*, **48**, 1795–1800 (*109*)

179 DUCKETT, J. G. (1973). An ultrastructural study of the differentiation of the spermatozoid of *Equisetum*. *J. Cell Sci.*, **12**, 95–129. (*146*)

180 DUPRAW, E. J. (1968). *Cell and Molecular Biology*. Academic Press, New York. (*149, 159*)

180a DUVE, C. DE (1973). Biochemical studies on the occurrence, biogenesis and life history of mammalian peroxisomes. *J. Histochem. Cytochem.*, **21**, 941–948. (*138*)

181 DYER, T. A., MILLER, R. H. and GREENWOOD, A. D. (1971). Leaf nucleic acids. I. Characteristics and role in the differentiation of plastids. *J. exp. Bot.*, **22**, 125–136. (*126*)

182 ECHLIN, P. (1971). The role of the tapetum during microsporogenesis of angiosperms. In *Pollen Development and Physiology* Ed. Heslop-Harrison, J., Butterworth, London, 41–61. (*22, 63*)

183 ECHLIN, P. and GODWIN, H. (1968). The ultrastructure and ontogeny of pollen in *Helleborus foetidus* L. II. Pollen grain development through the callose special wall stage. *J. Cell Sci.*, **3**, 175–186. (*20*)

184 ECKERT, W. A., FRANKE, W. W. and SCHEER, U. (1972). Actinomycin D and the central granules in the nuclear pore complex: Thin sectioning versus negative staining. *Z. Zellforsch.*, **127**, 230–239. (*56*)

184a EDDS, K. (1973). Particle movements in artificial axopodia of *Echinosphaerium nucleofilum*. *J. Cell Biol.*, **59**, 88a. (*153*)

185 EIGSTI, O. J. and DUSTIN, P. (1955). *Colchicine in Agriculture, Medicine, Biology and Chemistry*. Iowa State College Press, Ames. (*142*)

186 EKÉS, M. (1970). Electron—Microscopic—Histochemical demonstration of succinic—Dehydrogenase activity in root cells of Yellow Lupine. *Planta*, **94**, 37–46. (*91*)

187 ELIAS, H. (1971). Three dimensional structure identified from single sections. *Science, N.Y.*, **174**, 993–1000. (*4*)

188 EPSTEIN, E. (1973). Roots. *Scient. Am.*, **228**, 48–58. (*31*)

189 ERDELSKÁ, O. (1973). Vacuolar activity of nucleoli in the cells of the living embryo sacs of *Jasione montana* L. *Protoplasma*, **76**, 123–127. (*49*)

190 ERICKSON, H. P. (1974). Microtubule surface lattice and subunit structure and observations on reassembly. *J. Cell Biol.*, **60**, 153–167. (*142*)

191 ERIKSSON, G., KAHN, A., WALLES, B. and VON WETTSTIEN, D. (1961). Zur makromolekularen Physiologie der Chloroplasten III. *Ber. dt. bot. Ges.*, **74**, 221–232. (*100*)

192 ERIKSSON, L., SVENSSON, H. BERGSTRAND, A. and DALLNER, G. (1972). Physiochemical and enzymic properties of the endoplasmic reticulum in relation to membrane biogenesis. In *Role of Membranes in Secretory Processes*. Eds. Bolis, L., Keynes, R. D. and Wilbrandt, W. North-Holland, Amsterdam, 3–23. (*69*)

193 ESAU, K. (1969). The phloem. *Encyclopaedia of Plant Anatomy*, Vol. **5**, Pt. 1. Borntraeger. (*20*)

194 ESAU, K. (1972). Apparent temporary chloroplast fusions in leaf cells of *Mimosa pudica*. *Z. Pflanzenphysiol.*, **67**, 244–254. (*133*)

195 ESAU, K. and GILL, R. H. (1971). Aggregation of endoplasmic reticulum and its relation to the nucleus in a differentiating sieve element. *J. Ultrastruct. Res.*, **34**, 144–158. (*67*)

196 ESAU, K. and GILL, R. H. (1972). Nucleus and endoplasmic reticulum in differentiating root protophloem of *Nicotiana tabacum*. *J. Ultrastruct. Res.*, **41**, 160–175. (*67*)

197 ESCHRICH, W. (1970). Biochemistry and fine structure of phloem in relation to transport. *A. Rev. Pl. Physiol.*, **21**, 193–214. (*21*)

198 ESPONDA, P. and GIMÉNEZ-MARTIN, G. (1972). Ultrastructural morphology of the nucleolar organizing region. *J. Ultrastruct. Res.* **39**, 509–519. (*46*)

198a EVANS, L. V. and CALLOW, M. E. (1974). Polysaccharide sulphation in *Laminaria*. *Planta*, **117**, 93–95. (*77*)

199 EVANS, L. V. and CHRISTIE, A. O. (1970). Studies on the ship-fouling alga *Enteromorpha*. I. Aspects of the fine structure and biochemistry of swimming and newly settled zoospores. *Ann. Bot.*, **34**, 451–463. (*77*)

200 EVINS, W. H. and VARNER, J. E. (1971). Hormone-controlled synthesis of endoplasmic reticulum in barley aleurone cells. *Proc. natn. Acad. Sci. U.S.A.*, **68**, 1631–1633. (*62*)

201 EYMÉ, J. (1967). Nouvelles observations sur l'infrastructure de tissus nectarigènes floraux. *Le Botaniste*, série L, 169–183. (*64*)

202 FAHN, A. and RACHMILEVITZ, T. (1970). Ultrastructure and nectar secretion in *Lonicera japonica*. *Bot. J.*, **63**, Suppl. 1, 51–56. (*65*)

203 FAKAN, S., TURNER, G. N., PAGANO, J. S. and HANCOCK, R. (1972). Sites of replication of chromosomal DNA in a eukaryotic cell. *Proc. natn. Acad. Sci.*, **69**, 2300–2305. (*52*)

204 FALK, H. (1962). Zur physiologie der Golgi-Apparate in der Wurzelhaube der Zwiebel. *Z. Naturf.*, **17b**, 862–864. (*81*)

205 FALK, H. (1969). Rough thylakoids: polysomes attached to chloroplast membranes. *J. Cell Biol.*, **42**, 582–587. (*126*)

206 FAN, H. and PENMAN, S. (1971). Regulation of synthesis and processing of nucleolar components in metaphase-arrested cells. *J. molec. Biol.*, **59,** 27–42. *(161)*

207 FAWCETT, D. W. (1966). *An Atlas of Fine Structure. The Cell.* Saunders, W. B., London. *(89)*

208 FELDHERR, C. M. (1972). Structure and function of the nuclear envelope. *Adv. Cell and Molec. Biol.*, **2,** Ed. Dupraw. *(55, 56)*

209 FERDOUSE, M., RICKARD, P. A. D., MOSS, F. J. and BLANCH, H. W. (1972). Quantitative studies of the development of *S. cerevisiae* mitochondria. *Biotech. Bioengineering*, **15,** 1007–1026. *(94)*

210 FIGIER, J. (1968). Etude infrastructurale et cytochimique des glandes pétiolaires de *Mercurialis annua* L. Essai d'interprétation en rapport avec la sécrétion. *C. r. hebd. Acad. Sci., Paris*, **267,** 491–494. *(62)*

211 FIGIER, J. (1969). Incorporation de glycine-^{3}H chez les glandes pétiolaires de *Mercurialis annua* L. *Planta*, **87,** 275–289. *(62)*

212 FILNER, P. and BEHNKE, O. (1973). Stabilisation and isolation of brian microtubules with glycerol and dimethylsulfoxide (DMSO). *J. Cell Biol.*, **59,** 99a. *(143)*

213 FINDLAY, N. and MERCER, F. V. (1971). Nectar production in *Abutilon* I. Movement of nectar through the cuticle. *Aust. J. biol. Sci.*, **24,** 647–656. *(22)*

214 FINDLAY, N. and MERCER, F. V. (1971). Nectar production in *Abutilon*. II. Submicroscopic structure of the nectary. *Aust. J. biol. Sci.*, **24,** 657–664. *(65, 68)*

215 FINERAN, B. A. (1970). An evaluation of the form of vacuoles in thin sections and freeze-etch replicas of root tips. *Protoplasma*, **70,** 457–478. *(34)*

216 FINERAN, B. A. (1970). Organization of the tonoplast in frozen-etched root tips. *J. Ultrastruct. Res.*, **33,** 574–586. *(33)*

217 FINERAN, B. A. (1971). Ultrastructure of vacuolar inclusions in root tips. *Protoplasma*, **72,** 1–18. *(34)*

218 FINERAN, B. A. (1972). Fracture faces of the tonoplast in root tips after various conditions of pretreatment prior to freeze-etching. *J. Microsc.*, **96,** 333–342. *(33)*

219 FINERAN, B. A. (1973). Association between endoplasmic reticulum and vacuoles in frozen-etched root tips. *J. Ultrastruct. Res.*, **43,** 75–87. *(33, 65)*

220 FISHER, D. A. and BAYER, D. E. (1972). Thin sections of plant cuticles demonstrating channels and wax platelets. *Can. J. Bot.*, **50,** 1509–1511. *(21)*

221 FISHER, M. L., ANDERSON, A. J. and ALBERSHEIM, P. (1973). Host-pathogens interactions. VI A single plant protein efficiently inhibits endopolygalacturonases secreted by *Colletotrichum lindemuthianum* and *Aspergillus niger*. *Pl. Physiol., Lancaster*, **51,** 489–491. *(20)*

222 FLICKINGER, C. J. (1968). The effects of enucleation on the cytoplasmic membranes of *Amoeba proteus*. *J. Cell Biol.*, **37,** 300–315. *(67. 85)*

223 FLICKINGER, C. J. (1969). The development of Golgi complexes and their dependence upon the nucleus in *Amoebae*. *J. Cell Biol.*, **43,** 250–262. *(85)*

224 FLICKINGER, C. J. (1971). Decreased formation of Golgi bodies in *Amobae* in the presence of RNA and protein synthesis inhibitors. *J. Cell Biol.*, **49,** 221–226. *(81, 83, 85)*

225 FLICKINGER, C. J. (1971). Alterations in the Golgi apparatus of *Amobae* in the presence of an inhibitor of protein synthesis, *Exptl Cell Res.*, **68,** 381–387. *(81, 83)*

226 FLICKINGER, C. J. (1972). Influence of inhibitors of energy metabolism on the formation of Golgi bodies in *Amoebae*. *Exptl Cell Res.*, **73,** 154–160. *(81)*

227 FORER, A. (1969). Chromosome movements during cell division in *Handbook of Molecular Cytology*, Ed. Lima-de-Faria, A., North Holland, Amsterdam, 554–601. *(163, 165)*

228 FORER, A., EMMERSEN, J. and BEHNKE, O. (1972). Cytochalasin B: does it affect actin-like filaments. *Science*, **175,** 774–776. *(151)*

229 FOWKE, L. C. and SETTERFIELD, G. (1969). Multivesicular structures and cell wall growth. *Can. J. Bot.*, **47,** 1873–1877. *(23)*

230 FRANKE, W. W. (1969). On the universality of nuclear pore complex structure. *Z. Zellforsch.*, **105,** 405–429. *(54)*

231 FRANKE, W. W. (1970). Central dilations in maturing Golgi cisternae—a common structural feature among plant cells? *Planta*, **90,** 370–373. *(74)*

232 FRANKE, W. W. (1970). Nuclear pore flow rate. A characteristic for nucleocytoplasmic exchange of macromolecules and particles. *Naturwissenschaften*, **57,** 44–45. *(48, 56)*

233 FRANKE, W. W., DEUMLING, B., ERMEN, B., JARASCH, E.-D. and KLEINIG, H. (1970). Nuclear membranes from mammalian liver. I. Isolation procedure and general characterization. *J. Cell Biol.*, **46,** 379–395. *(52)*

234 FRANKE, W. W., DEUMLING, B., ZENTGRAF, H., FALK, H. and RAE, P. M. M. (1973). Nuclear membranes from mammalian liver. IV. Characterisation of membrane-attached DNA. *Exptl Cell Res.*, **81,** 365–392. *(52)*

235 FRANKE, W. W., HERTH, W., VAN DER WOUDE, W. J. and MORRE, D. J. (1972). Tubular and filamentous structures in pollen tubes: possible involvement as guide elements in protoplasmic streaming and vectorial migration of secretory vesicles. *Planta*, **105,** 317–341. *(146, 151, 153)*

236 FRANKE, W. W. and KARTENBECK, J. (1971). Outer mitochondrial membrane continuous with endoplasmic reticulum. *Protoplasma*, **73,** 35–41. *(89)*

237 FRANKE, W. W., KARTENBECK, J., KRIEN, S., VAN DER WOUDE, W. J., SCHEER, U. and MORRE, D. J. (1972). Inter- and Intracisternal elements of the Golgi complex. A system of membrane to membrane cross-links. *Z. Zellforsch.*, **132,** 365–380. *(72, 75)*

238 FRANKE, W. W., KARTENBECH, J., ZENTGRAF, H., SCHEER, U. and FALK, H. (1971). Membrane-tomembrane cross-bridges. A means to orientation and interaction of membrane faces. *J. Cell Biol.*, **51,** 881–888. *(174)*

239 FRANKE, W. W., MORRE, D. J., DEUMLING, B., CHEETHAM, R. D., KARTENBECK, J., JARASCH, E.-D. and ZENTGRAF, H.-W. (1971). Synthesis and turnover of membrane proteins in rat liver: an examination of the membrane flow hypothesis. *Z. Naturforsch.*, **26b,** 1031–1039. *(177)*

240 FRANKE, W. W. and SCHEER, U. (1972). Structural details of dictysomal pores. *J. Ultrastruct. Res.*, **40,** 132–144. *(75)*

241 FRANKE, W. W., TRENDELENBURG, M. F., and SCHEER, U. (1973). Natural segregation of nucleolar components in the course of a plant cell differentiation. *Planta*, **110,** 159–164. *(45)*

242 FRANTZ, C., ROLAND, J.-C., WILLIAMSON, F. A. and MORRÉ, J. D. (1973). Différenciation *in vitro* des membranes des dictyosomes. *C. r. hebd. Séanc. Acad. Sci., Paris*, **277,** 1471–1474. *(76)*

243 FRASER, T. W. and GUNNING, B. E. S. (1973). Ultrastructure of

the hairs of the filamentous green alga *Bulbochaete hiloensis* (Nordst.) Tiffany: an apoplastidic plant cell with a well developed Golgi apparatus. *Planta,* **113,** 1–19. (*50, 91, 97*)

244 FREDERICK, J. F. (1973). A primordial bifunctional polyglucan-forming enzyme. *Ann. N.Y. Acad. Sci.,* **210,** 254–264. (*118*)

245 FREDERICK, S. E., GRUBER, P. J. and TOLBERT, N. E. (1973). The occurrence of glycolate dehydrogenase and glycolate oxidase in green plants. An evolutionary survey. *Pl. Physiol. Lancaster,* **52,** 318–323. (*136*)

246 FREDERICK, S. E. and NEWCOMB, E. H. (1969). Cytochemical localization of catalase in leaf microbodies (peroxisomes). *J. Cell Biol.,* **43,** 343–353. (*136, 137*)

247 FREDERICK, S. E. and NEWCOMB, E. H. (1971). Ultrastructure and distribution of microbodies in leaves of grasses with and without CO_2-photorespiration. *Planta,* **96,** 152–174. (*137, 138*)

248 FREDERICK, S. E., NEWCOMB, E. H., VIGIL, E. L. and WERGIN, W. P. (1968). Fine structural characterization of plant microbodies. *Planta,* **81,** 229–252. (*135, 138*)

249 FREY-WYSSLING, A. (1969). On the molecular structure of starch granules. *Am. J. Bot.,* **56,** 696–701. (*118*)

249a FUGE, H. (1973). Verteilung der Mikrotubuli in Metaphase- und Anaphase-Spindeln der Spermatocytin von *Pales ferruginea. Chromosoma,* **43,** 109–143. (*168*)

250 FULCHER, R. G., O'BRIEN, T. P. and LEE, J. W. (1972). Studies on the aleurone layer. I. Conventional and flourescence microscopy of the cell wall with emphasis on phenol-carbohydrate complexes in wheat. *Aust. J. biol. Sci.,* **25,** 23–34. (*20*)

251 FUSELER, J. W. (1973). Temperature dependence of anaphase chromosome velocity and microtubule depolymerization rate. *J. Cell Biol.,* **59,** 106a. (*167*)

252 GAFF, D. F. and CARR, D. J. (1961). The quantity of water in the cell wall and its significance. *Aust. J. biol. Sci.,* **14,** 299–311. (*30*)

253 GALL, J. G. (1966). Chromosome fibres studied by a spreading technique. *Chromosoma,* **20,** 221–233. (*43*)

254 GALL, J. G. (1966). Microtubule fine structure. *J. Cell Biol.,* **31,** 639–643. (*142*)

255 GALL, J. G. and PARDUE, M. L. (1969). Formation and detection of RNA-DNA hybrid molecules in cytological preparations. *Proc. natn. Acad. Sci. U.S.A.,* **63,** 378–383. (*48*)

256 GAMBARINI, A. G., and MENEGHINI, R. (1972). Ribosomal RNA genes in salivary gland and ovary of *Rhynchosciara angelae. J. Cell Biol.,* **54,** 421–426. (*50*)

257 GARRETT, R. A. and WITTMANN, H.-G. (1973). Structure and function of the ribosome. *Endeavour,* **32,** 8–14. (*47*)

258 GASANOV, R. A. and FRENCH, C. S. (1973). Chlorophyll composition and photochemical activity of photosystems detached from chloroplast grana and stroma lamellae. *Proc. natn. Acad. Sci. U.S.A.,* **70,** 2082–2085. (*106*)

259 GAWLIK, S. R. and MILLINGTON, W. F. (1969). Pattern formation and the fine structure of the developing cell wall in colonies of *Pediastrum boryanum. Am. J. Bot.,* **56,** 1084–1093. (*147*)

260 GERGIS, M. S. (1972). Influence of carbon dioxide supply on the chloroplast structure of *Chlorella pyrenoidosa. Arch. Mikrobiol.,* **83,** 321–327. (*129*)

261 GERHARDT, B. (1973). Untersuchungen zur Funktionsänderung der Microbodies in den Keimblättern von *Helianthus annuus* L. *Planta,* **110,** 15–28. (*139*)

262 GERHARDT, B. and BERGER, C. (1971). Microbodies und Diaminobenzidin-Reaktion in den Acetat-Flagellaten *Polytomella caeca* und *Chlorogonium elongatum. Planta,* **100,** 155–166. (*136*)

262a GIBBONS, I. R. (1967). The organization of cilia and flagella. In *Molecular Organization and Biological Function,* Ed. Allen, J. M. Harper and Row, New York and London. 211–237. (*149*)

263 GIBBS, S. P. (1962). Nuclear envelope—Chloroplast relationships in algae. *J. Cell Biol.,* **14,** 433–444. (52)

264 GIBBS, S. P. (1967). Synthesis of chloroplast RNA at the site of chloroplast DNA. *Biochem. biophys. Res. Commun.,* **28,** 653–657. (*124, 126*)

265 GIBBS, S. P. (1971). The comparative ultrastructure of the algal chloroplast. *Ann. N.Y. Acad. Sci.,* **175,** 454–473. (*52, 65, 117, 123, 129, 130*)

266 GIBBS, S. P. and POOLE, R. J. (1973). Autoradiographic evidence for many segregating DNA molecules in the chloroplast of *Ochromonas danica. J. Cell Biol.,* **59,** 318–328. (*125*)

267 GIBSON, R. A. and PALEG, L. G. (1972). Lysosomal nature of hormonally induced enzymes in wheat aleurone cells. *Biochem. J.,* **128,** 367–375. (*62*)

268 GILES, K. L. (1971). The control of chloroplast division in *Fumaria hygrometrica.* II. The effects of kinetin and indoleacetic acid on nucleic acids. *Plant and Cell Physiol.,* **12,** 447–450. (*132*)

269 GILES, K. L. and SARAFIS, V. (1971). On the survival and reproduction of chloroplasts outside the cell. *Cytobios.,* **4,** 61–74. (*181*)

270 GILES, K. L. and SARAFIS, V. (1972). Chloroplast survival and division *in vitro. Nature New Biol.,* **236,** 56–58. (*181*)

270a GILES, K. L. and SARAFIS, V. (1974). Implications of rigescent integuments as a new structural feature of some algal chloroplasts. *Nature, Lond.,* **248,** 512–513. (*123*)

271 GILES, K. L. and TAYLOR, A. O. (1971). The control of chloroplast division in *Funaria hygrometrica.* I. Patterns of nucleic acid, protein and lipid synthesis. *Plant and Cell Physiol.,* **12,** 437–445. (*132*)

272 GIMENEZ-MARTIN, G., SOGO, J. M., ESPONDA, P. and STOCKERT, J. C. (1971). Morphology of the nucleolus and transverse expression of the organizing area. *Cytobiologie,* **3,** 343–350. (*46*)

273 GIMENEZ-MARTIN, G. and STOCKERT, J. C. (1970). Nucleolar structure during the meiotic prophase in *Allium cepa* anthers. *Z. Zellforsch.,* **107,** 551–563. (*45*)

274 GIRBARDT, M. (1968). Ultrastructure and dynamics of the moving nucleus. *Symp. Soc. exp. Biol.,* **22,** 249–259. (*153*)

275 GOLDBERG, I. and OHAD, I. (1970). Biogenesis of chloroplast membranes. V. A radioautographic study of membrane growth in a mutant of *Chlamydomonas reinhardi* y-1. *J. Cell Biol.,* **44,** 572–591. (*110*)

276 GOODAY, G. W. (1971). A biochemical and autoradiographic study of the role of the Golgi bodies in thecal formation in *Platymonas tetrathele. J. exp. Bot.,* **22,** 959–971. (*76*)

277 GOODCHILD, D. J. and BERGERSEN, F. J. (1966). Electron microscopy of the infection and subsequent development of soybean nodule cells. *J. Bact.,* **92,** 204–213. (*25*)

278 GOODE, D. (1973). Kinetics of microtubule formation after cold disaggregation of the mitotic apparatus. *J. molec. Biol.*, **80**, 531–538. (*144*)

279 GOODENOUGH, U. W., ARMSTRONG, J. J. and LEVINE, R. P. (1969). Photosynthetic properties of ac-31, a mutant strain of *Chlamydomonas reinhardi* devoid of chloroplast membrane stacking. *Pl. Physiol., Lancaster*, **44**, 1001–1012. (*107, 108*)

280 GOODENOUGH, U. W. and LEVINE, R. P. (1969). Chloroplast ultrastructure in mutant strains of *Chlamydomonas reinhardi* lacking components of the photosynthetic apparatus. *Pl. Physiol., Lancaster*, **44**, 990–1000. (*129*)

281 GOODENOUGH, U. W. and STAEHELIN, L. A. (1971). Structural differentiation of stacked and unstacked chloroplast membranes. Freeze-etch electron microscopy of wild-type and mutant strains of *Chlamydomonas*. *J. Cell Biol.*, **48**, 594–619. (*105, 108*)

282 GOODWIN, T. W. (1971). Biosynthesis by chloroplasts. In *Structure and Function of Chloroplasts*. Ed. Gibbs, M., Springer-Verlag, Berlin, Heidelberg, New York. 215–276. (*64, 176*)

282a GOOSEN-DE ROO, L. (1973). The relationship between cell organelles and cell wall thickenings in primary tracheary elements of the cucumber. I. Morphological aspects. *Acta Bot. Neerl.*, **22**, 279–300. (*148. 153*)

282b GOOSEN-DE ROO, L. (1973). The relationship between cell organelles and cell wall thickenings in primary tracheary elements of the cucumber. II. Quantitative aspects. *Acta Bot. Neerl.*, **22**, 301–320. (*148, 153*)

283 GORI, P., SARFATTI, G. and CRESTI, M. (1971). Development of spherical organelles from the endoplasmic reticulum in the nucellus of some *Euphorbia* species. *Planta.*, **99**, 133–143. (*138*)

284 GORSKA-BRYLASS, A. (1970). The 'callose stage' of the generative cells in pollen grains. *Grana*, **10**, 21–30. (*20*)

285 GOTTSCHALK, A. (1973). On the biosynthesis of Glycoproteins. *Z. Naturf.*, **28c**, 94–99. (*83*)

286 GRANETT, S., DENTLER, W., WITMAN, G. B. and ROSENBAUM, J. L. (1973). Characterization and directionality of chick brain tubulin assembly *in vitro*. *J. Cell Biol.*, **59**, 119a. (*144*)

287 GREEN, J. C. and JENNINGS, D. H. (1967). A physical and chemical investigation of the scales produced by the Golgi apparatus within and found on the surface of the cells of *Chrysochromulina chiton* Parke et Manton. *J. exp. Bot.*, **18**, 359–370. (*76*)

288 GREEN, P. B. (1963). On mechanisms of elongation. *Symp. Soc. dev. Biol.*, **21**, 203–234. (*148*)

289 GREEN, P. B. (1964). Cinematic observations on the growth and division of chloroplasts in *Nitella*. *Am. J. Bot.*, **51**, 334–342. (*109, 133*)

290 GREEN, P. B., ERICKSON, R. O. and RICHMOND, P. A. (1970). On the physical basis of cell morphogenesis. *Ann. New York Acad. Sci.*, **175**, 712–731. (*18*)

291 GREEN, T. R. and RYAN, C. A. (1972). Wound-induced proteinase inhibitor in plant leaves: a possible defense mechanism against insects. *Science, N.Y.* **175**, 776–777. (*34*)

292 GREGORY, S. C. and PETTY, J. A. (1973). Valve action of bordered pits in conifers. *J. exp. Bot.*, **24**, 763–767. (*19*)

293 GRIERSON, D. and LOENING, U. E. (1971). Distinct transcription products of ribosomal genes in two different tissues. *Nature New Biol.*, **235**, 80–82. (*47*)

294 GRNCAREVIC, M. and RADLER, F. (1967). The effect of wax components on cuticular transpiration—model experiments. *Planta*, **75**, 23–27. (*22*)

295 GRÖNEGRESS, P. (1971). The greening of chromoplasts in *Daucus carota* L. *Planta*, **98**, 274–278. (*132*)

296 GROVE, S. N., BRACKER, C. E. and MORRÉ, D. J. (1968). Cytomembrane differentiation in the endoplasmic reticulum —Golgi apparatus—vesicle complex. *Science, N.Y.*, **161**, 171–173. (*72, 75*)

297 GRUBER, P. J., TRELEASE, R. N., BECKER, W. M. and NEWCOMB, E. H. (1970). A correlative ultrastructural and enzymatic study of cotyledonary microbodies following germination of fat-storing seeds. *Planta*, **93**, 269–288. (*136, 138*)

298 GRUBER, P. J., BECKER, W. M. and NEWCOMB, E. H. (1973). The development of microbodies and peroxisomal enzymes in greening bean leaves. *J. Cell Biol.*, **56**, 500–518. (*138*)

299 GUALERZI, C. and CAMMARANO, P. (1969). Comparative electrophoretic studies on the protein of chloroplast and cytoplasmic ribosomes of spinach leaves. *Biochim. biophys. Acta*, **190**, 170–186. (*126*)

300 GUNNING, B. E. S. (1965). The greening process of plastids. I. The structure of the prolamellar body. *Protoplasma*, **60**, 111–130. (*112*)

301 GUNNING, B. E. S. and JAGOE, M. P. (1967). The prolamellar body. In *Biochemistry of Chloroplasts*. Goodwin, T. W., editor. Academic Press Inc., New York. **2**, 655–676. (*114, 131*)

301a GUNNING, B. E. S. and PATE, J. S. (1974). Transfer cells. In *Dynamic Aspects of Plant Ultrastructure*, Ed. Robards, A. W. McGraw-Hill, London, 441–476. (*30*)

302 GUNNING, B. E. S., PATE, J. S., MINCHIN, F. R. and MARKS, I. (1974). Quantitative aspects of transfer cell structure in relation to vein loading in leaves and solute transport in legume nodules. *Symp. Soc. exp. Biol.*, **28**, 87–124. (*27, 28, 29, 31*)

303 HABERLANDT, G. (1884). *Physiological Plant Anatomy*. Translation from the 4th edition by Drummond, M., 1914. Reprint edition published by Today and Tomorrow's Book Agency, New Delhi, 1965. (*1, 29*)

304 HAGGIS, G. H. (1966). *The Electron Microscope in Molecular Biology*. Longmans, Green & Co., Ltd., Henlow. (*2*)

304a HAIGH, W. G. FÖRSTER, H. J., BIEMANN, K., TATTRIE, N. H. and COLVIN, J. R. (1973). Induction of orientation of bacterial cellulose microfibrils by a novel terpenoid from *Acetobacter xylinum*, *Biochem. J.*, **135**, 145–149. (*148*)

305 HALL, D. M. (1967). The ultrastructure of wax deposits on plant leaf surfaces. II. Cuticular pores and wax formation. *J. Ultrastruct. Res.*, **17**, 34–44. (*21*)

306 HALL, D. O. and RAO, K. K. (1972). *Photosynthesis*, Edward Arnold, London. (*99, 104*)

307 HALLAM, N. D. (1970). Growth and regeneration of waxes on the leaves of *Eucalyptus*. *Planta*, **93**, 257–268. (*21*)

308 HALLBERG, R. L., and BROWN, D. D. (1969). Co-ordinated synthesis of some ribosomal proteins and ribosomal RNA in embryos of *Xenopus laevis*. *J. molec. Biol.*, **46**, 393–411. (*49*)

309 HAMKALO, B. A. and MILLER, O. L. Jr. (1973). Electronmicroscopy of genetic activity. *A. Rev. Biochem.*, **42**, 379–396. (*48*)

310 HANSON, J. B., LEONARD, R. T. and MOLLENHAUER, H. H. (1973). Increased electron density of tonoplast membranes in washed corn root tissue. *Pl. Physiol., Lancaster,* **52,** 298–300. *(35)*

311 HANZELY, L. and SCHJEIDE, O. A. (1971). Fine structural observations of mitochondria undergoing divison in *Allium sativum* root tip cells. *Cytobiologie,* **4,** 207–215. *(94)*

312 HARDIN, J. W., CHERRY, J. H., MORRE, D. J. and LEMBI, C. A. (1972). Enhancement of RNA polymerase activity by a factor released by auxin from plasma membrane. *Proc. natn. Acad. Sci. U.S.A.,* **69,** 3146–3150. *(26)*

313 HARRIS, H. (1970). *Nucleus and Cytoplasm.* Clarendon Press, Oxford. 2nd edition. *(51)*

314 HARRIS, P. J. and NORTHCOTE, D. H. (1971). Polysaccharide formation in plant Golgi bodies. *Biochim. biophys. Acta,* **237,** 56–64. *(77)*

315 HARRIS, W. M. (1970). Chromoplasts of tomato fruits. III. The high-delta tomato. *Bot. Gaz.,* **131,** 163–166. *(121)*

316 HARRIS, W. M. and SPURR, A. R. (1969). Chromoplasts of tomato fruits. I. Ultrastructure of low-pigment and high-beta mutants. Carotene analyses. *Am. J. Bot.,* **56,** 369–379. *(121)*

317 HARRIS, W. M. and SPURR, A. R. (1969). Chromoplasts of tomato fruits. II. The red tomato. *Am. J. Bot.,* **56,** 380–389. *(120, 121)*

318 HARRISON-MURRAY, R. S. and CLARKSON, D. T. (1973). Relationships between structural development and the absorption of ions by the root system of *Cucurbita pepo. Planta,* **114,** 1–16. *(31)*

319 HARTLEY, M. R. and ELLIS, R. J. (1973). Ribonucleic acid synthesis in chloroplasts. *Biochem. J.,* **134,** 249–262. *(126)*

320 HASCHKE, H.-P. and LÜTTGE, U. (1973). β—Indolylessigäure—(IES) — abhängiger K^+ — H^+ — Austauschmechanismus und Streckungswachstum bei *Avena*—Koleoptilen, *Z. Naturforsch,.* **28c,** 555–558 *(17)*

321 HASSID, W. Z. (1971). Biosynthesis of cellulose. A. Biosynthesis of cellulose and related plant cell-wall polysaccharides. *High Polymers* Vol. V. *Cellulose and Cellulose Derivatives Part IV.* Bikales, N. M. and Segal, L. *(25)*

322 HAVELKOVA, M. and MENSIK, P. (1966). The Golgi apparatus in the regenerating protoplasts of *Schizosaccharomyces. Naturwissenschaften,* **53,** 562. *(84)*

323 HAY, E. D. (1968). Structure and function of the nucleolus in developing cells. In *Ultrastructure in Biological Systems,* vol. 3, The nucleus. Eds. Dalton, A. J. and Haguenau. Academic Press, New York. 2–79. *(47)*

324 HEATH, I. B., GAY, J. L. and GREENWOOD, A. D. (1971). Cell wall formation in the Saprolegniales: cytoplasmic vesicles underlying developing walls. *J. gen. Microbiol.,* **65,** 225–232. *(77, 80)*

324a HEBER, U. (1974). Metabolite exchange between chloroplasts and cytoplasm. *A. Rev. Pl. Physiol.,* **25,** 393–421. *(122)*

325 HEBER, U. and SANTARIUS, K. A. (1970). Direct and indirect transfer of ATP and ADP across the chloroplast envelope. *Z. Naturforsch.* **25b,** 718–728. *(122)*

326 HEBER, U. and WILLENBRINK, J. (1964). Sites of synthesis and transport of photosynthetic products within the leaf cell. *Biochem. biophys. Acta,* **82,** 313–324. *(122)*

327 HEINRICH, G. (1970). Elektronenmikroskopische Beobachtungen an den Drüsenzellen von *Poncirus trifoliata;* zugleich ein Beitrag zur Wirkung ätherischer Öle auf Pflanzenzellen und eine Methode zur Unterscheidung flüchtiger von nichtflüchtigen lipophilen Komponenten. *Protoplasma,* **69,** 15–36. *(64)*

328 HEINRICH, G. (1973). Entwicklung, Feinbau und Ölgehalt der Drüsenschuppen von *Monarda fistulosa. Planta Medica,* **23,** 154–166. *(64)*

329 HEINRICH, G. (1973). Die Feinstruktur der Trichom-Hydathoden von *Monarda fistulosa. Protoplasma,* **77,** 271–278. *(79, 82)*

330 HELDT, H. W., SAUER, F. and RAPLEY, L. (1971). Differentiation of the permeability properties of the two membranes of the chloroplast envelope. *2nd Int. Congress on Photosynthesis,* Stresa, Proceedings, 1345–1355. *(122)*

331 HENDERSON, D. M., EUDALL, R. and PRENTICE, H. T. (1972). Morphology of the reticulate teliospores of *Puccinia chaerophylli. Trans. Br. mycol. Soc.,* **59,** 229–232. *(20)*

332 HENNINGSEN, K. W. (1970). Macromolecular physiology of plastids. VI. Changes in membrane structure associated with shifts in the absorption maxima of the chlorophyll pigments. *J. Cell Sci.,* **7,** 587–621. *(131)*

333 HENNINGSEN, K. W. and BOYNTON, J. E. (1969). Macromolecular physiology of plastids. VII. The effect of a brief illumination on plastids of dark-grown barley leaves. *J. Cell Sci.,* **5,** 757–793. *(131)*

334 HENNINGSEN, K. W. and BOYNTON, J. E. (1970). Macromolecular physiology of plastids. VIII. Pigment and membrane formation in plastids of barley greening under low light intensity. *J. Cell Biol.,* **44,** 290–304. *(115, 131)*

335 HEPLER, P. K. and FOSKET, D. E. (1971). The role of microtubules in vessel member differentiation in *Coleus. Protoplasma,* **72,** 213–236. *(147, 148)*

336 HEPLER, P. K. FOSKET, D. E. and NEWCOMB, E. H. (1970). Lignification during secondary wall formation in *Coleus:* an electron microscopic study. *Am. J. Bot.,* **57,** 85–96. *(19)*

337 HEPLER, P. K. and JACKSON, W. T. (1968). Microtubules and early stages of cell plate formation in the endosperm of *Haemanthus katherinae* Baker. *J. Cell Biol.,* **38,** 437–446. *(169)*

338 HEPLER, P. K. and JACKSON, W. T. (1969). Isopropyl N-phenylcarbamate affects spindle microtubule orientation in dividing endosperm cells of *Haemanthus katherinea* Baker. *J. Cell Sci.,* **5,** 727–743. *(164)*

339 HEPLER, P. K., MCINTOSH, J. R. and CLELAND, S. (1970). Intermicrotubule bridges in mitotic spindle apparatus. *J. Cell Biol.,* **45,** 438–444. *(163)*

340 HEPLER, P. K. and NEWCOMB, E. H. (1964). Microtubules and fibrils in the cytoplasm of *Coleus* cells undergoing secondary wall deposition. *J. Cell Biol.,* **20,** 529–533. *(141, 147)*

341 HEPLER, P. K. and NEWCOMB, E. H. (1967). Fine structure of cell plate formation in the apical meristem of *Phaseolus* roots. *J. Ultrastruct. Res.,* **19,** 498–513. *(169, 170)*

341a HEPLER, P. K. and PALEVITZ, B. A. (1974). Microtubules and microfilaments. *A. Rev. Pl. Physiol.,* **25,** 309–362. *(142, 150, 152)*

342 HEPLER, P. K., RICE, R. M. and TERRANOVA, W. A. (1972). Cytochemical localization of peroxidase activity in wound vessel members of *Coleus. Can. J. Bot.,* **50,** 977–983. *(19)*

343 HEREWARD, F. V. and NORTHCOTE, D. H. (1973). Fracture planes of the plasmalemma of some plants revealed by freeze-etch. *J. Cell Sci.,* **13,** 621–635. *(23)*

344 HERICH, R. (1969). Untersuchung über die Bedeutung der vegetativen Kerne und ihrer Nukleolen in den Pollenkörnern und Pollenschläuchen. *Theoret. Appl. Gen.*, **39**, 62–67. *(50)*

345 HERRMANN, R. G. (1970). Multiple amounts of DNA related to the size of chloroplasts. I. An autoradiographic study. *Planta*, **90**, 80–96. *(119, 125)*

346 HERRMANN, R. G. and KOWALLIK, K. V. (1970). Selective presentation of DNA-regions and membranes in chloroplasts and mitochondria. *J. Cell Biol.*, **45**, 198–202. *(124)*

347 HERRMANN, R. G. and KOWALLIK, K. V. (1970). Multiple amounts of DNA related to the size of chloroplasts. II. Comparison of electron-microscopic and autoradiographic data. *Protoplasma*, **69**, 365–372. *(124)*

348 HERTH, W., FRANKE, W. W., STADLER, J., BITTIGER, H., KEILICH, G. and BROWN, R. M., Jr. (1972). Further characterization of the alkali-stable material from the scales of *Pleurochrysis scherffelii*: a cellulosic glycoprotein. *Planta*, **105**, 79–92. *(76)*

349 HESLOP-HARRISON, J. (1962). Evanescent and persistent modifications of chloroplast ultrastructure induced by an unnatural pyrimidine. *Planta*, **58**, 237–256. *(101)*

350 HESLOP-HARRISON, J. (1963). Structure and morphogenesis of lamellar systems in grana-containing chloroplasts. I. Membrane structure and lamellar architecture. *Planta*, **60**, 243–260. *(101)*

351 HESLOP-HARRISON, J. (1966). Structural features of the chloroplast. *Sci. Prog. Oxf.*, **54**, 519–541. *(101, 104, 110)*

352 HESLOP-HARRISON, J. (1966). Cytoplasmic continuities during spore formation in flowering plants. *Endeavour*, **25**, 65–72). *(28, 30)*

353 HESLOP-HARRISON, J. (1968). Synchronous pollen mitosis and the formation of the generative cell in massulate orchids. *J. Cell Sci.*, **3**, 457–466. *(164, 170)*

354 HESLOP-HARRISON, J. (1971). The cytoplasm and its organelles during meiosis. In *Pollen development and physiology*. Ed. Heslop-Harrison, J. Butterworth, London. 16–31. *(20, 50)*

355 HESLOP-HARRISON, J. (1971). The pollen wall: structure and development. In *Pollen, Development and Physiology*, Ed. Heslop-Harrison, J. Butterworth, 75–98. *(22)*

356 HESLOP-HARRISON, J. (1971). Wall pattern formation in angiosperm microsporogenesis. *Symp. Soc. exp. Biol.*, **25**, 277–300. *(22, 66)*

357 HESLOP-HARRISON, Y. and KNOX, R. B. (1971). A cytochemical study of the leaf-gland enzymes of insectivorous plants of the genus *Pinguicula*. *Planta*, **96**, 183–211. *(78)*

358 HIGUCHI, T. (1971). Formation and biological degradation of lignins. *Adv. Enzymol.*, **34**, 207–283. *(20)*

359 HILL, A. E. and HILL, B. S. (1973). The *Limonium* salt gland: a biophysical and structural study. *Int. Rev. Cytol.*, **35**, 299–319. *(65)*

360 HILLIARD, J. H., GRACEN, V. E. and WEST, S. H. (1971). Leaf microbodies (peroxisomes) and catalase localization in plants differing in their photosynthetic carbon pathways. *Planta*, **97**, 93–105. *(138)*

361 HINCHMAN, R. R. (1972). The ultrastructural morphology and ontogeny of oat coleoptile plastids. *Am. J. Bot.*, **59**, 805–817. *(118, 132)*

362 HIRSCHAUER, M., REYSS, A., SARDA, C. and BOURDU, R. (1971). Effects of photoperiods on the development of chloroplast lamellae of *Lolium multiflorum*. *2nd International Congress on Photosynthesis. Stresa*. 2519–2523. *(111)*

363 HIXON, R. M. and BRIMHALL, B. (1968). Waxy starches and red iodine starches. Chapter 8 in *Starch and its Derivatives*. 4th Edition, Ed. Radley, J. A., Chapman and Hall, London. *(116, 117)*

364 HODGE, A. J., MCLEAN, J. D. and MERCER, F. V. (1955). Ultrastructure of the lamellae and grana in the chloroplasts of *Zea mays* L. *J. biophys. biochem. Cytol.*, **1**, 605–614 *(108)*

365 HOEFERT, L. L. (1969). Ultrastructure of *Beta* pollen. I. Cytoplasmic constituents. *Am. J. Bot.*, **56**, 363–368. *(147)*

366 HOFFMANN, H. P. and AVERS, C. J. (1973). Mitochondrion of yeast: ultrastructural evidence for one giant, branched organelle per cell. *Science, N.Y.*, **181**, 749–751. *(88, 94)*

367 HOKIN, L. E. (1968). Dynamic aspects of phospholipids during protein secretion. *Int. Rev. Cytol.*, **23**, 187–208. *(82)*

368 HOLDSWORTH, R. H. (1971). The isolation and partial characterization of the pyrenoid protein of *Eremosphaera viridis*. *J. Cell Biol.*, **51**, 499–513. *(130)*

369 HOLLIDAY, R. (1970). The organization of DNA in eukaryotic chromosomes. *Symp. Soc. gen. Microbiol.*, **20**, 359–380 *(44)*

370 HOLLOWAY, P. J. (1970). Surface factors affecting the wetting of leaves. *Pesticide Sci.*, **1**, 156–163. *(22)*

371 HOLLOWAY, P. J. and BAKER, E. A. (1970). The cuticles of some angiosperm leaves and fruits. *Ann. Appl. Biol.*, **66**, 145–154. *(21)*

372 HONDA, S. I., HONGLADAROM-HONDA, T. and KWANYUEN, P. (1971). Interpretations on chloroplast reproduction derived from correlations between cells and chloroplasts. *Planta*, **97**, 1–15. *(133)*

373 HOOBER, J. K. and STEGEMAN, W. J. (1973). Control of the synthesis of a major polypeptide of chloroplast membranes in *Chlamydomonas reinhardi*. *J. Cell Biol.*, **56**, 1–12. *(176)*

374 HOOKE, R. (1665). *Micrographia, or some physiological descriptions of Minute Bodies made by Magnifying Glasses* (especially Observation 18: of the Schematisme or Texture of Cork, and of the Cells and Pores of some other such frothy bodies). Published by the Royal Society. Facsimile reproduction by Dover Publications, Inc., New York, 1961. *(7)*

375 HOROWITZ, S. B. (1972). The permeability of the amphibian oocyte nucleus, *in situ*. *J. Cell Biol.*, **54**, 609–625. *(55)*

377 HOWELL, S. H. (1972). The differential synthesis and degradation of ribosomal DNA during the vegetative cell cycle in *Chlamydomonas reinhardi*. *Nature New Biology*, **240**, 264–267. *(50, 161)*

378 HRUBAN, Z. and RECHCIGL, M. Jr. (1969). Microbodies and related particles: Morphology, biochemistry and physiology. *Int. Rev. Cytol.*, Suppl., **1**, 1–296. *(135)*

379 HUANG, A. H. C. and BEEVERS, H. (1973). Localization of enzymes within microbodies. *J. Cell Biol.*, **58**, 379–389. *(136, 138)*

380 HUANG, J.-S., HUANG, P.-Y. and GOODMAN, R. N. (1973). Reconstitution of a membrane-like structure with structural proteins and lipids isolated from tobacco thylakoid membranes. *Am. J. Bot.*, **60**, 80–85. *(107)*

381 HUBERMAN, J. A. and RIGGS, A. D. (1966). Autoradiography of

chromosomal DNA fibers from chinese hamster cells. *Proc. natn. Acad. Sci. U.S.A.*, **55,** 599–606. *(43)*

382 HUBERMAN, J. A. and RIGGS, A. D. (1968). On the mechanism of DNA replication in mammalian chromosomes. *J. molec. Biol.*, **32,** 327–341. *(43)*

383 HUBERMAN, J. A., TSAI, A. and DEICH, R. A. (1973). DNA replication sites within nuclei of mammalian cells. *Nature, Lond.*, **241,** 32–36. *(52)*

384 HUGHES, A. (1959). *A History of Cytology.* Abelard-Schuman. *(7, 39, 40, 157)*

385 HUGHES-SCHRADER, S. (1947). The 'Pre-metaphase stretch' and kinetochore orientation in phasmids. *Chromosoma*, **3,** 1–21. *(166)*

386 HYDE, B. B. (1967). Changes in nucleolar ultrastructure associated with differentiation in the root tip. *J. Ultrastruct. Res.*, **18,** 25–54. *(46)*

387 HYDE, B. B., HODGE, A. J., KAHN, A. and BIRNSTIEL, M. L. (1963). Studies on phytoferritin. I. Identification and localization. *J. Ultrastruct. Res.*, **9,** 248–258. *(128)*

388 IHLE, J. N. and DURE, L. S. (1972). The developmental biochemistry of cottonseed embryogenesis and germination. III. Regulation of the biosynthesis of enzymes utilized in germination. *J. biol. Chem.*, **247,** 5048–5055. *(61)*

389 INFANTE, A. A., NAUTA, R., GILBERT, S., HOBART, P. and FIRSHEIN, W. (1973). DNA synthesis in developing sea urchins: role of a DNA-nuclear membrane complex. *Nature New Biol.*, **242,** 5–8. *(52)*

390 INGLE, J. (1973). The regulation of ribosomal RNA synthesis. *Biosynthesis and its Control in Plants. Phytochem. Soc. Sym. 9.* Ed. Milborrow, B. V., Academic Press, New York. *(47, 48, 50)*

391 INNAMORATI, M. (1966). Transformazioni strutturali e formazione degli strati nei granuli di amido *Triticum* coltivato in condizioni constanti e in ambiente naturale. *Caryologia*, **19,** 343–367. *(118)*

392 INOUE, S. (1952). The effect of colchicine on the microscopic and submicroscopic structure of the mitotic spindle. *Expl Cell Res. Suppl.*, **2,** 305–318. *(142, 158)*

392a INOUÉ, S., BORISY, G. G. and KIEHART, D. P. (1974). Growth and lability of *Chaetopterus* oocyte mitotic spindles isolated in the presence of porcine brain tubulin. *J. Cell Biol.*, **62,** 175–184. *(166)*

393 IZAWA, S. and GOOD, N. E. (1966). Effect of salts and electron transport on the conformation of isolated chloroplasts. II. Electron microscopy. *Pl. Physiol., Lancaster*, **41,** 544–552. *(107, 108)*

394 IZUTSU, K. (1961). Effects of ultraviolet microbeam irradiation upon division in grasshopper spermatocytes. II. Results of irradiation during metaphase and anaphase I. *Mie. Med. J.*, **11,** 213–232. *(166)*

395 JACKMAN, M. E. and VAN STEVENINCK, R. F. M. (1967). Changes in the endoplasmic reticulum of beetroot slices during aging. *Aust. J. biol. Sci.*, **20,** 1063–1068. *(61)*

395a JAROSCH, R. (1959). Die Protoplasmafibrillen der Characeen. *Protoplasma*, **50,** 93–108. *(152)*

396 JEFFREE, C. E. (1974). A method for recrystallizing selected components of plant epicuticular waxes as surfaces for the growth of micro-organisms. *Trans. Br. mycol. Soc.* **63,** 626–628 *(21)*

396a JENNINGS, R. C. and OHAD, I. (1972). Biogenesis of chloroplast membranes. XI. Evidence for the translation of extra chloroplast RNA on chloroplast ribosomes in a mutant of *Chlamydomonas reinhardi*, y-1. *Archs Biochem. Biophys.*, **153,** 79–87. *(175)*

397 JENSEN, C. and BAJER, A. (1973). Spindle dynamics and arrangement of microtubules. *Chromosoma*, **44,** 73–89. *(163, 168)*

397a JENSEN, C. and BAJER, A. (1974). Kinetochore microtubules in mitosis. *8th International Congress Electron Microscopy, Canberra*, **II,** 256–257. *(163, 167)*

398 JENSEN, W. A. (1964). Observations on the fusion of nuclei in plants. *J. Cell Biol.*, **23,** 669–672. *(41, 42, 53)*

399 JENSEN, W. A. (1969). Cotton embryogenesis: pollen tube development in the nucellus. *Can. J. Bot.*, **47,** 383–385. *(67)*

400 JENSEN, W. A. and FISHER, D. B. (1968). Cotton embryogenesis: the sperm. *Protoplasma*, **65,** 277–286. *(132)*

401 JENSEN, W. A. and FISHER, D. B. (1970). Cotton embryogenesis: the pollen tube in the stigma and style. *Protoplasma*, **69,** 215–235. *(20, 50)*

402 JOHN, P. C. L., MCCULLOUGH, W., ATKINSON, A. W. Jr., FORDE, B. G. and GUNNING, B. E. S. (1973). The cell cycle in *Chlorella. The Cell Cycle in Development and Differentiation.* Ed. Balls, M. and Billet, F. S. Cambridge University Press. *(94)*

403 JOHNSON, J. M. (1969). A study of nucleolar vacuoles in cultured tobacco cells using radioautography, actinomycin D. and electron microscopy. *J. Cell Biol.*, **43,** 197–206. *(49)*

404 JOHNSON, J. M. and JONES, L. E. (1967). Behaviour of nucleoli and contracting nucleolar vacuoles in tobacco cells growing in microculture. *Am. J. Bot.*, **54,** 189–198. *(49)*

405 JOHNSON, K. D. and KENDE, H. (1971). Hormonal control of lecinthin synthesis in barley aleuron cells: regulation of the CDP-choline pathway by gibberellin. *Proc. natn. Acad. Sci. U.S.A.*, **68,** 2674–2677. *(63)*

406 JOHNSON, R. and STREHLER, B. L. (1972). Loss of genes coding for ribosomal RNA in ageing brain cells. *Nature, Lond.*, **240,** 412–414. *(50)*

407 JOHNSON, U. G. and PORTER, K. R. (1968). Fine structure of cell division in *Chlamydomonas reinhardi J. Cell Biol.*, **38,** 403–425. *(169)*

408 JOKELAINEN, P. T. (1967). The ultrastructure and spatial organization of the metaphase kinetochore in mitotic rat cells. *J. Ultrastruct. Res.*, **19,** 19–44. *(159)*

409 JORDAN, E. G. and GODWARD, M. B. E. (1969). Some observations on the nucleolus in *Spirogyra. J. Cell Sci.*, **4,** 3–15. *(159)*

410 JULIANO, B. O. and VARNER, J. E. (1969). Enzymic degradation of starch granules in the cotyledons of germinating peas. *Pl. Physiol., Lancaster*, **44,** 886–892. *(118)*

411 JUNIPER, B. (1972). Mechanisms of perception and patterns of organization in root caps. *The Dynamics of Meristem Cell Populations.* Millar, M. W. and Kuehnert, C. C. Plenum, London. 119–131. *(30)*

412 JUNIPER, B. E. and BARLOW, P. W. (1969). The distribution of plasmodesmata in the root tip of maize. *Planta*, **89,** 352–360. *(28, 29)*

413 JUNIPER, B. E., COX, G. C., GILCHRIST, A. J. and WILLIAMS, P. R. (1970). *Techniques for Plant Electron Microscopy.* Blackwell Scientific Publications, Glasgow. *(2)*

415 JUNIPER, B. E. and FRENCH, A. (1973). The distribution and

redistribution of endoplasmic reticulum (ER) in geoperceptive cells. *Planta,* **109,** 211–224. (*68, 119*)

416 JUNIPER, B. E. and PASK, G. (1973). Directional secretion by the Golgi bodies in Maize root cells. *Planta,* **109,** 225–231. (*77*)

417 KAHN, A. (1968). Developmental physiology of bean leaf plastids. II. Negative contrast electron microscopy of tubular membranes in prolamellar bodies. *Pl. Physiol., Lancaster,* **43,** 1769–1780. (*114*)

417a KAMITSUBO, E. (1972). A 'window technique' for detailed observation of characean cytoplasmic streaming. *Exptl Cell Res.,* **74,** 613–616. (*152*)

417b KAMIYA, N. (1959). Protoplasmic streaming. In *Protoplasmatologia VIII/3a,* ed. Heilbrunn, L. V. and Weber, F. Springer-Verlag. (*151*)

418 KANAI, R. and EDWARDS, G. E. (1973). Enzymatic separation of mesophyll protoplasts and bundle sheath cells from C_4 plants. *Naturwissenschaften,* **60,** 157–158. (*108*)

419 KARTENBECK, J., JARASCH, E. D. and FRANKE, W. W. (1973). Nuclear membranes from mammalian liver. VI. Glucose-6-phospatase in rat liver, a cytochemical and biochemical study. *Exptl Cell Res.,* **81,** 175–194. (*52*)

420 KARTENBECK, J., ZENTGRAF, H., SCHEER, U. and FRANKE, W. W. (1971). The nuclear envelope in freeze-etching. *Advances in Anatomy, Embryology and Cell Biology,* **45,** 1. (*51*)

421 KAUSCHE, G. A. and RUSKA, H. (1940). Zur Frage der Chloroplastenstruktur. *Naturwissenschaften,* **28,** 303–304. (*100*)

422 KEEGSTRA, K., TALMADGE, K. W., BAUER, W. D. and ALBERSHIEM, P. (1973). The structure of plant cell walls. III. A model of the walls of suspension-cultured sycamore cells based on the interconnections of the macromolecular components. *Pl. Physiol., Lancaster,* **51,** 188–196. (*16, 17*)

423 KEENAN, T. W. and MORRÉ, D. J. (1970). Phospholipid class and fatty acid composition of Golgi apparatus isolated from rat liver and comparison with other cell fractions. *Biochemistry,* **9,** 19–25. (*75*)

424 KESSEL, R. G. (1973). Structure and function of the nuclear envelope and related cytomembranes. *Progr. in Surface & Membrane Sci.,* **6,** 243–329. (*51, 52, 54, 55*)

425 KIERMAYER, O. (1970). Electron microscopic studies on the problem of cytomorphogenesis in *Micrasterias denticulata* Breb. *Protoplasma.* **69,** 97–132. (*76*)

426 KIERMAYER, O. and DOBBERSTEIN, B. (1973). Dictyosome derived membrane complexes as templates for the extraplasmic synthesis and orientation of microfibrils. *Protoplasma,* **77,** 437–451. (*23, 76*)

427 KIND, J. and KRIEG, J. (1973). Unfixed cytomembranes in ultrathin frozen sections. *Cytobiologie,* **7,** 145–151. (*92*)

428 KIRK, J. T. O. (1971). Chloroplast structure and biogenesis. *A. Rev. Biochem.,* **40,** 161–196 (*104, 110, 124, 175*)

429 KIRK, J. T. O. (1972). The genetic control of plastid formation: Recent advances and strategies for the future. *Sub-Cell. Biochem.,* **1,** 333–361. (*133*)

430 KIRK, J. T. O. and TILNEY-BASSETT, R. A. E. (1967). *The Plastids.* W. H. Freeman and Co., London and San Francisco. (*97, 98, 110, 121, 123, 133, 175*)

431 KITAJIMA, E. W. and LAURITIS, J. A. (1969). Plant virions in plasmodesmata. *Virology,* **37,** 681–685. (*29*)

432 KITCHING, J. A. (1967). Contractile vacuoles, ionic regulation, and excretion. In *Research in Protozoology.* Ed. Chen, T.-T. **1,** 307–336. (*37*)

433 KLEIN, S. and SCHIFF, J. A. (1972). The correlated appearance of prolamellar bodies, protochlorophyll(ide) species, and the Shibata shift during development of bean etioplasts in the dark. *Pl. Physiol., Lancaster,* **49,** 619–626. (*115*)

434 KLOPPSTECH, K. and SCHWEIGER, H. G. (1973). Nuclear genome codes for chloroplast ribosomal proteins in *Acetabularia.* II. Nuclear transplantation experiments. *Expl Cell Res.,* **80,** 69–78. (*175*)

435 KNAPP, F. F., AEXEL, R. T. and NICHOLAS, H. J. (1969). Sterol biosynthesis in sub-cellular particles of higher plants. *Pl. Physiol., Lancaster,* **44,** 442–446. (*64*)

436 KNOX, R. B. (1973). Pollen wall proteins: pollen-stigma interactions in ragweed and *Cosmos* (Compositae). *J. Cell Sci.,* **12,** 421–443. (*23*)

437 KOEHLER, J. K. (1973). *Advanced Techniques in Biological Electron Microscopy.* Springer-Verlag, Berlin. (*2*)

438 KOLATTUKUDY, P. E. (1970). Plant waxes. *Lipids.* **5,** 259–275. (*21*)

439 KOLATTUKUDY, P. E. and BUCKNER, J. S. (1972). Chain elongation of fatty acids by cell-free extracts of epidermis from pea leaves (*Pisum sativum*). *Biochem. biophys. Res. Commun.,* **46,** 801–807. (*21, 64*)

440 KOLODNER, R. and TEWARI, K. K. (1972). Molecular size and conformation of chloroplast deoxyribonucleic acid from pea leaves. *J. biol. Chem.,* **247,** 6355–6364. (*124*)

441 KOWALLIK, K. V. (1971). The use of proteases for improved presentation of DNA in chromosomes and chloroplasts of *Prorocentrum micans (Dinophyceae). Arch. Mikrobiol.,* **80,** 154–165. (*124*)

442 KOWALLIK, K. V. and HABERKORN, G. (1971). The DNA-structures of the chloroplast of *Prorocentrum micans (Dinophyceae). Arch. Mikrobiol.,* **80,** 252–261. (*124, 125*)

443 KOWALLIK, K. V. and HERRMANN, R. G. (1972). Variable amounts of DNA related to the size of chloroplasts. IV. Three-dimensional arrangement of DNA in fully differentiated chloroplasts of *Beta vulgaris* L. *J. Cell Sci.,* **11,** 357–377. (*90, 124*)

444 KRETZER, F. (1973). Molecular architecture of the chloroplast membranes of *Chlamydomonas reinhardi* as revealed by high resolution electron microscopy. *J. Ultrastruct. Res.,* **44,** 146–178. (*104*)

445 KRIDER, H. M. and PLAUT, W. (1972). Studies on nucleolar RNA synthesis in *Drosophila melanogaster.* I. The relationship between number of nucleolar organizers and rate of synthesis. *J. Cell Sci.,* **11,** 675–687. (*50*)

446 KRIDER, H. M. and PLAUT, W. (1972). Studied on nucleolar RNA synthesis in *Drosophila melanogaster.* II. The influence of conditions resulting in a bobbed phentotype on rate of synthesis and secondary constriction formation. *J. Cell Sci.,* **11,** 689–697. (*50*)

447 KRISHAN, A. and HSU, D. (1971). Binding of colchicine-^{3}H to vinblastine- and vincristine-induced crystals in mammalian tissue culture cells. *J. Cell Biol.,* **48,** 407–410. (*143*)

448 KUBAI, D. F. and RIS, H. (1969). Division in the dinoflagellate *Gyrodinium cohnii* (Schiller). A new type of nuclear reproduction. *J. Cell Biol.,* **40,** 508–528. (*160, 161*)

449 KUHN, H. (1970). Chemismus, Struktur und Entstehung der Carotinkriställchen in der Nebenkrone von *Narcissus poeticus* L. var. 'La Riante'. *J. Ultrastruct. Res.,* **33,** 332–355. (*120*)

450 KUNG, S. D., THORNBER, J. P. and WILDMAN, S. G. (1972). Nuclear DNA codes for the photosystem II chlorophyll-protein of chloroplast membranes. *F.E.B.S. Letters*, **24**, 185–188. *(175)*

450a KUROIWA, T. (1973). Fine structure of interphase nuclei. VI. Initiation and replication sites of DNA synthesis in the nuclei of *Physarum polycephalum* as revealed by electron microscopic autoradiography. *Chromosoma*, **44**, 291–299. *(43, 52)*

450b KUROIWA, T. (1974). Fine structure of interphase nuclei. III. Replication site analysis of DNA during the S period of *Crepis capillaris*. *Exptl Cell Res.*, **83**, 387–398. *(52)*

451 KWIATKOWSKA, M. (1973). Polysaccharides connected with microtubules in the lipotubuloids of *Ornithogallum umbellatum* L. *Histochemie*, **37**, 107–112. *(154)*

452 LA COUR, L. F. (1966). The internal structure of nucleoli. *Chromosomes Today*, **1**, 150–160. *(45)*

453 LA COUR, L. F. and WELLS, B. (1967). The loops and ultrastructure of the nucleolus of *Ipheion uniflorum*. *Z. Zellforschung*, **82**, 25–45. *(45)*

453a LAETSCH, W. M. (1974). The C_4 syndrome: a structural analysis. *A. Rev. Pl. Physiol.*, **25**, 27–52. *(108, 122)*

454 LAFLECHE, D., BOVE, J. M. and DURANTON, J. (1972). Localization and translocation of the protochlorophyllide holochrome during the greening of etioplasts in *Zea mays* L. *J. Ultrastruct. Res.*, **40**, 205–214. *(131)*

454a LAFONTAINE, J. G. (1974). The nucleus. In *Dynamic Aspects of Plant Ultrastructure*, Ed. Robards, A. W. McGraw-Hill, London, 1–51 *(42, 45, 160)*

455 LAFONTAINE, J. G. and CHOUINARD, L. A. (1963). A correlated light and electron microscope study of the nucleolar material during mitosis in *Vicia faba*. *J. Cell Biol.*, **17**, 167–201. *(161)*

456 LAFOUNTAIN, J. R. Jr., (1973). Birefringence and fine structure of spindles in crane-fly spermatocytes. *J. Cell Biol.*, **59**, 182a. *(167)*

457 LAFOUNTAIN, J. R. Jr. and LAFOUNTAIN, K. L. (1973). Comparison of density of nuclear pores on vegetative and generative nuclei in pollen of *Tradescantia*. *Exptl Cell Res.*, **78**, 472–475. *(55)*

458 LAMBERT, A.-M. and BAJER, A. S. (1972). Dynamics of spindle fibers and microtubules during anaphase and phragmoplast formation. *Chromosoma*, **39**, 101–144. *(169)*

459 LAMPEN, J. O. (1974). Secretion of extracellular enzymes across cell membranes. *Symp. Soc. exp. Biol.*, **28**, 351–374. *(63)*

460 LAMPORT, D. T. A. (1970). Cell wall metabolism. *A. Rev. Pl. Phys.*, **21**, 235–270. *(17)*

461 LANCE-NOUGARÈDE, A. (1966). Présence de structures protéiques a arrangement périodique et d'aspect cristallin dans les mitochondries de l'epiderme des jeunes feuilles de lentille. *C. r. Séanc. Acad. Sci. Paris*. **263**, 246–249. *(90)*

462 LANG, W. and POTRYKUS, I. (1971). Licht mikroskopische und pigmentanalytische Untersuchungen zum Ablauf der Petalenentwicklung von *Torenia baillonii* (Scrophulariac). *Z. Pflanzenphysiol.*, **65**, 1–12. *(133)*

462a LATIES, G. G. (1969). Dual mechanisms of salt uptake in relation to compartmentation and long-distance transport. *A. Rev. Pl. Physiol.*, **20**, 89–116. *(34)*

463 LATIES, G. G. and TREFFRY, T. (1969). Reversible changes in conformation of mitochondria of constant volume. *Tissue and Cell*, **1**, 575–592. *(93)*

464 LAUDI, G. (1966). Ricerche infrastrutturali sui plastidi delle piante parassite II. *Lathraea squamaria*. *Caryologia*, **19**, 47–54. *(117)*

465 LAUDI, G. and ALBERTINI, A. (1967). Ricerche infrastrutturali sui plastidi delle piante parassite III. *Orobanche ramosa*. *Caryologia*, **20**, 207–216. *(117)*

466 LAUFFER, M. A. and STEVENS, C. L. (1968). Structure of the tobacco mosaic virus particle; polymerization of tobacco mosaic virus protein. *Adv. Virus Res.*, **13**, 1–63. *(143, 165)*

467 LAZAROW, A. and COOPERSTEIN, S. J. (1953). Studies on the enzymatic basis for the Janus Green B. staining reaction. *J. Histochem. Cytochem.*, **1**, 234–241. *(87)*

468 LAZAROW, P. B. and DE DUVE, C. (1973). The synthesis and turnover of rat liver peroxisomes. IV. Biochemical pathway of catalase synthesis. *J. Cell Biol.*, **59**, 491–506. *(138)*

469 LAZAROW, P. B. and DE DUVE, C. (1973). The synthesis and turnover of rat liver peroxisomes. V. Intracellular pathway of catalase synthesis. *J. Cell Biol.*, **59**, 507–524. *(138)*

470 LEAK, L. V. (1968). Intramitochondrial crystals in meristematic cells of *Pisum sativum*. *J. Ultrastruct. Res.*, **24**, 102–108. *(90)*

471 LEDBETTER, M. C. and PORTER, K. R. (1963). A 'microtubule' in plant cell fine structure. *J. Cell Biol.*, **19**, 239–250. *(141, 147)*

472 LEDBETTER, M. C. and PORTER, K. R. (1964). Morphology of microtubules of plant cells. *Science, N.Y.*, **144**, 872–874. *(142)*

473 LEDBETTER, M. C. and PORTER, K. R. (1970). *Introduction to the Fine Structure of Plant Cells*. Springer-Verlag, Berlin. *(10, 88, 120, 132)*

474 LEFORT-TRAN, M., COHEN-BRAZIRE, G. and POUPHILE, M. (1973). Les membrane photosynthétiques des algues à biliproteines observées apres Cryodecapage. *J. Ultrastruct. Res.*, **44**, 199–209. *(129)*

475 LEHMANN, H. and SCHULZ, D. (1969). Elektronenmikroskopische Untersuchungen von Differenzierungsvorgängen bei Moosen. II. Die Zellplatten—und Zellwandbildung. *Planta.*, **85**, 313–325. *(169)*

476 LEHNINGER, A. L. (1964). *The Mitochondrion*. Ed. Benjamin. 32–33. *(87)*

477 LEHNINGER, A. L. (1971). Metabolite carriers in mitochondrial membranes: the Ca^{2+} transport system. *The Dynamic Structure of Cell Membranes*. Ed. Hölzl Wallach, D. F. and Fischer, H. Springer, Berlin. *(90, 93)*

478 LEIGH, R. A., JONES, R. G. W. and WILLIAMSON, F. A. (1973). The possible role of vesicles and ATPases in ion uptake. In *Ion Transport in Plants*. Ed. Anderson, W. P., Academic Press Inc., London and New York. 407–418. *(25, 65)*

479 LEONARD, R. T., HANSEN, D. and HODGES, T. K. (1973). Membrane-bound adenosine triphosphatase activities of oat roots. *Pl. Physiol., Lancaster*, **51**, 749–754. *(24)*

480 LEONARD, R. T. and HODGES, T. K. (1973). Characterization of plasma membrane-associated adenosine triphosphatase activity of oat roots. *Pl. Physiol., Lancaster*, **52**, 6–12. *(24)*

482 LESKES, A. and SIEKEVITZ, P. (1969). Histochemical and biochemical studies of membrane differentiation: Glucose-

6-phosphatase as a marker enzyme. *J. Cell Biol.*, **43,** 80a. (*64, 69*)

483 LEVINE, R. P. (1971). Interactions between nuclear and organelle genetic systems. *Brookhaven Symp. Biol.*, **23,** 503–532. (*175, 176*)

484 LEVINE, R. P., ANDERSON, J. M. and DURAM, H. (1973). Polypeptides participating in chloroplast membrane stacking. *J. Cell Biol.*, **59,** 192a. (*107*)

485 LEWIS, D. (1970). Conclusions: organelles as membrane complexes. *Symp. Soc. exp. Biol.*, **24,** 497–501. (*176*)

486 LEYON, H. (1956). The structure of chloroplasts. *Svensk Kem. Tidskr.*, **68,** 70–88. (*100*)

487 LIBANATI, C. M. and TANDLER, C. J. (1969). The distribution of the water soluble inorganic orthophosphate ions within the cell: accumulation in the nucleus. *J. Cell Biol.*, **42,** 754–765. (*49*)

488 LICHTENTHALER, H. K. (1968). Plastoglobuli and the fine structure of plastids. *Endeavour,* **27,** 144–149. (*120, 127, 128*)

489 LICHTENHALER, H. K. (1970). Die Feinstruktur der Chromoplasten in plasmochromen Perigon-Blattern von *Tulipa. Planta,* **93,** 143–151. (*120*)

490 LICHTENTHALER, H. K. (1973). Regulation der Lipochinonsynthese in Chloroplasten. *Ber. dt. bot. Ges.*, **86,** 313–329. (*127*)

491 LIST, A. Jr. (1963). Some observations on DNA content and cell and nuclear volume growth in the developing xylem cells of certain higher plants. *Am. J. Bot.*, **50,** 320–329. (*41, 42*)

492 LITTAU, V. C., ALLFREY, V. G., FRENSTER, J. H. and MIRSKY, A. E. (1964). Active and inactive regions of nuclear chromatin as revealed by electron microscope autoradiography. *Proc. natn. Acad. Sci. U.S.A.*, **52,** 93–100. (*43*)

493 LOEWENSTEIN, W. R. (1970). Intracellular communication. *Scient. Am.*, **222,** 79–86. (*27, 31*)

494 LORD, J. M., KAGAWA, T., MOORE, T. S. and BEEVERS, H. (1973). Endoplasmic reticulum as the site of lecithin formation in castor bean endosperm. *J. Cell Biol.*, **57,** 659–667. (*64*)

495 LOTT, J. N. A., LARSEN, P. L. and WHITTINGTON, C. M. (1972). Frequency and distribution of nuclear pores in *Cucurbita maxima* cotyledons as revealed by freeze-etching. *Can. J. Bot.*, **50,** 1785–1787. (*55*)

496 LÖVLIE, A. (1964). Genetic control of division rate and morphogenesis in *Ulva mutabilis* Föyn. *C. r. Trav. Lab. Carlsberg,* **34,** 77–168. (*168*)

497 LUCK, D. J. L. (1963). Formation of mitochondria in *Neurospora crassa. J. Cell Biol.*, **16,** 483–499. (*93*)

498 LÜTTGE, U. (1971). Structure and function of plant glands. *A. Rev. Pl. Physiol.*, **22,** 23–44. (*22, 32, 36, 79*)

499 LUYKX, P. (1970). Cellular mechanisms of chromosome distribution. *Int. Rev. Cytol.* Suppl., **2.** (*160, 164, 167, 168*)

500 MACHE, R. and LOISEAUX, S. (1973). Light saturation of growth and photosynthesis of the shade plant *Marchantia polymorpha. J. Cell Sci.*, **12,** 391–401. (*111*)

501 MACHE, R., ROZIER, C., LOISEAUX, S. and VIAL, A. M. (1973). Synchronous division of plastids during the greening of cut leaves of maize. *Nature, New Biol.*, **242,** 158–160. (*133*)

502 MACKENDER, R. D. and LEECH, R. M. (1971). The isolation and characterization of plastid envelope membranes. *2nd Int. Congress on Photosynthesis,* Stresa, Proceedings, 1431–1440. (*121*)

502a MACROBBIE, E. A. C. (1971). Fluxes and compartmentation in plant cells. *A. Rev. Pl. Physiol.*, **22,** 75–96. (*34*)

503 MAGUN, B. (1973). Two actions of cyclic AMP on melanosome movement in frog skin. Dissection by cytochalasin B. *J. Cell Biol.*, **57,** 845–858. (*152*)

504 MAHER, E. P. and FOX, D. P. (1973). Multiplicity of ribosomal RNA genes in *Vicia* species with different nuclear DNA contents. *Nature New Biol.*, **245,** 170–172. (*51*)

505 MAHLBERG, P. (1973). Scanning electron microscopy of starch grains from latex of *Euphorbia terracine* and *E. tirucalli. Planta,* **110,** 77–80. (*117*)

506 MAHLBERG, P. (1972). Further observations on the phenomenon of secondary vacuolation in living cells. *Am. J. Bot.*, **59,** 172–179. (*34*)

507 MAHLBERG, P., OLSON, K. and WALKINSHAW, C. (1971). Origin and development of plasma membrane derived invaginations in *Vinca rosea* L. *Am. J. Bot.*, **58,** 407–416. (*23*)

508 MANGENOT, S. (1968). Sur la presence de leucoplastes chez les vegetaux vasculaires mycotrophes ou parasites. *C.r. hebd. Séanc. Acad. Sci. Paris,* **267,** 1193–1195. (*117*)

509 MANNERS, D. J. (1968). The biological synthesis of starch. Chapter 2 in *Starch and its Derivatives.* 4th edition, Ed., Radley, J. A. Chapman and Hall, London. (*118*)

510 MANNING, J. E. and RICHARDS, O. C. (1972). Synthesis and turnover of *Euglena gracilis* nuclear and chloroplast deoxyribonucleic acid. *Biochemistry,* **11,** 2036–2043. (*125*)

511 MANNING, J. E., WOLSTENHOLME, D. R. and RICHARDS, O. C. (1972). Circular DNA molecules associated with chloroplasts of spinach, *Spinacia oleracea. J. Cell Biol.*, **53,** 594–601. (*124*)

512 MANNING, J. E., WOLSTENHOLME, D. R., RYAN, R. S., HUNTER, J. A. and RICHARDS, O. C. (1971). Circular chloroplast DNA from *Euglena gracilis. Proc. natn. Acad. Sci. U.S.A.*, **68,** 1169–1173. (*125*)

513 MANTON, I. (1950). Demonstration of compound cilia in a fern spermatozoid by means of the ultra-violet microscope. *J. exp. Bot.*, **1,** 69–70. (*149*)

514 MANTON, I. (1957). Observations with the electron microscope on the cell structure of the antheridium and spermatozoid of *Sphagnum. J. exp. Bot.*, **8,** 382–400. (*141, 146*)

515 MANTON, I. (1959). Electron microscopical observations on a very small flagellate: the problem of *Chromulina pusilla* Butcher. *J. Mar. Biol. Ass. U.K.*, **38,** 319–333. (*88, 94*)

516 MANTON, I. (1959). Observations on the microanatomy of the spermatozoid of the bracken fern (*Pteridium aquilinum*). *J. biophys. biochem. Cytol.*, **6,** 413–417. (*141, 146*)

517 MANTON, I. (1966). Some possibly significant structural relations between chloroplasts and other cell components. In *Biochemistry of Chloroplasts* Vol. 1. Ed. Goodwin, T. W. Academic Press, London and New York. 23–47. (*123, 130*)

518 MANTON, I. (1966). Observations on scale production in *Pyramimonas amylifera* Conrad. *J. Cell Sci.*, **1,** 429–438. (*74*)

519 MANTON, I. (1967). Further observations on the fine structure of *Chrysochromulina chiton* with special reference to the haptonema, 'peculiar' Golgi structure and scale production. *J. Cell Sci.*, **2,** 265–272. *(74)*

520 MANTON, I. (1967). Further observations on scale formation in *Chrysochromulina chiton. J. Cell Sci.*, **2,** 411–418. *(73, 74)*

521 MANTON, I. and CLARKE, B. (1952). An electron microscope study of the spermatozoid of *Sphagnum. J. exp. Bot.*, **3,** 265–275. *(149)*

522 MANTON, I., CLARKE, B., GREENWOOD, A. D. and FLINT, E. A. (1952). Further observations on the structure of plant cilia, by a combination of visual and electron microscopy. *J. exp. Bot.*, **3,** 204–215. *(149)*

523 MANTON, I., KOWALLIK, K. and von STOSCH, H. A. (1969). Observations on the fine structure and development of the spindle at mitosis and meiosis in a marine centric diatom *(Lithodesmium undulatum)*. I. Preliminary survey of mitosis in spermatogonia. *J. Microsc.*, **89,** 295–320. *(165, 167)*

524 MANTON, I., KOWALLIK, K. and von STOSCH, H. A. (1969). Observations on the fine structure and development of the spindle at mitosis and meiosis in a marine centric diatom *(Lithodesmium undulatum)*. II. The early meiotic stages in male gametogenesis. *J. Cell Sci.*, **5,** 271–298. *(165, 167)*

525 MANTON, I., KOWALLIK, K. and von STOSCH, H. A. (1970). Observations on the fine structure and development of the spindle at mitosis and meiosis in a marine centric diatom *(Lithodesmium undulatum)*. III. The later stages of meiosis I in male gametogenesis. *J. Cell Sci.*, **6,** 131–157. *(165, 167)*

526 MANTON, I., KOWALLIK, K. and von STOSCH, H. A. (1970). Observations on the fine structure and development of the spindle at mitosis and meiosis in a marine centric diatom *(Lithodesmium undulatum)*. IV. The second meiotic division and conclusion. *J. Cell Sci.*, **7,** 407–443. *(165, 167)*

527 MANTON, I., OATES, K. and GOODAY, G. (1973). Further observations on the chemical composition of thecae of *Platymonas tetrathele* West (Prasinophyceae) by means of the X-ray microanalyser electron microscope (EMMA). *J. exp. Bot.*, **24,** 223–229. *(76)*

528 MANTON, I. and PARKE, M. (1960). Further observations on small green flagellates with special reference to possible relatives of *Chromulina pusilla* Butcher. *J. mar. biol. Ass. U.K.*, **39,** 275–298. *(88, 94)*

529 MANTON, I., RAYNS, D. G., ETTL, H. and PARKE, M. (1965). Further observations on green flagellates with scaly flagella: the genus *Heteromastix* Korshikov. *J. mar. biol. Ass. U.K.*, **45,** 241–255. *(74)*

530 MARČENKO, E. (1973). Plastids of the yellow y-1 strain of *Euglena gracilis. Protoplasma*, **76,** 417–433. *(129)*

530a MARCHANT, H. J. (1974). The role of microtubules in colony formation by the green alga *Pediastrum. 8th Int. Congress Electron Microscopy, Canberra*, **II,** 584–585. *(147)*

531 MARCHANT, R. and ROBARDS, A. W. (1968). Membrane systems associated with the plasmalemma of plant cells. *Ann. Bot. N.S.*, **32,** 457–471. *(23)*

531a MARGULIES, M. M. and MICHAELS, A. (1974). Ribosomes bound to chloroplast membranes in *Chlamydomonas reinhardtii. J. Cell Biol.*, **60,** 65–77. *(127)*

532 MARGULIS, L. (1970). *Origin of Eukaryotic Cells.* Yale University Press. New Haven. *(177)*

533 MARTIN, J. T. and JUNIPER, B. E. (1970). *The Cuticles of Plants.* Edward Arnold, *(21)*

534 MARTY, F. (1970). Les peroxysomes (microbodies) des laticiferes d'*Euphorbia characias* L. Une etude morphologique et cytochimique. *J. Microscopie*, **9,** 923–948. *(135, 136)*

535 MARTY, F. (1973). Mise en évidence d'un appareil provacuolaire et de son rôle dans l'autophagie cellulaire et l'origine des vacuoles. *C. r. hebd. Séanc. Acad. Sci., Paris*, **276,** 1549–1552. *(67)*

536 MARTY, F. (1973). Dissemblance des faces golgiennes et activite des dictysomes dans les cellules en cours de vacuolisation de la racine d'*Euphorbia characias* L. *C. r. hebd. Séanc. Acad. Sci., Paris*, **277,** 1749–1752. *(73, 178)*

537 MASCARENHAS, J. P. and LAFOUNTAIN, J. (1972). Protoplasmic streaming, cytochalasin B, and growth of the pollen tube. *Tissue and Cell*, **4,** 11–14. *(153)*

538 MATILE, PH. (1969). Plant lysosomes. In *Lysosomes in biology and pathology*. pp. 406–430. Ed. Dingle, J. T. and Fell. H. North Holland, Amsterdam. *(36)*

538a MATILE, PH. (1974). Lysosomes. In *Dynamic Aspects of Plant Ultrastructure*, Ed. Robards, A. W. McGraw-Hill, London, 178–218. *(36)*

539 MATILE, P., CORTAT, M., WIEMKEN, A. and FREY-WYSSLING, A. (1971). Isolation of Glucanase-containing particles from budding *Saccharomyces cerevisiae. Proc. natn. Acad. Sci.*, **68,** 636–640. *(66)*

540 MATILE, PH. and WINKENBACH, F. (1971). Function of lysosomes and lysomal enzymes in the senescing corolla of the morning glory *(Ipomoea purpurea)*. *J. exp. Bot.*, **22,** 759–771. *(37)*

541 MAUL, G. G., PRICE, J. W. and LIEBERMAN, M. W. (1971). Formation and distribution of nuclear pore complexes in interphase. *J. Cell Biol.*, **51,** 405–418. *(54)*

542 MCCLINTOCK, B. (1934). The relation of a particular chromosomal element to the development of the nucleoli in *Zea mays. Z. Zellforsch Mikroskop. Anat.*, **21,** 294–328. *(46)*

543 MCGUIRE, J. and MOELLMANN, G. (1972). Cytochalasin B: effects on microfilaments and movement of melanin granules within melanocytes. *Science, N.Y.*, **175,** 642–644. *(152)*

544 MCINTOSH, J. R., HEPLER, P. K. and VAN WIE, D. G. (1969). Model for mitosis. *Nature, Lond.*, **224,** 659–663. *(167, 168)*

545 MCINTOSH, J. R. and LANDIS, S. C. (1971). The distribution of spindle microtubules during mitosis in cultured human cells. *J. Cell Biol.*, **49,** 468–497. *(168)*

545a MCINTOSH, J. R. (1974). Bridges between microtubules. *J. Cell Biol.*, **61,** 166–187. *(145)*

546 MEEUSE, B. J. D. and HALL, D. M. (1973). Studies on the cell wall starch of *Hericium. Ann. N.Y. Acad. Sci.*, **210,** 39–45. *(117)*

547 MEIDNER, H. and MANSFIELD, T. A. (1968). *Physiology of Stomata.* McGraw-Hill, London. *(35)*

548 MENKE, W. (1962). Structure and chemistry of plastids. *A. Rev. Physiol.*, **13,** 27–44. *(100)*

549 MEPHAM, R. H. and LANE, G. R. (1969). Nucleopores and polyribosome formation. *Nature, Lond.*, **221,** 288–289. *(56)*

550 MESQUITA, J. F. (1969). Electron microscope study of the

origin and development of the vacuoles in root-tip cells of *Lupinus albus* L. *J. Ultrastruct. Res.*, **26,** 242–250. *(37)*

551 MESQUITA, J. F. (1972). Ultrastructure de formations comparables aux vacuoles autophagiques dans les cellules des racines de l'*Allium cepa* L. et du *Lupinus albus* L. *Cytologia*, **37,** 95–110. *(37, 67)*

552 MEYLAN, B. A. and BUTTERFIELD. B. G. (1972). *Three-dimensional Structure of Wood, a Scanning Electron Microscope Study.* Chapman & Hall, London. *(19)*

553 MIGNOT, J.-P., BRUGEROLLE, G. and METENIER, G. (1972). Complements a l'étude des mastigonèmes des protistes flagelles. Utilisation de la technique de Thiéry pour la mise en évidence des polysaccharides sur coupes fines. *J. Microscopie*, **14,** 327–342. *(74)*

554 MILLER, K. R. and STAEHELIN, A. (1973). Direct identification of photosynthetic enzymes on membrane surfaces revealed by deep -etching. *J. Cell. Biol.*, **59,** 226a. *(107)*

555 MILLER, K. R. and STAEHELIN, L. A. (1973). Fine structure of the chloroplast membranes of *Euglena gracilis* as revealed by freeze-cleaving and deep-etching techniques. *Protoplasma*, **77,** 55–78. *(104, 121, 122)*

555A MILLER, K. R. and STAEHELIN, L. A. (1974). The coupling factor: direct localization on chloroplast membranes. *8th International Congress Electron Microscopy, Canberra*, **II,** 204–205. *(107)*

556 MILLER, L. and KNOWLAND, J. (1971). The number and activity of ribosomal RNA genes in *Xenopus laevis* embryos carrying partial deletions in both nucleolar organisers. *Biochem. Genetics*, **6,** 65–73. *(50)*

557 MILLER, O. L., Jr. and BEATTY, B. R. (1968). Amphibian oocyte nucleoli: structure and function. *J. Cell Biol.*, **39,** 156a. *(44)*

558 MILLER, O. L., Jr. and BEATTY, B. R. (1969). Visualization of nucleolar genes. *Science, N.Y.*, **164,** 955–957. *(44, 48)*

559 MILLINGTON, W. F. and GAWLIK, S. R. (1970). Ultrastructure and initiation of wall pattern in *Pediastrum boryanum*. *Am. J. Bot.*, **57,** 552–561. *(147)*

560 MILLONIG, G. and MARINOZZI, V. (1968). Fixation and embedding in electron microscopy. *Adv. Opt. Elect. Micr.*, **2,** 251–341. *(4)*

561 MISHRA, A. K. and COLVIN, J. R. (1970). On the variability of spherosome-like bodies in *Phaseolus vulgaris*. *Can. J. Bot.*, **48,** 1477–1480. *(135)*

562 MOESTRUP, ϕ. (1970). The fine structure of mature spermatozoids of *Chara corallina*, with reference to microtubules and scales. *Planta*, **93,** 295–308. *(146)*

563 MOESTRUP, ϕ. (1970). On the fine structure of the spermatozoids of *Vaucheria sescuplicaria* and on the later stages in spermatogenesis. *J. mar. biol. Ass. U.K.*, **50,** 513–523. *(146)*

563a MOESTRUP, ϕ. and THOMSEN, H. A. (1974). An ultrastructural study of the flagellate *Pyramimonas orientalis* with particular emphasis on Golgi apparatus activity and the flagellar apparatus. *Protoplasma*, **81,** 247–269. *(74)*

564 MOLLENHAUER, H. H. (1971). Fragmentation of mature dictyosome cisternae. *J. Cell Biol.*, **49,** 212–214. *(82)*

565 MOLLENHAUER, H. H. and MORRÉ, D. J. (1966). Golgi apparatus and plant secretion. *A. Rev. Pl. Physiol.*, **17,** 27–46. *(71, 72)*

566 MOLLENHAUER, H. H., MORRÉ, D. J. and TOTTEN, C. (1973). Intercisternal substances of the Golgi apparatus. Unstacking of plant dictyosomes using chaotropic agents. *Protoplasma*, **78,** 443–459. *(72)*

567 MONNERON, A., LIEW, C. C. and ALLFREY, V. G. (1971). Isolation and biological activity of mammalian helical polyribosomes. *J. molec. Biol.*, **57,** 335–350. *(63)*

568 MOOR, H. (1967). Endoplasmic reticulum as the initiator of bud formation in yeast *(S. cerevisiae)*. *Arch. Mikrobiol.*, **57,** 135–146. *(66)*

569 MOOR, H. (1967). Der Feinbau der Mikrotubuli in Hefe nach Gefrierätzung. *Protoplasma*, **64,** 89–103. *(141)*

570 MOORE, T. S., LORD, J. M., KAGAWA, T. and BEEVERS, H. (1973). Enzymes of phospholipid metabolism in the endoplasmic reticulum of castor beam endosperm. *Pl. Physiol., Lancaster*, **52,** 50–53. *(64)*

571 MORRÉ, D. J., CHEETHAM, R. D. and NYQUIST, S. E. (1972). A simplified procedure for isolation of Golgi apparatus from rat liver. *Preparative Biochemistry*, **2,** 61–69. *(72)*

572 MORRÉ, D. J., JONES, D. D. and MOLLENHAUER, H. H. (1966). Golgi apparatus mediated polysaccharide secretion by outer root cap cells of *Zea mays*. I. Kinetics and secretory pathway. *Planta*, **74,** 286–301. *(77, 80)*

573 MORRÉ, D. J., MERLIN, L. M. and KEENAN, T. W. (1969). Localisation of glycosyl transferase activities in a Golgi apparatus-rich fraction isolated from rat liver. *Biochem. biophys. Res. Commun.*, **37,** 813–819 *(78)*

573a MORRÉ, D. J. and MOLLENHAUER, H. H. (1974). The endomembrane concept: a functional integration of endoplasmic reticulum and Golgi apparatus. In *Dynamic Aspects of Plant Ultrastructure*, Ed. Robards, A. W. McGraw-Hill, London, 84–137. *(85, 177)*

574 MORRÉ, D. J., MOLLENHAUER, H. H. and BRACKER, C. E. (1971). Origin and continuity of Golgi apparatus. In *Origin and Continuity of Cell Organelles*. Ed. Reinert, J. and Ursprung, H. Springer-Verlag, Berlin, Heidelberg, New York. 82–126. *(71, 84, 85, 177)*

575 MUGHAL, S. and GODWARD, M. B. E. (1973). Kinetochore and microtubules in two members of *Chlorophyceae, Cladophora fracta* and *Spirogyra majuscula*. *Chromosoma*, **44,** 213–229. *(159)*

575a MURAKAMI, S. and PACKER, L. (1971). The role of cations in the organization of chloroplast membranes. *Archs Biochem. Biophys.*, **146,** 337–347. *(107)*

575b MUSCATINE, L. (1973). Chloroplasts and algae as symbionts in molluscs. *Int. Rev. Cytol.*, **36,** 137–169. *(123)*

576 NAGL, W. (1962). 4096—Ploidie und 'Riesenchromosomen' im Suspensor von *Phaseolus coccineus*. *Natürwissenschaften*, **49,** 261–262. *(42)*

577 NAGL, W. (1972). Molecular and structural aspects of the endomitotic chromosome cycle in angiosperms. In *Chromosomes Today*, volume **3,** 17–23. *(42)*

578 NANNINGA, N. (1973). Structural aspects of ribosomes. *Int. Rev. Cytol.*, **35,** 135–188. *(47)*

579 NASS, M. M. K. (1969). Mitochondrial DNA I. Intramitochondrial distribution and structural relations of single- and double-length circular DNA. *J. molec. Biol.*, **42,** 521–528. *(90)*

580 NASS, M. M. K. (1969). Mitochondrial DNA: Advances, problems and goals. *Science, N.Y.*, **165,** 25–35. *(90, 93)*

581 NELMES, B. J. and PRESTON, R. D. (1968). Wall development in apple fruits: a study of the life history of a parenchyma cell. *J. exp. Bot.*, **19,** 496–518. *(17)*

582 NEUSHUL, M. (1971). Uniformity of thylakoid structure in a red, a brown, and two blue-green algae. *J. Ultrastruct. Res.*, **37,** 532–543. *(121)*

583 NEUSHUL, M. and DAHL, A. L. (1972). Ultrastructural studies of brown algal nuclei. *Am. J. Bot.*, **59,** 401–410. *(55)*

584 NEUTRA, M. and LEBLOND, C. P. (1969). The Golgi apparatus. *Scient. Am.*, **220**, 100–107. (*78*)

585 NEWCOMB, E. H. (1969). Plant microtubules. *A. Rev. Pl. Physiol.*, **20**, 253–288. (*141, 147, 148, 169*)

586 NIR, I. and PEASE, D. C. (1973). Chloroplast organisation and the ultrastructural localisation of photosystems I and II. *J. Ultrastruct. Res.*, **42**, 534–550. (*104, 107*)

587 NISSEN, H.-U. (1969). Crystal orientation and plate structure in echinoid skeletal units. *Science, N.Y.*, **166**, 1150–1152. (*114*)

588 NOBEL P. S. (1970). *Plant Cell Physiology*. Chaps. 1–3 (Cells, Water, Solutes), Freeman, San Francisco. (*24, 34*)

589 NOEL, J. S., DEWEY, W. C., ABEL, J. H., Jr. and THOMPSON, R. P. (1971). Ultrastructure of the nucleolus during the chinese hamster cell cycle. *J. Cell Biol.*, **49**, 830–847. (*161*)

590 NORSTOG, K. (1972). Early development of the barley embryo: fine structure. *Am. J. Bot.*, **59**, 123–132. (*28*)

591 NORTHCOTE, D. H. (1969). Fine structure of cytoplasm in relation to synthesis and secretion in plant cells. *Proc. R. Soc. B*, **173**, 21–30. (*25, 64, 65, 66*)

592 NORTHCOTE, D. H. (1972). Chemistry of the plant cell wall. *A. Rev. Pl. Physiol.*, **23**, 113–132. (*16, 17, 82, 83*)

593 NORTHCOTE, D. H. and LEWIS, D. R. (1968). Freeze-etched surfaces of membranes and organelles in the cells of pea root tips, *J. Cell Sci.*, **3**, 199–206. (*23*)

594 NORTHCOTE, D. H. and PICKETT-HEAPS, J. D. (1966). A function of the Golgi apparatus in polysaccharide synthesis and transport in the root-cap cells of wheat. *Biochem. J.*, **98**, 159–167. (*77*)

595 NOUGAREDE, A. (1964). Evolution infrastructurale des chromoplastes au cours de l'ontogenese des petales chez le *Spartium junceum* L. (Papilionacees). *C. r. Seanc. Acad. Sci. Paris*, **258**, 683–685. (*120*)

596 NOVIKOFF, A. B. and HOLTZMAN, E. (1970). *Cells and Organelles*. Holt, Rinehart & Winston, New York. (*7*)

597 O'BRIEN, T. P. (1970). Further observations on hydrolysis of the cell wall in the xylem. *Protoplasma*, **69**, 1–14. (*19*)

598 O'BRIEN, T. P. (1972). The cytology of cell-wall formation in some eukaryotic cells. *Bot. Rev.*, **38**, 87–118. (*78, 79, 169*)

598a O'BRIEN, T. P. (1974). Primary vascular tissues. In *Dynamic Aspects of Plant Ultrastructure*, Ed. Robards, A. W. McGraw-Hill, London, 414–440. (*19*)

599 O'BRIEN, T. P. and CARR, D. J. (1970). A suberised layer in the cell wall of the bundle sheath of grasses. *Aust. J. biol. Sci.*, **23**, 275–287. (*29, 31*)

600 O'BRIEN, T. P. and MCCULLY, M. E. (1969). *Plant Structure and Development, a pictorial and physiological approach*. Collier-Macmillan. London. (*7*)

600a O'BRIEN, T. P. and MCCULLY, M. E. (1970). Cytoplasmic fibres associated with streaming and saltatory-particle movement in *Heracleum mantegazzianum*. *Planta*, **94**, 91–94. (*152*)

601 O'BRIEN, T. P. and THIMANN, K. V. (1967). Observations on the fine structure of the oat coleoptile. II. The parenchyma cells of the apex. *Protoplasma*, **63**, 417–442. (*29*)

602 OCKLEFORD, C. D. and TUCKER, J. B. (1973). Growth, breakdown, repair, and rapid contraction of microtubular axopodia in the heliozoan *Actinophrys sol*. *J. Ultrastruct. Res.*, **44**, 369–387. (*144*)

603 OHAD, I. (1972). Biogenesis and modulation in chloroplast membranes. In *Role of Membranes in Secretory Processes*, Eds. Bolis, L., Keynes, R. D. and Wilbrandt, W., North Holland, Amsterdam, London, 24–51. (*110*)

604 OHAD, I., FRIEDBERG, L., NE'EMAN, Z. and SCHRAMM, M. (1971). Biogenesis and degradation of starch. I. The fate of the amyloplast membranes during maturation and storage of potato tubers. *Pl. Physiol., Lancaster*, **47**, 465–477. (*118*)

606 OLIVER, P. T. P. (1973). Influence of cytochalasin B on hyphal morphogenesis of *Aspergillus nidulans*. *Protoplasma*, **76**, 279–281. (*153*)

607 OLMSTED, J. B. and BORISY, G. G. (1973). Microtubules. *A. Rev. Biochem.*, **42**, 507–540. (*147*)

608 OPIK, H. (1968). Structure, function and development changes in mitochondria of higher plant cells. *Plant Cell Organelles*. Ed. Pridham, J. B. Academic Press, London. (*94*)

609 OPIK, H. (1973). Effect of anaerobiosis on respiratory rate, cytochrome oxidase activity and mitochondrial structures in coleoptiles of rice *Oryza sativa* L. *J. Cell Sci.*, **12**, 725–739. (*94*)

609a OPIK, H. (1974). Mitochondria. In *Dynamic Aspects of Plant Ultrastructure*, Ed. Robards, A. W. McGraw-Hill, London, 52–83. (*88*)

610 ORY, R. L. and HENNINGSEN, K. W. (1969). Enzymes associated with protein bodies isolated from ungerminated barley seeds. *Pl. Physiol., Lancaster*, **44**, 1488–1498. (*37*)

611 OSAFUNE, T. (1973). Three dimensional structures of giant mitochondria, dictyosomes and 'concentric lamellar bodies' formed during the cell cycle of *Euglena gracilis* (Z) in synchronous culture. *J. Electron Micr.*, **22**, 51–61. (*88*)

612 OSAFUNE, T., MIHARA, S., HASE, E. and OHKURO, I. (1972). Electron microscope studies on the vegetative cellular life cycle of *Chlamydomonas reinhardii* Dangeard in synchronous culture. I. Some characteristics of changes in subcellular structures during the cell cycle, especially in formation of giant mitochondria. *Plant & Cell Physiol.*, **13**, 211–227. (*88*)

613 OSCHMAN, J. L., WALL, B. J. and GUPTA, B. L. (1974). Cellular basis of water transport. *Symp. Soc. exp. Biol.*, **28**, 305–350. (*24*)

614 OSMOND, C. B. (1971). Metabolite transport in C_4 photosynthesis. *Aust. J. biol. Sci.*, **24**, 159–163. (*108*)

615 ÖSTERGREN, G. (1951). The mechanism of co-orientation in bivalents and multivalents. The theory of orientation by pulling. *Hereditas*, **37**, 85–156. (*167*)

616 OUTKA, D. E. and WILLIAMS, D. C. (1971). Sequential coccolith morphogenesis in *Hymenomonas carterae*. *J. Protozool.*, **18**, 285–297. (*73*)

617 OVTRACHT, L. and THIÉRY, J. P. (1972). Mise en évidence par cytochimie ultrastructurale de compartiments physiologiquement différents dans un même saccule Golgien. *J. Microscopie*, **15**, 135–170. (*74*)

618 OWELLEN, R. J., OWENS, A. H., Jr. and DONIGIAN, D. W. (1972). The binding of vincristine, vinblastine and colchicine to tubulin. *Biochem. biophys. Res. Commun.*, **47**, 685–691. (*143*)

619 PAASCHE, E. (1968). Biology and physiology of coccolithophorids. *A, Rev. Microbiol.*, **22**, 71–86. (*73*)

620 PACINI, E., CRESTI, M. and SARFATTI, G. (1972). Incorporation of integumentary nuclei in *Eranthis hiemalis* endosperm and ther disaggregation by the endoplasmic reticulum. *J. Submicr. Cytol.*, **4**, 19–31. (*67*)

621 PACKER, L., WILLIAMS, M. A. and CRIDDLE, R. S. (1973). Freeze-fracture studies on mitochondria from wild-type and respiratory-deficient yeasts. *Biochem. biophys. Acta*, **292**, 92–104. *(92)*

622 PALADE, G. E. (1952). The fine structure of mitochondria. *Anat. Record.* **114,** 427–451. *(89)*

623 PALADE, G. E. and PORTER, K. R. (1954). Studies on the endoplasmic reticulum. I. Its identification in cells *in situ*. *J. exp. Med.*, **100,** 641–656. *(59)*

623a PALEVITZ, B. A., ASH, J. F. and HEPLER, P. K. (1974). Actin in the green alga, *Nitella. Proc. natn. Acad. Sci. U.S.A.*, **71,** 363–366. *(152)*

624 PALEVITZ, B. A. and NEWCOMB, E. H. (1970). A study of sieve element starch using sequential enzymatic digestion and electron microscopy. *J. Cell Biol.*, **45,** 383–398. *(116, 118)*

625 PALMER, J. M. and HALL, D. O. (1972). The mitochondrial membrane system. *Pro. Biophys. & Molec. Biol.*, **24,** 127–175. *(91, 92, 93)*

626 PAOLILLO, D. J. Jr. (1970). The three-dimensional arrangement of intergranal lamellae in chloroplasts. *J. Cell Sci.*, **6,** 243–255. *(102, 111)*

626a PAOLILLO, D. J. (1974). Motile gametes of plants. In *Dynamic Aspects of Plant Ultrastructure*, Ed. Robards, A. W. McGraw-Hill, London, 504–531. *(147, 150, 164)*

626b PAPSIDERO, L. D. and BRASELTON, J. P. (1973). Ultrastructural localization of ribonucleoprotein on mitotic chromosomes of *Cyperus alternifolius. Cytobiologie*, **8,** 118–129. *(161)*

627 PARK, R. B. and BIGGINS, J. (1964). Quantasome: size and composition. *Science, N.Y.*, **144,** 1009–1011. *(105, 106)*

628 PARK, R. B. and PFEIFHOFER, A. O. (1969). Ultrastructural observations on deep-etched thylakoids. *J. Cell Sci.*, **5,** 299–311. *(105)*

629 PARK, R. B. and PFEIFHOFER, A. O. (1969). The effect of ethylenediaminetetra-acetate washing on the structure of spinach thylakoids. *J. Cell Sci.*, **5,** 313–319. *(106)*

630 PARK, R. B. and SANE, P. V. (1971). Distribution of function and structure in chloroplast lamellae. *A. Rev. Pl. Physiol.*, **22,** 395–430. *(105, 106)*

631 PARKE, M. and MANTON, I. (1962). Studies on marine flagellates. VI. *Chrysochromulina pringsheimii* sp. nov. *J. mar. biol. Ass. U.K.*, **42,** 391–404. *(73)*

632 PARSONS, J. A. and RUSTAD, R. C. (1968). The distribution of DNA among dividing mitochondria of *Tetrahymena pyriformis*. *J. Cell Biol.*, **37,** 683–693. *(93)*

633 PATE, J. S. and GUNNING, B. E. S. (1972). Transfer cells. *A. Rev. Pl. Physiol.*, **23,** 173–196. *(30, 65)*

633a PARTHASARATHY, M. V. and MUHLETHALER, K. (1972). Cytoplasmic microfilaments in plant cells. *J. Ultrastruct. Res.*, **38,** 46–62. *(150)*

634 PAYNE, E. S., BROWNRIGG, A., YARWOOD, A. and BOULTER, D. (1971). Changing protein synthetic machinery during development of seeds of *Vicia faba*. *Phytochemistry*, **10,** 2299–2303. *(61)*

635 PAYNE, P. I. and BOULTER, D. (1969). Free and membrane bound ribosomes of the cotyledons of *Vicia faba* (L.). I. Seed development. *Planta*, **84,** 263–271. *(61)*

636 PAZUR, J. H. and ARONSON, N. N. Jr. (1972). Glycoenzymes: enzymes of glycoprotein structure. *Adv. carbohydrate Chem. Biochem.*, **27,** 301–341. *(78)*

637 PEACHEY, L. D. (1964). Electron microscopic observations on the accumulation of divalent cations in intramitochondrial granules. *J. Cell Biol.*, **20,** 95–111. *(90)*

638 PELLING, C. (1964). Ribonukleinsäure-synthese der Riesenchromosomen. Autoradiographische Untersuchungen an *Chironomus tentans. Chromosoma*, **15,** 71–122. *(43)*

639 PENDLAND, J. C. and ALDRICH, H. C. (1973). Ultrastructural organisation of chloroplast thylakoids of the green alga *Oocystis marssonii*. *J. Cell Biol.*, **57,** 306–314. *(104)*

640 PETERMANN, M. L. (1964). *The Physical and Chemical Properties of Ribosomes*. Published by Elsevier, Amsterdam, London and New York. *(49)*

641 PHILIPPOVICH, I. I., BEZSMERTNAYA, I. N. and OPARIN, A. I. (1973). On the localisation of polyribosomes in the system of chloroplast lamellae. *Exptl Cell Res.*, **79,** 159–168. *(126)*

642 PHILLIPS, D. M. and PHILLIPS, S. G. (1973). Repopulation of postmitotic nucleoli by preformed RNA. II. Ultrastructure. *J. Cell Biol.*, **58,** 54–63. *(161)*

643 PHILLIPS, R. L., KLEESE, R. A. and WANG, S. S. (1971). The nucleolus organizer region of Maize (*Zea mays* L.): Chromosomal site of DNA complementary to ribosomal RNA. *Chromosoma*, **36,** 79–88. *(46, 48)*

644 PICKETT-HEAPS, J. D. (1967). Further observations on the Golgi apparatus and its functions in cells of the Wheat seedling. *J. Ultrastruct. Res.*, **18,** 287–303. *(37, 78, 83)*

645 PICKETT-HEAPS, J. D. (1967). The effects of colchicine on the ultrastructure of dividing plant cells, xylem, wall differentiation and distribution of cytoplasmic microtubules. *Devl. Biol.*, **15,** 206–236. *(147, 148, 169)*

646 PICKETT-HEAPS, J. D. (1968). Xylem wall deposition. Radioautographic investigations using lignin precursors. *Protoplasma*, **65,** 181–205. *(66)*

647 PICKETT-HEAPS, J. D. (1969). The evolution of the mitotic apparatus: an attempt at comparative ultrastructural cytology in dividing plant cells. *Cytobios*, **1,** 257–280. *(145, 161, 169)*

648 PICKETT-HEAPS, J. D. (1969). Preprophase microtubules and stomatal differentiation in *Commelina cyanea*. *Aust. J. biol. Sci.*, **22,** 375–391. *(170)*

649 PICKETT-HEAPS, J. D. (1969). Preprophase microtubule bands in some abnormal mitotic cells of wheat. *J. Cell Sci.*, **4,** 397–420. *(170)*

650 PICKETT-HEAPS, J. D. (1970). The behaviour of the nucleolus during mitosis in plants. *Cytobios*, **2,** 69–78. *(160, 161)*

651 PICKETT-HEAPS, J. D. (1971). 'Bristly' cristae in algal mitochondria. *Planta*, **100,** 357–359. *(91)*

652 PICKETT-HEAPS, J. D. (1971). The autonomy of the centriole: fact or fallacy? *Cytobios*, **3,** 205–214. *(164)*

653 PICKETT-HEAPS, J. D. (1972). Reproduction by zoospores in *Oedogonium*. III. Differentiation of the germling. *Protoplasma*, **74,** 169–193. *(84)*

654 PICKETT-HEAPS, J. D. (1974). Plant microtubules. In *Dynamic Aspects of Plant Ultrastructure*. Ed. Robards, A. W. McGraw-Hill, London. 219–255. *(170)*

655 PICKETT-HEAPS, J. D. and FOWKE, L. C. (1970). Cell division in *Oedogonium*. III. Golgi bodies, wall structure, and wall formation in *O. cardiacum*. *Aust. J. biol Sci.*, **23,** 93–113. *(79)*

656 PICKETT-HEAPS, J. D. and NORTHCOTE, D. H. (1966). Cell division in the formation of the stomatal complex of the young leaves of wheat. *J. Cell Sci.*, **1,** 121–128. *(170)*

657 PLOWE, J. Q. (1931). Membranes in the plant cell. I. Morphological membranes at protoplasmic surfaces. *Protoplasma*, **12,** 196–220. *(33)*

658 PLOWE, J. Q. (1931). Membranes in the plant cell. II. Localisation of differential permeability in the plant protoplast. *Protoplasma,* **12,** 221–240. *(33)*

659 POLLARD, E. (1963). Collision kinetics applied to phage synthesis, messenger RNA, and glucose metabolism. *J. theor. Biol.,* **4,** 98–112. *(173, 176)*

660 PORTER, K. R. (1955). Fine structure of cells. *Proc. Symp. Int. Congr. Cell Biol.,* 8th, London, 1954, pp. 236–250. *(158)*

661 PORTER, K. R. (1963). Diversity at the subcellular level and its significance. *The Nature of Biological Diversity.* Ed. Allen, J. M. McGraw Hill, London. *(59, 64)*

662 PORTER, K. R. (1966). Cytoplasmic microtubules and their functions. *Principles Biomol. Ciba Found. Symp.* (1965), 308–345. *(146, 153)*

663 PORTER, K. R. and BONNEVILLE, M. A. (1966). *Fine Structure of Cells and Tissues.* Published by Lea and Febiger, Philadelphia. *(27, 63, 65)*

664 PORTER, K. R. and MACHADO, R. D. (1960). Studies on the endoplasmic reticulum. IV. Its form and distribution during mitosis in cells of onion root tip. *J. biophys. biochem. Cytol.,* **7,** 167–180. *(169)*

665 POSSINGHAM, J. V. (1973). Chloroplast growth and division during the greening of spinach leaf discs. *Nature, New Biol.,* **245,** 93–94. *(133)*

666 POSSINGHAM, J. V. and SAURER, W. (1969). Changes in chloroplast number per cell during leaf development in spinach, *Planta,* **86,** 186–194. *(132, 133)*

667 POSSINGHAM, J. V. and SMITH, J. W. (1972). Factors affecting chloroplast replication in spinach. *J. exp. Bot.,* **23,** 1050–1059. *(132)*

668 POUX, N. (1972). Localisation d'activités enzymatiques dans le méristème radiculaire de *Cucumis sativus* L. IV. Réactions avec la diaminobenzidine mise en évidence de peroxysomes. *J. Microscopie,* **14,** 183–218. *(136)*

669 PREISS, J., OZBUN, J. L., HAWKER, J. S., GREENBERG, E. and LAMMEL, C. (1973): ADPG synthetase and ADPG-glucan 4-glucosyl transferase: enzymes involved in bacterial glycogen and plant starch synthesis. *Ann. N.Y. Acad. Sci.,* **210,** 265–278. *(118)*

670 PRESTON, R. D. (1968). Plants without cellulose. *Scient. Am.,* **218,** 102–108. *(16)*

670a PRESTON, R. D. (1974). Plant cell walls. In *Dynamic Aspects of Plant Ultrastructure,* Ed. Robards, A. W. McGraw-Hill, London, 256–309. *(16)*

671 PRESTON, R. D. and GOODMAN, R. N. (1968). Structural aspects of cellulose microfibril biosynthesis. *Jl R. microsc. Soc.,* **88,** 513–527. *(25)*

672 PRIOUL, J.-L. and BOURDU, R. (1968). Utilisation de l'analyse biométrique a l'étude de la dynamique des infrastructures chloroplastiques lors de la levée d'une carence en azote. *J. Microscopie,* **7,** 419–439. *(111)*

673 PUSZKIN, E., PUSZKIN, S., LO, L. W. and TANENBAUM, S. W. (1973). Binding of cytochalasin D to platelet and muscle myosin. *J. biol. Chem.,* **248,** 7754–7761. *(151)*

674 QUINCEY, R. V. and WILSON, S. H. (1969). The utilization of genes for ribosomal RNA, 5S RNA, and transfer RNA in liver cells of adult rats. *Proc. natn. Acad. Sci. U.S.A.,* **64,** 981–988. *(48)*

675 RACHMILEVITZ, T. and FAHN, A. (1973). Ultrastructure of nectaries of *Vinca rosea* L., *Vinca major* L. and *Citrus sinensis* Osbeck cv. *Valencia* and its relation to the mechanism of nectar secretion. *Ann. Bot.,* **37,** 1–9. *(65)*

676 RACUSEN, D. and FOOTE, M. (1971). The major glycoprotein in germinating bean seeds. *Can. J. Bot.,* **49,** 2107–2111. *(78)*

677 RADLEY, J. A. (1968). *Starch and its derivatives.* 4th edition. Chapter 1. (General survey of starch chemistry to 1950.) Chapman and Hall, London. *(116)*

678 RAFF, R. A., GREENHOUSE, G., GROSS, K. W. and GROSS, P. R. (1971). Synthesis and storage of microtubule proteins by sea urchin embryos. *J. Cell Biol.,* **50,** 516–527. *(144)*

679 RAFF, R. A. and KAUMEYER, J. F. (1973). Soluble microtubule proteins of the sea urchin embryo: partial characterisation of the proteins and behaviour of the pool in early development. *Devl. Biol.,* **32,** 309–320. *(144)*

680 RAFF, R. A. and MAHLER, H. R. (1972). The non-symbiotic origin of mitochondria. *Science, N.Y.,* **177,** 575–582. *(176)*

681 RAMBOURG, A. (1969). L'appareil de Golgi: examen en microscopie électronique de coupes épaisses (0.5–1μ). colorées par le melanage chlorhydrique-phosphotungstique. *C. r. hebd. Séanc. Acad. Sci., Paris,* **269,** 2125–2127. *(73)*

682 RAMBOURG, A. and CHRÉTIEN, M. (1970). L'appareil de Golgi: examen en microscopie électronique de coupes epaisses (0.5–1μ), après imprégnation des tissus par le tétroxyde d'osmium. *C. r. hebd. Séanc. Acad. Sci., Paris,* **270,** 981–983. *(73)*

683 RAMIREZ-MITCHELL, R., JOHNSON, H. M. and WILSON, R. H. (1973). Strontium deposits in isolated plant mitochondria. *Exptl Cell Res.,* **76,** 449–451. *(90)*

684 RAWSON, J. R. Y. and HASELKORN, R. (1973). Chloroplast ribosomal RNA genes in the chloroplast DNA of *Euglena gracilis. J. molec. Biol.,* **77,** 125–132. *(126)*

685 RAY, P. M., SHININGER, T. L. and RAY, M. M. (1969). Isolation of β-glucan synthetase particles from plant cells and identification with Golgi membranes. *Proc. natn. Acad. Sci. U.S.A.,* **64,** 605–612. *(78)*

686 REAVEN, E. P. and AXLINE, S. G. (1973). Subplasmalemmal microfilaments and microtubules in resting and phagocytizing cultivated macrophages. *J. Cell Biol.,* **59,** 12–27. *(151)*

686a REBHUN, L. I., ROSENBAUM, J., LEFEBVRE, P. and SMITH, G. (1974). Reversible restoration of the birefringence of cold-treated, isolated mitotic apparatus of surf clam eggs with chick brain tubulin. *Nature, Lond.,* **249,** 113–115. *(166)*

687 REITH, E. J. and JOKELAINEN, P. T. (1973). Cytokinesis in the stratum intermedium of the rat molar enamel organ. *J. Ultrastruct. Res.,* **42,** 51–65. *(170)*

688 REST, J. A. and VAUGHAN, J. G. (1972). The development of protein and oil bodies in the seed of *Sinapis alba* L. *Planta,* **105,** 245–262. *(36)*

689 RICHARDSON, M. (1974). Microbodies (glyoxysomes and peroxisomes) in plants. *Sci. Prog. Oxf.,* **61,** 41–61. *(137, 138)*

690 RICHTER, H. and SLEYTER, U. (1971). Gefrierätzung des assimilationsparenchyms von *Asparagus sprengeri* Regel. *Mikroskopie,* **26,** 329–346. *(23)*

691 RICKSON, F. R. (1971). Glycogen plastids in Mullerian body cells of *Cecropia peltata*—a higher green plant. *Science, N.Y.,* **173,** 344–347. *(116)*

692 RICKSON, F. R. (1973). Review of glycogen plastid differentiation in Mullerian body cells of *Cecropia peltata. Ann. N.Y. Acad. Sci.,* **210,** 104–114. *(16)*

693 RIS, H. and PLAUT, W. (1962). Ultrastructure of DNA-containing areas in the chloroplast of *Chlamydomonas*. *J. Cell Biol.*, **13**, 383–391. *(124)*

694 ROBARDS, A. W. (1969). Particles associated with developing plant cell walls. *Planta*, **88**, 376–379. *(25)*

695 ROBARDS, A. W. (1970). *Electron Microscopy and Plant Ultrastructure*, McGraw-Hill, London. *(2)*

696 ROBARDS, A. W. (1971). The ultrastructure of plasmodesmata. *Proroplasma*, **72**, 315–323. *(27, 29)*

697 ROBARDS, A. W. and KIDWAI, P. (1969). A comparative study of the ultrastructure of resting and active cambium of *Salix fragilis* L. *Planta*, **84**, 239–249. *(68)*

698 ROBARDS, A. W. and KIDWAI, P. (1972). Microtubules and microfibrils in xylem fibres during secondary cell wall formation. *Cytobiologie*, **6**, 1–21. *(18, 147)*

699 ROBARDS, A. W. and ROBB, M. E. (1972). Uptake and binding of uranyl ions by barley roots. *Science, N.Y.*, **178**, 980–982. *(31)*

700 ROBERTS, K. and NORTHCOTE, D. H. (1971). Ultrastructure of the nuclear envelope; structural aspects of the interphase nucleus of sycamore suspension culture cells. *Microscopia Acta*, **71**, 102–120. *(54)*

701 ROBINSON, D. G. and PRESTON, R. D. (1972). Plasmalemma structure in relation to microfibril biosynthesis in *Oocystis*. *Planta*, **104**, 234–246 *(25)*

701a ROBINSON, D. G., WHITE, R. K. and PRESTON, R. D. (1972). Fine structure of swarmers of *Cladophora* and *Chaetomorpha*. III. Wall synthesis and development. *Planta*, **107**, 131–144 *(148)*

702 RODKIEWICZ, B. (1970). Callose in cell walls during megasporogenesis in angiosperms. *Planta*, **93**, 39–47 *(20)*

703 ROELOFSEN, P. A. (1959). The plant cell wall. In *Encyclopaedia of Plant Anatomy*, Vol. III, Pt. 4. Borntraeger. *(16)*

704 ROELOFSEN, P. A. (1965). Ultrastructure of the wall in growing cells and its relation to the direction of the growth. *Adv. Bot. Res.*, **2**, 69–149. *(16, 18)*

705 ROGERS, H. J. and PERKINS, H. R. (1963). *Cell Walls and Membranes*. E. & F. N. Spon Ltd, London. *(19)*

706 ROLAND, J. C., LEMBI, C. A. and MORRE, D. J. (1972). Phosphotungstic acid—chromic acid as a selective electron dense stain for plasma membranes of plant cells. *Stain. Technol.*, **47**, 195–200. *(23)*

707 ROLAND, J. C. and VIAN, B. (1971). Réactivité du plasmalemme végétal. Etude cytochimique. *Protoplasma*, **73**, 121–137. *(23)*

708 ROLLESTON, F. S. (1972). The binding of ribosomal subunits to endoplasmic reticulum membranes. *Biochem. J.*, **129**, 721–731. *(63)*

709 ROLLESTON, F. S. and MAK, D. (1973). The binding of polyribosomes to smooth and rough endoplasmic reticulum membranes. *Biochem. J.*, **131**, 851–853. *(63)*

710 ROODYN, D. B. and WILKIE, D. *The Biosynthesis of Mitochondria*. Methuen, London. *(93)*

711 ROOS, U.-P. (1973). Light and electron microscopy of rat kangaroo cells in mitosis. II. Kinetochore structure and function. *Chromosoma*, **41**, 195–220. *(159)*

711a ROSE, R. J. (1974). Changes in nucleolar activity during the growth and development of the wheat coleoptile. *Protoplasma*, **79**, 127–143. *(50)*

712 ROSE, R. J. and SETTERFIELD, G. (1971). Cytological studies on the inhibition by 5-fluorouracil of ribosome synthesis and growth in Jerusalem artichoke tuber slices. *Planta*, **101**, 210–230. *(45)*

713 ROSE, R. J., SETTERFIELD, G., and FOWKE, L. C. (1972). Activation of nucleoli in tuber slices and the function of nucleolar vacuoles. *Expl Cell Res.*, **71**, 1–16. *(49, 50)*

714 ROSEN, W. G. (1968). Ultrastructure and physiology of pollen. *A. Rev. Pl. Physiol.*, **19**, 435–462. *(79)*

715 ROSENBAUM, J. L., MOULDER, J. E. and RINGO, D. L. (1969). Flagellar elongation and shortening in *Chlamydomonas*. The use of cycloheximide and colchicine to study the synthesis and assembly of flagellar proteins. *J. Cell Biol.*, **41**, 600–619. *(144)*

716 ROSINSKI, J. and ROSEN, W. G. (1972). Chloroplast development: fine structure and chlorophyll synthesis. *Q. Rev. Biol.*, **47**, 160–191. *(111, 115)*

716a ROTH, L. E. and DANIELS, E. W. (1962). Electron microscopic studies of mitosis in Amebae. II. The giant Ameba *Pelomyxa carolinensis*. *J. Cell Biol.*, **12**, 57–78 *(159)*

717 ROTH, L. E., PIHLAJA, D. J. and SHIGENAKA, Y. (1970). Microtubules in the heliozoan axopodium. I. The gradion hypothesis of allosterism in structural proteins. *J. Ultrastruct. Res.*, **30**, 7–37 *(145, 146)*

718 ROTHFIELD, L. I. (1971). Some aspects of the structure and assembly of bacterial membranes in *The Dynamic Structure of Cell Membranes*. Ed. Holzl Wallach, D. V. and Fischer, H. Springer, 1971. *(78)*

719 ROUGIER, M. (1971). Étude cytochimique de la secrétion des polysaccharides végétaux a l'aide d'un matériel de choix: les cellules de la coiffe de *Zea mays*. *J. Microscopie*, **10**, 67–82. *(77)*

720 ROUGIER, M. (1972). Etude cytochimique des squamules d'*Elodea canadensis*. Mise en évidence de leur sécrétion polysaccharidique et de leur activité phosphatasique acide. *Protoplasma*, **74**, 113–131. *(77)*

721 RYAN, C. A. and SHUMWAY, L. K. (1970). Differential synthesis of Chymotrypsin inhibitor 1 in variegated leaves of a cytoplasmic mutant of tobacco. *Pl. Phys., Lancaster*, **45**, 512–514. *(34)*

722 RYSER, U., FAKAN, S. and BRAUN, R. (1973). Localization of ribosomal RNA genes by high resolution autoradiography. *Expl Cell Res.*, **78**, 89–97. *(48)*

723 SABNIS, D. D. and JACOBS, W. P. (1967). Cytoplasmic streaming and microtubules in the coenocytic marine alga, *Caulerpa prolifera*. *J. Cell Sci.*, **2**, 465–472. *(152)*

724 SAGER, R. (1972). *Cytoplasmic Genes and Organelles*. Published by Academic Press, New York and London. *(175)*

724a SAKANO, K., KUNG, S. D. and WILDMAN, S. G. (1974). Identification of several chloroplast DNA genes which code for the large subunit of *Nicotiana* fraction 1 proteins. *Molec. gen. Genet.*, **130**, 91–97. *(176)*

725 SALEMA, R. and BADENHUIZEN, N. P. (1967). The production of reserve starch granules in the amyloplasts of *Pellionia daveauana* N. E. Br. *J. Ultrastruct. Res.*, **20**, 383–399. *(132)*

726 SALEMA, R. and BADENHUIZEN, N. P. (1969). Nucleic acids in plastids and starch formation. *Acta Bot. Neerl.*, **18**, 203–215. *(175)*

727 SALMON, E. D. (1973). Chromosome movement coupled to pressure-induced microtubule depolymerisation. *J. Cell Biol.*, **59**, 300a. *(167)*

728 SANGER, J. M. and JACKSON, W. T. (1971). Fine structure study of pollen development in *Haemanthus katherinae* Baker. II. Microtubules and elongation of the generative cells. *J. Cell. Sci.*, **8**, 303–315. *(147)*

729 SATIR, B. (1974). Membrane events during the secretory process. *Symp. Soc. exp. Biol.*, **28**, 399–418. *(65)*

730 SCHEER, U. (1972). The ultrastructure of the nuclear envelope of amphibian oocytes. IV. On the chemical nature of the nuclear pore complex material. *Z. Zellforsch.*, **127**, 127–148. *(54)*

731 SCHEER, U. (1973). Nuclear pore flow rate of ribosomal RNA and chain growth rate of its precursor during oogenesis in *Xenopus laevis*. *Devl. Biol.*, **30**, 13–28. *(47, 56)*

732 SCHEER, U. and FRANKE, W. W. (1972). Annulate lamellae in plant cells. Formation during microsporogenesis and pollen development in *Canna generalis*. *Planta*, **107**, 145–159. *(55)*

733 SCHENKEIN, I. and UHR, J. W. (1970). Immunoglobulin synthesis and secretion. I. Biosynthetic studies of the addition of the carbohydrate moieties. *J. Cell Biol.*, **46**, 42–51. *(78)*

734 SCHMID, G. and GAFFRON, H. (1966). Chloroplast structure and the photosynthetic unit. In *Energy Conservation by the Photosynthetic Apparatus*. Brookhaven Symposia in Biology, **19**, 380–392. *(107)*

735 SCHNARRENBERGER, C., OESER, A. and TOLBERT, N. E. (1972). Isolation of protein bodies on sucrose gradients. *Planta*, **104**, 185–194. *(37)*

736 SCHNEPF, E. (1960). Quantitative Zusammenhänge zwischen der sekretion des Fangschleimes und den Golgi Strukturen bei *Drosophyllum lusitanicum*. *Z. Naturforschung*, **16b**, 605–610. *(81)*

737 SCHNEPF, E. (1965). Licht- und elektronenmikroskopische Beobachtungen an den Trichom-Hydathoden von *Cicer arietinum*. *Z. Pflanzenphysiol.* **53**, 245–254. *(79)*

738 SCHNEPF, E. (1968). Zur Feinstruktur der schleimsezernierenden Drüsenhaare auf der Ochrea von *Rumex* und *Rheum*. *Planta*, **79**, 22–34. *(77, 79)*

739 SCHNEPF, E. (1969). Uber den Feinbau von Öldrüsen. I Die Drüsenhaare von *Arctium lappa*. *Protoplasma*, **67**, 185–194. *(64)*

740 SCHNEPF, E. (1969). Über den Feinbau von Öldrüsen. II. Die Drüsenhaare in *Calceolaria*–Blüten. *Protoplasma*, **67**, 195–203. *(64)*

741 SCHNEPF, E. (1969). Über den Feinbau von Öldrüsen. III. Die Ölgänge von *Solidago canadensis* und die Exkretschläuche von *Arctium lappa*. *Protoplasma*, **67**, 205–212. *(64)*

742 SCHNEPE, E. (1969). Uber den Feinbau von Öldrusen. IV. Die Ölgänge von Umbelliferen: *Heracleum sphondylium* und *Dorema ammoniacum*. *Protoplasma*, **67**, 375–390. *(64)*

743 SCHNEPF, E. (1969). Sekretion und Exkretion bei Pflanzen. *Protoplasmatologia*, **8**, 1–181. *(62, 72, 79, 81)*

744 SCHNEPF, E. (1971). Über die Wirkung von Hennstoffen der Proteinsynthese auf die Sekretion des Kohlenhydrat-Fangschleimes von *Drosophyllum lusitanicum*. *Planta*, **103**, 334–339. *(85)*

745 SCHNEPF, E. (1972). Tubuläres endoplasmatisches Reticulum in Drüsen mit lipophilen Ausscheidungen von *Ficus*, *Ledum* und *Salvia*. *Biochem. Physiol. Pflanzen*, **163**, 113–125. *(64)*

746 SCHNEPF, E. (1973). Mikrotubulus-Anordnung und -Umordnung, Wandbildung und Zellmorphogenese in jungen *Sphagnum*-Blättchen. *Protoplasma*, **78**, 145–173. *(146, 149)*

746a SCHNEPF, E. (1974). Gland cells. In *Dynamic Aspects of Plant Ultrastructure*, Ed. Robards, A. W. McGraw-Hill, London, 331–357. *(62, 79, 81)*

747 SCHNEPF, E. and DEICHGRÄBER, G. (1969). Über die Feinstruktur von *Synura petersenii* unter besonderer Berücksichtigung der Morphogenese ihrer Kieselschuppen. *Protoplasma*, **68**, 85–106. *(66, 73)*

748 SCHNEPF, E. and KLASOVA, A. (1972). Zur Feinstruktur von Öl- und Flavon-Drüsen. *Ber. dt. bot. Ges.*, **85**, 249–258. *(64)*

749 SCHNEPF, E. and NAGL, W. (1970). Über einige Strukturbesonderheiten der Suspensorzellen von *Phaseolus vulgaris*. *Protoplasma*, **69**, 133–143. *(64)*

750 SCHNEPF, E. and VON TRAITTEUR, R. (1973). Über die traumatotaktische Bewegung der Zellkerne in *Tradescantia*-Blättern. *Z. Pflanzenphysiol.*, **69**, 181–184. *(154)*

751 SCHOEFL, G. I. (1964). The effect of actinomycin D on the fine structure of the nucleolus. *J. Ultrastruct. Res.*, **10**, 224–243. *(45)*

752 SCHOEN, A. H. (1969). Infinite periodic minimal surfaces without self-intersections. *N.A.S.A.*, Technical Note C-98, 1–52. *(114)*

752a SCHÖNBOHM, E. (1973). Kontraktile Fibrillen als aktive Elemente bei der Mechanik der Chloroplastenverlagerung. *Ber. dt. bot. Ges.*, **86**, 407–422. *(154)*

753 SCHÖTZ, F. (1970). Effects of the disharmony between genome and plastome on the differentiation of the thylakoid system in *Oenothera*. *Symp. Soc. exp. Biol.*, **24**, 39–54. *(176)*

754 SCHÖTZ, F., BATHELT, H., ARNOLD, C.-G. AND SCHIMMER, O. (1972). The architecture and organisation of the chlamydomonas cell. Results of serial-section electron microscopy and a three dimensional reconstruction. *Protoplasma*, **75**, 229–254. *(88)*

755 SCHREIL, W. H. (1964). Studies on the fixation of artificial and bacterial DNA plasms for the electron microscopy of thin sections. *J. Cell Biol.*, **22**, 1–20. *(90)*

756 SCHRÖTER, K., RODRIGUEZ-GARCIA, M. I. and SIEVERS, A. (1973). Die Rolle des endoplasmatischen Retikulums bei der Genese der *Chara*-statolithen. *Protoplasma*, **76**, 435–442. *(119)*

757 SCHULZ, P. and JENSEN, W. A. (1971). *Capsella* embryogenesis: the chalazal proliferating tissue. *J. Cell Sci.*, **8**, 201–227. *(28)*

758 SCHULZ, R. and JENSEN, W. A. (1968). *Capsella* embryogenesis: the synergids before and after fertilization. *Am. J. Bot.*, **55**, 541–552. *(67)*

759 SECKBACH, J. (1972). Remarks on ferritin from iron loaded plants. *Planta Medica*, **21**, 267–273. *(128)*

760 SECKBACH, J. (1972). Electron microscopical observations of leaf ferritin from iron-treated *Xanthium* plants: localization and diversity in the organelle. *J. Ultrastruct. Res.*, **39**, 65–76. *(129)*

761 SEITZ, U. and SEITZ, V. (1972). Transport of newly synthesised rRNA from the nucleus to the cytoplasm in freely suspended cells of parsley (*Petroselinium sativum*). *Planta*, **106**, 141–148. *(49)*

762 SHANNON, J. C. and CREECH, R. G. (1973). Genetics of storage polyglucosides in *Zea mays* L. *Ann. N.Y. Acad. Sci.*, **210**, 279–289. *(116)*

763 SHELANSKI, M. L., GASKIN, F. and CANTOR, C. R. (1973). Microtubule assembly in the absence of added nucleotides. *Proc. natn. Acad. Sci. U.S.A.*, **70**, 765–768. *(143)*

764 SHEPHARD, D. C. and BIDWELL, R. G. S. (1973). Photosynthesis and carbon metabolism in a chloroplast preparation from *Acetabularia*. *Protoplasma*, **76**, 289–307. *(123)*

765 SHIMONY, C., FAHN, A. and REINHOLD, L. (1973). Ultrastructure and ion gradients in the salt glands of *Avicennia marina* (Forssk.) Vierh. *New Phytol.*, **72,** 27–36. (*65*)

766 SIEKEVITZ, P. (1972). Biological membranes: the dynamics of their organization. *A. Rev. Physiol.*, **34,** 117–139. (*63, 68*)

767 SIEVERS, A. (1965). Elektronenmikroskopische Untersuchungen zur geotropischen Reaktion. I. Über Besonderheiten im Feinbau der Rhizoide von *Chara foetida. Z. Pflanzen physiol.*, **53,** 193–213. (*72*)

768 SIMON, E. W. and CHAPMAN, J. A. (1961). The development of mitochondria in *Arum* spadix. *J. exp. Bot.*, **12,** 414–420. (*90*)

769 SINGER, S. J. and NICOLSON, G. L. (1972). The fluid mosaic model of the structure of cell membranes. *Science, N.Y.*, **175,** 720–732. (*12, 104*)

770 SINGH, S. and WILDMAN, S. G. (1973). Chloroplast DNA codes for the ribulose diphosphate carboxylase catalytic site on fraction I proteins of *Nicotiana* species. *Molec. gen. Genet.*, **124,** 187–196. (*175, 176*)

771 SLATYER, R. O. (1967). *Plant-Water Relationships.* Academic Press, New York. (*30*)

772 SLAUTTERBACK, D. B. (1963). Cytoplasmic microtubules. I. Hydra. *J. Cell Biol.*, **18,** 367–388. (*141*)

772a SLEIGH, M. (Ed.) (1974). *Cilia and Flagella.* Academic Press, New York and London. (*149*)

773 SMITH, D. G. and SVOBODA, A. (1972). Golgi apparatus in normal cells and protoplasts of *Schizosaccharomyces* Pombe. *Microbios,* **5,** 177–182. (*84*)

774 SMITH, E. L., DELANGE, R. J. and BONNER, J. (1970). Chemistry and biology of the histones. *Physiol. Rev.*, **50,** 159–170. (*43*)

775 SMITH, M. and BUTLER, R. D. (1971). Ultrastructural aspects of petal development in *Cucumis sativus* with particular reference to the chromoplasts. *Protoplasma,* **73,** 1–13. (*120*)

776 SOROKIN, H. P. (1966). The intercellular spaces of the *Avena* coleoptile. *Physiol. Plant.*, **19,** 691–701. (*21*)

777 SOROKIN, H. P. (1967). Distribution of intercellular material in the apical part of pea seedlings. *Physiol. Plant.*, **20,** 643–654. (*21*)

778 SOROKIN, H. P. and SOROKIN, S. (1966). The spherosomes of *Campanula persicifolia* L. *Protoplasma,* **62,** 216–236. (*135*)

779 SPADARI, S. and RITOSSA, F. (1970). Clustered genes for ribosomal RNA in *Escherichia coli. J. molec. Biol.*, **53,** 357–367. (*48*)

780 SPANSWICK, R. M. (1972). Electrical coupling between cells of higher plants: A direct demonstration of intercellular communication. *Planta,* **102,** 215–227. (*29*)

781 SPENCER, F. S. and MACLACHLAN, G. A. (1972). Changes in molecular weight of cellulose in the pea epicotyl during growth. *Pl. Physiol., Lancaster,* **49,** 58–63. (*16*)

782 SPOONER, B. S., YAMADA, K. M. and WESSELLS, N. K. (1971). Microfilaments and cell locomotion. *J. Cell Biol.*, **49,** 595–613. (*151*)

783 SPREY, B. and GIETZ, N. (1973). Isolierung von Etioplasten und elektronenmikroskopische Abbildung membranassoziierter Etioplasten-DNA. *Z. Pflanzenphysiol.*, **68,** 397–414. (*124, 125*)

784 SPUDICH, J. A. and LIN, S. (1972). Cytochalasin B, its interaction with actin and actomyosin from muscle. *Proc. natn. Acad. Sci. U.S.A.*, **69,** 442–446. (*151*)

785 SPURR, A. R. and HARRIS, W. M. (1968). Ultrastructure of chloroplasts and chromoplasts in *Capsicum annuum.* I. Thylakoid membrane changes during fruit ripening. *Am. J. Bot.*, **55,** 1210–1224. (*120*)

786 SRERE, P. A. (1972). Is there an organization of Krebs cycle enzymes in the mitochondrial matrix? *Energy Metabolism and the Regulation of Metabolic Processes in Mitochondria.* Ed. Mehlman, M. A., and Hanson, R. W., Academic Press, London. (*90, 91*)

787 SRIVASTAVA, L. M. (1966). On the fine structure of the cambium of *Fraximus americana* L. *J. Cell Biol.*, **31,** 79–93. (*68*)

788 STADLER, J. and FRANKE, W. W. (1974). Characterisation of the colchicine binding of membrane fractions from rat and mouse liver. *J. Cell Biol.*, **60,** 297–303. (*148*)

789 STAEHELIN, L. A. (1968). Ultrastructural changes of the plasmalemma and the cell wall during the life cycle of *Cyanidium caldarium. Proc. Roy. Soc., B,* **171,** 249–259. (*23*)

790 STAEHELIN, L. A. and KIERMAYER, O. (1970). Membrane differentiation in the Golgi complex of *Micrasterias denticulata* Breb. visualized by freeze-etching. *J. Cell Sci.*, **7,** 787–792. (*73, 76, 85*)

790a STAEHELIN, L. A. and MILLER, K. R. (1974). Particle movements associated with unstacking and restacking of chloroplast membranes *in vitro.* A freeze-cleave and deep-etch study. *8th International Congress Electron Microscopy, Canberra,* **II,** 202–203. (*105, 107*)

791 STEARNS, M. E. and WAGENAAR, E. B. (1971). Ultrastructural changes in chloroplasts of autumn leaves. *Can. J. Genet. Cytol.*, **13,** 550–560. (*133*)

792 STEBBINS, G. L. and PRICE, H. J. (1971). The developmental genetics of the *calcaroides* gene in barley. I. Divergent expression at the morphological and histological level. *Genetics,* **68,** 527–538. (*168*)

793 STEBBINGS, H. and WILLISON, J. H. M. (1973). Structure of microtubules: a study of freeze-etched and negatively stained microtubules from the ovaries of *Notonecta. Z. Zellforsch.*, **138,** 387–396. (*142*)

793a STEER, M. W. (1974). The development of tapetal cells in *Avena sativa* L. *8th Int. Congress Electron Microscopy, Canberra,* **II,** 594–595. (*59, 63*)

794 STEPHENS, R. E. (1972). Studies on the development of the sea urchin *Strongylocentrotus droebachiensis.* II. Regulation of mitotic spindle equilibrium by environmental temperature. *Biol. Bull., Woods Hole,* **142,** 145–159. (*165*)

795 STEPHENS, R. E. (1973). A thermodynamic analysis of mitotic spindle equilibrium at active metaphase. *J. Cell Biol.*, **57,** 133–147. (*165*)

796 STERLING, C. (1968). The structure of the starch grain. Chapter 4 in *Starch and its Derivatives.* 4th edition, Ed. Radley, J. A., Chapman and Hall, London. (*116, 117, 118*)

797 STERN, C. (1970). The continuity of genetics. *Proc. Am. Acad. Arts Sci.*, **99,** 882–908. (*40*)

798 STOCKERT, J. C., COLMAN, O. D., RISUENO, M. C. and FERNANDEZ-GOMEZ, M. E. (1971). Nucleolar segregation by cold 2-4 dinitrophenol or anoxia in plant cells. *Cytologia,* **36,** 499–503. (*45*)

799 STOCKERT, J. C., SOGO, J. M., DIEZ, J. L. and GIMENEZ-MARTIN, G. (1969). Alteration in the SAT-chromosome of *Allium cepa* affecting the size of the nucleolus. *Experentia,* **25,** 773–774. (*46*)

800 STOCKING, C. R. and LARSON, S. (1969). A chloroplast cytoplasmic shuttle and the reduction of extraplastid NAD. *Biochem. biophys. Res. Commun.*, **37**, 278–282. *(122)*

801 STRUGGER, S. (1950). Über den Bau der Proplastiden und Chloroplasten. *Naturwissenschaften*, **37**, 166–167. *(98)*

802 SURZYCKI, S. J. (1969). Genetic functions of the chloroplast of *Chlamydomonas reinhardi* : effect of rifampin on chloroplast DNA dependent RNA polymerase. *Proc. natn. Acad. Sci. U.S.A.*, **63**, 1327–1334. *(176)*

803 SURZYCKI, S. J. and ROCHAIX, J. D. (1971). Transcriptional mapping of ribosomal RNA genes of the chloroplast and nucleus of *Chlamydomonas reinhardi*. *J. molec. Biol.*, **62**, 89–109. *(126)*

804 SUTCLIFFE, J. (1968). *Plants and Water*. Edward Arnold, London. *(24, 34)*

805 SWIFT, J. G. and BUTTROSE, M. S. (1972). Freeze-etch studies of protein bodies in wheat scutellum. *J. Ultrastruct. Res.*, **40**, 378–390. *(36)*

806 SWIFT, J. G. and BUTTROSE, M. S. (1973). Protein bodies, lipid layers and amyloplasts in freeze-etched pea cotyledons. *Planta*, **109**, 61–72. *(36)*

807 SYRETT, P. J., MERRETT, M. J. and BOCKS, S. M. (1963). Enzymes of the glyoxylate cycle in *Chlorella vulgaris*. *J. exp. Bot.*, **14**, 249–264. *(136)*

808 TALMADGE, K. W., KEEGSTRA, K., BAUER, W. D. and ALBERSHEIM, P. (1973). The structure of plant cell walls. I. The macromolecular components of the walls of suspension-cultured sycamore cells with a detailed analysis of the pectic polysaccharides. *Pl. Physiol., Lancaster*, **51**, 158–173. *(16)*

809 TANDLER, B. and HOPPEL, C. L. (1972). *Mitochondria*. Academic Press, New York and London. *(116)*

810 TANDLER, C. J., and SOLARIA, A. J. (1969). Nucleolar orthophosphate ions. Electron microscope and diffraction studies. *J. Cell Biol.*, **41**, 91–108. *(49)*

811 TARTOF, K. D. (1971). Increasing the multiplicity of ribosomal RNA genes in *Drosophila melanogaster*. *Science, N.Y.*, **171**, 294–297. *(50)*

812 TATEWAKI, M. and NAGATA, K. (1970). Surviving protoplasts *in vitro* and their development in *Bryopsis*. *J. Phycol.*, **6**, 401–403. *(180)*

813 TAYLOR, D. L., CONDEELIS, J. S., MOORE, P. J. and ALLEN, R. D. (1973). The contractile basis of amoeboid movement. I. The chemical control of mobility in isolated cytoplasm. *J. Cell Biol.*, **59**, 378–394. *(151)*

814 TELFORD, J. N. and RACKER, E. (1973). A method of increasing contrast of mitochondrial inner membrane spheres in thin sections of epon-araldite embedded tissue. *J. Cell Biol.*, **57**, 580–586. *(92)*

815 TEWARI, J. P., MALHOTRA, S. K. and TU, J. C. (1973). Ultrastructural transformations as a function of conformational states studied in freeze-fractures of rat liver mitochondria. *Cytobios*, **7**, 15–33. *(91)*

816 TEWARI, K. K. and WILDMAN, S. G. (1970). Information content in the chloroplast DNA. *Symp. Soc. exp. Biol.*, **24**, 147–179. *(125, 176)*

817 THAIR, B. W. and WARDROP, A. B. (1971). The structure and arrangement of nuclear pores in plant cells. *Planta*, **100**, 1–17. *(54)*

818 THIÉRY, J.-P. (1969). Role de l'appareil de Golgi dans la synthese des mucopolysaccharides etude cytochimique. I. Mise en évidence de mucopolysaccharides dans les vésicules de transition entre l'ergastoplasme et l'appareil de Golgi. *J. Microscopie*, **8**, 689–708. *(78)*

819 THOMAS, J. B., MINNAERT, K. and ELBERS, P. F. (1955). Chlorophyll concentrations in plastids of different groups of plants. *Acta Bot. Neerl.*, **5**, 315–321. *(102)*

820 THOMSON, W. W., FOSTER, P. and LEECH, R. M. (1972). The isolation of proplastids from roots of *Vicia faba*. *Pl. Physiol., Lancaster*, **49**, 270–272. *(99)*

821 THOMSON, W. W., LEWIS, L. N. and COGGINS, C. W. (1967). The reversion of chromoplasts to chloroplasts in Valencia oranges. *Cytologia*, **32**, 117–124. *(132)*

822 THOMSON, W. W., RAISON, J. K. and LYONS, J. M. (1972). The induction of energized configurational changes in plant mitochondria, *in vivo*. *Bioenergetics*, **3**, 531–538. *(93)*

823 TILNEY, L. G. (1968). Ordering of subcellular units. The assembly of microtubules and their role in the development of cell form. *Develop. Biol.*, Suppl., **2**, 63–102. *(143)*

824 TILNEY, L. G., BRYAN, J., BUSH, D. J., FUJIWARA, K., MOOSEKER, M. S., MURPHY, D. B. and SNYDER, D. H. (1973). Microtubules; evidence for 13 protofilaments. *J. Cell Biol.*, **59**, 267–275. *(142, 150)*

825 TOLBERT, N. E. (1971). Microbodies – peroxisomes and glyoxysomes. *A. Rev. Pl. Physiol.*, **22**, 45–74. *(137)*

826 TOLBERT, N. E. (1973). Compartmentation and control in microbodies. In *Rate Control of Biological Processes. Symp. Soc. exp. Biol.*, **27**, 215–239. *(137, 138)*

827 TOLBERT, N. E., OESER, A., KISAKI, T., HAGEMAN, R. H. and YAMAZAKI, R. K. (1968). Peroxisomes from spinach leaves containing enzymes related to glycolate metabolism. *J. biol. Chem.*, **243**, 5179–5184. *(137)*

828 TOLBERT, N. E., OESER, A., YAMAZAKI, R. K., HAGEMAN, R. H. and KISAKI, T. (1969). A survey of plants for leaf peroxisomes. *Pl. Physiol., Lancaster*, **44**, 135–147. *(137)*

829 TORREY, J. G., FOSKET, D. E. and HEPLER, P. K. (1971). Xylem formation: a paradigm of cytodifferentiation in higher plants. *Am. Scient.*, **59**, 338–352. *(19, 148)*

830 TREFFRY, T. (1970). Phytylation of chlorophyllide and prolamellar-body transformation in etiolated peas. *Planta*, **91**, 279–284. *(131)*

831 TREFFRY, T. (1973). Chloroplast development in etiolated peas: reformation of prolamellar bodies in red light without accumulation of protochlorophyllide. *J. exp. Bot.*, **24**, 185–195. *(116, 131)*

832 TRELEASE, R. N., BECKER, W. M., GRUBER, P. J. and NEWCOMB, E. H. (1971). Microbodies (glyoxysomes and peroxisomes) in cucumber cotyledons. Correlative biochemical and ultrastructural study in light- and dark-grown seedlings. *Pl. Physiol., Lancaster*, **48**, 461–475. *(138)*

832a TRELEASE, R. N., BECKER, W. M. and BURKE, J. J. (1974). Cytochemical localization of malate synthase in glyoxysomes. *J. Cell Biol.*, **60**, 483–495. *(136)*

833 TRENCH, R. K., BOYLE, J. E. and SMITH, D. C. (1973). The association between chloroplasts of *Codium fragile* and the mollusc *Elysia viridis*. I. Characteristics of isolated *Codium* chloroplasts. *Proc. R. Soc., B.*, **184**, 51–61. *(123)*

834 TRENCH, R. K., BOYLE, J. E. and SMITH, D. C. (1973). The association between chloroplasts of *Codium fragile* and the mollusc *Elysia viridis*. II. Chloroplast ultrastructure and photosynthetic carbon fixation in *E. viridis*. *Proc. R. Soc., B.*, **184**, 63–81. *(123)*

835 TRIBE, M. and WHITTAKER, P. (1972). *Chloroplasts and Mitochondria*. Edward Arnold, London. *(102)*

836 TRUCHET, G. and COULOMB, P. (1973). Mise en évidence et évolution du système phytolysosomal dans les cellules

des différentes zones de nodules radiculaires de pois (*Pisum sativum* L.). Notion d'hétérophagie. *J. Ultrastruct. Res.*, **43,** 36–57. (*67*)

838 TUCKER, J. B. (1970). Initiation and differentiation of microtubule patterns in the ciliate *Nassula. J. Cell Sci.*, **7,** 793–821. (*144, 145*)

839 TUKEY, H. B., Jr. (1970). The leaching of substances from plants. *A Rev. Pl. Physiol.*, **21,** 305–324. (*22*)

840 TYREE, M. T. (1968). Determination of transport constants of isolated *Nitèlla* cell walls. *Can. J. Bot.*, **46,** 317–327. (*23, 30*)

841 TYREE, M. T. (1970). The Symplast Concept. *J. theoret. Biol.*, **26,** 181–214. (*28, 29*)

842 TZAGOLOFF, A., RUBIN, M. S. and SIERRA, M. F. (1973). Biosynthesis of mitochondrial enzymes. *Biochem. biophys. Acta*, **301,** 71–104. (*175*)

843 URBAN, P. (1969). The fine structure of pronuclear fusion in the coenocytic marine alga *Bryopsis hypnoides* Lamouroux. *J. Cell Biol.*, **42,** 606–611. (*53*)

844 URL, W. G. (1971). The site of penetration resistance to water in plant protoplasts. *Protoplasma*, **72,** 427–447. (*35*)

845 VANDERWOUDE, W. J., LEMBI, C. A. and MORRE, D. J. (1972). Auxin (2,4-D) stimulation (*in vivo* and *in vitro*) of polysaccharide synthesis in plasma membrane fragments isolated from onion stems. *Biochem. biophys. Res. Commun.*, **46,** 245–253. (*25*)

846 VANDERWOUDE, W. J., MORRÉ, D. J. and BRACKER, C. E. (1971). Isolation and characterization of secretory vesicles in germinated pollen of *Lilium longiflorum*. *J. Cell Sci.*, **8,** 331–351. (*77, 80, 82*)

847 VAN STEVENINCK, M. E. and VAN STEVENINCK, R. F. M. (1971). Effect of protein synthesis inhibitors on the formation of crystalloid inclusions in the endoplasmic reticulum of beetroot cell. *Protoplasma*, **73,** 107–119. (*61*)

848 VAN STEVENINCK, R. F. M. and CHENOWETH, A. R. F. (1972). Ultrastructural localization of ions. I. Effect of high external sodium chloride concentration on the apparent distribution of chloride in leaf parenchyma cells of barley seedlings. *Aust. J. biol. Sci.*, **25,** 499–516. (*29*)

849 VARNER, J. E. and MENSE, R. M. (1972). Characteristics of the process of enzyme release from secretory plant cells. *Pl. physiol., Lancaster*, **49,** 187–189. (*62*)

849a VESK, M. and JEFFREY, S. W. (1974). The effect of blue light on chloroplast number and thylakoid stacking in the marine diatom, *Stephanopyxis turris. 8th Int. Congress Electron Microscopy, Canberra*, **II,** 586–587. (*132*)

850 VIAN, B. (1972). Aspects, en cryodecapage, de la fusion des membranes des vesicules cytoplasmiques et du plasmalemme lors des phénonmènes de sécrétion végétale. *C. r. hebd. Séanc. Acad. Sci., Paris*, **275,** 2471–2474. (*65, 76*)

851 VIAN, B. and ROLAND, J. C. (1972). Différenciation des cytomembranes et renouvellement du plasmalemme dans les phénomènes de sécrétions végétales. *J. Microscopie*, **13,** 119–136. (*76, 80*)

852 VIGIL, E. L. (1970). Cytochemical and developmental changes in microbodies (glyoxysomes) and related organelles of castor bean endosperm. *J. Cell Biol.*, **46,** 435–454. (*136, 138*)

852a VIGIL, E. L. (1973). Structure and function of plant microbodies. *Sub-Cell. Biochem.*, **2,** 237–285. (*135*)

853 VIGIL, E. L. and RUDDAT, M. (1973). Effect of gibberellic acid and actinomycin D on the formation and distribution of rough endoplasmic reticulum in barley aleurone cells. *Pl. Physiol., Lancaster*, **51,** 549–558. (*62*)

854 VILLIERS, T. A. (1972). Cytological studies in dormancy. II. Pathological ageing changes during prolonged dormancy and recovery upon dormancy release. *New Phytol.*, **71,** 145–152. (*37, 67*)

855 VINCENT, W. S. and MILLER, O. L., Jr. (1966). International Symposium. *The nucleolus, its structure and function.* National Cancer Institute Monograph 23. Published by U.S. Dept. of Health, Education, and Welfare, Bethesda. (*44, 46, 50*)

856 VOELLER, B. R. (1964). The plant cell: aspects of its form and function. In *The Cell*, vol. 6, pp. 245–312. Ed. Brachet, J. Mirsky, A. E., Academic Press. London. (*34*)

857 VON WETTSTEIN, D., HENNINGSEN, K. W., BOYNTON, J. E., KANNANGARA, G. C. and NIELSEN, O. F. (1971). The genic control of chloroplast development in barley. In *Autonomy and Biogenesis of Mitochondria and Chloroplasts*, Eds. Boardman, N. K., Linnane, A.W. and Smillie, R. M. North Holland, Amsterdam. 205–223. (*175*)

857a VON WETTSTEIN-KNOWLES, P. (1974). Ultrastructure and origin of epicuticular wax tubes. *J. Ultrastruct. Res.*, **46,** 483–498. (*21*)

858 WAGNER, G., HAUPT, W. and LAUX, A. (1972). Reversible inhibition of chloroplast movement by cytochalasin B in the green alga *Mougeotia. Science, N.Y.*, **176,** 808–809. (*154*)

859 WAGNER, R. P. (1969). Genetics and phenogenetics of mitochondria. *Science, N.Y.*, **163,** 1026–1031. (*93*)

860 WAGNER, R. R. and CYNKIN, M. A. (1971). Glycoprotein biosynthesis. Incorporation of glycosyl groups into endogenous acceptors in a Golgi apparatus-rich fraction of liver. *J. biol. Chem.*, **246,** 143–151. (*78*)

861 WALLES, B. (1971). Chromoplast development in a carotenoid mutant of maize. *Protoplasma*, **73,** 159–175. (*121*)

862 WALLES, B. (1972). Plastid inheritance and mutations. In *Structure and Function of Chloroplasts.* Edited by Gibbs, M. published by Springer-Verlag, Berlin, Heidelberg, New York. 51–88. (*123*)

863 WALNE, P. L. and ARNOTT, H. J. (1967). The comparative ultrastructure and possible function of eyespots: *Euglena granulata* and *Chlamydomonas eugametos. Planta*, **77,** 325–353. (*128*)

864 WARNER, F. D. and SATIR, P. (1973). The substructure of ciliary microtubules. *J. Cell Sci.*, **12,** 313–326. (*142, 149, 150*)

865 WATSON, J. D. (1970). *Molecular Biology of the Gene.* 2nd Edition. Benjamin, New York. (*40, 41*)

866 WATSON, J. D. and CRICK, F. H. C. (1953). Molecular structure of nucleic acids. A structure for deoxyribose nucleic acid. *Nature, Lond.*, **171,** 737–738. (*40*)

866a WAYMAN, M. and OBIAGA, T. I. (1974). The modular structure of lignin. *Can. J. Chem.*, **52,** 2102–2110. (*19*)

867 WEHRMEYER, W. (1964). Zur Klärung der strukturellen Variabilität der Chloroplastengrana des Spinats in Profil und Aufsicht. *Planta*, **62,** 272–293. (*110*)

868 WEHRMEYER, W. (1965). Zur Kristallgitterstruktur der sogenannten Prolamellarkörper in Proplastiden etiolierter Bohnen. I. Pentagondodekaeder als Mittelpunkt konzentrischer Prolamellarkörper. *Z. Naturf.*, **20b,** 1270–1278. (*113*)

869 WEHRMEYER, W. (1965). Zur Kristallgitterstruktur der sogenannten Prolamellarkörper in Proplastiden etiolierter Bohnen. II. Zinkblendegitter als Muster tubulärer

Anordnungen in Prolamellarkörpern. *Z. Naturf.*, **20b**, 1278–1288. (*112*)

870 WEHRMEYER, W. (1965). Zur Kristallgitterstruktur der sogenannten Prolamellarkörper in Proplastiden etiolierter Bohnen. III. Wurtzitgitter als Muster tubulärer Anordnungen in Prolamellarkörpern. *Z. Naturf.*, **20b**, 1288–1296. (*112*)

871 WEHRMEYER, W. (1967). Prolamellar bodies, structure and development. In *Croissance et Viellissement des Chloroplastes.* Masson et Cie., Paris, p. 62–68. (*112, 113*)

871a WEIBEL, E. R. (1974). Selection of the best method in stereology. *J. Microsc.*, **100**, 261–269. (*59*)

872 WEIER, E. (1938). The structure of the chloroplast. *Bot. Rev.*, **4**, 497–530. (*99, 100*)

873 WEIER, T. E. and BENSON, A. A. (1967). The molecular organisation of chloroplast membranes. *Am. J. Bot.*, **54**, 389–402. (*104, 108*)

874 WEIER, T. E. and BROWN, D. L. (1970). Formation of the prolamellar body in 8-day, dark-grown seedlings. *Am. J. Bot.*, **57**, 267–275. (*112, 115*)

875 WEIER, T. E., SHUMWAY, K. L. and STOCKING, C. R. (1968). The organisation of chloroplast membranes of *Vicia faba* and *Zea mays* after processing to retain chlorophyll and hydrophobic lipids in the chloroplasts. *Protoplasma*, **66**, 339–355. (*104*)

876 WEIER, T. E., SJOLAND, R. D. and BROWN, D. L. (1970). Changes induced by low light intensities on the prolamellar body of 8-day, dark-grown seedlings. *Am. J. Bot.*, **57**, 276–284. (*115, 131*)

877 WEIER, T. E. and STOCKING, C. R. (1952). The chloroplast: structure, inheritance, and enzymolgy. II. *Bot. Rev.*, **18**, 14–75. (*98*)

878 WEIER, T. E., STOCKING, C. R., THOMSON, W. W. and DREVER, H. (1963). The grana as structural units in chloroplasts of mesophyll of *Nicotiana rustica* and *Phaseolus vulgaris. J. Ultrastruct. Res.*, **8**, 122–143. (*100, 101, 104*)

879 WEINBACH, E. C. and VON BRAND, T. (1967). Formation, isolation and composition of dense granules from mitochondria. *Biochim. biophys. Acta*, **148**, 256–266. (*90*)

880 WEINSTOCK, A. (1970). Cytotoxic effects of puromycin on the Golgi apparatus of pancreatic acinar cells, hepatocytes and ameloblasts. *J. Histochem. Cytochem.*, **18**, 875–886. (*85*)

881 WEISENBERG, R. C. (1972). Microtubule formation *in vitro* in solutions containing low calcium concentrations. *Science, N.Y.*, **177**, 1104–1105. (*143*)

881a WEISENBERG, R. C. (1973). Regulation of tubulin organization during meiosis. *Amer. Zool.*, **13**, 981–987 (*145*)

882 WEISENBERG, R. C., BORISY, G. G. and TAYLOR, E. W. (1968). The colchicine-binding protein of mammalian brain and its relation to microtubules. *Biochemistry*, **7**, 4466–4479. (*143*)

883 WELLS, R. and BIRNSTIEL, M. (1969). Kinetic complexity of chloroplast deoxyribonucleic acid and mitochondrial deoxyribonucleic acid from higher plants. *Biochem. J.*, **112**, 777–786. (*125*)

884 WELLS, R. and SAGER, R. (1971). Denaturation and renaturation kinetics of chloroplast DNA from *Chlamydomonas reinhardi. J. molec. Biol.*, **58**, 611–622. (*125*)

885 WESSELLS, N. K., SPOONER, B. S., ASH, J. F., BRADLEY, M. O., LUDUENA, M. A., TAYLOR, E. L., WRENN, J. T. and YAMADA, K. M. (1971). Microfilaments in cellular and developmental processes. *Science, N.Y.*, **171**, 135–143. (*150, 151, 153*)

886 WHALEY, W. G., DAUWALDER, M. and KEPHART, J. E. (1972). Golgi apparatus: influence on cell surfaces. *Science, N.Y.*, **175**, 596–599. (*83*)

887 WHALEY, W. G. and MOLLENHAUER, H. H. (1963). The Golgi apparatus and cell plate formation—a postulate. *J. Cell Biol.*, **17**, 216–221. (*169*)

888 WHEELDON, L. W. (1973). Products of mitochondrial protein synthesis. *Biochimie*, **55**, 805–814. (*176*)

889 WHETSELL, W. O. Jr. and BUNGE, R. P. (1969). Reversible alterations in the Golgi complex of cultured neurons treated with an inhibitor of active Na and K transport. *J. Cell Biol.*, **42**, 490–500. (*80*)

890 WHITTAKER, R. H. and FEENY, P. P. (1971). Allelochemics: chemical interactions between species. *Science, N.Y.*, **171**, 757–770. (*36*)

891 WILBUR, F. H. and RIOPEL, J. L. (1971). The role of cell interaction in the growth and differentiation of *Pelargonium hortorum* cells *in vitro*. II. Cell interaction and differentiation. *Bot. Gaz.*, **132**, 193–202. (*30*)

892 WILDER, B. M. and ALBERSHEIM, P. (1973). The structure of plant cell walls. IV. A structural comparison of the wall hemicellulose of cell suspension cultures of sycamore *(Acer pseudoplatanus)* and of red kidney bean *(Phaseolus vulgaris). Pl. Physiol., Lancaster*, **51**, 889–893. (*16*)

892a WILDMAN, S. G., HONGLADAROM, T. and HONDA, S. I. (1962). Chloroplasts and mitochondria in living plant cells: cinephotomicrographic studies. *Science, N.Y.*, **138**, 434–436. (*89, 100*)

893 WILKINS, M. H. F., STOKES, A. R. and WILSON, H. R. (1953). Molecular structure of deoxypentose nucleic acids. *Nature, Lond.*, **171**, 738–740. (*40*)

894 WILLIAMS, E., HESLOP-HARRISON, J. and DICKINSON, H. G. (1972). The activity of the nucleolus organizing region and the origin of cytoplasmic nucleoloids in meiocytes of *Lilium. Protoplasma*, **77**, 79–93. (*50*)

895 WILLIAMS, N. E. and FRANKEL, J. (1973). Regulation of microtubules in *Tetrahymena.* I. Electron microscopy of oral replacement. *J. Cell Biol.*, **56**, 441–457. (*144*)

896 WILLIAMS, N. E. and NELSEN, E. M. (1973). Regulation of microtubules in *Tetrahymena.* II. Relation between turnover of microtubule proteins and microtubule dissociation and assembly during oral replacement. *J. Cell Biol.*, **56**, 458–465. (*144*)

897 WILLIAMSON, D. H. (1970). The effect of environmental and genetic factors on the replication of mitochondrial DNA in yeast. *Symp. Soc. exp. Biol.*, **24**, 247–276. (*94*)

898 WILSON, E. B. (1928). *The Cell in Development and Heredity*. 3rd edition, Macmillan, New York. (*39, 157*)

899 WILSON, H. J. (1968). The fine structure of the kinetochore in meiotic cells of *Tradescantia. Planta*, **78**, 379–385. (*159*)

900 WILSON, H. J. (1969). Arms and bridges on microtubules in the mitotic apparatus. *J. Cell Biol.*, **40**, 854–859. (*146, 163*)

901 WILSON, L. and MEZA, I. (1973). The mechanism of action of colchicine. Colchicine binding properties of sea urchin sperm tail outer doublet tubulin. *J. Cell Biol.*, **58**, 709–719. (*143*)

902 WILSON, R. H., THURSTON, E. L. and MITCHELL, R. (1973). Ultrastructural transformations in bean inner mitochondrial membranes. *Pl. Physiol., Lancaster*, **51**, 26–30. (*93*)

903 WISCHNITZER, S. (1970). *Introduction to Electron Microscopy.* 2nd. Edition. Pergamon Press., Oxford. (*2*)

904 WISCHNITZER, S. (1973). The submicroscopic morphology of the interphase nucleus. *Int. Rev. Cytol.*, **34**, 1–48. (*51, 54*)

905 WISE, G. E. and FLICKINGER, C. J. (1971). Patterns of cytochemical staining in Golgi apparatus of amebae following enucleation. *Exptl Cell Res.*, **67**, 323–328. (*73, 76*)

906 WISE, G. E. and PRESCOTT, D. M. (1973). Initiation and continuation of DNA replication are not associated with the nuclear envelope in mammalian cells. *Proc. natn. Acad. Sci. U.S.A.*, **70**, 714–717. (*52*)

907 WOJTCZAK, L., BARANSKA, J., ZBOROWSKI, J. and DRAHOTA, Z. (1971). Exchange of phospholipids between microsomes and mitochondrial outer and inner membranes. *Biochim. biophys. Acta*, **249**, 41–52. (*89*)

908 WOLFE, S. L. (1968). The effect of prefixation on the diameter of chromosome fibres isolated by the Langmuir trough—critical point method. *J. Cell Biol.*, **37**, 610–620. (*43*)

909 WOLFE, S. L. (1972). *Biology of the Cell.* Wadsworth Pub. Co. (*7*)

910 WOLLENWEBER, E. and SCHNEPF, E. (1970). Vergleichende Untersuchungen über die flavonoiden Exkrete von 'Mehl'- und 'Öl'-Drüsen bei Primeln und die Feinstruktur der Drüsenzellen. *Z. Pflanzenphysiol.*, **62**, 216–227. (*64*)

911 WOODCOCK, C. L. F. and BOGORAD, L. (1970). Evidence for variation in the quantity of DNA among plastids of *Acetabularia. J. Cell Biol.*, **44**, 361–375. (*124*)

912 WOODCOCK, C. L. F. and BOGORAD, L. (1971). Nucleic acids and information processing in chloroplasts. In *Structure and Function of Chloroplasts.* Ed. Gibbs, M. Springer-Verlag, Berlin, Heidelberg, New York. 89–128. (*124, 175*)

913 WOODCOCK, C. L. F. and FERNÁNDEZ-MORÁN, H. (1968). Electron microscopy of DNA conformations in spinach chloroplasts. *J. molec. Biol.*, **31**, 627–631. (*125*)

914 WOODCOCK, C. L. F. and MILLER, G. J. (1973). Ultrastructural features of the life cycle of *Acetabularia mediterranea.* II. Events associated with the division of the primary nucleus and the formation of cysts. *Protoplasma*, **77**, 331–341. (*152*)

915 WRIGGLESWORTH, J. M., PACKER, L. and BRANTON, D. (1970). Organisation of mitochondrial structure as revealed by freeze-etching. *Biochem. biophys. Acta*, **205**, 125–135. (*91*)

916 WUNDERLICH, F. (1972). The macronuclear envelope of *Tetrahymena pyriformis* GL in different physiological states. V. Nuclear pore complexes—a controlling system of protein biosynthesis? *J. memb. Biol.*, **7**, 220–230. (*55, 56*)

917 WUNDERLICH, F., MÜLLER, R. and SPETH, V. (1973). Direct evidence for a colchicine-induced impairment in the mobility of membrane components. *Science, N.Y.*, **182**, 1136–1138. (*148*)

918 YAGO, N., SEKI, M., SEKIYAMA, S., KOBAYASHI, S., KUROKAWA, H., IWAI, Y., SATO, F. and SHIRAGAI, A. (1972). Growth and differentiation of mitochondria in the regenerating rat adrenal cortex. *J. Cell Biol.*, **52**, 503–513. (*94*)

919 YATSU, L. Y. and JACKS, T. J. (1968). Association of lysosomal activity with aleurone grains in plant seeds. *Archs Biochem. Biophys.*, **124**, 466–471. (*37*)

920 YATSU, L. Y. and JACKS, T. J. (1972). Spherosome membranes. Half unit-membranes. *Pl. Physiol., Lancaster*, **49**, 937–943. (*135*)

921 YEOMAN, M. M., TULETT, A. J. and BAGSHAW, V. (1970). Nuclear extension in dividing vacuolated plant cells. *Nature, Lond.*, **226**, 557–558. (*42*)

922 YOO, B. Y. (1968). Some observations on chromatin fibres of isolated pea nuclei. *Can. J. Bot.*, **46**, 1111–1114. (*43*)

923 YUNGHANS, W. N., KEENAN, T. W. and MORRÉ, D. J. (1970). Isolation of Golgi apparatus from rat liver. III. Lipid and protein composition. *Exptl. & Molecular Pathology*, **12**, 36–45. (*75*)

924 YUNIS, J. J. and YASMINEH, W. G. (1971). Heterochromatin, satellite DNA, and cell function. *Science, N.Y.*, **174**, 1200–1209. (*43, 44*)

925 ZAAR, K. and SCHNEPF, E. (1969). Membranfluss und Nucleosiddiphosphatase-Reaktion in Wurzelhaaren von *Lepidium sativum. Planta*, **88**, 224–232. (*73, 75, 78*)

926 ZAGURY, D., UHR, J. W., JAMIESON, J. D. and PALADE, G. E. (1970). Immunoglobulin synthesis and secretion. II. Radioautographic studies of sites of addition of the carbohydrate moieties and intracellular transport. *J. Cell Biol.*, **46**, 52–63. (*78*)

927 ZELITCH, I. (1973). Plant productivity and the control of photorespiration. *Proc. natn. Acad. Sci. U.S.A.*, **70**, 579–584. (*109*)

928 ZELLWEGER, A., RYSER, U. and BRAUN, R. (1972). Ribosomal genes of *Physarum*: their isolation and replication in the mitotic cycle. *J. molec. Biol.*, **64**, 681–691. (*161*)

929 ZIEGLER, H. (1968). La sécrétion du nectar. In *Traité de Biologie de l'Abeille.* Vol. 3. Ed. Chauwin, R. 218–248. Masson et Cie., Paris. (*65*)

930 ZIEGLER, H. and LÜTTGE, U. (1967). Die Salzdrüsen von *Limonium vulgare.* II. Mitteilung. Die Lokalisierung des Chlorids. *Planta*, **74**, 1–17. (*29*)

931 ZILKEY, B. F. and CANVIN, D. T. (1972). Localisation of oleic acid biosynthesis enzymes in the proplastids of developing castor endosperm. *Can. J. Bot.*, **50**, 323–326. (*99*)

932 ZIMMERMANN, H.-P. (1973). Electron microscopic investigations on the spermatogenesis of *Sphaerocarpos donnellii* Aust. I. Mitochondria and the plastid. *Cytobios*, **7**, 42–54. (*93*)

933 ZIRKLE, C. (1927). The growth and development of plastids in *Lunularia vulgaris, Elodea canadensis,* and *Zea mays. Am. J. Bot.*, **14**, 429–445. (*98*)

934 ZIRKLE, C. (1937). The plant vacuole. *Bot. Rev.*, **3**, 1–30. (*33, 37*)

Subject Index

Numbers in **heavy type** refer to major sections, numbers *in italics* refer to Plate captions.